August 2013

Kendall Hunt
publishing company
www.kendallhunt.com
Send all inquiries to:
4050 Westmark Drive
Dubuque, IA 52004-1840

TABLE OF CONTENTS

Acknowledgments

This book is a compendium of sources derived from the literature at Kendall Hunt Publishing and is intended as far as possible to present a Christian World View towards the ancient but unstructured discipline of geography.

This book is intended to be used at Liberty University; the world's largest Christian university. In particular, I would like to thank Dr. Emily Heady the brilliant vice-provost and Dean of the College of General Studies for her encouragement in both this profession and particular endeavor. Dr. Carey Roberts, the Associate Dean of the College of Arts and Sciences likewise has been a terrific influence and an outstanding academic example.

Sam Borgia assisted immeasurably with the organization of the transcript, as did Curtis Ross at Kendall Hunt. Thank you to the Plaid Avenger for his willingness to assist Captain Geo in this endeavor.

I would like to dedicate this work and express my thanks to my beautiful wife, Shannon Cavin Ritchie who for 35 years has stayed with me through the entire adventure.

Robert F. Ritchie IV
Liberty University

PART ONE
INTRODUCTION TO GEOGRAPHY

Introduction to Physical Geography

1

"And He has made from one blood every nation of men to dwell on all the face of the earth, and has determined their preappointed times and the boundaries of their dwellings, so that they should seek the Lord, in the hope that they might grope for Him and find Him, though He is not far from each one of us." (Acts 17:25–27)

CHAPTER OUTLINE:

→ Introduction: Main Ideas and Thematic Structure
→ The Nature of Geography
→ Unifying Themes in Geography
→ World-Systems Analysis
→ Semi-Peripheral States
→ International Organizations
→ Economic Entitities—Show Me the Money
→ Defense
→ The United Nations

→ New Kids on the Bloc!!!
→ Cultural Organizations
→ International Oddballs
→ The Nuke Group
→ Key Terms to Know
→ Further Reading
→ Web Sites
→ Study Questions

ACCORDING to the book of Genesis, God created the world. God's creation consists of time and space and is studied as history and geography, respectively. God's love for this Earth is epitomized by his son Jesus Christ, who gave his life for it. As followers and servants of Jesus, we shall continue to seek opportunities to use our individual talents vocationally. Geography offers an excellent tool for the development of future service strategies. What then is geography? This definition will be discussed in detail later, but for now, geography is basically the study of the earth's surface and is concerned with spatial concepts. More specifically, the subject of human geography is concerned with various aspects of human activity over space. Any proper study of human geography, however, must begin with an understanding of the planet's physical properties.

This book will survey geography from a regional and human perspective. The regions chosen are somewhat arbitrary and extremely subjective.

The physical properties of planet Earth that we shall focus upon in this introduction to physical geography are shape, spin, tilt, proximity to the sun's energy, and surface relief. These phenomena work in tandem to create the various climates observed. These climates, in turn, continue to shape the surface of the earth, thereby affecting human activities.

Do you think the earth is round? The earth is indeed a generally spherically shaped planet. The **sphere** is an ideal form and varies slightly from the reality of creation. The earth is actually an oblate ellipsoid, but for the purposes of this course, it will be considered a sphere. A sphere is a round, solid form with the surface equally distant from the center. Interestingly, many geologists believe the earth has actually changed in shape over time, and this change in shape has resulted in both changes to the atmosphere and perhaps even to gravity. Our planet is a sphere suspended in space, it spins on an axis with a 23.5-degree tilt, and it revolves around the sun.

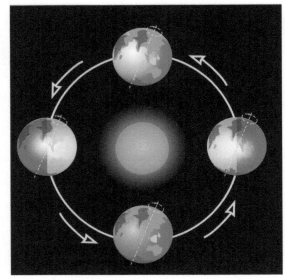

Image © Matthew Cole, 2011. Used under license from Shutterstock, Inc.

The spherical nature of planet Earth is of great importance. The curvature on the surface is approximately 16 feet for every five miles of distance travelled. This curvature is an important consideration in understanding the nature of the earth and its surrounding atmosphere. Often, students of geography mistakenly envision the earth in a two-dimensional plane, such as that seen on a GPS screen or a map. In fact, it is through the study of geometry we begin to understand the exciting characteristics of the sphere and see how it affects our planet. By looking at a sphere extrinsically (from a distance), we can see that any attempt to divide the sphere results in a lack of symmetry unless we cut the sphere exactly through the center. The nature of a sphere is such that when it is divided at the center, the circle formed by passing a plane through the exact center is the largest circle that can be drawn on its surface. Longitude lines and the equator are examples of **great circles**. A small circle is any place on a sphere where a plane does not go through the center. Latitudes reflect small circles at various angles to the earth's center, with the angle from the center of the sphere increasing in size when moving from the equator to the poles. The small circle's lack of symmetry offers a way of understanding the circular patterns associated with the atmosphere.

By virtue of its spherical shape, the earth is witness to a phenomenon called the **Coriolis effect**. A way of understanding the Coriolis effect is through a simple illustration. For example, the shortest distance from point A to point B is not necessarily over a mountain because of its height; but rather, the shortest distance can be around the mountain. Dr. David W. Henderson, in his book, *Experiencing Geometry*, demonstrates the importance of symmetry by illustrating a toy car rolling across a sphere. The toy car would remain on the great circle, but it would fall off a small circle. If you were walking on a small circle, for example, one leg would need to do more work than the other just to stay on a straight line, whereas the symmetry of a great circle would allow you to move your feet at the same speed.

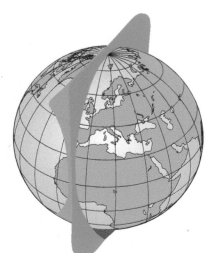

Now consider the earth's **spin**. Spinning in roughly a 24-hour period, a small circle (latitude) is actually moving faster than a great circle as the earth moves on its axis because the axis of rotation is decreasing in size as one heads toward the poles. This fact may seem counterintuitive, but if one sees the earth as a circle on a Cartesian coordinate system and the equator is the X line, a Sine function at X (equator) is zero, whereas the Y (poles) would be 1. This illustration numerically reflects the increased angle towards the poles with a decreasing axis of rotation.

If this seems a bit confusing, just remember that the earth is spinning at great speed and, as a sphere, the angle of momentum is perpendicular to the direction of spin. Additionally, the atmosphere is affected by the friction of the surface. Since the shortest distance between two points on a sphere is a great circle, and the angles from the earth's center (latitude) grow, motion toward the poles picks up speed.

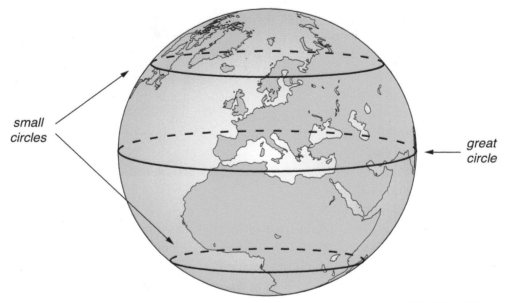

small
circles

great
circle

Image © Kendall Hunt Publishing

To an observer on Earth, these characteristics of sphere and spin appear to create wind and ocean currents moving in a clockwise manner (generally) in the northern hemisphere and counterclockwise (generally) in the southern hemisphere. Imagine a trail of smoke behind a locomotive. Because the train is moving, the smoke, which in actuality is rising straight up, appears to be in a trail behind it. Similarly, our atmosphere is deflected from the earth's surface as the planet is moving under it and appears to be deflecting at an angle. When you think of spin, reflect upon how an ice skater pulls his/her arms in to increase the speed of spin—a function of angular momentum and conservation. The skater is producing a smaller axis of rotation, and, because the sum total of an objects momentum cannot change, the skater's speed increases. Planetary bodies exhibit the same features as they move in circular paths while spinning on an axes. If you can find a globe, look at it from the poles as you spin it. Doing so will help you visualize the aforementioned concepts of great and small circles and axis of rotation; the sum ingredients of the Coriollis effect. If you spin the globe, you will find you are hitting it with your hands at an angle because of its tilt.

The **tilt** of the earth is about 23.5 degrees. In about 24 hours, a complete **rotation** occurs on this axis. This tilt, also a function of the conservation of angular momentum, holds the earth in space where its momentum is shared with the moon, as evidenced by the tides. A way of picturing this tilt as a function of the attraction between the earth and the moon is to think of how your bicycle stays upright when you ride it down the road. The perfect balance of gravitational attraction between the earth and moon, and to a lesser extent, the other planets in the solar system, help explain the earth's revolution around the sun. The combination of tilt and rotation produce differing levels of incoming solar radiation upon the earth's surface. This alternation between cool night and warm day, combined with the annual revolution, results in the unequal distribution of the sun's energy. The earth's **revolution** around the sun creates the seasons and is of importance to us because this revolution creates a situation where heat is absorbed at different rates between the northern and southern hemispheres. The tendency of the earth's atmosphere is to attempt to achieve levels of equilibrium. The differing amounts of radiation the hemispheres absorb from the sun during the different seasons contribute to the climate and weather patterns. How does this tilt cause different levels of absorption of energy?

Many people think the earth experiences different extremes of temperature because of the changing proximity to, or distance from, the sun. Actually, our atmosphere is similar to a microwave oven in that the heat produced is a function of the absorption of solar energy. **Insolation**, or incoming solar radiation, hits the earth's surface at different **angles of incidence** and is then reflected back through the atmosphere at various angles at different levels of latitude called angles of incidence. The differing angles of incidence at various latitudes result in distinct levels of energy absorbed into the atmosphere as short-wave radiation and then, subsequently, reflected off the earth's surface into the atmosphere (into outer space) as long-wave radiation. Depending on the altitude, humidity, and **albedo**—or level of solar radiation reflected from the earth's surface—of the side of the earth facing the sun, the resulting reflective long-wave energy will heat up the atmosphere at different rates. The atmo-

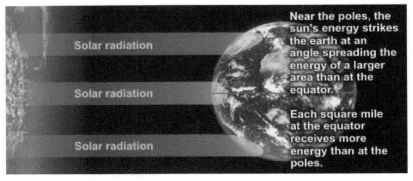

Near the poles, the sun's energy strikes the earth at an angle spreading the energy of a larger area than at the equator.

Each square mile at the equator receives more energy than at the poles.

Courtesy of NOAA

sphere also serves to protect the earth from certain types of harmful solar radiation through a unique layer of oxygen molecules arranged in threes (triatomic molecules), which is called the **ozone layer**. The different seasons, and even the time of day and latitude determine the different levels of insolation covering the earth.

The surface of the earth is a final determining aspect of the different climates we confront. The surface is divided between land and sea. Over 70 percent of the earth's surface is water, and therefore, there is more variance in the levels of absorption and reflection of the sun's radiation here than compared to on land. This varying albedo is also instrumental in accounting for different temperatures and levels of precipitation. Like a giant machine, the earth is churning and chugging through space, subject to various forces and phenomena, like a spherical battleground attempting to achieve equilibrium. The energy that is absorbed is cooled in the atmosphere, while the sun's rays warm other places. This spinning sphere's surface releases and absorbs enormous amounts of energy, and the result is the generally circular patterns of wind and ocean currents.

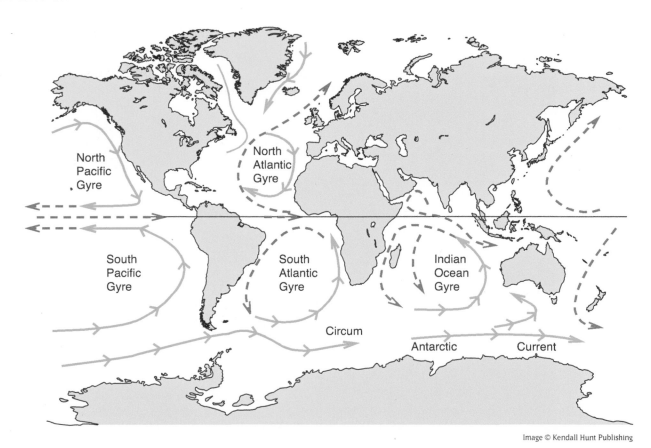

Image © Kendall Hunt Publishing

The **ocean currents** of the world are consistent patterns of flow across the earth's hydrosphere. The currents, in large measure, are also a function of the moving atmosphere. In the northern hemisphere, the winds and ocean currents follow a clockwise direction. In the southern hemisphere, these take on the characteristics of a counter-clockwise pattern in keeping with the Coriolis effect. In summary, the oceans and seas of the planet tend to absorb and release, not retain, energy at slower rates than the land and air around them. These currents attempt to redistribute the energy absorbed from the sun more evenly over the earth's surface. Hopefully, you are beginning to see how the earth's shape, its spin, and its topography contribute greatly to the creation of the climates of the world and will see how these climates have affected culture.

The atmosphere, like the oceans themselves, also tends to cool as one goes higher up in elevation. The phenomenon of **orographic lifting** causes air to cool as it rises. As it cools, moisture assumes a lower state of energy, and precipitation may result. After the release of energy, the result is a cooler and dryer air pattern. This sinking phenomenon can occur both locally and globally and is best demonstrated by the global patterns established by **Hadley cells**. These global cells of varying air pressure account for the general patterns of climate in the world, such as moisture at the equator, dry conditions at 30 degrees north and south, and the global temperate climates where warmer equatorial currents interface with cold, dry, and polar fronts. These colliding cells of varying heat and moisture are turning on the sphere, alternately rising and sinking as they warm and cool. When topography is introduced into the discussion of currents and atmosphere, we begin to be able to define climate and see how the world can be divided into general patterns of weather.

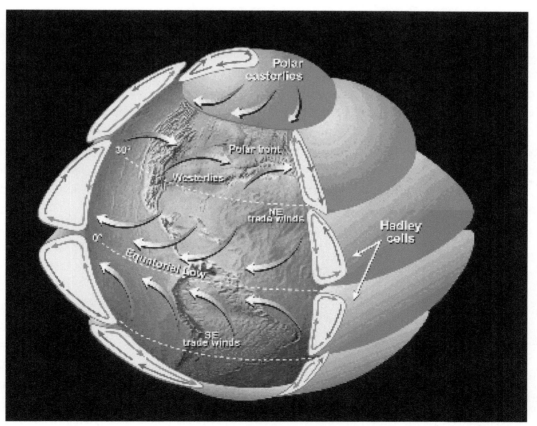

Courtesy of NASA

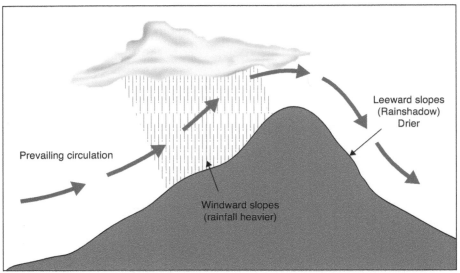

Leeward slopes
(Rainshadow)
Drier

Prevailing circulation

Windward slopes
(rainfall heavier)

Image © Kendall Hunt Publishing

Orographic uplift may prevent moisture-laden clouds from rising above mountains with their moisture. As the clouds rise, they cool, and energy is released in the form of winds and heat as precipitation occurs. The result of an ocean-borne current when coming ashore and encountering high terrain can result in a **windward side**, or wet side, of the mountain. Just as we saw with the rise of air in the Hadley cells at 30 degrees north and south, the wind absorbs the moisture often creating a dry side, known as the **leeward** side. The factors of latitude, tilt, and spin on a global scale create high-pressure and low-pressure systems rising and falling, and as they come into contact with surface relief, create climates.

The different climates of the world therefore describe consistently measured weather patterns and are the result of geostrophic winds blowing ocean currents, resulting in part from the unequal distribution of the sun's energy on the earth's surface, combined with the spherical nature of the planet and different latitudes to produce the different climates. **Climates** are defined for the purpose of this class as generalized weather conditions at a given location consistent over long periods of recorded time. For example, the climate in Lynchburg, Virginia, is generally consistent for warmer temperatures and drier conditions in July compared to January. **Weather** is much more variable and can change quickly over time, such as when the sun comes out at the end of a rainstorm. An easy way to recognize different climates is by their biomes. A **biome** is defined as shared characteristics in animal and plant species within a particular ecosystem on the earth's surface and are generally compatible within a particular climate zone.

Rainforest Biome. Tropical Wet Climate
Image © STILLFX, 2011. Used under license from Shutterstock, Inc.

Brief descriptions of the various world climates and biomes are given as one travels from the equator towards the poles:

Equatorial climates generally occur between the tropics of Cancer and Capricorn located at 23.5 degrees north and south latitude. These tropical climates are characterized by extreme amounts of heat and moisture, in part because they receive the lowest angles of insolation, resulting in a greater release of energy and consequent excitement of atmospheric molecular activity. The moisture may be seasonal in areas, and the result is a high contrast between wet and dry seasons. The biome often associated with this climate is the **rainforest**.

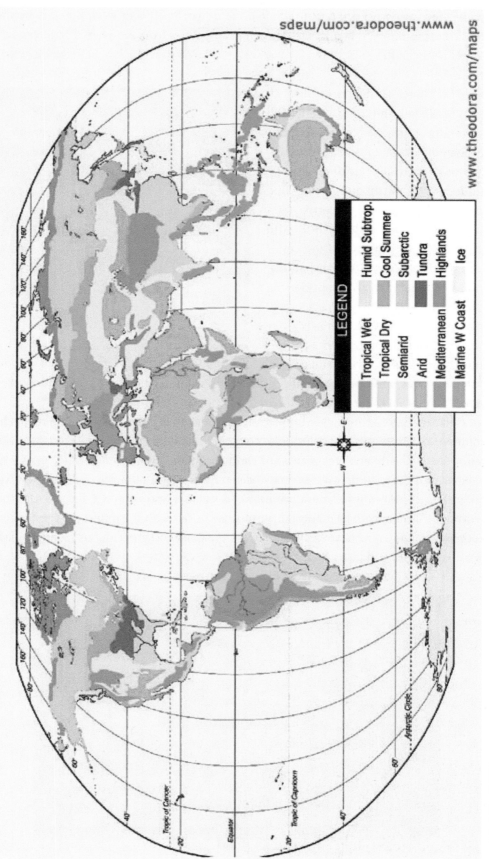

www.theodora.com/maps

LEGEND

Tropical Wet | Humid Subtrop.
Tropical Dry | Cool Summer
Semiarid | Subarctic
Arid | Tundra
Mediterranean | Highlands
Marine W Coast | Ice

Generally, plants within the rainforest are adapted to survive in extremely wet environments by the use of large, thin leaves to enable rapid transpiration and leaf tips to direct moisture away from the plant. Plant roots are notoriously shallow and near the surface. Animals are most noted for their extreme diversity. One log, for example, may be home to numerous species.

The **dry climates** vary between semi-arid and arid, depending on amounts of rainfall. The biomes associated with this category generally are marked by **desertification**—an increased degradation or loss of land usage. The deserts of the world are often located between the 15-degree latitude marks north and south, and between rising and falling Hadley cells. Desertification might result from colder ocean currents, such as with the Atacama Desert in Chile, or perhaps be the result of relief-blocking moisture through orographic lifting, such as happens in Eastern Australia. We consider the biome here to be that of a **desert**—an area generally recognized by animals able to survive without much water because of physiological adaptations to a dry environment. Plant life generally consists of shrubs and grasses

Desert Biome/Arid Climate
Image © agap, 2011. Used under license from Shutterstock, Inc.

having fewer and smaller leaves and trees becoming noticeably absent. In the New World, the predomination of cacti is similar to the pattern of Euphorbia in the Old World.

The **temperate climates** exist at the higher latitudes from about 30 degrees north and south toward the poles in both the northern and southern hemispheres. The collision of cold polar air sinking towards the equator with warm tropical air rising from the equator partially explains the creation of various climates. These temperate climates are often cooler and drier as one goes farther into the continent where they occur. For example, the proximity of New York to the Canadian land mass results in greater extremes of temperature between summer and winter. Contrasted to this is Virginia, where the proximity of the Atlantic Ocean, with its moderating influence, is felt with the differences between winter and summer being much less pronounced. The animals are variable, as are the plants. The deciduous forest typifies the biomes of this climate zone.

Deciduous Forest/Temperate Climate
Image © Inga Nielsen, 2011. Used under license from Shutterstock, Inc.

Coniferous Forest/Highland Climate
Image © Nikolay Stefanov Dimitrov, 2011. Used under license from Shutterstock, Inc.

The cool, dry air of the high mountainous areas of the earth is typical of the **highland climate**. These intensely cold areas create an environment counter-intuitively similar to a desert due to their extremely dry nature. Any moisture in the air is crystallized and may exist for long periods in a solid condition. A highland biome is characterized by the plants that have adapted to this climate with needles and a shape generally associated with Christmas trees as the limbs slant down in an imbricated manner resembling the pitch of a roof. This design protects the trees from the harmful effects of ice and wind.

Tundra/Semi Arctic
Image © Nadezhda Bolotina, 2011. Used under license from Shutterstock, Inc.

The **polar climates** are characterized by extremely dry and cold climatic features and a biome where animals and plants have adapted to these extremes of low temperatures. The most obvious characteristic of the biome, also known as the **tundra**, is the low diversification of the animals. The same amount of animals or biomass may exist in these areas as in other climates and biomes, but generally, the creatures tend to be more homogenous and of the same species.

Mediterranean Climate
Image © Dmitriy Yakovlev, 2001. Used under license from Shutterstock, Inc.

The final two climates briefly mentioned here are important because they demonstrate the importance of contrasting ocean and land temperatures and air pressure. A cold ocean current next to a warm land mass tends to draw the lower-pressured and warm rising air off the land, and this result is a dry, warm land mass. California and southern Italy are characteristic of this pleasant and somewhat rare **Mediterranean climate**. The opposite phenomenon, however, occurs when the ocean is warm and the land is cool. This **maritime climate** can result in a temperate rainforest biome where ferns and other plants abound; plants in this climate are capable of living in a moist environment. Western Oregon and Washington state are typical of this climate and, along with the Cascade mountain chain, are examples of the windward and leeward effects on climate. Much of Northern Europe epitomizes this type of climate, which is characterized by moderate winters and rainy weather patterns.

Temperate Rainforest/Maritime Climate
Image © 2009fotofriends, 2011. Used under license of Shutterstock, Inc.

This brief survey is not intended to ignore other forces that tend to act upon the earth's surface. The limited space here prohibits an in-depth analysis, but the mysterious **magnetic field** surrounding the earth also may serve to protect the earth from harmful effects from outer space. Some experts believe the magnetic fields emanating from other planets in the solar system may serve to further protect the earth from the harmful effects of space matter. Some scientists also think currents of molten rock or convection currents beneath the earth's surface may also be responsible for this mysterious phenomenon.[1] Attempting to reconcile tectonic plate activity with the creation of the oceans and seas with the great flood described in Genesis 7:11 offers opportunities for studying the changing surface of the planet. Further exciting opportunities for study exist in the amazing influences of tectonic, hydrospheric, and atmospheric affects upon the earth's surface. Our planet is truly vibrant and is being shaped and eroded daily under the sovereignty of a living God.

Magnetic Field Around the Earth
Image © Andrea Danti, 2011. Used under license from Shutterstock, Inc.

God has given us a beautiful and enchanted world in which to live. By understanding the various forces and effects that result from the earth's shape, spin, tilt, and revolution about the sun, we can determine the reasons for the variations of climate and weather patterns on the earth as functions of location and topography. Appreciating the physical geography of the earth is a wonderful starting point for beginning to understand the themes of human geography and the differences in development in the world economy.

INTRODUCTION: MAIN IDEAS AND THEMATIC STRUCTURE*

"Beloved, you do faithfully whatever you do for the brethren and for strangers." (3 John 1:5)

God created the world. Despite the many problems, despair, and fears we may confront, we can rest assured His Love for us is unceasing. In Genesis 9:1–3, God blessed Noah's descendants and instructed them "to be fruitful and multiply, and replenish the earth." Furthermore, God provided mankind the resources necessary to spread across the earth. In Mark 14:62 and Rev 19:11–16, Jesus promises to return, thereby giving us both hope and a reason to be faithful in attempting to reach the world with the truth and to serve the needs of a lost and suffering world. A spatial perspective in general, and knowledge of human geography in particular, can better enable us to serve the various peoples of the earth, armed with the truth of the Great Commission. Any goal for service must, however, have as its basis a strategy. Any strategy should be compatible with concepts of human geography. To understand the concepts of human geography, we will examine the nature of geography and the themes governing a geographic perspective. A world system analysis can delineate the areas of the world most capable of serving others, as well as identify areas particularly in need.

* Pages 13–20 from *Introduction to Human Geography: A World-Systems Approach,* 4th Edition by Timothy G. Anderson. Copyright © 2012 by Timothy G. Anderson. Reprinted by permission of Kendall Hunt Publishing Company.

---

[1] http://184.154.224.5/creatio1/index.php?option=com_content&task=view&id=38

THE NATURE OF GEOGRAPHY

Most Americans relate the term geography to the rote memorization of trivial facts associated with various countries of the world: capitals, imports and exports, highest mountains, largest lakes, most populous cities, climates and the like. This trivialized impression of the subject matter with which geography is concerned most likely stems from the fact that until recently, geography was not taught as a distinctive subject in most American elementary schools and high schools. Even today, it is often taught as a marginalized part of "social science" courses in elementary schools and civics or history courses in high schools. As a result, most Americans think of geography as a collection of trivial factoids rather than a distinct academic subject such as history, economics, mathematics or physics are. But within academia, geography is a thriving, diverse discipline with its own set of theories, research methods, terminology and subject matter and with a unique way of looking at the world. Indeed, the Association of American Geographers, the largest professional society of geographers in the world, counts over 5,000 professors, students and professionals as members.

So what is it that geographers do? What is their method of analysis? What questions do geographers ask about the world? These questions, as it turns out, are not easily answered. Indeed, any two geographers might answer them in very different ways, for geography is not unified by subject, but rather method. This is yet another reason for misconceptions about what geography is on the part of the lay public and, indeed, on the part of many academicians themselves. Geography does not seem to "fit" into academe the same way that other subjects do because the discipline is not unified or defined by the subject matter with which it is concerned. Rather, it is unified and defined by its method and approach, its mode of analysis. This mode of analysis (or approach or perspective) involves a **spatial perspective**. In this way geography can be conceptualized as a way of thinking about the world, about space and about places. This spatial (geographical) perspective involves the fundamental question as to how both cultural and natural phenomena vary spatially (geographically) across the earth's surface. As such, it is quite literally possible to study the geography of almost any phenomenon that occurs on the earth's surface. Employing these ideas, we may define the academic discipline of geography as *the study of the spatial variation of phenomena across the earth's surface.* So, geographers do not study a certain "thing" or subject. Instead, they study all sorts or "things" and subjects in a specific way—spatially.

The 18th century German philosopher Immanuel Kant understood very well the unique position that geography held within academia. Kant argued that human beings make sense of and organize the world in one of three ways, and human knowledge, as well as academic departments in the modern university, is arranged or organized in the same way. First, Kant wrote, we make sense of the world *topically* by organizing knowledge according to specific subject matter: biology as the study of plant and animal life, geology as the study of the physical structure of the earth, sociology as the study of human society, and so on. Each of these fields is unified by the subject matter with which it is concerned. Second, according to Kant, we make sense of the world *temporally* by organizing phenomena according to time, periods and eras. This is the sole domain of the discipline of history. Finally, Kant argued that we make sense of the world *chorologically* (geographically) by organizing phenomena according to how they vary across space and from place to place. This is the sole domain of the discipline of geography. The disciplines of history and geography are similar in that they are both unified by a method rather than the study of a specific subject matter. The region (discussed below) in geography is analogous to the era or period in history—they are the main units of analysis in each of the respective disciplines.

Figure 1.1 represents a generalized model for understanding the nature of the academic discipline of geography. This figure illustrates the field's three main subfields: physical geography, human geography and environmental geography. The description and analysis of patterns on the landscape unites each of these subfields. When most people think of the term landscape they think of something depicted in a painting, or perhaps a garden. But when geographers employ the term landscape, they are referring to the totality of our surroundings. In this sense, the **physical landscape** refers to the patterns created on the earth's surface as a result of natural or physical processes. For example, tectonic forces create continents and mountain chains. Long term climatic processes create varying vegetative realms. Wind and water shape and modify landforms. **Physical geography**, then, is the subfield of geography that is concerned with the description and

FIGURE 1.1 THE GEOGRAPHIC WORLD-VIEW

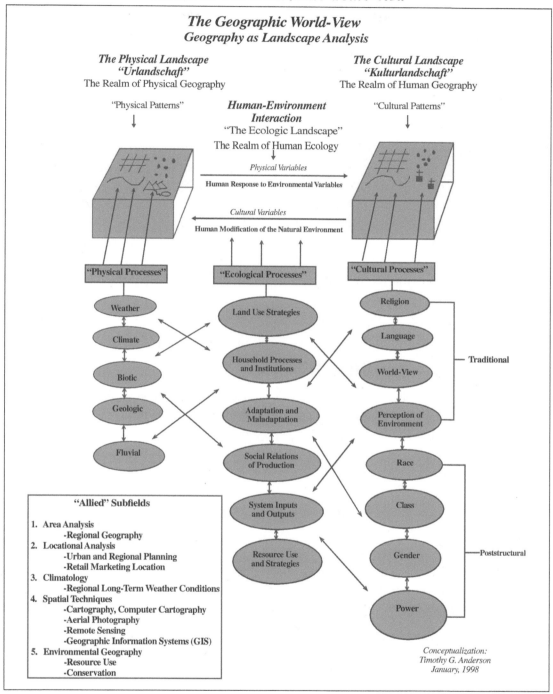

analysis of the physical landscape and the processes that create and modify that landscape. The various subfields of physical geography such as biogeography, climatology and geomorphology are natural sciences, allied with such fields as botany, meteorology and geology, and, like these allied fields, research in physical geography is undertaken largely according to the scientific method.

Human geography is the subfield of geography that is concerned with the description and analysis of cultural landscapes and the social and cultural processes that create and modify those landscapes. While physical geography is concerned with the natural forces that shape the earth's physical landscapes, human geography analyzes the social and cultural forces that create cultural landscapes. The **cultural landscape**, then, may be conceptualized as the human "imprint" on the physical landscape resulting from modification of the physical landscape by human social and cultural forces. Given the power and influence of human technology and institutions, many human geographers see human beings as the ultimate modifiers of the physical landscape. The various subfields of human geography, such as historical geography, economic geography and political geography are social sciences, allied with such fields as history, economics and political science. In addition to their own research methodologies, human geographers often employ research methods from these allied fields. Traditionally, human geography has been concerned with how cultural processes such as religion, language and world-view affect the cultural landscape. More recently, however, cultural geographers have begun to question traditional ideas about how cultural landscapes are created by focusing on so-called post-structural processes to reevaluate the nature of cultural landscapes. In such research, the cultural landscape is conceptualized as a stage upon which social struggles dealing with such concepts as race, class, power and gender are played out. In this sense, the cultural landscape is the product of such struggles.

Although all geographers do not recognize it as such, it can be argued that **environmental geography** is a third major subfield of geography. Environmental geography is concerned with the interrelationships between humans and the natural environments in which they live, or nature-society relationships. Environmental geographers study the patterns created on the landscape by such interactions and the processes involved, such as social relations of production, cultural adaptations or maladaptations to particular environments, modifications to the environment wrought by human economies and technologies, and the use (or misuse) of natural resources.

UNIFYING THEMES IN GEOGRAPHY

The discussion in the previous section illustrates the diverse nature of the academic discipline of geography. In spite of this diversity, however, we can identify some common themes with which most geographical studies are concerned and which unify the field. Three common concepts have already been identified: pattern, process and landscape. In addition, we can delineate five unifying themes in geographic research:

LOCATION

This theme addresses the question, "where is it at?" We can think of location in two primary ways. **Absolute location** refers to a specific point on the globe—latitude and longitude, for example. On the other hand, **relative location** refers to the location of a place or phenomenon with respect to other places or phenomena around it. For example some places may occupy a more central role with respect to the economy of a country. Such places may be said to occupy a more central relative location.

PLACE AND SPACE

This theme refers to the description and analysis of the patterning of phenomena in a certain place or across space. This patterning can be analyzed in terms of the distribution, density and concentration of phenomena.

HUMAN-ENVIRONMENT INTERACTION

This theme explores the complex relationships between human beings and their natural environment, how people interact with the environment around them, the nature of that interaction and the influence of the natural environment on human culture. Interpretations of the role played by the environment in human affairs and the nature of human-environment interaction have changed substantially over time. The ancient Greeks, for example, attributed observed cultural differences around the world to the climates in which people lived. It was argued that African cultures differed from European cultures because African societies evolved in a hot climate. That is, they argued that the environment plays a deciding role in terms of shaping cultural values, ideals and traditions—certain environments and climates create certain cultures. This supposition is known as **environmental determinism**. Surprisingly, this argument continued to be used by many social scientists to account for cultural differences around the world well into the 20th century. Some prominent American geographers in the early 20th century, for example, argued that a unique subculture punctuated by poverty, subsistence agriculture and a distinctive folk culture (food, music, social systems, etc.) developed in Appalachia because of the region's mountainous terrain and geographical remoteness. But under close scrutiny, of course, environmental determinism is an untenable supposition that does not explain reality. Culture is not created by the climate of a place or by mountains—culture is learned behavior passed on from generation to generation. By the 1940s the use of environmental deterministic arguments to explain culture began to wither under close scientific scrutiny. Led by French human geographers beginning in the 1930s, the idea of **possibilism** began to replace determinism. This supposition argued that the environment surrounding us offers certain possibilities, but what people do with those possibilities is determined by cultural values and traditions which are learned and which evolve over time. Today, geographers have modified possibilism into a workable hypothesis that may be termed **environmental perception**. This hypothesis argues that different cultures perceive or conceptualize the environment in which they live in different ways. These differences in perception may be attributed to differences in cultural values and traditions, which result in the creation of varying cultural landscapes. In this way, it is argued, any cultural landscape can be "read," analyzed and deconstructed to understand the values, ideals and traditions (the "culture") of the people that created that landscape.

MOVEMENT

This theme seeks to understand how and why observed phenomena move through space—"how did this pattern or distribution come to be?" The central concept employed by geographers to explain the movement of people, ideas and innovations from place to place is **diffusion**. In this sense, cultural diffusion may be thought of as the movement of ideas, traditions and innovations through space. Three different kinds of movement can be employed to explain how diffusion takes place. *Hierarchical diffusion* describes the movement of ideas and innovations in stair-step fashion from person to person: one person tells two people and they tell two people and they tell two people, and so on. This kind of diffusion is relatively slow and it may take weeks, months or years for ideas to move from one place to another. Before the advent of mass communication technology in the early 20th century, most diffusion was hierarchical in nature. A second type of diffusion is *contagious diffusion*. Contagious diffusion refers to the rapid movement of innovations and ideas through space in the same manner that a contagious disease spreads through a population. Many people (perhaps in the millions) become aware of an idea or innovation at the same time. In today's age of globalization, the internet and highly-evolved methods of mass communication, contagious diffusion is the primary way in which ideas move from place to place. A third type of diffusion is known as *relocation diffusion*. Relocation diffusion describes the movement of ideas over long distances in a relatively short time, usually attributed to mass migrations of people from one place to another place. The fact that the majority of North Americans speak English as a mother tongue and practice some form of Christianity can most likely be attributed to the migration of millions of Europeans to North America over some 400 years (relocation diffusion) and the subsequent movement across the continent (contagious diffusion).

REGIONS

The **region** is the primary unit of analysis for the geographer. A region is simply an area within which there is homogeneity of a certain phenomenon or certain phenomena—that is, homogeneity through space. Both physical and human geographers use the concept of region to describe and analyze places and subject matter. In this sense, a **culture region** is a place within which a certain culture is predominant. Geographers identify three primary types of regions. A **formal region** is a region that is easily identified through the use of verifiable data. The distribution of German speakers in Europe or the region in the United States in which winter wheat is grown is an example of a formal region. A **functional region** is a place within which there is movement. The predominance of a certain kind of movement in that place defines the region itself. The best example of a functional region is a trade area or a market area. Finally, a **perceptual region** is a region that is not easily identifiable, the parameters of which may vary from person to person. The "Midwest" and the "South" in the United States are examples of perceptual regions. The location of both of these regions is not exact and may vary from person to person based on their own perceptions.

WORLD-SYSTEMS ANALYSIS

The main goals of this text are to give students a basic understanding of the human geography of the world, to delineate the world's major culture regions, to illustrate how human culture and cultural landscapes vary across space, and to examine the phenomenon of economic "development" and "underdevelopment." In order to accomplish these goals effectively we will employ the concepts of pattern, process and landscape, as well as the five themes outlined above. In addition, **world-systems analysis** is used as a thematic context around which the book is structured. World-systems analysis is best described as a distinctive approach to the study of social change developed by the American sociologist Immanuel Wallerstein in the early 1970s. This approach combines economic, political, sociological, geographical and historical aspects in a holistic historical social science that is based on three different research traditions: dependency theory, the French *Annales* School of history (especially the work of Fernand Braudel), and Marxist theory.

World-systems analysis is chiefly concerned with analysis of the nature, development and structure of the capitalist world economy and provides a useful context within which to compare and contrast regions of relative "development" and "underdevelopment" in the world's economy through time and space. In his three-volume work *The Modern World-System*, Wallerstein maintains that there have been only three fundamental ways, or **modes of production**, in which societies throughout history have been organized in order to sustain production. These modes of production, these societies, can be distinguished by determining the division of labor in production that is dominant in these societies.

The *reciprocal-lineage mode of production* refers to a society in which production is differentiated mainly by age and gender, and exchange is merely reciprocal in nature (that is, barter exchange dominates the economy). Wallerstein calls such systems **mini-systems**, and there have been countless numbers of these throughout history. Such societies are usually small in terms of population and geographical area and there are few, if any, class divisions (anthropologists refer to such societies as *tribes*).

The *redistributive-tributary mode of production* describes a society that is class-based and in which production is performed by a large agricultural underclass that pays tribute to a small ruling class (anthropologists refer to such societies as *chiefdoms*). Wallerstein calls these societies **world-empires**, and there have been many of them since the Neolithic Revolution around 10,000 B.C. (examples would include the ancient kingdoms of Mesopotamia and the Nile Valley, the Mayan and Aztec societies in Middle America, various kingdoms in western and southeastern Africa, the ancient dynasties of China, and the Ottoman and Roman empires). World-empires were very large in terms of population and geographical size.

Finally, the *capitalist mode of production* also refers to a class-based society, but is distinguished by the goal of a ceaseless accumulation of capital operating within a market logic and structure (capitalism). Wallerstein calls such societies **world-economies**. According to Wallerstein, there has been only one successful world-economy. It originated in Western Europe around the middle of the 15th century and spread to encompass the entire world through long-distance sea trade, colonialism and world conflicts by the beginning of the 20th century. World-empires and world-economies are called *world-systems* by Wallerstein because the divisions of labor operating within them are larger than any one local grouping.

In world-systems analysis, then, the world is seen as a single entity—a capitalist world-economy. Wallerstein argues that in order to meaningfully understand the nature of the global economy and of social change one must not consider the role of individual countries so much as the entire world-system. To do so is to commit the fundamental **error of developmentalism**, which dominates current liberal studies of development and Marxist analyses, both of which see individual countries progressing through stages of "development."

SEMI-PERIPHERAL STATES

The **semi-peripheral states** today include many of the countries of Middle and South America (e.g. Mexico, Costa Rica, Chile, Brazil, and Argentina), Eastern and Southeastern Europe (e.g. Russia, Poland, Bulgaria, and Hungary), Southwest Asia (e.g. Turkey, Saudi Arabia, and Iran) and Southeast and East Asia (e.g. Indonesia, Malaysia, Thailand, China and South Korea). Other semi-peripheral "outliers" include such countries as South Africa and much of Saharan Africa (e.g. Tunisia, Algeria, and Libya). As the term implies, the semi-periphery occupies a "middle" place in the hierarchy of "development" in the world-economy. If one considers socio-economic indicators and demographic data, it is clear that the data for semi-peripheral countries are midway between the extremes of the core and periphery—per capita incomes, birth and fertility rates, and rates of natural increase in the semi-periphery are neither the highest nor the lowest in the world, but rather somewhere in between. Accordingly, many world-systems analysts characterize semi-peripheral countries as not the richest, but certainly not the "poorest" countries in the world.

Many, but not all, of these countries are also former colonies. But since the 1960s most of them have managed to achieve some amount of economic and political stability, largely with loans from the World Bank and foreign aid from the core countries. But as a result, many are saddled with large debts to the World Bank and banks in the core. Economic stability in these countries was largely achieved through a focus on heavy industry and manufacturing as central features of the economy under the authority of rather strong (and sometimes corrupt and heavy handed) central governments. At the same time, most semi-peripheral economies are still heavily dependent upon the agricultural sector. Indeed, one of the most characteristic aspects of the semi-periphery is a mixed economy dependent upon agriculture (largely for export to the core), heavy industry and manufacturing, and a small but growing service sector. Social stratification in semi-peripheral societies reflects this mixed economy: a large rural agricultural lower class; a relatively large urban blue collar manufacturing class; and a small, wealthy urban professional class.

This intense social stratification—a vast difference between the richest and poorest members of society—is one of the hallmarks of the semi-periphery. According to Wallerstein, the most pronounced and acute class struggle occurs in the semi-periphery. This class struggle is often accompanied by chronic political and economic instability. The semi-periphery is also the focus of periodic restructurings of the world-economy during times of economic stagnation, which provide the necessary conditions for this restructuring. For example, it is usually the semi-periphery that is most adversely affected by crises in the world-economy. During the Industrial Revolution in the late 18th and early 19th centuries, for example, traditional agricultural and artisan economies in places such as Germany and Ireland (part of the semi-periphery at that time) were upset by the changes wrought by industrialism in Great Britain. Many farmers and artisans who could no longer make a living at home moved to core regions like Great Britain and the United States in order to take jobs in urban factories, thus supplying the core with a needed industrial workforce. In today's world-economy a similar situation is occurring in the semi-periphery. This time around traditional economies are being reordered by the current restructuring usually referred to as "globalization." This restructuring involves the outsourcing of manufacturing jobs by multi-national firms from the core to the semi-periphery, especially in the textile industry (the manufacturing of clothes, shoes and the like). At the same time, this economic reordering has again resulted in traditional economies being upset and phased out. One of the consequences of this has been a renewed large-scale migration of low skilled farmers and laborers from the semi-periphery (Latin America, East and Southeast Asia, Southwest Asia) to the core (Western Europe, North America, Australia).

This has resulted in a pronounced international division of labor characterized by economic specialization in each of the three regions of the world-economy. Peripheral economies are dominated by subsistence agriculture, plantation agriculture and natural resource extraction, all mainly for export to the semi-periphery and core. While local, low-wage labor is employed in the production of these resources and products, the capital and management is often controlled from or by the core in the form of multi-national corporations. Semi-peripheral economies specialize in small-scale commercial agriculture, heavy industry and manufacturing (steel, chemicals, etc.) and textile production, the latter for export primarily to the core. Core economies are highly diversified but are primarily service-based. That is, most workers are employed in the service sector of the economy, which includes everything from retain sales to real estate, banking, health care, education, government and high-tech industries such as computer software production. Commercial agriculture is an important part of the economy in all of the core countries, but relatively few people make a living wholly as farmers (typically less than 10% of the population). While heavy industry and manufacturing was the mainstay of the industrial economies of the core from World War II until the 1970s, employment in this sector of the economy and its overall importance to the economies of the core have declined dramatically over the past twenty years.

SUMMARY

This section has discussed the nature of the discipline of geography, outlining its main themes, its place in academia, and its distinctive way of looking at the world around us. It has also introduced some of the main themes with which this text is concerned and outlined its primary goals, namely general patterns in the human geography of the world, with an emphasis on the following:

→ A basic understanding of the major culture regions of the world, where these regions are located, their general characteristics, and their attendant cultural landscapes

→ An appreciation of how populations vary from place to place around the world, what populations "look" like, how they are structured, and various problems and policies related to population in different parts of the world

→ A comprehension of the various ways in which people make a living around the world, how economies and societies are structured in different parts of the world, and the nature of the world-economy today and in the past

→ An understanding of the concepts of economic "development" and "underdevelopment" and how critically important these are in understanding the nature of economies, societies, and ways of life around the world both today and in the past

This section has also introduced the basic premises of world-systems analysis and presented it as a general model for understanding the development of capitalism in the world-economy and for comprehending its nature and structure today. The three-tiered spatial and economic hierarchy of core, semi-periphery, and periphery will be used as a thematic context and as a model, within which various regions of the world might be placed. It is argued throughout the text that virtually all aspects of differences in the human geography of the world, especially with respect to the goals outlined above, can be more fully understand within the context of this world-systems model. As is the case with any model, it must be understood that this model simplifies reality to a certain extent. The model is useful in understanding *general*, global patterns and differences, especially at the regional level. It is hoped that through the application of this model, students will gain a fuller comprehension of the myriad ways in which the human geography of the world varies from place to place and from region to region.

INTERNATIONAL ORGANIZATIONS*

And now we get to the trendiest world trend of globalization. This section consists of brief explanations of some entities that fall outside, or rather across, state boundaries—global players in a global age. We call them supranationalist organizations. Above and beyond the national level, these organizations play an increasingly important role in what is happening

*Pages 20–41 from *The Plaid Avenger's World #7 Ukranian Unraveling Edition*. Copyright © 2008, 2009, 2010, 2011, 2012, 2014 by John Boyer. Copyright © 2006 by Kendall Hunt Publishing Company. Reprinted by permission.

across our planet. But who are they? Where did they come from? How are we supposed to know this stuff? I don't know, friends. If the Plaid Avenger doesn't tell you these things, who will?

Supranationalist organizations are groups of states working together to achieve a common, or outlined, objective. This is another fairly new concept in human history, as states or nations have spent most of their time doing the opposite: beating the daylights out of each other or undercutting each other at every available opportunity. So why do countries now work together? The Plaid Avenger sees order in the universe; we can classify cooperation into three main classes: economic, defensive, and cultural.

I'll introduce you to the more important and happening entities here, but by no means is this list exhaustive. This section will also serve as a functioning reference for you as you progress through the rest of the book; come back often to refresh your memory when you see these acronyms appear in the regional chapters.

ECONOMIC ENTITIES—SHOW ME THE MONEY

Money. Who doesn't want it? Not any of the states of the world, that's for sure. A great way to make more money, if you are a country, is to make some trade deals with other countries. I'll buy all my bananas from you if you buy all of your wheat from me—sound good? On top of that, I won't put an import tax on your bananas, but if any other countries try to sell bananas here, I'll tax the heck out of them. Deal? This is the essence of **trade blocks** which are, as you might have guessed, a dandy vehicle for increasing trade between two countries . . . or perhaps among a whole bunch of countries, depending upon how many new kids are in your bloc.

Many economists believe that **free trade** between countries increases competition, which decreases prices for consumers, which in turn increase consumption of products . . . which ultimately benefits producers and consumers alike! Get governments out of the way, and let the market rule! For this reason, both the United States and the European Union are trying their hardest to promote "trade blocs" or "free trade zones" with neighboring countries, so that they can improve economic performance and increase sales. Even countries in Latin America, Africa and Asia have caught the bug.

However, there is a tug-of-war going on. Independent sovereign states naturally want to protect their own economies, so they are reluctant to sign up for free trade when they think that their local industries may lose the trade game. For example, if Chinese shoe companies make cheaper and better shoes than French shoe companies, France is not going to want free trade in shoes with China. Everybody in France might want to buy the less expensive Chinese shoes, so the French shoe companies would go out of business. Historically, the cheaper imported products are hit with a **tariff**, an import tax, which subsequently makes the price of the product more expensive, and thus the local products can compete better.

Many times, poor countries accuse rich countries of trying to take advantage of them by using free trade agreements. These poor countries argue that free trade isn't equally beneficial for both sides; that it's mostly beneficial for the fully developed, industrialized country because their companies are more competitive. Furthermore, they accuse developed countries of cheating, and they point to agricultural **subsidies** in these rich countries as an example. Farmers in Europe and America produce crap tons of food using lots of big equipment and fertilizers, thus their costs are high, and subsequently the food they make costs more. Farmers in poor countries don't use that expensive stuff and have cheaper labor, therefore their food should cost much less, giving them a competitive advantage in the world market. However, the rich farmers still "win" on the international market because Europe and America give their farmers huge subsidies to offset the higher costs of production they face. Uncle Sam gives American farmers money just to be farmers, and the farmers can turn around and sell their food for cheaper prices and still make money. You dig? And if you dig a lot, maybe you should become a farmer.

Poor countries argue that the only reason that rich countries became rich in the first place is by protecting their domestic industries by using things like tariffs and other forms of **protectionism**. Also, it can be argued that fully developed mega-rich companies from mega-rich countries are so technologically superior that they have a competitive advantage that can never be overcome . . . meaning that the less developed states will always be stuck buying finished goods and selling primary level commodities, thus always losing money. On the other hand, free trade usually does mean more trade, so the less developed countries do stand to sell much more oil or lima beans or flip-flops or beef lips. Poorer countries are torn as to whether or not it is in their best interest to join these trade blocks with the fully developed states.

Perhaps it's on these grounds that we are seeing many new trade blocks springing up that are comprised solely of states in "developing status," with no "rich kids" invited to the party. It certainly is the reason for the foot-dragging with the FTAA, but once again, I have gotten ahead of the story.

Check out these economic entities that you will be hearing a hell of a lot more about, as they will play an increasingly larger role in the way the global economy operates:

NAFTA

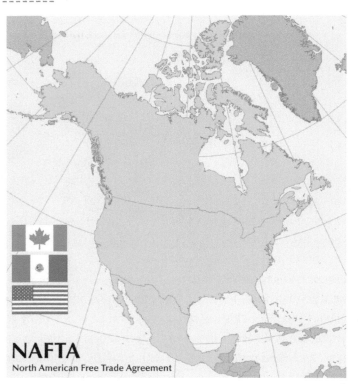

NAFTA
North American Free Trade Agreement

Members: United States, Canada, and Mexico

Summary: NAFTA, which stands for North American Free Trade Agreement, is between the United States, Mexico, and Canada, enacted in 1994. This agreement is meant to gradually eliminate all duties and tariffs on all goods and services between these three countries. However, the three nations are resisting lowering specific barriers that would hurt specific components of their economies. For example, the United States and Canada have been bickering because the United States imposes a duty on Canadian lumber that goes to the United States. The Canadians are accusing the Americans of not sticking to the treaty and are considering imposing duties on American goods to retaliate.

NAFTA has been very controversial in other ways too. Generally, multinational corporations support it because lower tariffs mean higher profits for them. Labor unions in the United States and Canada have opposed it because they believe that jobs will go from the United States and Canada to Mexico because of lower wages there. They were all correct: manufacturing jobs, particularly from the auto industry, migrated rapidly south of the border where wages were significantly lower, but this did make the costs of products cheaper to the consumer as well. American jobs were lost, but Americans pay lower prices for goods. Also, farmers in Mexico oppose it because agricultural subsidies in the United States have forced them to lower the prices on their goods. But make no bones about it: NAFTA has been an incredible plus to the 3 countries, as trade has exploded; more goods and more services now flow between the North American titans than ever before, and it looks set to expand into the future.

Fun Plaid Fact: Chapter 11 of the NAFTA treaty allows private corporations to sue federal governments in the NAFTA region if they feel like that government is adversely affecting their investments.

DR-CAFTA

Members: United States, Costa Rica, Dominican Republic, El Salvador, Guatemala, Honduras, Nicaragua. Currently, the US Administration is pushing hard to get Colombia and Panama into this club as well.

Summary: DR-CAFTA stands for Dominican Republic—Central America Free Trade Agreement and is an international treaty to increase free trade. It was ratified by the Senate of the United States in 2005. Like NAFTA, its goal is to privatize public services, eliminate barriers to investment, protect intellectual property rights, and eliminate tariffs between the participating nations. Many people see DR-CAFTA as a stepping stone to the larger, more ambitious, FTAA (Free Trade Agreement of the Americas).

The controversy regarding DR-CAFTA is very much like the controversy regarding NAFTA. Many people are concerned about America losing jobs to poorer countries where the minimum wage is lower and environmental laws are more lax. Also, some people are concerned that regional trade blocs like DR-CAFTA undermine the project of creating a worldwide free trade zone using organizations like the WTO.

Fun Plaid Fact: Many Washington insiders see DR-CAFTA as a way of reducing the influence of China in Central America.

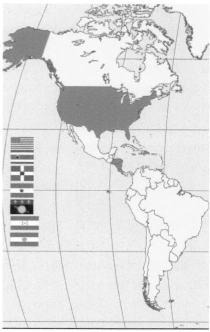

CAFTA members

FTAA (PROPOSED ONLY!)

Members: PROPOSED All of the nations in North and South America, except Cuba. 'Cause the US hates commies. Dirty pinko commies.

Summary: The FTAA, which stands for Free Trade Area of the Americas, is a proposed agreement to end trade barriers between all of the countries in North and South America. It hasn't been ratified yet, because there are some issues that need to be worked out by the participating countries. The developed (rich) countries, such as the United States, want more free trade and increased intellectual property rights. The developing (poorer) countries, especially powerhouse Brazil, want an end to US/Canadian agricultural subsidies and more free trade in agricultural goods.

The key issue here for poor countries is agricultural subsidies. Farmers in the United States, and in rich countries generally, produce agricultural goods at a higher price than poor countries do. However, to keep their goods cheap, and thus competitive on the world market, the government of the United States pays their farmers subsidies. These subsidies make developing countries very angry because they believe subsidies give American farmers an unfair advantage. For this reason, some Latin American leaders have stalled the agreement. Former Venezuelan President Hugo Chavez called the agreement "a tool of imperialism" and proposed an alternative agreement called the Bolivarian Alternative for the Americas.

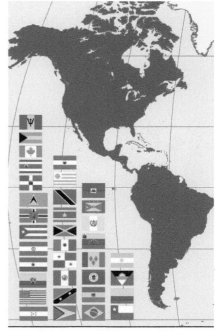

Proposed FTAA members

For rich countries, the issue is intellectual property rights, which is best exemplified by copyright laws. Less developed countries sometimes oppose these rights because they believe that if they are enacted, they will stifle scientific research in Latin America and widen the gap between the rich and poor countries in the Americas.

It should be noted, as of this writing in 2014, that this tentative agreement is still stalled by more independent and "leftist" Latin American states that don't want to join a group that the US would likely dominate . . . especially Brazil, which sees itself as the natural true leader of an economically united Latin America. The "leftward swing" of Latin America, that we'll talk about in a later chapter, has seriously squashed the US administration's agenda on this issue.

Fun Plaid Fact: The only country that would not be included in the FTAA is Cuba, because the United States has an economic embargo that prohibits all trade with the communist regime. For this reason, Cuba more fully supports the Bolivarian alternative for the Americas, which is MERCOSUR . . . we will get to that on the next page.

EU

Members: Belgium, Bulgaria, France, Germany, Italy, Luxembourg, The Netherlands, Denmark, Ireland, United Kingdom, Greece, Portugal, Spain, Austria, Finland, Sweden, Cyprus, Czech Republic, Estonia, Hungary, Latvia, Lithuania, Malta, Poland, Romania, Slovakia, Slovenia

Summary: For years, European philosophers and political observers have recognized that the best way to ensure peace on the European continent while also increasing trade is to politically and economically integrate the nations. After the destruction and loss of life caused by World War II, European nations finally began taking small steps toward interdependence. They started by integrating their coal and steel industries in the ECSC (European Coal and Steel Community). What a long way they have come since then! Currently, the European Union, which has 27 member states, has a common market, a common European currency (the euro), a European Commission, a European Parliament, and a European Court of Justice. The nations of the EU have also negotiated treaties to have common agricultural, fishing, and security policies. More so than any other free trade agreement, the European Union covers way more areas other than just trade.

Consequently, the EU is the most evolved supranationalist organization the world has ever seen—perhaps a "United States of Europe." Free movement of people across international borders of the member states makes it unique in the trade block category. Of greater importance is an evolving EU armed force, a single environmental policy, and increasingly, a single foreign policy voice. That is a very big deal!

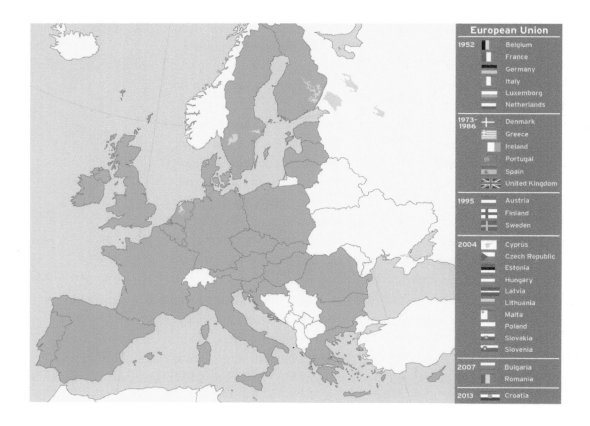

However, there has been some resistance to integration within European countries. Some countries, like Norway and Switzerland, have refused to join, and others, like the United Kingdom, have refused to fully adopt the euro: the Brits do use the euro, but maintain their traditional pound as well. Also, in 2005, a constitution for the European Union was rejected by French and Dutch voters, putting the future of European integration into question. Many Europeans simply don't care about the European Union and others see it as a secretive, undemocratic organization that is taking away power from their home countries. Some of the richer countries in Europe are afraid that adding nations with weaker economies will take away money from them just to give it to less productive economies.

But make no bones about it, the EU unification and expansion has made Europe a global power once again. Individually, these European countries are rich places, but none could compete on their own with the likes of the US or even China. However, as a unit, the EU has the largest GDP on the planet. After a serious growth spurt in the last decade, it appears that the expansionism may be played out. Future candidates include Croatia, Bosnia, Macedonia, Albania, and possibly Turkey . . . but Ukraine and Georgia are now seen as too hot to handle for the EU given Russian resurgence of influence in these areas. Russia itself is usually invited to big EU talks as kind of an associate member already, although the idea of Russian ascension into the EU will never happen. Not everybody likes it, but the EU makes Europe a player in the world economy and world political terms. Divided, they are not much. United, Europe still has a big voice.

MERCOSUR

Members: Brazil, Argentina, Uruguay, Paraguay, with the newest addition: Venezuela!

Associate Members: Bolivia, Chile, Colombia, Ecuador, Peru, Guyana, & Surinam.

Summary: MERCOSUR is a free trade agreement between several South American countries that was created in 1991 by the Treaty of Asuncion. Like other free trade agreements, its purpose is to promote free trade and the fluid movement of goods and currency between the member countries. Many people see MERCOSUR as a counterweight to other global economic powers such as the European Union and the United States.

MERCOSUR's combined GDP is only 1/12 of the United States', standing at 1 trillion dollars. But watch out! Brazil is becoming a serious global player with a booming economy, and actually the entire group is prospering. They are further integrating on economic and political decision, and even talking about free movement of workers and a standard labor law throughout the block. Good stuff! However, keep an eye on this one as leftist events unfold in Latin America. It could become a viable force. Already, MERCOSUR has now added all members of the Andean Community (another economic bloc) as "observer states" with the grander goal of uniting the entire continent in super-block called the Union of South American Nations (UNASAR)!!! Wow!

**MERCOSUR
(MERCOSUL)**

Member states
Associate members

Fun Plaid Fact: I have been watching to see how serious this UNASAR concept is, and the concrete steps taken just in the last few years indicate that it will likely soon supercede MERCOSUR itself, and thus be an added entry in the next edition of this book!!!!

ASEAN

Members: Brunei, Cambodia, Indonesia, Laos, Malaysia, Burma, Philippines, Singapore, Thailand, Vietnam

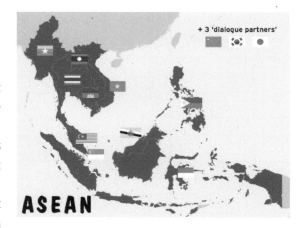

Summary: ASEAN, which stands for Association of Southeast Asian Nations, is a free trade bloc of Southeast Asian countries. Like the European Union, ASEAN may possibly evolve into more than just a free-trade zone; it aims for political, cultural, and economic integration. It was formed in 1967 as a show of solidarity against expansion of Communist Vietnam and insurgency within their own borders. During that time, many countries around Vietnam were turning communist, and capitalist governments were extremely worried that communism might infect them as well. However, even Vietnam has joined ASEAN since then.

ASEAN is significant because of the heterogenous nature of its constituent countries. ASEAN countries are culturally diverse, including Muslims, Buddhists, and other religions. Governments in ASEAN range from democracy to autocracy. The economies of ASEAN countries are also very diverse, but they mainly focus on electronics, oil, and wood.

These guys are increasingly modeling themselves after the EU experiment, too. Even though they remain much more nationalistic than the European countries, the ASEAN group has much bigger goals than to simply be trade block. A common electric grid across the member countries has been proposed; an "open-sky" arrangement is soon taking effect (free movement of all aircraft among member states); and common environmental policies are being adopted across the region.

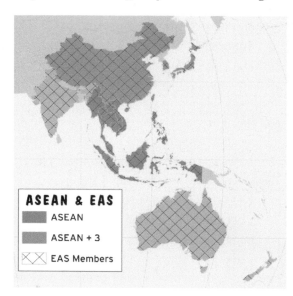

The most interesting ASEAN prospect is now focused on, of all things, human rights! They actually wrote and adopted a legitimate constitution/charter which mentions the idea of protection of equal rights for all. This from a club which, at the time, had the dictatorship of Burma as a member! Interesting Asian times ahead for the ASEAN.

In addition, during annual ASEAN meetings, the three 'Dialogue Partners' (China, Japan, and South Korea) meet with ASEAN leaders . . . and when this happens, it is referred to as **ASEAN+3**. Wow! Those 3 economic titans as semi-associate members of ASEAN? That is a serious economic situation!

And I can do you one better: now ASEAN has become the core of the annual **East Asia Summit** which is the ASEAN+3 plus India, Australia, and New Zealand! The summit discusses issues including trade, energy, security and regional community building. Is this the early formation of an Asian EU? Who knows! But it's a fascinating development!

PLAID ALERT! Holy trade tips! Keep your eyes on this one as well. This could turn into the largest, richest, baddest block on the planet in your lifetime. Booming economies in most member states with more growth in the future. Bigger deal than that: China, South Korea, Japan and India as "associates" members? Are you kidding me? That's like over half of the entire planet's population under one economic umbrella. Watch out! They are going to be hot!

APEC

Members: Australia, Brunei, Canada, Chile, China, Indonesia, Japan, Korea, Malaysia, Mexico, New Zealand, Papua New Guinea, Peru, Philippines, Russia, Singapore, Taiwan, Thailand, United States, Vietnam—pretty much all the guys with Pacific coastline.

Summary: The Asia-Pacific Economic Cooperation trade bloc is a group of Pacific Rim countries that meets with the aim of improving economic and political ties. Like most free-trade blocs, the goal of APEC is to eventually reduce tariffs to nothing. Also, like many free trade agreements, agricultural subsidies have become a point of controversy. The leaders of all APEC countries meet annually in a summit called "APEC Economic Leaders' Meeting" which meets in a different location every year. The first of these meetings was in 1993 and was organized by US president Bill Clinton.

The countries in APEC are responsible for the production of about 80 percent of the world's computer and high-tech components. The countries in many of the Pacific Rim are also significant because the population in many of these countries is increasing dramatically. This trade bloc could possibly become a huge force in the global economy in the near future. Or it could totally be replaced by a proposed Pacific pact being pushed hard by the USA named the TPP. The Trans-Pacific Partnership: a very similar economic union that makes even deeper ties between the states, while simultaneously excluding China from the group. Currently only proposed, but being put together at such a frantic pace that it will likely be included in the next edition of this book . . . in which case APEC gets deleted!

OECD

Members: Austria, Belgium, Canada, Denmark, France, Germany, Greece, Iceland, Ireland, Italy, Luxembourg, Netherlands, Norway, Portugal, Spain, Sweden, Switzerland, Turkey, United Kingdom, United States, Japan, Finland, Australia, New Zealand, Mexico, Czech Republic, Hungary, South Korea, Poland, Slovakia

Summary: The Organisation for Economic Co-operation and Development (OECD) is an international organization of countries that accept the principles of democracy and free markets. After World War II, when Europe was in ruins, the United States gave European countries aid in the form of **the Marshall Plan** to rebuild the continent and repair the economy, while also ensuring that European countries remain democracies. The Organisation for European Economic Co-operation (OEEC) was formed in 1948 to help administer the Marshall Plan. In 1961, membership was extended to non-European countries and renamed the OECD. Because it contains most of the richer, more developed states around the world which are all democracies to boot, the OECD is really kind of the core of what I refer to as "Team West" throughout this text.

Like many trade agreements, the purpose of the OECD is to promote free trade, economic development, and coordinate policies. The OECD also does a lot of research on trade, environment, agriculture, technology, taxation, and other areas. Since the OECD publishes its research, it has become one of the world's best sources for information and statistics about the world.

Fun Plaid Fact: While you will still see many references to OECD in news/literature, it has never really been a really pro-active group like the other ones discussed so far, and is of decreasing significance altogether here in the 21st century as regional (specifically Asian) trade blocks have proliferated and dominated.

DEFENSE

Why should countries get together defensively? If they are all on the same team, then they won't fight—right? Well, that's the emphasis of the UN. But perhaps more pertinent are regional defense blocks that have cropped up between countries throughout history. Their thinking is more along the lines of: "I'll help you if you get attacked by an outsider, if you help me if I'm attacked by an outsider." If this sounds like trivial schoolyard thinking, don't laugh; the basis for World War I was a whole host of such pacts between European countries—once one country was attacked, virtually every other country was immediately pulled in as a consequence of defense agreements. Here are the big three that are pertinent in today's world—even though one of them is now gone—plus a fascinating newcomer with tremendous future potential growth. But first, the easy ones . . .

THE UNITED NATIONS

UN Secretary General Ban Ki-Moon

Members: All the sovereign states in the world except Vatican City. There are currently 193 of them. Even the Swiss finally joined a few years ago.

Summary: The United Nations, or UN, was founded in 1945 as a successor to the League of Nations. Like the League, the goal of the UN is to maintain global peace. Unlike the League, no major world wars have happened on the UN's watch. This is not to say that the United Nations has achieved global peace. In fact, UN "peacekeepers" have been on hand to witness some of the most egregious violations of human rights in recent history.

The UN is made up of several bodies, the most important of which is the Security Council (see Security Council section). The second most important body in the UN is the General Assembly where each of the 193 member nations has a representative and a vote. The General Assembly has produced gems such as the *Universal Declaration of Human Rights* and the lesser known *International Convention on the Protection of the Rights of All Migrant Workers and Members of Their Families*. The General Assembly is clearly the home of utopian thinkers, but not of any real international power. This leaves the major world powers like the US and China free to ignore everything that the General Assembly says, without even having to waste the time vetoing it.

The UN also includes hundreds of sub-agencies that you've heard of before, such as the World Health Organization (WHO) and UNICEF. The WHO is in charge of coordinating efforts in international public health. UNICEF (The United Nation's Children Fund) provides health, educational, and structural assistance to children in developing nations. Both agencies are supported by member nations and private donors. UNICEF also receives millions of pennies collected each year by children on Halloween. Just a handful of the hundreds of other UN agency acronyms you may have heard of include the FAO, IAEA, UNESCO, IMF, WMO, and the WTO.

Critics often charge the UN with being ineffective. This is largely true, but the United Nations was never really intended to be a global government. The best way to view the UN is a forum in which nations can communicate and work together. The UN is ill-equipped to punish any strong member for violations. If a member is especially naughty, a strongly worded UN resolution might recommend voluntary diplomatic or economic sanctions. Perhaps after World War III, the United Nations will be once again renamed and given stronger international authority. If there is anything left of us.

THE REAL POWER AT THE UN: THE UN PERMANENT SECURITY COUNCIL

Members: US, UK, Russia, China, and France and 10 other rotating positions.

The Security Council is composed of five permanent members (the United States, the United Kingdom, Russia, China, and France) and ten other elected members serving rotating two year terms. The Security Council is charged with responding to threats to peace and acts of aggression. Basically, for anything to get done, the Security Council has to do it. But things rarely get done because each of the five permanent members has the power to veto and prevent any resolution that they do not like. A single veto from any one of the permanent members kills the resolution on the spot. This group of rag-tag veto-wielding pranksters is currently the ultimate source in interpreting international law. Most of the Cold War saw little to no consensus on anything, as Team US/UK faced off against Team Russia/China. Whatever one team tried to push, the other team generally would veto. The Frenchies vetoed according to mood and lighting of the room. Even today, votes tend to fall along these same alliance lines.

The other ten rotating members of the Council do not have veto power, but are often used as a coalition building tool to get things done. E.g.: During the build-up to the most recent war in Iraq, the US worked very hard to get as many members of the Council as possible to back the resolution to invade Iraq, knowing full well that China and Russia would veto it. This was a strategic move to show broad support for the war, even though the US accepted up front that the resolution would not be passed.

There is currently speculation that new members may be added to the UN Permanent Security Council. The prime candidates are Germany and Japan. The United States supports their candidacies; maybe because they have over 270,000 military personnel (including dependents of military) in Germany and Japan combined, and they are staunch US allies. There is also talk of including Brazil or India, or even more remotely, an "Islamic member" or an "African member." But seriously, what incentive does the Security Council have to dilute their powers? Remember, all five would have to agree to let a new member in, so while the United States would certainly support the incorporation of Japan, China would be more likely to tell Japan to go commit **Seppuku**, veto-style. However, the four most likely members (Japan, Germany, Brazil, and India) have released a joint statement saying that they will all support the others' entry bids. The best argument for enlargement is that Japan and Germany are the second and third largest contributors to the UN general fund, and thus deserve more power. Regardless, don't count on the Security Council getting any bigger unless serious global strife starts going down, which it will, sooner or later.

NATO

Members: Bulgaria, Estonia, Latvia, Lithuania, Romania, Slovakia, Slovenia, the United States of America, France, the United Kingdom, Iceland, Spain, Portugal, Germany, Italy, Belgium, Switzerland, Luxembourg, Finland, Poland, the Czech Republic, Hungary, Greece, Turkey, Norway, the Netherlands, Denmark, and Canada. New members inducted in 2009: Albania and Croatia!

NATO, which stands for the North Atlantic Treaty Organization, is a military alliance between certain European countries, Canada, and America. It was originally created in 1949 to serve as a discouragement to a possible attack from the Soviet Union (which never occurred). The most important part of NATO is Article V of the NATO Treaty, which states, *"The Parties agree that an armed attack against one or more of them in Europe or North America shall be considered an attack against them all. . . ."* This is called a **mutual defense clause** and basically means that the United States must treat an attack on Latvia the same as it would treat an attack on Tennessee.

NATO Secretary General Fogh Rasmussen. Don't make him angry. You wouldn't like NATO when he's angry.

Although NATO is a multilateral organization, the United States is clearly the captain of the ship. As a rule, US troops are never under the command of a foreign general. NEVER. Because of this, NATO troops (mainly American) are ALWAYS under American command. The United States also uses NATO countries to base its own troops and station nuclear weapons. Many historians blame the United States for provoking the Cuban Missile Crisis, saying that the Russians only wanted to put nukes in Cuba because the United States had at that time stationed nukes in Turkey (a NATO member).

Since the Cold War, NATO has been looking for a new role in the world. Many of the former Soviet republics have since been admitted to NATO—which, by the way, really ticks off Russia. NATO expansion was promoted as an expansion of democracy and freedom into Eastern Europe. More likely, it was to make sure Russia would never be able to regain

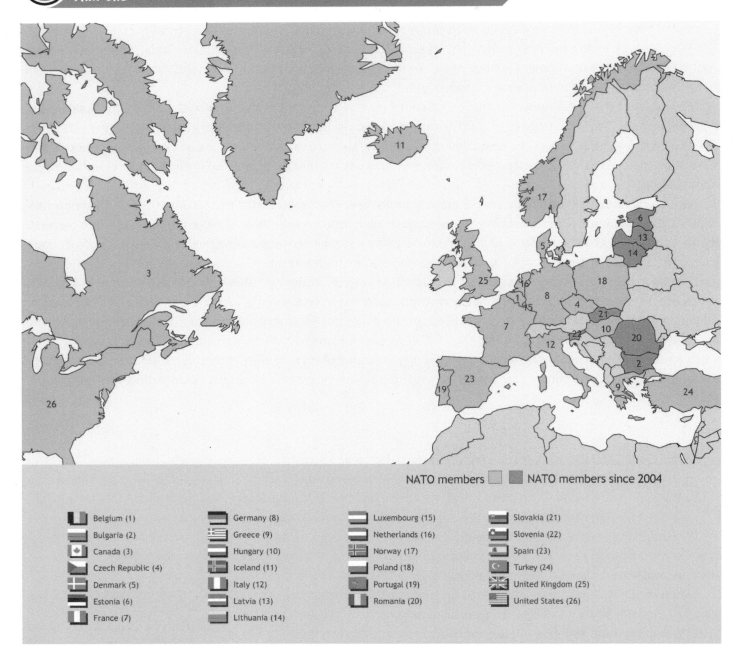

NATO members ☐ ☐ NATO members since 2004

Belgium (1)	Germany (8)	Luxembourg (15)	Slovakia (21)
Bulgaria (2)	Greece (9)	Netherlands (16)	Slovenia (22)
Canada (3)	Hungary (10)	Norway (17)	Spain (23)
Czech Republic (4)	Iceland (11)	Poland (18)	Turkey (24)
Denmark (5)	Italy (12)	Portugal (19)	United Kingdom (25)
Estonia (6)	Latvia (13)	Romania (20)	United States (26)
France (7)	Lithuania (14)		

the territory. NATO has also been increasingly active in international police work, although there is no real justification for this in the NATO charter. NATO forces were heavily involved in the Bosnia conflict in 1994 and the Yugoslavia conflict in 1999, although in reality these were just American troops under a multinational flag. After September 11th attacks on the US, NATO has also become involved in the anti-terrorism game, even invoking Article V for the first time with regard to Afghanistan. Remember, the war in Afghanistan is a NATO mission, not a US mission. But let's be honest here; the US does most of the heavy lifting, as usual.

And the NATO role continues to become broader and more bullish lately: NATO was the central organizing entity in the 2011 invasion of Libya. Say what? What the heck did Libya do to any NATO country? Answer: Nothing, which goes to show how the entity is rapidly redefining itself here in the 21st century. It appears that NATO is fast becoming the military muscle for the objectives of 'Team West,' wether those objectives are defense, economic, or purely political. Interesting stuff, eh? And infuriating stuff to those not aligned with the NATO countries.

Fun Plaid Fact: The only country in NATO without a military force is Iceland. The Icelandic Defense Force is an American military contingent stationed permanently on the island.

WARSAW PACT—DEFUNCT!

PAST Members: Soviet Union (club president), Albania (until 1968), Bulgaria, Czechoslovakia, East Germany (1956–1990), Hungary, Poland, and Romania.

The Warsaw Pact, or if you prefer the more Orwellian Soviet name—the "Treaty of Friendship, Co-operation and Mutual Assistance," was the alliance formed by the Soviet Union to counter the perceived threat of NATO. The Warsaw Pact was established in 1955 (six years after NATO) and lasted officially until 1991 (two years post-Berlin Wall). Much like the Bizarro-Superman to the United State's real Superman, the Warsaw Pact never had the teeth or the organizational strength of NATO. Perhaps this is because many of the members actually hated the dominance of the Soviet Union. Two countries, Hungary (1956) and Czechoslovakia (1961), tried to assert political independence and were subsequently crushed by Soviet military forces in exercises that would make **Tiananmen Square** uprising look like an after school special.

The main idea behind the Warsaw Pact was mutual protection. If the United States attempted to invade any of the Warsaw members, it would guarantee a Soviet response. In this way, the Warsaw countries acted like a tripwire against the expansion of Western-style capitalism and democracy, firmly establishing the location of the Iron Curtain. Shortly after the Cold War, most Warsaw Pact countries either ceased to exist or defected to NATO.

Fun Plaid Fact: The Soviet Union despised American acronyms. Instead of taking the first letter from each word, the Soviets preferred taking the entire first sound. For example, Communist International was "Comintern."

NEW KIDS ON THE BLOC!!!

SCO

Members: China, Russia, Kazakhstan, Kyrgyzstan, Tajikistan, and Uzbekistan

Observer States: India, Pakistan, Mongolia, and Iran

Goodbye, Warsaw Pact! Hello, SCO! Watch this new BLOC! It is evolving fast, with gigantic repercussions for international relations and world balance of power in the future of Asia!

Summary: The Shanghai Cooperation Organization grouping was originally created by 5 member states in 1996 with the signing of the *Treaty on Deepening Military Trust in Border Regions* . . . however with the addition of Uzbekistan in 2001, which brought them up to 6 members, they can't use that wicked cool nickname of "Shanghai 5" anymore. Too bad. But they are are the hottest defense block to keep an eye on, as they are the newest on the scene; so new, in fact, that they have not exactly quite figured out what they are yet. Part military, part economic, part cultural . . . but 100% important to know. Let's just focus on military aspects for now.

SHANGHAI COOPERATION ORGANISATION

■ Member States ■ Observer States ■ Dialogue Partners

Created with that *Deepening Military Trust* issue, they have also agreed to a *Treaty on Reduction of Military Forces in Border Regions* in 1997 and in July 2001, Russia and China, the organization's two leading nations, signed the *Treaty of Good-Neighborliness and Friendly Cooperation* . . . wow, could they get any more sickening sweet with the descriptors? What an Eurasian love-fest. The SCO is primarily centred on its member nations' security-related concerns, often citing its main threats/focuses as: terrorism, separatism, and extremism. You can easily read into this as a great vehicle for all these governments to help each other crack down on any internal political dissidents as well.

They work together thwarting terrorism and stuff, but they are also quickly absorbing other avenues of cooperation, like in domestic security, crime, and drug trafficking. Over the past few years, the organization's activities have expanded to include increased military cooperation, intelligence sharing, and counterterrorism. Of significant note: there have been a number of SCO joint military exercises, and while it is very early to think that the SCO is a serious strategic military power, please keep in mind that the group is still very young and has a long way to go and to grow. We may just be seeing the beginning of their military prowess. Dig this: both Russia and China are nuke powers, Russia has a ton o' of weaponry, and China has the largest standing army on Earth. So this club could become a serious defensive force, if they put their minds to it.

Is this thing sizing up to be a counterbalance to US power or as an overt anti-NATO? Perhaps. But one thing is for sure: its six full members account for 60% of the land mass of Eurasia and its population is a third of the world's . . . and if you include the observer states, they collectively account for half of the human race. The SCO has now initiated over two dozen large-scale projects related to transportation, energy and telecommunications and held regular meetings of security, military, defense, foreign affairs, economic, cultural, banking and other officials from its member states. No multinational organization with such far-ranging and comprehensive mutual interests and activities has ever existed on this scale before. A combo EU/NATO of Eurasia for this century? Could be! So you gotsta' know the SCO! They are fast becoming a playa'!

CULTURAL ORGANIZATIONS

Some supranationalist organizations form out of a desire to maintain a cultural coherence with like countries, or to promote certain aspects of their culture among their member states. In other words, monetary gain is not the driving force behind the organization, although economics usually sneaks in there as well. Here are three very different such organizations to compare and contrast.

ARAB LEAGUE

Members: Egypt, Iraq, Jordan, Lebanon, Saudi Arabia, Syria, Yemen, Libya, Sudan, Morocco, Tunisia, Kuwait, Algeria, United Arab Emirates, Bahrain, Qatar, Oman, Mauritania, Somalia, Palestine, Djibouti, Comoros

The Arab League is an organization designed to strengthen ties among Arab member states, coordinate their policies, and promote their common interests. The league is involved in various political, economic, cultural, and social programs, including literacy campaigns and programs dealing with labor issues. The common bond between the countries in the Arab League is that they speak a common language, Arabic, and they practice a common religion, Islam. The Charter of the Arab League also forbids member states from resorting to force against each other. In many ways, the Arab League can be seen as a regional UN. It was formed in 1945.

However, the Arab League is better known for their lack of coherence and in-fighting more so than any unifying activities they have had so far to date. In fact, Libyan leader Muammar Qaddafi threatened to withdraw from the League in 2002, because of "Arab incapacity" in resolving the crises between the United States and Iraq and the Israeli-Palestinian conflict. If the Arab League ever gets its act together, it could be a powerful force in the world. Right now, it is not.

Let's get this Arab party started!

Radical Arab League Update for 2012: Holy Middle Eastern Madness! The League finally agreed to something, and that something significantly impacted world events! I am referring to the international invasion of Libya in the early months of 2011, a move that was supported by the Arab League! They voted against one of their own too . . . all the other Arab leaders so despised Muammar Qaddafi that they supported outside intervention to help have him deposed! The Arab League's support of this measure was crucial because US/UK/French/NATO/UN intervention would have likely not happened at all if the League would have voted against it. No western power would have wanted to be seen

as a foreign, anti-Islamic, imperialist force invading Libya . . . but once all the other Arab states backed it, then game on! And now the Arab League has united once more: they group is standing with a single opposition voice against fellow Arab Bashar al-Assad, President of Syria currently cracking down brutally on his own people. Saudi Arabia in particular is making plans to arm the Syrian rebels, and the rest of the League will likely support that action! A new era for the Arab group may be dawning.

Fun Plaid Fact: Egypt was suspended from the Arab League from 1979 to 1989 for signing a peace treaty with Israel. Libya has sporadically quit and re-joined the group multiple times depending on how Muammar Qaddafi was feeling at the time. (see more on that in the AU section on next page.) Syria has currently been suspended as well, due in part to the government mass killing of Arabs, aka Syrian citizens.

OAS

Members: Argentina, Bolivia, Brazil, Chile, Colombia, Costa Rica, Cuba, Dominican Republic, Ecuador, El Salvador, Guatemala, Haiti, Honduras, Mexico, Nicaragua, Panama, Paraguay, Peru, United States, Uruguay, Venezuela, Barbados, Trinidad and Tobago, Jamaica, Grenada, Suriname, Dominica, Saint Lucia, Antigua and Barbuda, Saint Vincent and the Grenadines, Bahamas, Saint Kitts and Nevis, Canada, Belize, Guyana

Summary: The OAS, which stands for Organization of American States, is an international organization headquartered in Washington, DC. According to Article 1 of its Charter, the goal of the member nations in creating the OAS was "to achieve an order of peace and justice, to promote their solidarity, to strengthen their collaboration, and to defend their sovereignty, their territorial integrity, and their independence." Other goals include economic growth, democracy, security, the eradication of poverty, and a means to resolve disputes. Historically, the first meeting to promote solidarity and cooperation was held in 1889 and was called the First International Conference of American States. Since then, the OAS has grown, through a number of small steps, to become the organization it is today. Like the Arab League, this organization is a kind of regional UN.

Unlike free-trade blocs, the OAS encompasses many areas other than just trade. For example, it oversees elections in all of its member countries. However, it has also been criticized as a means for America to control the countries in Latin America. For example, when America wanted Cuba kicked out of the OAS, the organization quickly did so. However, many dictatorships that America has supported have remained within the OAS.

But this "US as imperialist" attitude is changing fast in the OAS. US President Barack Obama personally addressed the group in April 2009, pledging to bring more unity to the hemisphere and even hinting at thawing relations with the Cubans. At that meeting, Obama even shook hands with legendary US-hating Venezuelan President Hugo Chavez! The OAS may actually be gaining ground as a legit forum for problem-solving in the Americas. As of this writing, Cuba is being reconsidered for membership, and it appears that things are moving forward rapidly.

AU: AFRICAN UNION

Members: Algeria, Angola, Benin, Botswana, Burkina Faso, Burundi, Cameroon, Cape Verde, Central African Republic, Chad, Comoros, Democratic Republic of the Congo, Republic of the Congo, Côte d'Ivoire, Djibouti, Egypt, Equatorial Guinea, Eritrea, Ethiopia, Gabon, Gambia, Ghana, Guinea, Guinea-Bissau, Kenya, Lesotho, Liberia, Libya, Madagascar, Malawi, Mali, Mauritius, Mozambique, Namibia, Niger, Nigeria, Rwanda, & Zimbabwe.

Suspended Members: Mauritania

Summary: The Organization of African Unity was established in 1963 at Addis Ababa, Ethiopia, by 37 independent African nations to promote unity and development; defend the sovereignty and territorial integrity of members; eradicate all forms of colonialism and promote international cooperation. This organization changed its name to the African Union in 2002. Institutionally, the AU is very much like the EU with a parliament, a commission, a court of justice, and a chairmanship which rotates between the member countries. The AU is also beginning to deploy peacekeepers, and it has sent over 2500 soldiers to the Darfur region of the Sudan. Every country in Africa is a member of the AU except for Morocco, which withdrew in 1985. Also, Mauritania was suspended in 2005 after a coup d'etat occurred and a military government took power. The new government has promised to hold elections within two years, but many observers are doubtful that it will.

African military problems to be increasingly solved by African troops. How novel!

There are many problems facing Africa such as civil war, disease, undemocratic regimes, poverty, and the demographic destabilization caused by the AIDS epidemic . . . these issues are more than even a rich, well-established regional block could handle, and the AU has thus far proved incapable of positively impacting any of these problems successfully. However, its most successful component to date is their military wing. Major powers from around the globe have hailed the common AU military as the most awesome thing the continent could do, and have supported it whole-heartedly. Why? Because the AU military can thus be tasked with solving African conflicts, without the outside powers becoming directly involved. After the US debacle in Somalia (go watch Blackhawk Down) no one wants to send actual troops to stop African conflicts, even in the face of titanic humanitarian disasters like the Rwandan genocide. And maybe now they don't have to! Because the AU has become the choice d'jour for outside support to deal with these issues. While the AU may not get outside support for anything else, look for the military to be bolstered with funding and training from the US, the EU and maybe even China and Russia, in lieu of actual participation in conflicts on the ground.

"Goodbye, Arab League. I'm African now!"

Fun Fact: The idea of an African Union separate from the OAU came from Muammar Qaddafi, who wanted to see a "United States of Africa." He was sick of developments in the Arab world and publicly gave up on being an Arab. In 2011 he publicly gave up on living, when he was assassinated during the Libyan uprising.

INTERNATIONAL ODDBALLS

In addition to these, other entities have been formed at the international level for specific functions. Many of these organizations are frequently in the news, so I feel a brief introduction to them is merited. Haven't you ever wondered who the heck is the G7, the G8, the IMF, the World Bank, or the WTO, and what is an NGO?

G-7 GROUP OF SEVEN

Members: Canada, France, Germany, Italy, Japan, UK, and the United States

The **Group of Seven** (or G-7) is the "country club" of international relations. It's a place where the richest industrialized nations go to talk about being rich, form strategies for staying rich, and—like any other country club—figure out how to keep everyone else poor. The leaders from G-7 countries meet each year for a summit. The summits are often widely protested for reasons such as global warming, poverty in Africa, unfair trade policies, unfair medical patent laws, and basically, for a general sense of arrogance.

Originally, the group was called the G-6, which was the G-7 minus Canada. They formed out of the **Library Group** because rich countries were enraged about the **1973 oil crisis**. Then Canada joined the group in 1976, making it the G-7. Even though this group has since evolved into the G-8, the G-7 group still meets annually to discuss financial issues.

G-8 GROUP OF EIGHT

Members: G-7 plus Russia

Same as above, with Russia in attendance. Russia is still a nuclear power, still the largest territorial state on the planet, and perhaps most significantly a major energy resource provider to the world—especially to the G-7 countries. They have to invite the big boy to the party every now and again, or the big boy might feel neglected and go play with China. The G-7 doesn't want that. They need their energy! In 1991, Russia's entry into the group turned it from G-7 to the G-8. Who knows how much longer the G-7 or G-8 will even last given the rapidly changing face of our planet. What do I mean? Well, in our day and age, what point is there to having a meeting of the "biggest" world economies that doesn't include China? Or India? Or Brazil? It's getting to be kind of outdated and pathetically sad for all the old white-boy countries to get together to chat about world affairs without these other up-and-coming-non-white states having input, so we are probably witnessing the end of days of both the G-7 and G-8, although old traditions die hard and they will survive for a bit longer. And the G-8 as a whole maybe be down-graded back to the G-7, as Russia continues to lean more back towards Asian ties over Western ones . . . as evidenced by their more staunch support for the SCO and BRICS (more on that in a minute).

G-20 GROUP OF TWENTY

PLAID ALERT!!!

Great googly global groups! We have a new international organization titan in our midst, my friends! The G-20 (formally, the Group of Twenty Finance Ministers and Central Bank Governors).

Members: Technically, it is comprised of the finance ministers and central bank governors of the G7, 12 other key countries, and the European Union Presidency (if not a G7 member). Also at the meetings are the heads of the European Central Bank, the IMF, and the World Bank.

The literal translation: the G-8 rich states + Argentina, Australia, Brazil, China, India, Indonesia, Mexico, Korea, Saudi Arabia, South Africa, Turkey + all those aforementioned bank heads. Collectively, the G-20 economies comprise 85% of global GDP, 80% of world trade and two-thirds of the world population. Whoa! That's all the money!

Summary: The G-20 is a forum for cooperation and consultation on matters pertaining to the international financial system which was created in 1999. It studies, reviews, and promotes discussion among key industrial and emerging market countries of policy issues pertaining to the promotion of international financial stability, and seeks to address issues that go beyond the responsibilities of any one organization. The key here is to "promote international financial stability," but I'm thinking they are quickly turning into a whole hell of a lot more. Why would I think that?

Because all the prime ministers and presidents are now showing up for these meetings! A Washington, D.C. summit was held in November 2008 and a London summit in April 2009, which were attended by all the heads of the respective states. Barack was there baby, chillin' with other heads of states! And these folks are truly transforming the power structure

of the planet: the G-8 guys realize that they can't fix nothing without the cooperation of titans like China and India. In London, the G-20 adopted a collective approach to solving the global recession, but also are reorganizing the IMF and World Bank to incorporate other non-G8 voices, which have dominated global financial institutions for years. This is exciting stuff!

I truly believe this G-20 unit has now fully superseded the G-7, the G-8, and any other G's as the premier problem-solving institute on earth. The UN is too big and bureaucratic, and the G-7 too white and too exclusive, but the 20 seems to be a great balance of peoples and powers from across the planet. I would speculate that this much more representative and simultaneously streamlined group will be where all sorts of international treaties and agreements will have their foundation stones set; from economic crises to global warming issues to nuclear proliferation problems, look for the talking points to be brought up at these G-20 pow-wows.

This star is new, but shining already. And its future is bright indeed!

Fun Plaid Fact: There actually has been another G-20 gang comprised of an array of developing nations from Africa, Latin America and Asia. Their primary goal has been fighting for trade rights, particularly in the agricultural sector. You probably won't hear much from that group anymore, but just be aware in case you do . . . they are the poor G-20, not this huge rich G-20 described above.

BRICS

Members: Brazil, China, India & Russia

The Avenger has his plaid panties in a bunch about this unofficial "group" because it represents a serious global shift that has started in the last several years: a move to more representative economic and political power on the world stage. You can possibly look at the BRIC as an alternative center of power to the "Team West" rich countries that have run the global show for a century. It is a new, dynamic, and growing coalition which you will hear a lot more about in the future . . . . just as you will increasingly hear less and less about the G-7 or G-8. So what is this BRIC house all about?

An acronym for the states of Brazil, Russia, India and China combined, BRIC was first coined in 2003 by a financial analyst at Goldman Sachs who speculated that by 2050, these four economies would be wealthier than most of the current

BRICS 2014

major economic powers put together. For sure, China and India will become the world's biggest suppliers of manufactured goods and services, with Russia and Brazil becoming some of the biggest suppliers of natural resources. All four are developing and industrializing rapidly and making bank. Because of their emerging statuses, their increasingly educated populations, and their lower labor costs, the BRIC has also become a magnet for foreign investment, outsourcing, and development of R&D centers.

But this thing has already grown well beyond a simple classification of future rich countries, and this is the deal I really want you to know. It is consistently pointed out that these four countries have not formed a structured political alliance like the EU, NAFTA or NATO . . . . . yet! I'm telling you friends, these guys are thinking about it! They started having face-to-face summits in 2009, and are increasingly agreeing on a whole lot of economic AND foreign policy issues which they then announce to the world under a common voice. Craziness!

What am I spouting about? At the 2011 meeting of the G-20, the BRIC countries demanded, and received, more voting power in global financial institutions like the IMF and World Bank (which they should, since they are increasingly funding them). And what is certainly the boldest and most telling strategic political initiative they have taken to date, all four BRIC countries have jointly declared that they do not support harsh sanctions against Iran and that they support Iran's right to develop nuclear power. What's that got to do with economic policy? Nothing! But it certainly does fly in the face of Team West's opinion of Iran, and that is why I think its important that you know that the BRIC is fast evolving into something much more than a simple economic group . . . . they are becoming a global entity on par with Team West. It's not official, it's not on paper, and it's not heavily coordinated yet, but for sure the BRIC is going to be a significant entity of the 21st century. Get hip to the BRIC!

BODACIOUS BRIC 2012 UPDATE: Exciting expansion has been enacted! The BRIC has just become the BRICS my friends! Say what? A new 'S' in the bric-house? That's right, the S is for South Africa which just joined the ranks of this increasing important global group. Now the gang includes the biggest economy of Africa, to join the biggest economy of South America, along with the 3 biggest economies of Asia.

At their last meeting, the members worked on a slew of trade issues, promised to invest heavily in each other's economies, and also made a public statement that the use of force "should be avoided" in Libya. What's that got to do with economics? Nada, which is why its important to note. This group, while still young, is already making unified political statements about world events . . . world events on which they share common cause, a common cause that is distinctly NOT the Team West view. Hmmmm . . . do I see the formation of a bi-polar world forming? Let's keep a close eye on this entity to find out.

WTO

WORLD TRADE ORGANIZATION
ORGANISATION MONDIALE DU COMMERCE
ORGANIZACIÓN MUNDIAL DEL COMERCIO

Dudes In Club—157 total members
    Dudes Observing the Club—Iran, Iraq, Sudan, Vietnam, Vatican City . . .
    Dudes NOT In the Club—Palestine, Somalia, North Korea . . .
    The World Trade Organization (WTO) is an international, multilateral organization that makes the rules for the global trading system and resolves disputes between its member states. The stated mission of the WTO is to increase trade by promoting lower trade barriers and providing a platform for the negotiation of trade. In principle, each member of the WTO is a privileged trading partner with every other member. This means that if one member gives another a special deal, he's got to give it to everyone else—like in elementary school if you were caught with candy.

Things that make the WTO sigh in delight are "open markets," "tariff reductions," and "long walks on the beach at sunset." The WTO is basically the application of capitalism on a global scale. The key idea is that competition creates efficiency and growth. Any country should be able to sell anything it can, anywhere it wants, at any price that anyone will pay. The WTO can boast some successes in growing international trade, but these have also been accompanied by increased wealth disparity between rich and poor nations AND between the rich and poor within nations.

Formally established in 1995, the WTO is structured around about 30 different trade agreements, which have the status of international legal texts. Member countries must ratify all WTO agreements to join the club. Many of the agreements are highly criticized including the Agreement on Agriculture, which reduces tariffs hurting small farms in developing countries. One of the most famous anti-globalization protests occurred around the 1999 WTO meeting in Seattle that galvanized various groups like pro-labor, pro-environment, anti-fur, etc. under a single anti-WTO banner.

Fun Plaid Fact: The Kingdom of Tonga became the 150th member in 2005. The oldest animal ever recorded, a tortoise named Tu'i Malila, died in Tonga in 1965. Okay, this has nothing to do with the WTO, but turtles are cool.

IMF

Members: Everyone. Seriously. Okay, you got me! Everyone except North Korea, Cuba, Liechtenstein, Andorra, Monaco, Tuvalu and Nauru.

The primary responsibility of the International Monetary Fund (IMF) is to monitor the global financial system. The IMF works at stabilizing currency exchange rates. Doing this, they provide security to overseas investors and help promote international trade. The IMF's policies are also aimed at reducing the phenomenon of "boom and bust," where economies grow rapidly, then stagnate, then grow rapidly, then stagnate, then grow rapidly, et cetera. The main tool of the IMF is "financial assistance" (aka loans), which they provide to countries with "balance of payment problems" (aka big time debt). As a condition of the loans, the IMF mandates "structural adjustment programs." These programs are designed to turn a cash profit, allowing the borrowing country to repay its debt to the IMF. Here is a short glossary of "structural adjustment" terms and how they are interpreted by the locals:

IMF sez . . .	Locals sez it means . . .
Austerity	cutting social programs
User Fees	charging for stuff like education and health care and water
Resource Extraction	selling stuff out of the ground that rich countries want
Privatization	selling state owned stuff to rich companies (usually foreigner-owned)
Deregulation	removing domestic control over stuff
Trade liberalization	allowing foreigners to open sweatshops and exporting stuff made in sweatshops

Much like the WTO and World Bank, the IMF is often congratulated for growth in the global trade and production and simultaneously scorned for increasing the poverty gaps within countries and between countries.

THE WORLD BANK

Similar to the IMF, the mission of the World Bank (actually the World Bank Group) is to encourage and safeguard international investment. All the while, the World Bank attempts to help reduce poverty and spawn economic development. The World Bank works primarily with "developing countries" helping them develop in a Westerly fashion. Also, like the IMF, the World Bank loans are contingent on adopting "structural adjustments." While the IMF deals primarily with currency stabilization, the World Bank is primarily like a real bank, loaning countries money for very specific development projects like a hydropower plant or a disease-eradication program.

Unlike the IMF, which is headquartered in Switzerland, the World Bank Group is headquartered in Washington, D.C. and the US plays a very heavy hand in both its leadership and loaning activities. At least for now . . .

It should be noted that the World Bank is breaking new ground on the redefinition of the sovereignty issue. Due to government corruption in many states, particularly Africa, the World Bank is now putting further stipulations and oversight on all the loans it makes to countries. They are going in and making sure that the money they give a country to build an AIDS clinic doesn't end up getting used to buy weaponry for the state. The Plaid Avenger is proud as punch at such a bold move, but many states are hopping mad that this type of intense oversight violates their sovereignty—I'm with the World Bank on this one. Suck it up, sovereign states. If you want the jack, prove it's going to be used responsibly.

Watch for this issue to gain world attention soon. As of this writing, it's causing consternation in Chad as we speak . . .

NGOS

NGOs are Nongovernmental Organizations. These are basically every private organization that is not directly affiliated with a government, like Amnesty International or Greenpeace. We are talking about a growing number of highly influential international groups that are playing an ever important role in transnational politics. These groups transcend borders and unite common interests. NGOs are set up to represent a diverse array of special interests (environmental protection, human rights, et cetera). Here are a few examples of influential NGOs:

→ Human Rights Watch—With a budget of $20 million a year, this NGO aims to document violations of international humanitarian law by sponsoring fact-finding research. HRW recently waged a successful campaign against the use of land-mines, which the U.S. government opposed.

→ Freedom House—This NGO supports research for democracy promotion. Also, each year FH ranks countries on a scale from "Free" to "Not Free." Luckily, as of 2014, the United States is still "Free."

→ Greenpeace—As the name suggests, this NGO hopes to achieve greenness using peaceful means. Greenpeace is also the only NGO to own a ship (*The Rainbow Warrior*) that the French government intentionally sunk. Funny story, actually; google it and find out. The greatest French military victory since the Napoleonic era.

→ Amnesty International—This NGO is committed to protecting the human rights enshrined in the UN Universal Declaration of Human Rights.

→ International Red Cross (and Red Crescent)—The sole function of the Red Cross is to protect the life and dignity of victims in armed conflict. The Red Cross is independent, neutral, and all that other crap. They help anyone and everyone.

We will discuss some of these NGOs in more detail in later chapters. Party on.

THE NUKE GROUP

Members: US, Russia, UK, France, China, India, Pakistan, Israel

Questionable Members: North Korea, Iran

Last, but certainly not least, is a most important gang of states on planet earth with this homogenous trait: they got nuclear bombs! Or perhaps maybe they do. Or perhaps maybe they are trying to get them. While not an "official" entity in the spirit of the other groups discussed in this chapter, I would be remiss without sticking this information in somewhere . . . . and what better place than a chapter on clubs, as this particular club has enough firepower to blow up all of the other clubs on the planet? The club with the biggest clubs! But it's much more important than that if you want to understand how the world works and how geopolitical power is actually wielded in real life in our day and age. What do I mean?

Well, who is in the club? Let's start with the original declared nuclear powers: US, Russia, UK, France, and China. Hey! Wait a minute! That's the exact same group that as the UN Permanent Security Council! How true, my quick-witted friends, and that is no coincidence! These 5 countries were the first to develop, test, and therefore prove that they possessed

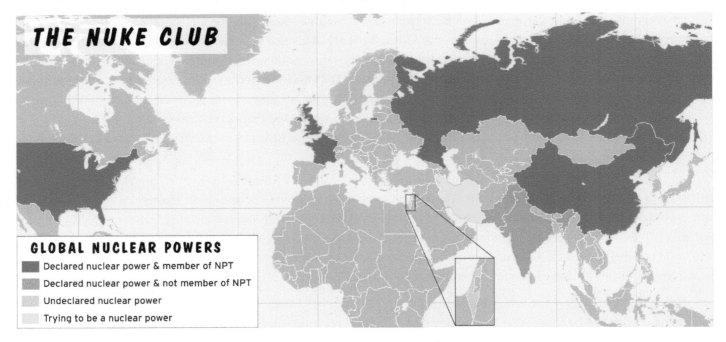

THE NUKE CLUB

GLOBAL NUCLEAR POWERS
- Declared nuclear power & member of NPT
- Declared nuclear power & not member of NPT
- Undeclared nuclear power
- Trying to be a nuclear power

nuclear weapons. The fact that they admitted this to the world means that they have "declared" their status. Kind of like coming out of the nuclear closet. Because of their "first-ness" and openness, these nuke powers have the veto powers at the UN, as per their permanent status on the security council.

Soon after the development of nuclear weapons, most everyone agreed it would be a horrible idea for all countries on earth to have access to these weapons, so a movement to limit them emerged, culminating in the 1970 **Nuclear Non-Proliferation Treaty** (or **NPT**). This treaty has been signed by 189 countries, including the aforementioned big 5 declared nuke powers . . . only four countries are not signed up for it (more on that in a sec). The treaty has three basic pillars it strives to deal with:

1. non-proliferation, or non-spread, of nuclear weapons and/or nuclear weapon information;

2. disarmament, or getting rid of existing nuclear weapons; and

3. the right to peacefully use nuclear technology, which means all signers of the treaty are allowed to have nuclear energy, but not nuclear bombs.

As you can see, this treaty is mostly to ensure that nobody is making nuke weapons, but nuke power technology for energy production is allowed by the NPT, which makes enforcement of it tricky and a hot potato in current events. But before we get to that, what about the other possible nuke powers not yet named? That would be India, Pakistan, North Korea, and Israel. Ah! Yes! Those would be the exact four countries that have not signed the NPT (or dropped out of it in the case of North Korea). What a coincidence! Not! Here is the deal:

India developed and tested its nuclear weapons in 1974, and not to be outdone, Pakistan followed suit in the late 1990's. Why would these countries need nuclear weapons? Because they hate each other and have already fought three wars so far, with more to come. Once India became nuclear, Pakistan could not rest until it got the bomb as well. That's the way rivalries work. The tension between these two countries is so great that neither will give up their right to possess, and even create more, nuclear weapons, which is why both countries have refused to sign the NPT. But they have declared and proven that they have bombs, so they make the list of nuke powers, even though they cannot legally pursue nuclear energy industries since all that stuff is also regulated by the NPT.

Who is left? Israel, of course! With no reservations, everyone on the planet knows that Israel is also a nuclear power, but they have never (openly) tested, proven, or declared it. The US and many others never want them to declare it either, for fear it would spark a regional **arms race**. Israel has probably been a nuclear power since the 1960's, and it wants to have that nuclear edge mostly in order to ensure its survival, were it to ever be attacked by surrounding states again. Because of this "secret" status, Israel has refused to sign the NPT, because to do so would mean it would have to open up for inspections and declare their stockpile . . . which, again, might cause a regional arms race.

Finally, there are those who might be trying to get into this nuclear club, but aren't quite there yet. North Korea originally signed the NPT but has subsequently quit, mostly because having the elusive illusion of possibly having nuclear material is the only bargaining chip they have left to play to get international attention as their state nears collapse. The koo-koo North Koreans exploded something underground a few years back and claimed it was a nuclear device, but no one is really quite sure what the heck they actually have. Including themselves. And I'm sure you have heard plenty about Iran recently too . . . they actually are a signatory of the NPT, which gives them the right to develop nuclear energy. Which they are, while claiming peaceful intent. But some of Iran's semi-crazy leaders have already made reference to themselves as a "nuclear power," which of course they are technically not. Yet. Maybe. Wowsers, this is confusing!

Why am I throwing you all this info on the Nuke Group? Well, unless you have been hiding out in an underground nuclear bunker since the 1950's, you know that the issue of nuclear energy/nuclear weapons has become an extremely hot topic in today's world. Like radioactive hot. Like 1.21 gigawatts hot. Crazy North Korea may be developing something which it then may use to blow up a neighbor. And there is great fear of terrorist organizations obtaining nuclear material and doing something nasty with it. But most of all, the world is currently at arms with itself over what to do with Iran: half the world thinks Iran is just developing nuclear power, the other half think Iran is trying to get nuclear weapons . . .

it is possibly the most divisive issue of current times, and may result in war. That's why the UN permanent Security Council is debating these issues non-stop, why Team West is at odds with the BRICS, and why current US President Barack Obama held the first-ever "Nuclear Security Summit" in Washington DC in 2010, at which all these issues and more were debated. No state currently with nuclear weapons (and the lion's share of states that don't) wants to see more states go nuclear; these things are dangerous and deadly and could spell the end of humanity! Dudes! I've seen *The Road Warrior* and *The Matrix*! It's scary! At the same time, there are those states that want nuclear weapons. With no question, nukes are the absolute best deterrent that would

prevent your country from being attacked. Who wouldn't want a nuke? I mean, let's be honest here: the main reason that the Cold War never turned into a hot war was because both sides had nuclear weapons; therefore, neither side could attack without suffering the same nuclear annihilation itself. Would the US have dropped atomic weapons on Japan in World War II if Japan was equally armed? Would the US administration be openly talking about invading/bombing Iran if Iran was already a nuclear power? Having a nuclear weapon is a game-changer. No wonder some countries may still want one.

This brief section was only to alert you to the nuclear status of our planet, not to fully explain and engage all the complicated topics surrounding these weapons of ultimate destruction. Hopefully, you now at least have a handle on the hotness and how it plays out in current events and the world regions. Regions? Oh yeah! That's what this book is about! Let's get to the regions now!

KEY TERMS TO KNOW

Absolute Location

Adiabatic lapse rate

Albedo

Climate

Core States

Coriollis effect

Cultural Landscape

Culture region

Currents

Desert

Desertification

Diffusion

Dry

Environmental Determinism

Environmental Geography

Environmental Perception

Error of Developmentalism

Formal Region

Functional Region

Geostrophic winds

Great circle

Hadley cells

Hegemonic Powers

Highland

Human geography

Insolation

International Division of Labor

Leeward

Magnetic Field

Maritime

Mediterranean

Mini-Systems

Modes of Production

Orographic uplift

Ozone

Perceptual Region

Peripheral States

Physical geography

Physical Landscape

Polar/Sub-polar

Possibilism

Region

Relative Location

Revolution

Rotation

Semi-Peripheral States

Spatial Perspective

Sphere

Spin

Temperate

Tilt

Tropical/Equatorial Wet

Tropical/Equatorial Wet/Dry

Tundra

Weather

Windward

World-Economy

World-Empires

World-Systems Analysis

FURTHER READING

Alex MacGillivray, *Globalization.* (London: Constable & Robinson Ltd., 2006), 288.

Henry Morris, *The Long War Against God.* (Grand Rapids, MI: Baker Books, 1989), 344.

Henderson, *Experiencing Geometry: In Euclidean, Spherical, and Hyperbolic Spaces.* (Upper Saddle River, NJ:Prentice Hall, 2001), 352.

Christopher Chase-Dunn and Thomas D. Hall, eds., *Rise and Demise: Comparing World-Systems* (Boulder: Westview Press, 1997).

Saul Cohen, *Geopolitics of the World System* (Lanham: Rowman & Littlefield, 2003).

Harm De Blij, *The Power of Place: Geography, Destiny, and Globalization's Rough Landscape* (Oxford: Oxford University Press, 2009).

James S. Duncan, Nuala C. Johnson, and Richard H. Schein, eds., *A Companion to Cultural Geography* (Oxford: Blackwell, 2004).

Derek Gregory et al, eds., *The Dictionary of Human Geography* (Oxford: Blackwell, 2009).

Thomas D. Hall, ed., *A World-Systems Reader* (Lanham: Rowman & Littlefield, 2000).

Geoffrey J. Martin, *All Possible Worlds: A History of Geographical Ideas* (New York: Oxford University Press, 2005).

Donald Mitchell, *Cultural Geography: A Critical Introduction* (Oxford: Blackwell, 2000).

Peter Taylor, *Political Geography: World-Economy, Nation-State and Locality* (Essex: Longman, 1993).

Immanuel Wallerstein, *World-Systems Analysis: An Introduction* (Durham: Duke University Press, 2004).

Peter N. Stearns, *Globalization in World History.* (New York: Routledge, 2010).

WEB SITES

Population Reference Bureau (http://www.prb.org).

United Nations (http://data.un.org).

The World Bank (http://data.worldbank.org).

North America: An Economic and Cultural Powerhouse

2

You are the salt of the earth, but if salt has lost its taste, how shall its saltiness be restored? It is no longer good for anything except to be thrown out and trampled under people's feet. You are the light of the world. A city set on a hill cannot be hidden. Nor do people light a lamp and put it under a basket, but on a stand, and it gives light to all in the house. In the same way, let your light shine before others, so that they may see your good works and give glory to your Father who is in heaven. (Matthew 5:13–16)

CHAPTER OUTLINE:

A VERSE commonly heard throughout the university is that from Luke 12:48b: "For everyone to whom much is given, from him much will be required; and to whom much has been committed, of him they will ask the more." To understand how we can help others or share our hope it is sometimes necessary to honestly assess our strengths and weaknesses. Just as the joy we experience in doing a job well results from diligently reflecting upon both our successes and failures in the past, America's successes are likewise in large measure a function of learning from our past spatially—through geography! Let's see what we can learn about ourselves and how our history has been shaped spatially for the purpose of better serving and providing hope to others!

North America is uniquely placed to fulfill an important role in the world today. From the first footprints left by explorers to the New World, the geography of the continent lent itself to a potentially vibrant culture and one served by the bounties of location and geographic resources. A faithful people clawed an existence out of a few small colonies and set the foundation for a culture destined to be one of the greatest powers on Earth. Understanding how God's creation physically intertwined with the creation of a unique culture and history begins with the study of the earth's surface.

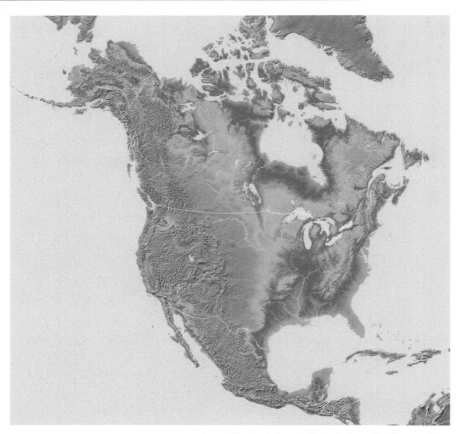

Image © Vitoriano Junior, 2013. Used under license from Shutterstock, Inc.

North America is a mid-sized continent in size somewhere between that of Africa and South America. This land area is bordered on two sides by oceans and a gulf. North America's location and topography have both complemented and facilitated the expansion of European settlement at a time in history when transportation routes demanded harbors and rivers—both of which the continent has in abundance. The placement of North America at middle latitudes ensures it is largely temperate in nature with a unique interaction of topography and climate ideal for agriculture. With oceans bordering both sides of the continent and a gulf to the south, moisture exists in abundance.

FUNNELING AND SUNNING

The topography of North America interacts with the warm sub-tropical atmosphere to the south by funneling the moisture between the two major mountain chains—the Rockies and the Appalachians, both generally running from north to south. The somewhat funneled and steady rainfall pattern helps to create some of the most productive farmlands in the world. The rivers also follow a pattern of north and south with the major rivers thereby flowing in directions generally advantageous to settlement and transport while uniting the nation **longitudinally**. The mouths of these rivers create some of the world's finest **harbors** and further contribute to the ability to transport goods to international markets. An abundance of well-watered forests of valuable lumber and mature soils have sustained much of the world's population in recent centuries.

The generally latitudinal or east to west movement patterns of European settlers across the North American continent over time lent itself to what appeared to be a divided culture from the founding of the colonies until the American Civil War. The diffusion of these differing agricultural patterns and their accompanying economies are examples of **cultural diffusion**. A southern feudalistic and agrarian society based on servile labor stood in stark contrast to a northern manufacturing center dependent on various transportation outlets or **nodes**. This ability to connect different networks still is a characteristic of the

American economic system and gave the northern or Union views of freedom and the fears of a southern slavocracy an ultimate advantage in movement. Violent war ultimately reconciled different views of freedom and the end of slavery while initiating increased westward expansion in an attempt to spread freedom. Today the cities of the United States still demonstrate patterns of movement often in generally western directions utilizing the latest offerings in various means of transportation.

America's relief affords an excellent opportunity to review some of the aspects of physical geography discussed earlier in this work. The mountains of North America have actually been instrumental in reinforcing from a cultural standpoint the different regions of the North American continent. The rivers likewise have been important in the settlement patterns of North America, and the harbors have served as gateways to the world for the purposes of shipping and the projection of military power. How these factors of relief and the themes of movement have intertwined to create a narrative will be discussed in this section.

The mountains of eastern North America are believed by many scientists to be among the oldest in the world. Possibly some of the finest soils for farming in the United States are in large measure a result of erosion from these mountains. Soils vary and are broken down by physical weathering and the chemicals in the acids from the residue of living organisms such as leaves. Large enough to buffer the extremes of continental climates from the north, the topography did not serve as an impediment to movement for those self-sufficient settlers seeking to support themselves and their families.

SECTIONALISM AND THE CITY

As the first Americans arrived off of the **easterly** currents at Jamestown, Virginia in 1607, they sought a place of refuge that was near the famed gold of the Spanish but not so near that they could suffer the catastrophic conditions that afflicted Fort Caroline in 1565 (near modern day Jacksonville, Florida). Here the Spanish destroyed a French-speaking colony for venturing too close to their supplies of New World riches. Nevertheless, despite carefully choosing their settlements, the colonists at Jamestown would find themselves geographically challenged. The first English-speaking colonists chose a location relatively near the headquarters of the Indian nation called the Powhatan. Imagine colonists from another culture camping near Washington, D.C. today! To make matters worse, they set up camp in an area that could be compared both to the local dump of the capital city and in a disputed frontier area— a place of contention between two rival nations. Needless to say, the first English-speaking colonists faced spatial challenges, indeed!

To the north in Canada and a short time later, the French disembarked their boats at Quebec City in 1608 at what today we would call a **break-bulk point** where the river shallows and prevents further travel by ship. A break-bulk point is where we see one form of transport changed to another such as from a ship to a wagon. Like the patterns of movement to the south, the Canadians followed a generally western pattern and tended to populate the lower regions within 100 miles of what is today the US border. This distribution of population

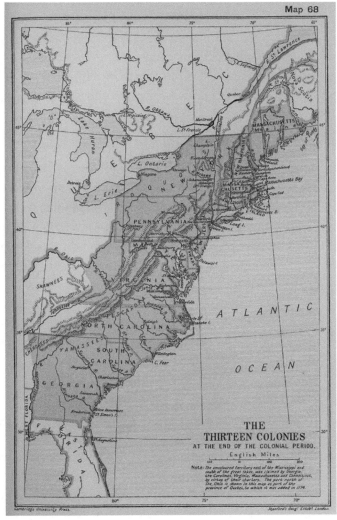

Map 68

THE
THIRTEEN COLONIES
AT THE END OF THE COLONIAL PERIOD.

Map source: *The Cambridge Modern History Atlas 1912*, courtesy of the University of Texas Libraries, Perry-Castaneda Library Map Collection.

is referred to as an **ecumene** or more habitable region. As a result of the variety in climates to the south, distinct cultures in the English colonies from New England to Virginia emerged with fearful implications for freedom.

The frontier dynamic as the settlers moved west created two different types of agricultural economies. The successful diffusion of these cultures and subsequent population growth created unique American experiences. As you probably learned in school, tobacco's introduction to the public tended to wear out the soil in the Virginia river lands, and this created a need for more rich soils and led to the subsequent deforestation of the eastern deciduous forest. As the settlers moved into the interior of the country, the local courthouses'

Art by John C. McRae (ca. 1620), from Library of Congress

recording and maintenance of drawn borders reinforced a culture that was both individually competitive and placed a premium on private property rights. Here is where we can see that the intertwining of the topography of North America would ultimately shape a unique and vibrant culture.

While Canadians crossed the windswept prairie, Americans began moving in two different directions and into two different regions, also a function of latitude and moisture. Many of the compromises for union you may have studied in US history also appear to be related to topographic relief and the two different climates seen on the map on page 9. Settlers moved west through the gaps in the Appalachian Mountains into southern states such as Kentucky and Tennessee. Simultaneously to the north, immigrants continued to arrive into coastal harbors and traveled to the Great Lakes through the Erie Canal beginning in the 1820s. This ensured the rapid growth of northern cities. The clearing of the northwest bisected the southern culture through the central river valley of the Mississippi and ultimately doomed the southern slave culture.

DIFFERENT LATITUDE—DIFFERENT ATTITUDE

The mighty Mississippi River flowed generally north to south and was fed by the powerful tributaries of the Ohio, Tennessee, Cumberland and Missouri Rivers. The river system served not only to support inexpensive transport, creating a **gateway city** in New Orleans as American crops and manufacturing spread to the world, but the existence of a river running almost perpendicular to the spread of settlers and culture west ultimately ensured the dominance of a northern culture over a southern one. It also served to inextricably bind the North and the South together through trade and transport. Northern industry benefited from the physical geography.

The cultural diffusion of southern agriculture encouraged the use of slaves as a labor source for picking cotton as far west as Texas while Americans in the North who were entering into a modern industrialized era experienced a series of religious revivals that embodied a **dualistic** or **Manichean perspective** and placed a premium on doing good work. This eventually initiated a crusade of sorts to free the slaves in the South. Interestingly, it would be the American West that would bring this issue to full warfare.

HEAD WEST, YOUNG PEOPLE!

When California came into the Union it did so in an expedited manner that upset the national balance. Why the big rush? One word—gold. With an international economy based on **specie** since the age of exploration, the possession of this metal was the fastest ticket to wealth and power of its time. As both the northern and southern cultures eyed each other over this precious resource, the paucity or lack of plantation agriculture in

Image © Richard Peterson, 2013. Used under license from Shutterstock, Inc.

the western region ensured California would tend towards a free culture rather than relying on slave labor. The subsequent railroad connections and the continued westward movement of a population increasingly marked by immigrants further inflamed passions of freedom. A diffusion of a free and more unionized industrial culture in the north intersected with that of a slave culture based on state's rights in the south. The Civil War was won because of geography by the unionization of the North over the decentralized conservative South. It would seem the American experience was bound to be one of strong unity after the conflagration of war due to the effect of topography and culture.

INTERIOR

The North American continent is blessed with both incredible resources and a culture reflecting a solid and intense Christian work ethic. By the time the American people closed the western frontier in the 1890s, Americans had established important agricultural contributions. The climate in Jamestown produced the native crops of corn; the hopefulness of the Puritans in New England enabled capitalism to distribute the crop. Ease of connecting resources across the lakes to the emerging cities of New York, Pittsburgh and Chicago kept relative transport rates down. It has been said transportation costs on water range historically from 1:30 to 1:100 relative cost of tonnage to transport on land. This reunited nation of northerners and southerners continued to influence the world through the nineteenth and twentieth centuries.

Meanwhile, the movement of Americans into the Pacific coast continued unabated as America marched into the global scene as an economic powerhouse. What lured people to California, for example? The answer is jobs and ultimately an opportunity to participate in industries that would change the entire world—such as aviation and Hollywood. World War II presented increased opportunities, in particular, with the proliferation of technologies that enabled people to survive the harsh deserts and to cross them quickly. These technologies included the automobile and air conditioning. One could argue the southwestern United States and the southeastern part of the deep-south, Florida, actually presented some of the last frontiers of the continental United States and was finally settled because of the invention of cars and air conditioning!

CITY LIGHTS

The same world that recognizes McDonald's drive-through food would recognize the automobile culture that poured people west into California where these fast food restaurants would originate. In a sense, the **cultural landscape** of American cities reflects historical movement patterns. American westward travel from Jamestown to the Pacific may still be seen when studying America's cities. The first stage of the growth of American cities has generally been from a transportation node such as a river or road. After the arrival of the Industrial Revolution in the early nineteenth century, people walked to the factories, often taking the trolley cars to work. Later, the automobile made the suburbs within reach to the upwardly mobile. Today, fiber optic networks have created corridors in tandem with interstate highways and airports. The vast majorities of these global networks cross or interconnect in North America still. The modern city demonstrates the impact of the cultural landscape on the history of the country.

Resources abound in North America but none are as valuable as oil. First discovered in Pennsylvania in the 1850's, the United States would begin to actually need imports of oil by the end of the Second World War. Nevertheless, the potential for oil production still exists in various forms and has without question directed America's interest in the world since the fall of the Soviet Union in the 1900s.

California, with its terrific harbors, and the valuable Washington, with Puget Sound on the border of the United States and Canada offered priceless gateways to the Pacific—the very same goal of the ancients as they sought the headwaters of the Euphrates River, attempting to circumvent the silk road! But the American story doesn't end on the Pacific coast; rather the western movement patterns of America would extend into the Pacific. In fact, America would expand across the islands

We Like Ike! What's the Deal with the Interstate System?

The formal name of the United States Interstate System is the "Dwight D. Eisenhower National System of Interstate and Defense Highways." The US interstate system was created by the Federal-Aid Highway Act of 1956, which, as the name suggests, was championed by Eisenhower. The US interstate system was built for both civilian and military purposes. Some of you might be thinking, "Yeah! One in every five miles of road must be straight so that military aircraft can land and take off on them!" Well, maybe. The true military aspects of the Interstate System are primarily to facilitate troop movement and to allow for the evacuation of major cities in the event of nuclear war.

The civilian aspects of the Interstate Highway System have helped shape American culture in more ways than we can possibly imagine. Most US interstates pass through the center of cities, which allows people to live outside the city and commute in for work, which has also played a huge role in the creation of suburbs and urban sprawl. The highway system also gives the federal government power over state governments. The US government can withhold interstate highway funds, which are huge amounts of money, from uncooperative state governments. The US government used this tactic to increase the national drinking age to 21 and to lower the blood alcohol level for intoxication to 0.08%. According to the Constitution, both of these issues should be decided by states. However, if they choose to disobey, they don't get their highway money.

Also, the US highway system allows high speed transportation of consumer products. This makes the prices of everything, from bananas to concrete, cheaper. The result: the US is an automobile culture heavily reliant on fossil fuels for the movement of everything and for virtually all aspects of our lives. Americans are quite unique in this respect. It's also why they are the best stock car racers in the world. Who else has NASCAR? Who is more worried that the price of oil is reaching $150 a barrel?

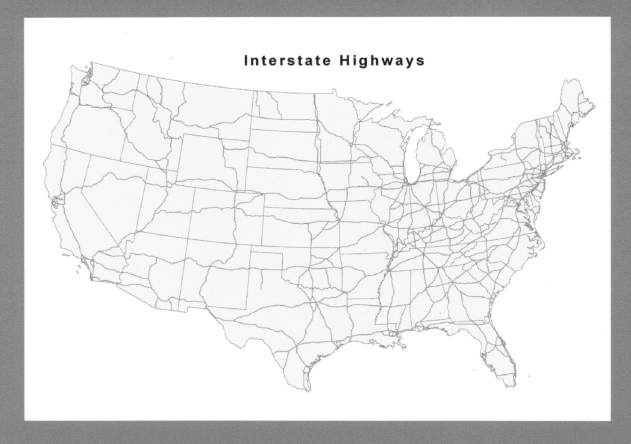

Interstate Highways

USA GDP =

Top 5 economies	
State	**GDP (USD)**
USA	$13,843,825,000,000
Japan	$4,383,762,000,000
Germany	$3,322,147,000,000
China	$3,250,827,000,000
UK	$2,772,570,000,000

Map courtesy of Katie Pritchard

All images © 2007 JupiterImages Corporation

Photo by M.B. Marcell (ca. 1911), from Library of Congress

Country	Military Expenditures 2011 (USD)
US	$741,200,000,000
China	$380,000,000,000
India	$92,000,000,000
Russia	$82,500,000,000
Saudi Arabia	$59,090,000,000
France	$54,444,000,000
UK	$50,952,000,000
Turkey	$46,634,700,000
Germany	$42,255,000,000
South Korea	$36,774,000,000
Brazil	$34,170,000,000
Japan	$33,192,000,000
World Total	$2,157,172,000,000

http://www.globalsecurity.org/military/world/spending.htm

From *The Plaid Avenger's World #6 Nuclear Insecurity Edition.* Copyright © 2008, 2009, 2010, 2011, 2012 by John Boyer. Copyright © 2006 by Kendall Hunt Publishing Company.

of the Pacific and into the Philippines. With two coasts and more good harbors than many continents have, America's role in the Philippines as a military power seemed almost predestined. Many historians in fact believed the high-water mark of the American westward expansion or "Manifest Destiny"—or a perception of the rights to occupy spatially westward—would not come until the Vietnam War. Regardless, overseas commitments and treaties would be instrumental in America's growth as a military power with the vast majority of aircraft carriers. America's role as a global power seemed assured by the end of the bipolar world of the 1990's with the fall of communism.

U.S. Air Force photo by SSGT Jacob N. Bailey

Department of Defense photo by U.S. Navy

U.S. Navy photo by Mass Communication Specialist 3rd Class Kathleen Gorby

2011 TOP SALES OF MULTINATIONAL COMPANIES

Company	Home Base	Sales (USD)
Wal-Mart Stores	USA	421.8 billion
Royal Dutch Shell	Netherlands	369.1 billion
ExxonMobil	USA	341.6 billion
BP	UK	297.1 billion
Sinopec-China Petroleum	China	284.8 billion
PetroChina	China	222.3 billion
Toyota Motor	Japan	202.8 billion
Chevron	USA	189.6 billion
Total	France	188.1 billion
ConocoPhillips	USA	175.8 billion
Volkswagen Group	Germany	168.3 billion
General Electric	USA	162.4 billion

Fannie Mae USA 54.3 billion From Special Report: The World's Biggest Companies, edited by Scott DeCarlo 04.02.08, Forbes.com

From *The Plaid Avenger's World #6 Nuclear Insecurity Edition.* Copyright © 2008, 2009, 2010, 2011, 2012 by John Boyer. Copyright © 2006 by Kendall Hunt Publishing Company.

A powerful United States prospered as international networks of fiber optics and space satellites in various levels of Earth orbit maintained an ability to navigate the globe through GPS and to maintain the security of American interests. Almost like the British Empire, which was based on trade and utilized the waterways to accomplish its business, the United States has overseen suzerainty over much of the world somewhat unintentionally utilizing the electromagnetic spectrum. With **multinational corporations**, or MNCs, overseeing an invisible empire of sorts, the American people often stand to benefit both publicly and privately by their profits.

Bravely walking into a new world order and perhaps blunderingly assisting in the rise of powerful autocracies in Russia and China with well-intentioned military interventions, America's importance in the future is undoubtedly assured do to the blessings of relief, currents, the particulars of climate and their effects on agriculture as well as the rich soils of the east which served to lure and direct the movement of the first colonists west—a pattern that has continued almost to the present. Hopefully Americans have not forgotten the geographic basis for the prosperity that exists today nor the author of this land, which in large measure has shaped the culture of freedom we so enjoy today.

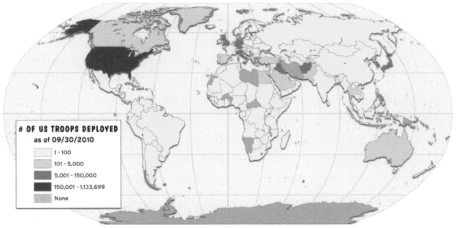

Map courtesy of Katie Pritchard

FOUR REVOLUTIONS: THE EVOLUTION AND DYNAMICS OF GLOBAL CORE-PERIPHERY RELATIONSHIPS*

¹Now the whole earth had one language and one speech.² And it came to pass, as they journeyed from the east, that they found a plain in the land of Shinar, and they dwelt there.³ Then they said to one another, "Come, let us make bricks and bake them thoroughly." They had brick for stone, and they had asphalt for mortar.⁴ And they said, "Come, let us build ourselves a city, and a tower whose top is in the heavens; let us make a name for ourselves, lest we be scattered abroad over the face of the whole earth." (Genesis 11:1–5)

Since the goal of this text is to aid the student in understanding the basic characteristics of the human geography of the world today, especially with respect to development and underdevelopment and global core-periphery relationships, then it is altogether appropriate that we understand the evolution of these relationships over time. We must start at the beginning; we must know from where we have come before we can know where we are today. What are the historical processes that have resulted in present global geographical patterns? Toward these ends, this chapter introduces general, long-term, global-scale trends in the development of the world-economy.

The chapter argues that there have been four significant "revolutions" in human history that have successively shaped and reordered the world's economies, political geographies, cultures and landscapes. Each revolution represents a clear break with what had come before with respect to dominant modes of production, forms of labor control, social relations of production, dominant technologies in use, and the types of natural resources exploited. Each revolution resulted in a significant increase in the amount of power that could be harnessed per person per year. Further, each revolution resulted in substantial changes in the ways in which human beings conceived of the world around them, how humans perceived of themselves in the world and their place in it, and how certain natural resources could be used and exploited. Over time, the effects of these revolutions resulted in the alteration of the world's cultural landscapes into what they are today. The cultural landscapes of today reveal these past changes. They reveal changing cultural traditions, ideals, and values. They reveal social, political, and economic struggles that have taken place over time and space. The world's cultural landscapes are, in short, a palimpsest of the last 12,000 years of human history.

THE NEOLITHIC REVOLUTION

WHAT WAS IT?

The first revolution to significantly alter the ways in which human beings viewed and used the natural world around them was what anthropologists refer to as the **Neolithic Revolution**. This term refers to the period in human history about 10,000–12,000 B.C. when the first large-scale urban settlements began to appear, together with concomitant changes in the structure and nature of societies, modes of production, and social relations of production. The most important feature of the Neolithic Revolution, however, and what engendered most of these changes, was the domestication of plants and animals. **Plant and animal domestication** refers to the gradual genetic change of plants and animals through selective breeding such that they become dependent upon human intervention for their reproduction. Certain plants and animals or certain varieties of plants and animals were selected for particular qualities such as taste or caloric value or nutritional value, while other varieties were left behind. These selected varieties of plants and animals were nurtured, protected and cared for in gardens and fields and continuously reproduced. Over many generations this caused genetic change, resulting in plants and animals with distinctive characteristics found to be helpful to human beings (again, characteristics such as

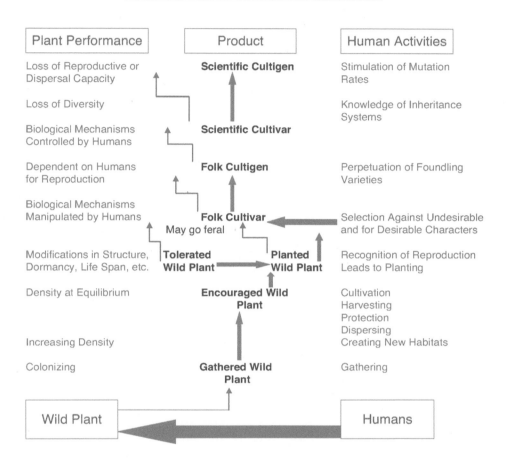

Stages of Plant Domestication

Plant Performance	Product	Human Activities
Loss of Reproductive or Dispersal Capacity	**Scientific Cultigen**	Stimulation of Mutation Rates
Loss of Diversity		Knowledge of Inheritance Systems
Biological Mechanisms Controlled by Humans	**Scientific Cultivar**	
Dependent on Humans for Reproduction	**Folk Cultigen**	Perpetuation of Foundling Varieties
Biological Mechanisms Manipulated by Humans	**Folk Cultivar** May go feral	Selection Against Undesirable and for Desirable Characters
Modifications in Structure, Dormancy, Life Span, etc.	**Tolerated Wild Plant** → **Planted Wild Plant**	Recognition of Reproduction Leads to Planting
Density at Equilibrium	**Encouraged Wild Plant**	Cultivation Harvesting Protection Dispersing Creating New Habitats
Increasing Density		
Colonizing	**Gathered Wild Plant**	Gathering

Wild Plant ← Humans

Conceptualization by Dr. Clarissa T. Kimber, 1973

taste or resistance to pests and drought). Today, this process occurs in the form of scientific genetic breeding of plants and animals carried out at laboratories and universities.

Without a doubt, the most important and far-reaching result of plant and animal domestication was the advent of agriculture on a scale that had not been seen before. Before the advent of large-scale agriculture, the vast majority of societies all over the world could be described, in world-systems analysis language, as mini-systems, with social relations of production and social and political organizations that were tribal in nature. Exchange in these mini-systems was basically reciprocal, social classes were weakly developed, and their geographic extent was very small, amounting to little more than a claimed "homeland" or hunting territory. Agriculture, however, wrought social and economic changes that changed the world forever. Where it occurred, much more "complex" societies quickly developed, with highly stratified societies, vast increases in food production, and despotic forms of political organization that claimed and defended large territories by force. That is, where plant and animal domestication first occurred, where agriculture first originated, mini-systems gave way to the development of world-empires.

WHY AND WHERE DID IT OCCUR?

There are two primary theories about when, where, and why agriculture first originated. The first, what we might call the orthodox theory of domestication, is agreed upon by most historians and anthropologists. The second was proposed by the geographer Carl O. Sauer in the mid-20th century. The orthodox theory, that which is most widely accepted by social scientists and for which there is the most archaeological evidence, argues that plant and animal domestication occurred independently in several areas of the world at about the same time in history—from about 8,000 to 12,000 B.C.—through a process of **independent invention**. That is, populations in different parts of the world "invented" agriculture on their own without contact from outside populations, without "learning" it from others. Most experts argue that some sort of ecological stress—overpopulation or climate change, for example—forced these populations to move from a nomadic hunting and gathering type of lifestyle to a more sedentary lifestyle more and more dependent upon agriculture to feed burgeoning populations.

This revolutionary change did not happen overnight, or even in a few generations, but rather most likely occurred over hundreds of years and over many human generations. But where it did occur, in the following so-called **culture hearths** of innovation and invention, societies began to look very different, the hallmark of which was the first urban civilizations. Table 2.1 lists some of the earliest culture hearths and gives dates by which world-empire types of civilizations had developed. The dates for the domestication of various plants and animals are not given here, but radiocarbon dating of plant and animal remains suggests that plant domestication occurred as early as 12,000 to 15,000 years ago in Southwest Asia and the Nile Valley and 5,000 years ago in Middle America and the Andean Highlands of South America.

TABLE 2.1 MAJOR CULTURE HEARTHS

Culture Hearth	Location	Date	Domesticates
Mesopotamia	Southwest Asia	by 5,500 B.C.	wheat, barley, rye, grapes, oats, cattle, horses, dogs, sheep
East Africa	Nile Valley	by 3,300 B.C.	barley, coffee, cotton, millet, wheat
Incan	Andean Highlands	by 2,500 B.C.	potato, tomato, llama, guinea pig, alpaca
Mediterranean	Crete, Greece	by 2,500 B.C.	barley, grapes, goats olives, dates, garlic
Indus Valley	Pakistan	by 2,300 B.C.	wheat, cattle, dog, rye, sheep, horse
North China	Huang He Valley	by 2,200 B.C.	soybeans, buckwheat, cabbage, barley, plum
Southeast Asia	Vietnam, Cambodia, Thailand	by 1,500 B.C.	bananas, chicken, pig, tea, dog, rice, taro, water buffalo, yams
Meso-America	Mexico, Guatemala	by 1,250 B.C.	maize, beans, taro, chile peppers, dog
West Africa	Ghana, Mali	by 400 B.C.	arrowroot, millet, pigs, rice, oil palm

Another less widely accepted theory of the development of agriculture was proposed by the American geographer Carl O. Sauer in his book *Agricultural Origins and Dispersals* in 1952. Sauer suggested that agricultural first developed in Southeast Asia as early as 20,000 years ago and from there diffused to the other areas of the world. This theory is thus one based on the idea of **cultural diffusion** from an original hearth area to all other regions of the world. Sauer reasoned that agriculture would have developed first in an area where the richest array of different kinds of useful plants naturally occurred, most likely a tropical region, and in a place where people were not under any kind of ecological stress. Such populations, Sauer argued, would have not been *forced* to invent agriculture, but rather would have had ample time over many generations to experiment with growing different varieties of useful plants. This theory, however, is at most a thought experiment and there is little archaeological evidence to support it. It also presupposes that populations in certain parts of the world were not "advanced" enough to invent agriculture on their own and that they therefore must have been shown how by invading populations with such knowledge. For these reasons Sauer's theory is not widely held to be true.

WHAT CHANGES DID IT ENGENDER?

An agricultural lifestyle had the effect of radically altering the nature and structure of societies and economies in each of the various culture hearths. Aside from new innovations and techniques that increased crop yields and food production, such as crop rotation and large-scale irrigation projects, the most revolutionary change was the development of highly stratified, agriculturally-based societies with codified class structures, a class-based division of labor, and a redistributive-tributary mode of production. Anthropologists refer to such societies as *chiefdoms*, and in essence this describes a *feudal* social and economic order. In the world-systems perspective, Wallerstein refers to these societies as *world-empires*.

At one end of the socio-economic spectrum in these societies an agricultural underclass comprised up to 90% of the population. This peasant underclass produced agricultural surpluses, the storage and redistribution of which was controlled and directed by an aristocratic (title by blood birth) royalty at the other extreme of the socio-economic spectrum. These rulers also directed the building of vast agricultural projects such as irrigation schemes (in North China and the Nile Valley) and the construction of monumental structures (such as the Great Wall and the pyramids). Such rulers (known as kings, emperors, pharaohs and the like) are often called **god-kings** by anthropologists because they often claimed to possess supernatural powers and to have achieved communion with the gods. These rulers surrounded themselves with other elite classes such as scribes and priests. These ruling elite lived in a central urban location known as a city-state, the basic political unit of world-empires. The presence of distinctive elite classes points to the development of other revolutionary changes first witnessed in each of these culture hearth that set the stage for the development of vastly more complex societies and economies that forever changed the ways in which humans interacted with the natural world around them:

→ Written languages to keep records of agricultural surpluses and the redistribution of surpluses

→ The development and use of calendars to track the seasons and to predict planting and harvesting times

→ The development of an organized military under the command of the ruling elite in order to defend a claimed political territory

→ Significant population increases as a result of more stable food supplies

THE MERCHANT CAPITALIST "REVOLUTION"

WHAT WAS IT?

Although it is not always referred to as a "revolution," the development of merchant capitalism in northwestern Europe in the mid-15th century and its diffusion around the world over the ensuing 400 years engendered changes in societies, economies and cultures that shaped the modern world-economy more than any other event. If we define **capitalism** as the production and exchange of goods and services for private money profit, then most historians would agree that it has existed for millennia in many places around the world. **Merchant capitalism**, however, arose in only one region of the world at a specific point in history—Holland and England in the mid-15th century. Merchant capitalism was an altogether new version of capitalist production and exchange because it was based heavily on long-distance sea trade in exotic products from the tropics and subtropics.

Because such products were rare and costly to acquire (it took five years to travel by ship from Holland to the southeast Asian spice islands in the 16th century) the high relative value of such trade meant great wealth to the countries and individuals that controlled it. Before the advent of merchant capitalism, most trade was merely regional in nature and used land-based caravans or shipping routes that plied coastal regions. Merchant capitalism involved the use of new sailing and navigation technologies to strike out into the oceans out of sight of land on risky overseas ventures in search of rare tropical goods for which Europeans were willing to pay high prices (such as sugar, cotton, tea, coffee, pepper, and other spices). In short, merchant capitalism ushered in the era of colonialism and global trade that radically altered the world forever. It was the first step toward creating the "globalized" world that we live in today.

WHY AND WHERE DID IT OCCUR?

Why did merchant capitalism develop in a relative backwater spot in the world at the particular time that it did? Compared to the great civilizations of China, South Asia and Southwest Asia, Europe in the early Middle Ages lagged far behind in terms of technology and scientific know-how. Earlier, traditional theories attributed the development of merchant capitalism in northwest Europe to such things as Protestant Christianity (Weber's "Protestant work ethic") or European racial superiority. While Weber's theory is still held by some to be of explanatory value, nobody still believes that Europeans were somehow "better" or "smarter" than other world civilizations and "invented" capitalism due to this superiority. Instead, most historians today argue that a series of events happened first in Europe, effecting societies there in more drastic ways than others and resulting in a fundamental reordering of European societies and economies.

The most important of these events was most likely the **Black Death**, the Bubonic plague that ravaged Europe from 1340–1440. The Black Death struck elsewhere around the world, but nowhere with such far-reaching consequences. Anywhere from one-third to one-half of the population of Europe died during this period. Most of those that died came from the peasant classes, in a feudal society that part of the population with the lowest caloric intake and the highest susceptibility to such respiratory diseases. This had drastic repercussions for European economies because with so much of the peasantry gone, labor was now in very short supply. As a result, those peasants that remained acquired something that they had never possessed before—bargaining power over their aristocratic landlords. Peasants began to demand something for their labor, either payment in kind or money wages. In short, the Black Death precipitated a **crisis of feudalism** in Europe. The ancient feudal organization of labor and production could no longer keep up with the demand for food, especially as populations began to increase rapidly again after the plagues subsided. World-systems analysts argue that the replacement of a feudal organization to societies and economies with a different form of production and exchange based on individual money profit—capitalism—was the solution to this crisis in the feudal order.

Along with the reordering of economies came a fundamental reordering of societies as well, the most important aspect of which was the emergence of a new class of people—a middle class. Because production was now based on profit and competition, those that could produce the most reaped the most profit. This spurred technological innovations, especially in agricultural production. The use of steel plows, the draining of marshlands and crop rotation schemes, for example, led to higher agricultural surpluses than ever before. This also meant that more and more of the European population did not have to farm because agriculture had become more efficient and more productive. Many people began to move to towns and cities and became shopkeepers, merchants, and artisans (weavers, blacksmiths, millers, tailors, bakers, etc.). Over ensuing generations this new middle class rose in social, economic and political importance and they began to pass wealth on to successive generations. Such dynastic, wealthy urban merchant families came to dominate the social and economic life of coastal towns in northwestern Europe, especially in Holland and England, by the 16th and 17th centuries. By the late

16th century, these powerful urban merchants began to sponsor overseas trading ventures to newly "discovered" tropical regions in order to supply the new and growing demand for rare, expensive tropical agricultural products by the expanding middle class. Colonialism—the political, economic and social control of tropical peripheral regions by European core powers using coerced labor—was invented by northwestern European urban merchant capitalists in order to more efficiently supply this demand. It was this new merchant capitalist colonial order that created the European-centered capitalist world-economy with its characteristic tripartite global economic geography of core, semi-periphery and periphery.

WHAT CHANGES DID IT ENGENDER?

In summary, the development and expansion of merchant capitalism based on long-distance sea trade resulted in the following revolutionary developments:

- → The replacement of feudal, agricultural societies with economies dominated by merchant capital interests involved in long-distance sea trade
- → The invention and domination of colonialism
- → The birth and dominance of a middle class of urban merchant entrepreneurs
- → The creation of a capitalist world-economy dominated by a few nation-states in Europe
- → The creation of a "core" of relatively high economic and technological development and a "periphery" of relatively low economic and technological development
- → The creation of the use of varying labor control systems in the different part of the world-economy: slavery in the periphery, wage labor in the core, tenant farming in the semi-periphery
- → The increased use of non-human and non-animal sources of energy; a drastic increase in the amount of energy harnessed per capita, per year
- → The initiation of a capitalistic logic in the world-economy; since capitalism rewards innovations that make production cheaper, it engenders constant technological innovation

THE INDUSTRIAL REVOLUTION

WHAT WAS IT AND WHERE DID IT OCCUR?

The Industrial Revolution represents a third major break with the past with respect to far-reaching societal changes around the world, especially in the core countries of the world-economy during the 19th and 20th centuries, whose wealth was built on the backs of economies focused on heavy industry and manufacturing. Most significantly, the Industrial Revolution ushered in vastly different methods and modes of industrial production, systems of labor control, and social relations of production. These changes occurred first in the Midlands of central England beginning in the early- to mid-18th century but spread very quickly to the rest of Europe (Belgium, Holland, France, Germany, and Russia) and North America (especially the northeast United States) by the mid- to late-19th century. The Industrial Revolution ushered in the Machine Age.

The Industrial Revolution radically altered societies and economies in two important ways. First, modes of production were revolutionized with the introduction and application of new labor-saving technologies and machines. The earliest of these new technologies was the steam engine. Steam engines were first put to use in the British Midlands running pumps to remove excess water from coal mines, but the technology was soon put to use running all types of machines that heretofore had employed human or animal power such as mills and looms. The first industry to be completely revolutionized by the machine age was the textile industry. Steam engines that spun and wove yarn into textile goods replaced traditional cottage textile industries in the rural countryside of Europe, where such goods had for millennia been produced by hand. The mode of production that came to dominate regions that became industrialized was the urban factory staffed by low-wage, low-skilled workers running steam-driven machines.

Many of these workers were women and children from the rural hinterlands of emerging industrial cities who came to the cities in search of work as traditional ways of making a living were being radically altered. This represents the second major change wrought by the Industrial Revolution—a new form of labor control characterized by the factory organization of low-wage, low-skilled laborers. Such laborers lived in housing, often built by the factory owners, very near to the factories in which they worked. Long working hours (production could take place around the clock with the advent of gas lighting and, later, electricity), poor living conditions, and exposure to hazardous environmental pollutants (coal smoke, chemicals) characterized the lives of the earliest factory workers. Workers often performed the same menial tasks hour after hour in cramped and hot conditions. In the United States, the industrialist Henry Ford mastered the factory mode of production and labor control with the innovation of the assembly line, where workers performed the same task hour after hour as cars traveled down an assembly line (this mode of production is often referred to as **Fordism**).

The factory mode of production built around urban factories staffed by low-skilled wage laborers and powered by petroleum-based fuels (first coal, then oil products like gasoline) revolutionized production. The countries that industrialized first, like Great Britain, Germany, and the United States, soon outpaced and out produced their competitors. Although industrialization came to different countries at different times, the economies of all of the core powers of today were built upon heavy industrial production between the mid-18th century and 1960. Industrialization brought immense wealth and power to these core countries, especially in the early 20th century when steel production, shipbuilding and automobile manufacturing became the hallmark of the core economies. Many semi-peripheral areas did not "industrialize" until the 1960s and 1970s, while large-scale industrialization has yet to appear in most areas of the periphery. The semi-peripheral economies hope that a strong manufacturing and heavy industry base will also bring the same wealth that it brought to the core economies, but it remains to be seen whether this will happen or not.

THE POST-INDUSTRIAL (POST-MODERN) REVOLUTION

The **Post-Industrial Revolution** describes revolutionary changes in societies, economies and cultures that have taken place since the early 1970s, predominately in the core regions of the world-economy. But because changes that take place in the core affect all regions of the world-economy, this revolutionary period in which we now live is fundamentally transforming societies, economies, and cultures all over the world. This period has seen the development of new modes of production, the advent of so-called "new information technologies" such as the internet (these have ushered in the "computer age"), and a pronounced global division of labor in the world-economy.

Perhaps the most significant development during this latest revolution is the process of **globalization** in the world-economy, directed from the core by multi-national conglomerates employing rapidly-evolving forms of information technology to increasingly expand the scope of *interconnectedness* among the parts of the world-economy. Such interconnectedness has led not only to a fundamental restructuring of the world-economy. This post-industrial or "post-modern" age has also ushered in an era in which people and places all over the world are linked and tied together to a degree never before seen in human history. Rapidly-evolving forms of mass communication, faster and cheaper forms of travel between continents, and large-scale international migrations are all in part responsible for such international linkages.

In terms of modes of production, a central feature of the post-industrial era has been a restructuring of the world-economy characterized by globalization and a marked international division of labor. In the core regions of the world-economy this restructuring involved a move from economies based on heavy industry and manufacturing between World War II and the early 1970s, to service-based economies from the 1970s until the present. Manufacturing jobs and manufacturing-based industries (such as steel production and automobile assembly plants), once the cornerstone of core economies, have increasingly moved to the semi-periphery and in some instances to peripheral locations. This process many be referred to as **global outsourcing**, and it has occurred primarily due to lower costs of labor (wages) in semi-peripheral and peripheral locations such as Mexico, Indonesia, Malaysia, and much of Latin America. This geographical outsourcing has led to the movement of manufacturing-based industries and jobs (especially those in which the costs of labor are very high, such as textile production and automobile assembly) to such semi-peripheral locations. In the core regions themselves a new mode of production, called "just-in-time" or **post-Fordist production**, has replaced traditional factory organizations of labor. So-called "blue collar" low-wage and low-skilled jobs have rapidly declined in number with the decline in the number of manufacturing industries. In their place, a new kind of worker has emerged: highly-skilled, high-wage labor in which brains rather than brawn matter the most. This is especially true in the new information technologies like software production and computer assembly that are an increasingly important part of the core economies in the post-industrial era.

In the end, the post-industrial era has led to a reinforcement of the core—semi-periphery—periphery structure of the world-economy and made the differences between them greater than ever before. That is, life in the core for the average person is vastly different than it is in the periphery. Today, the core controls the flow of wealth and information in the post-industrial world-economy. The core contains the nodes, or nerve centers, of the world-economy (places such as New York, London, and Tokyo) and it is here that the largest multi-national companies in the world are located. Core economies are primarily service-based and information-based in nature. The semi-peripheral regions today are the places where much of the world's manufacturing now takes place. Agriculture is still an important part of these economies, especially the production of plantation products for export to the core. A small but expanding service sector is also characteristic of many of the semi-peripheral economies. Peripheral economies remain largely dependent upon traditional subsistence agricultural economies and plantation agriculture for export to the core. The service and manufacturing sectors of the economy in the periphery are both still rather weakly developed.

Finally, **post-modernism** has significantly altered traditional ways of life and forms of artistic expression such as art, literature, music, and poetry, as well as philosophy and other scientific endeavors. These alterations have been most conspicuous in the core regions of the world-economy, but their effects have trickled down to the semi-periphery and periphery as well, albeit to a lesser extent. Post-modern expression is characterized by a lack of faith in absolute truths, a mélange of forms and styles, a rejection of order, and a deconstructionist ideology in which traditional ways of articulation are continually questioned. This post-modern "condition" has resulted in a reordering of societies, economies, and modes of production, especially in the core, and is characterized by the following conditions:

→ Increasing globalization of the world-economy

→ The development of a "frenetic" international financial system

→ The development of, and reliance upon, new information technologies

→ A world-economy more and more reliant upon the flow of information

→ A world-economy that is increasingly *illegible* to the average person; interconnections are so complex that the world is harder to comprehend, global capitalism is harder to "locate"; a world of confused senses and order

→ A world that is increasingly "*hyper-mobile*"; a world-wide informational economy with telecommunication technology as its foundation; a "space of flows" that dominates sense of place; a perception of the world through the medium of information technologies

→ A world increasingly effected by *time-space compression*; a marked increase in the pace of life; a seeming collapse of time and space that affects our abilities to grapple with and comprehend the world

Unfortunately, after the flood, mankind disobeyed God's instructions to Noah to spread over the Earth. In Genesis 11:1, we begin to see the marks of civilization on the cultural landscape and the beginning of civilization with its consequent characteristics. Without adequate food supplies, however, urban growth would have been as impossible then as it would be now, and civilization thwarted. Adequate agricultural production is likewise inseparably connected to the rise of technology. A series of agricultural and industrial revolutions has resulted in an increasingly globalized, technologically connected world today. An increasingly efficient planet-wide "assembly line" of production requiring practically zero inventories, and accompanied by time-space compression, will probably result in decreased loyalties. There has never been a greater opportunity for leadership through service than today. Our role model during these hectic times is the Savior. Jesus said, in John 12:26, "If any man serve me, let him follow me; and where I am, there shall also my servant be: if any man serve me, him will my Father honor."

KEY TERMS TO KNOW

"Crisis of Feudalism"	Ecumene	Manifest Destiny
"god-kings"	Fordism	Merchant Capitalism
Break-bulk Point	Gateway City	Multinational Corporations
Capitalism	Global Outsourcing	Neolithic Revolution
Colonialism	Globalization	Nodes
Cultural Diffusion	Harbors	Post-Fordist Production
Cultural Landscape	Independent Invention	Post-Industrial Revolution
Culture Hearth	Industrial Revolution	Post-Modernism
Domestication	Longitudinally	Specie
Easterly	Manichean	The Black Death

FURTHER READING

Hans Bertens, *Literary Theory: The Basics* (New York: Routledge, 2001).

J. M. Blaut, *The Colonizer's Model of the World: Geographical Diffusionism and Eurocentric History* (New York: Guilford Press, 1993).

Fernand Braudel, *Civilization and Capitalism, 15th–18th Century*, Vol. I, *The Structures of Everyday Life* (New York: Harper & Row, 1981).

Carlo Cippola, *The Fontana Economic History of Europe: The Industrial Revolution, 1700–1914* (London: Fontana, 1976).

Phyllis Deane, *The First Industrial Revolution* (Cambridge: Cambridge University Press, 1979).

Jared Diamond, *Guns, Germs, and Steel: The Fates of Human Societies* (New York: W. W. Norton & Co., 1997).

Michel Foucault, *The Order of Things: An Archaeology of the Human Sciences* (New York: Vintage Books, 1994).

Marvin Harris, *Good to Eat: Riddles of Food and Culture* (Long Grove: Waveland Press, 1985).

David Harvey, *The Condition of Postmodernity* (Oxford: Blackwell, 1989).

J. David Hoeveler, Jr., *The Postmodernist Turn: American Thought and Culture in the 1970s* (Lanham: Roman & Littlefield, 1996).

David Landes, *The Unbound Prometheus: Technological Change and Industrial Development in Western Europe from 1750 to the Present* (Cambridge: Cambridge University Press, 1969).

William McNeill, *The Rise of the West: A History of the Human Community* (Chicago: University of Chicago Press, 1963).

Sidney W. Mintz, *Sweetness and Power: The Place of Sugar in Modern History* (New York: Penguin, 1985).

Lewis Mumford, *Technics and Civilization* (San Diego: Harcourt Brace & Co., 1963).

Carl O. Sauer, *Agricultural Origins and Dispersals* (New York: The American Geographical Society, 1952).

Wolfgang Schivelbusch, *Tastes of Paradise: A Social History of Spices, Stimulants, and Intoxicants* (New York: Vintage Books, 1992).

Peter Stearns, *The Industrial Revolution in World History*, 3rd ed. (Boulder: Westview Press, 2007).

Immanuel Wallerstein, *The Modern World-System I: Capitalist Agriculture and the Origins of the European World-Economy in the Sixteenth Century* (San Diego: Academic Press, 1981).

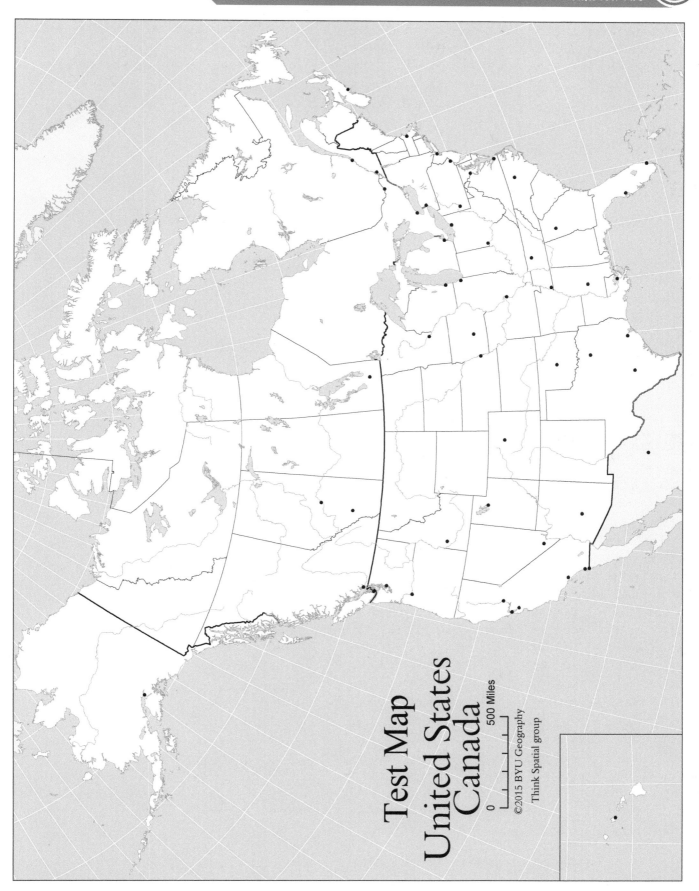

Test Map
United States
Canada

0   500 Miles

©2015 BYU Geography
Think Spatial group

Middle America and Population Growth

3

But godliness with contentment is great gain, for we brought nothing into the world, and we cannot take anything out of the world. But if we have food and clothing, with these we will be content. But those who desire to be rich fall into temptation, into a snare, into many senseless and harmful desires that plunge people into ruin and destruction. For the love of money is a root of all kinds of evils. It is through this craving that some have wandered away from the faith and pierced themselves with many pangs. (1 Timothy 6:6-10)

CHAPTER OUTLINE:

MOST Christians are aware of what Jesus meant when he said, "Therefore I say to you, do not worry about your life, what you will eat or what you will drink; or about your body, what you will put on. Is not life more than food and the body more than clothing?" (Matthew 6:25 NKJV) But fewer actually realize that worrying about these material things choke out our lives like weeds do crops in the garden. Though geographically close to us, the loveliness of the Caribbean landscape and culture can hide its spiritual wounds; just below the border we can find desperation that needs more than a material solution.

The area we call Middle America consists of a triangular shaped landform that has acted as a narrowing land bridge largely located in the tropics between North America and South America. This landform is flanked on the west by the warm Caribbean Sea and the Gulf of Mexico. On the east is the enormous Pacific Ocean. As do most tropical areas, Middle America faces both a great deal of challenges and exciting opportunities. Because of the diversity of cultures existing in this area, cultural terms such as Latin America are actually inaccurate when describing this region. Where exactly is Middle America? For the purpose of this work it is considered to be all of the areas with a coastline bordering the Pacific and Atlantic Oceans south of the United States and above the continent of South America. Let us examine how the physical geography of this region has affected many aspects of the human geography.

CAREFUL WITH THOSE PLATES

Besides an **attenuated** landform, much of Middle America is mountainous. In fact, without the mountains resulting from tectonic activity, many of the islands in the Caribbean Sea would not exist at all! Geologists tell us the Caribbean plate is actually in contact with four other plates. The resulting stresses of the converging and diverging of these plates are instrumental in producing volcanoes and earthquakes as well as the mountain ranges running down the spine of Middle America. The climate, as one might expect, is warm. This warmth and the resulting precipitation can be a deadly mix; erosion has long plagued farmers of Middle America especially since the high relief, or elevation, has been over grazed in large measure because of the arrival of European domesticated animals.

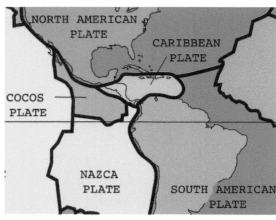

From *This Dynamic Earth: The Story of Plate Tectonics* (online edition) by W. Jacqueline Kious and Robet I. Tilling, prepared by U.S. Geological Survey

Do not be deceived, however; though the majority of Middle America is tropical, the mountains and high relief can indeed lead to cooling. As you might expect in a largely tropical area, a majority of the climate, however, is warm and experiences heavy rainfall. The cooling and the subsequent orographic effect of precipitation that are due to the high relief of the mountains and volcanoes result in **windward** and leeward effects. The variance in precipitation has given Middle America a range of climatic conditions—many of which are ideal for particular crops. Previous discussions of North America in Chapter 2 revealed how important differing agricultural patterns can be to the development of societal culture and even ideology. Similarly, Middle America is an amazingly complex area with a diversity of agriculture.

THE COAST WITH THE MOST

One can go from rainforests and warm-moist areas along the coast called the **tierra caliente** (sea level to 2500 feet elevation) to the moderate climates of the **tierra templada** (2500 to 6000 feet elevation) and then to the dryer and cold climates in the **tierra fría** (6000 to 1150 feet elevation) within relatively short distances, merely by taking a trip up the mountains. Each of these areas has the potential to yield a fantastic array of crops. The tierra caliente area near the coast is famous for its bananas and other tropical fruits. In fact, some geographers believe the tomato originated in the area between the tierra caliente and the templada regions. For many years, Europeans and North Americans were afraid to eat a member of the nightshade family because this fruit had an infamous reputation as being a plant used to poison royalty. A great example of a type of inverted **hierarchical diffusion** actually occurred here when the lower classes would eat the fruit and eventually it made its way up to the higher classes. Usually hierarchical diffusion works the other way around. Like when the author dresses up like his boss in another failed attempt to be promoted. Anyway, the tierra templada is famous for its coffee, wheat, and maize (native corn), whereas in the harsher, drier and much colder highlands one can grow the tough potato from South America and many of the various grains such as barley and wheat that we are so accustomed to.

The biomes in Middle America lend themselves to a great deal of diversity in terms of animals and plants. In fact, of the millions of species on earth many of those yet to be discovered are in this region. Despite this wonderful diversity, however, risks exist. The cutting of the rainforests contributes greatly to the loss of precious soils through erosion and by the percolation or **leaching** of valuable nutrients resulting from the tropical rains taking nutrients down into the soil away from the shallow roots of the tropical trees. Since we all want to be good stewards of God's resources, it is important to understand the nature of this area and the importance of good soil and forest management techniques.

Another **human-environmental** difficulty associated with the arrival of Europeans into the Caribbean resulted from a combination of mountains and tropical conditions that have unfortunately caused much **erosion** after the Europeans brought animals such as goats with them. The grazing of these animals too closely to the ground because of the shapes of their mouths resulted in the grass dying and the rain eroding the soil. The poverty in this area is largely due to the arrival of these Old World animals during the Columbian Exchange. The damage done by these animals has, in the long run, made it difficult for farmers everywhere, including Middle America, to feed a growing tropical population.

HANG THOSE FLAGS!

Vexillology is the study of flags. The flags of the Caribbean bear depictions of the stars, the sun, and the sea which reflect strong symbolic meanings associated with the landscapes of this region. The cultural legacies of Africa are manifested often in green, yellow and black, and these combinations offer interesting insights into the origins and history of many of the cultures here. Jamaica is a great example; though relatively close to Spanish-speaking Cuba and Hispaniola in the Greater Antilles of the Caribbean, Jamaica is an English-speaking nation with the predominant ethnicity being black. Looking at the flag of Jamaica, one can trace the African roots by viewing the green and yellow flag with the black colors. Flags can be a terrific addition to any classroom and by definition should be able to be drawn by a child from memory accurately because

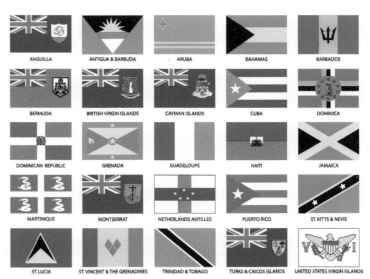

Image © Basheera Designs, 2013. Used under license from Shutterstock, Inc.

children are normally indoctrinated into nationalism by the happy recognition of their flag. For these reasons, flags symbolically depict a nation's values and its story and speak volumes of a culture's views and history. Many of the republics of the mainland in Middle America, for example, use varying degrees of blue and white in their flags like the French and Spanish flags.

You may not be aware that the unfortunate legacy of slavery that occurred in the South extends beyond the borders of the United States. The first slaves who were brought to the New World were probably bound for the Caribbean. South America and the Caribbean are where the preponderance of these slaves arrived because of the agricultural demands. The Portuguese initially used mostly west-African slaves as had Arabs previously in east Africa. The West Africans demonstrated a valuable ability to grow sugar cane initially on Sao Tome and the Principe Islands. And as we saw with Jamestown, the easterly currents brought them to the New World. Under terrible conditions, these unwilling travelers on slave ships crossed the **Middle Passage** of the Atlantic. The islands of the Caribbean sat waiting, perfect for growing sugar. Eventually, the plantation form of agriculture arose with its emphasis on profits. Ultimately, the Barbadian slave code, which created perpetual slavery would diffuse north and be adopted into the slave markets of Charleston, South Carolina, and would manifest itself into the codified laws of Virginia and the other colonies of the South. Climate and currents worked together to create the largest forced movement of people in the world against their wills.

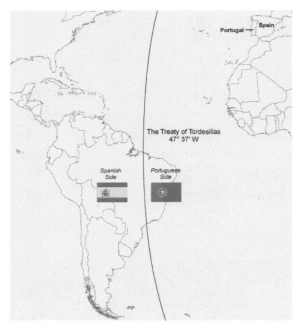

The Treaty of Tordesillas
47° 37' W

Spanish Side

Portuguese Side

Map courtesy of Katie Pritchard

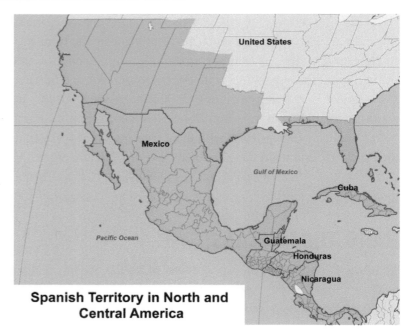

Spanish Territory in North and Central America

The World Factbook

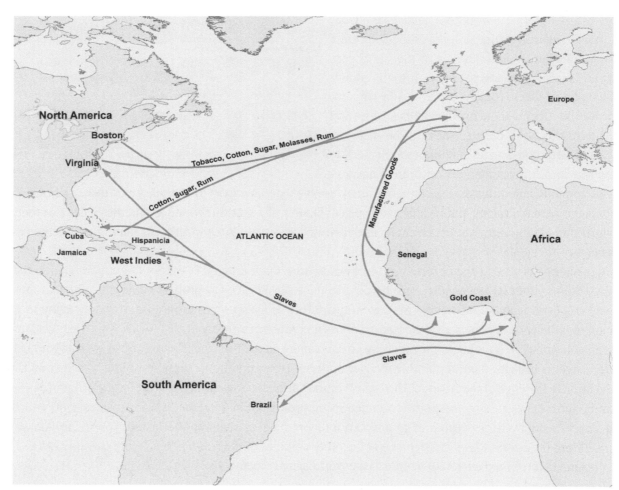

The World Factbook

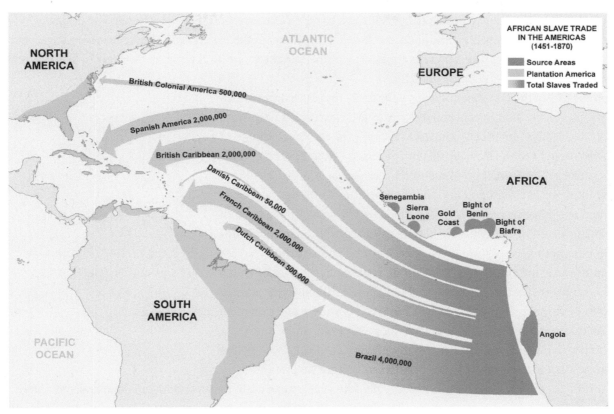

AFRICAN SLAVE TRADE
IN THE AMERICAS
(1451-1870)

Source Areas
Plantation America
Total Slaves Traded

The World Factbook

WHOSE HOUSE IS THIS?

The **hacienda**, or "big house," is a characteristic item on the cultural landscape of Middle America. This livestock farm, or precursor of American ranches, represented an independent and self-sufficient form of agriculture that in many ways exemplified the individual spirit of the region. This diffusion of Spanish culture into the agricultural realm of America is associated with the **viceroy system** of colonial administration. The drawn borders in Central and South America reflect the centralized political entities that were authoritarian regimes in this region. These borders are deep divisions that keep the peoples apart and enable the political leadership to control them. Both the hacienda and the viceroy system according to some authorities represented a breakup of Spanish rule and were characteristically absolute in the authority delegated to local leaders and their ability to rule over their subordinates. The Spanish used the **encomienda system** as a means of employing the labor of Native-Americans in Middle America; however, these natives especially suffered under the immense power of the state which had the power of life and death over them. The transition from an encomienda to a hacienda system of agriculture, a self-supporting body raising all crops and livestock needed for survival, stands in marked contrast to the plantation system, a capitalistic endeavor designed to specialize in one or two staple crops thereby maximizing profits, located usually where the English-speaking colonists settled in this region. The nature of the Spanish political system described above also shaped agriculture which in itself reflects the independent culture of the southern latitudes as opposed to the more capitalistic or unified north.

The **plantation** is marked by a large-scale endeavor usually based on one or two cash crops. The plantation system is more reflective of the industrial age with its emphasis on production and efficiency. Both

Image © Elena Kalistratova, 2013. Used under license from Shutterstock, Inc.

the hacienda systems represented attempts by capitalists to receive returns on their investments. The unique ways in which each system was profitable reveals cultural diversity which originated in the agricultural attempts toward sugar and tobacco production in the New World. The agricultural patterns of the Caribbean are as diverse as the French, Spanish, and English peoples who inhabit it.

The ethnicities of the Caribbean are of a wide array ranging from Spanish-speaking people in the Dominican Republic, who are descended from both the Spanish and Native Americans known as **Creole**, to the descendants of African slaves brought to the New World with modified British speech accents such as in Jamaica. The French presence in the rich "sugar islands" manifests itself today in places like Haiti and into the Leeward Islands where the islands of Guadalupe and Martinique are still dependent on France. The mixture of African and Europeans are often referred to as **mulatto** and generally the differences in population are of little consequence to the peoples there. Notice on the map towards the south near the Gulf of Venezuela, the former Dutch islands of Aruba and Curacao sit strategically offshore of the oil rich areas of Maracaibo in Venezuela. Though you might think of Venezuela and Colombia as South American countries, in many ways they culturally share the same sunny and cheerful culture of the Caribbean, as does the Gulf Coast of the United States. What are the sub-regions, or areas, of Middle America?

© Peter Hermes Furian/Shutterstock.com

Middle America can easily be separated into mainland and islands. **Mexico** has tended to be stereotyped into a uniform climate and culture by Americans as a result of their experiences with the border, but beneath the border is an amazing array of different biomes and climates. Contrary to the views of most North Americans, Mexico is actually a quite diverse landscape. Somewhat similar to the United States in that two mountain chains run basically north and south, it differs by virtue of possessing a huge plateau between them in the center of the nation. The rich legacy of the Spanish can be seen in the **plazas** and landscapes of Mexico. The movement of peoples and culture into the United States is reflected in large measure by the flat-roofed landscapes and the use of fountains whose tradition is rooted in the plaza. Revolution has shaped Mexico and generally replaced the hacienda system previously mentioned. The control of public lands on the **ejidos** continues, however, and normally are public lands controlled by the government and run by 20 families or more. To the south in Mexico we find the state of **Chiapas** where centrifugal effects of ethnicity and even language pose a modest threat to the Mexican union. Farther to the north in Mexico we see marked social-cultural differences.

From *Historical Atlas* (1911) by William Shepherd, courtesy of University of Texas Libraries

MOVE OUT!

To the north we find the **maquiladoras**, or the factory areas, along the border. Unlike the ecumene in Canada, this area is somewhat impoverished and tends to be a less permanent region. Initially, many of the "border towns" along the US and Mexican border were small, but, with the change in laws affecting manufacturing decades ago, factories became profitable on the border due to the relatively inexpensive labor available to the US companies. With the displacement of

The World Factbook

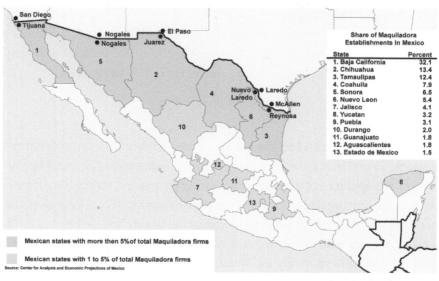

Share of Maquiladora Establishments in Mexico	
State	Percent
1. Baja California	32.1
2. Chihuahua	13.4
3. Tamaulipas	12.4
4. Coahuila	7.9
5. Sonora	6.5
6. Nuevo Leon	5.4
7. Jalisco	4.1
8. Yucatan	3.2
9. Puebla	3.1
10. Durango	2.0
11. Guanajuato	1.8
12. Aguascalientes	1.8
13. Estado de Mexico	1.5

Mexican states with more then 5% of total Maquiladora firms

Mexican states with 1 to 5% of total Maquiladora firms

Source: Center for Analysis and Economic Projections of Mexico

Map source: *The World Factbook*

SOME MAQUILADORA COMPANIES YOU MAY KNOW

3 Day Blinds	Eli Lilly Corporation	Honeywell, Inc.	Pioneer Speakers
20th Century Plastics	Ericsson	Hughes Aircraft	Samsonite Corporation
Acer Peripherals	Fisher Price	Hyundai Precision America	Samsung
Bali Company, Inc.	Ford	IBM	Sanyo North America
Bayer Corp./Medsep	Foster Grant Corporation	Matsushita	Sony Electronics
BMW	General Electric Company	Mattel	Tiffany
Canon Business Machines	JVC	Maxell Corporation	Toshiba
Casio Manufacturing	GM	Mercedes Benz	VW
Chrysler	Hasbro	Mitsubishi Electronics Corp.	Xerox
Daewoo	Hewlett Packard	Motorola	Zenith
Eastman Kodak/Verbatim	Hitachi Home Electronics	Nissan	
Eberhard-Faber	Honda	Philips	

the American manufacturing sector largely to Asia in the last generation, many of these towns faced massive unemployment and the result was the beginning of a movement pattern known as **chain migration** into North America. The movement into North America consisted of young males initially seeking work—**pull factor**—and many desperate to leave an area where little work existed—**push factor**—and then eventually as wives moved to be with their husbands, the eventual movement of entire towns resulted. These patterns of movement are generally a global phenomenon, with most migrations occurring today from southern to northern latitudes. Many of these longitudinal travelers represent a **counter-movement** or reverse of earlier colonial era patterns. Interestingly, it is usually the intention of the migrants to travel only short distances when such moves are made as families. Usually young, single male settlers travel a much longer distance. The colonization process described earlier in

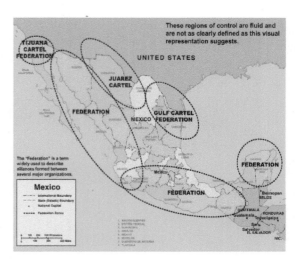

Source: U.S. Drug Enforcement Administration, adapted by CRS (P. McGrath 3/2/2007)

North America is a prime example of what has occurred in much of Middle America in the last generation. Although Mexico faces numerous challenges, including seemingly unending drug wars and gang activities, it is a mineral-rich nation with limitless potential. Tourism has been a leading industry in Mexico for obvious reasons. The beauty of God's creation in the people and in His beautiful natural world is hard to miss in the region of Middle America!

FUN IN THE SUN

Incredibly beautiful, but at times violent and harsh, the islands of the Caribbean Sea are sometimes buffeted cruelly by hurricanes and other tropical storms. Traditionally, these areas consisted of highly contested areas sought by the empires of Europe for their enormous profit potential in the form of sugar and other crops such as tobacco. Nevertheless, life could be hard and fast with tropical diseases in large measure a function of the increased movement of the age of exploration and the Colombian Exchange of animals and plants. Perhaps Cuba is a good example of the beauty and tumult that is a part of the Caribbean.

© 2007 JupiterImages Corporation

The island members of the Greater Antilles, which include Cuba, Jamaica, Puerto Rico, and Hispaniola, generally lay latitudinally across the areas that see many of the seasonal storms that wash them with moisture. The great nation of Cuba has seen itself the battleground of efforts by American force in the Spanish American War and the Cold War. Due to its proximity to the United States, Cuba both chafed under and received an advantage from its allies in the former Soviet Union. To American eyes, revolution seems to be the lay of the land in this land of extremes, but to the people attempting to survive it represents just another autocracy in the long history of totalitarian power to our south. Recent attempts by the son of the former dictator Fidel Castro to serve only two terms are being watched carefully as an opportunity for this formerly great culture to shift back into a free market mode. Venezuela's growing relationship with increasingly autocratic Russia and China may slow any transition to capitalism, but Cuba's future appears brighter in terms of relations with the U.S. and the world perhaps due to its longer history of international trade and more successful legacy of agricultural management.

Most Christians are familiar with the earthquakes that hit Haiti in 2010, but very few statistics are available to document the total generosity displayed toward this poor island nation which is on the eastern half of the island of Hispaniola; a divided island like New Guinea in Indonesia. Millions of dollars and countless hours of time have been given to provide essential needs to the Haitians in recent years. The Dominican Republic next door, actually occupied by US Marines for years earlier in the last century, has so far escaped many of the challenges facing the former French colony of Haiti, but the Spanish-speaking nation of the Dominican Republic will continue to be persuaded toward economic changes by developments between the United States and Cuba. The island of Puerto Rico (the last major island group in the Greater Antilles) is in fact a territory of the United States as are the Virgin Islands next door.

Image © Art_girl, 2013. Used under license from Shutterstock, Inc.

Photo by C.B. Waite, from Frank and Frances Carpenter Collection, Library of Congress

CONTRIBUTION OF TOURISM

Country/Territory	Economy to GDP (2002) Percent of GDP
Anguilla	58
Antigua and Barbuda	72
Aruba	47
Bahamas	46
Barbados	37
Bermuda	26
British Virgin Islands	85
Cayman Islands	31
Cuba	11
Dominica	22
Dominican Republic	18
Grenada	23
Guadeloupe	33
Jamaica	27
Martinique	10
St. Kitts and Nevis	25
St. Lucia	51
St. Vincent and the Grenadines	29
Virgin Islands (U.S.)	42

From *The Plaid Avenger's World #6 Nuclear Insecurity Edition.* Copyright © 2008, 2009, 2010, 2011, 2012 by John Boyer. Copyright © 2006 by Kendall Hunt Publishing Company.

Puerto Rico has influenced America through migration and its contributions to the nation through military service and political support for continued membership with commonwealth status in the United States. Primarily, sugar is grown and much of the American industry has moved to the island to take advantage of inexpensive labor. Jamaica historically provided aluminum ore (bauxite) to a world entering the age of air and space. The nation of Denmark actually sold the neighboring Virgin Islands to the United States. Since the US dollar is the medium of exchange in the Virgin Islands, it can be considered to be an insular part of the United States. The Virgin Islands is the first of the Lesser Antilles we

© 2007 JupiterImages Corporation

will discuss in the Caribbean. Unfortunately, due to high demand in the United States, drugs are a valuable resource in the Caribbean. Narcotics trafficking can make people rich as they try to enter the United States. **Offshore banking** is a means of escaping the tax burden that has grown in the United States, for example, while maintaining close proximity to American businesses. Compared to the banks of Switzerland, drugs, tourism and offshore banking are all heavily influenced by the American economy.[1]

Two island chains coming from off the northern coast of South America—the Windward Islands that eventually become the Leeward Islands to the north, form the Lesser Antilles. Aptly named for their exposure to the sometimes violent storms emerging from the Atlantic Ocean, these beautiful islands actually share more than their local cultures. The entire region

[1] Plaid Avenger's World (Dubuque, Iowa: Kendall Hunt), 325.

Caribbean Colonial Holdings

Puerto Rico

1636 1672 1682 1690 1696 1781 1784 1810 1616 — **St. Eustatius**

1623 17th Century 1702 1938 — **St. Kitts**

1628 1706 1706 1782 1784 1983 — **Nevis**

1635 1666 1667 1981 — **Antigua**

1632 1667 1667 1782 1784 — **Montserrat**

1635 1759 1763 1810 1814 1815 1816 — **Guadeloupe**

1632 1761 1778 1783 1978 — **Dominica**

Caribbean Sea

1625 1762 1763 1794 1802 1809 1814 — **Martinique**

France 9 times / British 6 times / Neutral 2 times — 1803 — 1803 1979 — **St. Lucia**

1627 1762 1779 1783 1979 — **St. Vincent**

1627 1966 — **Barbados**

1650 1762 1779 1783 1974 — **Grenada**

1658 1677 1763 1781 1793 1802 1803 1962 — **Tobago**

Legend:
- British
- French
- Dutch
- Dominica
- St. Kitts & Nevis
- Antigua
- St. Vincent
- St. Lucia
- Barbados
- Tobago
- Grenada

Map source: *The World Factbook*

of the Gulf of Mexico and the Atlantic coasts of Venezuela and the countries of northern South America share a Caribbean-type regional culture. A visit to Charleston, South Carolina, or New Orleans will give a hint of this atmosphere.

Sugar has been an important resource of this area causing colonial competition throughout history and is partially responsible for an amazing diversity of people and cultures. The triangle of trade previously mentioned actually made the Caribbean the center of world trade (early globalization?) and in many ways was responsible for a series of world wars in the 18th century that were felt in North America. Perhaps inaccurately characterized as relaxed and laid back, the rhythms of this area seem to strongly suggest an indomitable joy despite the apparent challenges of the environment. Caribbean coastal cultures differing from the changing landscapes of Middle America and Mexico make this area potentially rich and important for the United States as we continue to view the areas in terms of opportunities for missions and enjoy the security of this southern area.

THE GEOGRAPHY OF WORLD POPULATION*

> "Then they also brought infants to Him that He might touch them; but when the disciples saw it, they rebuked them. But Jesus called them to Him and said, "Let the little children come to Me, and do not forbid them; for of such is the kingdom of God. Assuredly, I say to you, whoever does not receive the kingdom of God as a little child will by no means enter it." (Luke 18:15–17)

"Jesus loves the little children" is how the song goes. If the Lord Jesus Christ did not see children as interruptions; and compared them to the Kingdom of God so should we. The study of population geography is of vital interest to the human geographer. We see a terrific opportunity during this time of unprecedented population growth to fulfill a purpose of eternal consequence. In John 4:34–35 it says, "Do you not say, 'There are still four months and *then* comes the harvest'? Behold, I say to you, lift up your eyes and look at the fields, for they are already white for harvest!" Life is indeed short, let's get busy!

We begin this section with how the human population varies across the planet, especially with respect to distributions, structures, and core-periphery relationships. We are often reminded by the news media and various international organizations that population "problems" confront our world today. Invariably these problems are presented as having something to do with either overpopulation or how the growing world population affects the supply and use of various natural resources. If there are "problems" related to population, what are they? Do such issues vary between the core and periphery? How do the structures of populations differ in the zones of the world-economy? How is population distributed around the world?

WORLD POPULATION DISTRIBUTION

If one examines a map of global population distribution one of the first things that is readily apparent is that the world's population is not evenly distributed. While some regions are very densely populated (Europe and much of Asia, for example), there are large parts of the earth that are very lightly populated (such as the arctic regions, Australia, and Siberia). In general, if indeed there are "problems" relating to overpopulation, those problems are not found everywhere around the world; it is not that there are too many people on the planet (that is a value judgment), but that there are too many people in certain places. The densest population clusters tend to be located in two main types of natural environments around the world. The first is the fertile river valleys of the tropics and sub-tropics and the second is the coastal plains of the mid-latitudes which are in general temperate regions. More precisely, it is possible to identify four major concentrations of population:

→ East Asia (China, the Koreas, Vietnam, and Japan)

→ South Asia (India, Pakistan, and Bangladesh)

→ Europe (The British Isles to western Russia)

→ North America (Boston to Washington, D.C.; West Coast)

Roughly 75 percent of the world's population lives in these four areas; four out of ten people in the world live in just two: East and South Asia. Other smaller concentrations of dense populations occur on the island of Java (part of Indonesia), the Nile Valley of Egypt, central Mexico, and southeastern South America (southern Brazil and eastern Argentina).

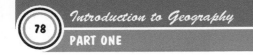

FACTORS IN WORLD POPULATION

DENSITY

We have alluded to the fact that the way populations "look," the way they are structured and their rates of growth, vary significantly between the core and the periphery. It is possible to compare and contrast different populations by comparing various statistics (examples of these statistics for various countries are listed in Table 3.1). One of the most elementary of these factors is population density. Density can be measured as **crude population density**, the total number of people per unit area of land in a place or region, or as **physiologic population density**, the total number of people per unit area of *arable* (agriculturally productive) land. The latter is actually a more telling figure because it measures the density of populations with respect to how much of the land on which they are living is productive enough to produce enough food for that population. When the difference between a country's crude and physiologic densities is very large, it is a sure sign that that country has quite a bit of marginally productive agricultural areas. This can be observed in the following list:

TABLE 3.1

Country	Crude Density/km²	Physiologic Density/km²
Japan	862	6,637
Bangladesh	2,124	3,398
Egypt	142	7,101
Netherlands	1,041	4,476
USA	67	335

GROWTH RATES

A second method for comparing populations is by examining population growth rates. Since the Industrial Revolution, the world's population has been growing at an exponential rate (2, 4, 8, 16, 32, 64 . . .). Currently, the world rate of natural increase is about 1.8 percent per annum. This means that 1.8 percent of the current population is being added each year. This figure translates into a current **doubling rate** of 40 years, but this doubling rate will decrease with added population each year. Given this growth rate, the world's population likely reached 7 billion in 2011. Peripheral populations are growing at a much faster rate than those in the core and semi-periphery. That is, at a broad global scale, there is a direct correlation between economic "development" and population growth rates:

TABLE 3.2

Country	Rate of Natural Increase	Doubling Time
Poland	0.5%	141 years
Australia	0.75%	94 years
China	1.5%	46 years
Kenya	4.0%	17 years

STRUCTURE

A third way of comparing populations around the world is by observing differences in the "structure" of populations. By structure we mean the relative number of men and women in different age cohorts in a population. A population's structure is most clearly seen by constructing a **population pyramid** that charts both male and female populations in five-year age cohorts on a y-axis and percentage of total population on an x-axis. The term population pyramid is used to refer to such age-sex diagrams because the shape of these diagrams is pyramidal in developing countries, that is, in the peripheral regions of the world-economy. This pyramidal shape indicates a population that is "young" and growing. Birth rates and fertility rates (discussed below) are relatively high and life expectancies are relatively low. Thus, a substantial proportion of the population in the peripheral countries is very young, under 15 years of age, while the number of people in higher age cohorts, above 60, is very low. By contrast, in the core and parts of the semi-periphery, age-sex diagrams tend to have a rectangular shape. These countries have low birth and fertility rates, higher life expectancies, and thus the population is more evenly distributed among age cohorts. These populations are "old" and stable.

DEMOGRAPHIC CYCLES

If we examine past patterns of population growth rates in different parts of the world it is possible to identify demographic "stages" through which populations tend to pass. Where a country is at with respect to this cycle (that is, what "stage" the country is in) tends to mirror economic "development." These cycles and stages can be discerned by examining the relationship between three major indictors: the **crude birth rate**, defined as the number of live births per 1000 persons per year; the **crude death rate**, defined as the number of death per 1000 persons per year; and the **total fertility rate**, defined as the average number of children born to women of childbearing age (roughly 15–45) during their lifetimes. It is possible to compute the **rate of natural increase** for a given country by subtracting the crude death rate from the crude birth rate. For example, if the crude death birth rate of a country is 20/1000 and the crude birth rate is 5/1000, then the natural increase is 15/1000, or a rate of 1.5% per annum. Death rates do not vary substantially from core to periphery. Indeed, some core countries have higher death rates than some of the poorest countries in the world (see Table 3.1). Only in areas of famine or economic and political unrest (and such occurrences are usually short-lived) are death rates inordinately high. This is largely due to advancements in medical technology, especially immunizations for diseases that used to kill millions of people every year. Such technology has become available even in some of the poorest countries in the world in the last 50 years. On the other hand, as can be seen in Table 3.2, birth rates and fertility rates, and thus rates of natural increase, vary substantially between the core and periphery.

CORE-PERIPHERY POPULATION PATTERNS

As mentioned above, populations tend to pass through stages, and these are revealed by comparing long-term historical patterns with respect to the relationship between birth rates, death rates, and rate of natural increase. This historical model of population change is usually referred to as the **Demographic Transition Model**, of which there are four stages (Figure 3.1). In Stage 1, birth rates and death rates are both very high, resulting in relatively low or fluctuating rates of natural increase. Until the Industrial Revolution, when societies around the world were still agricultural in nature, all populations around the world were in Stage 1, but today there are virtually no populations in this stage. In Stage 2, death rates fall off substantially but birth rates remain high, resulting in very high rates of natural increase. In Europe, this stage began around the middle of the 18th century, as economies and societies began to industrialize. New medical technology greatly reduced death rates, but birth rates and fertility rates remained very high due to advances in medicine and improved agricultural yields as a result of more efficient agricultural techniques and tools. With enough food to go around, most people saw little need to alter traditional conceptualizations and norms with respect to reproduction. This rapid and exponential growth in population in Stage 2 is the "transition" in populations that is referred

FIGURE 3.1 THE DEMOGRAPHIC TRANSITION MODEL

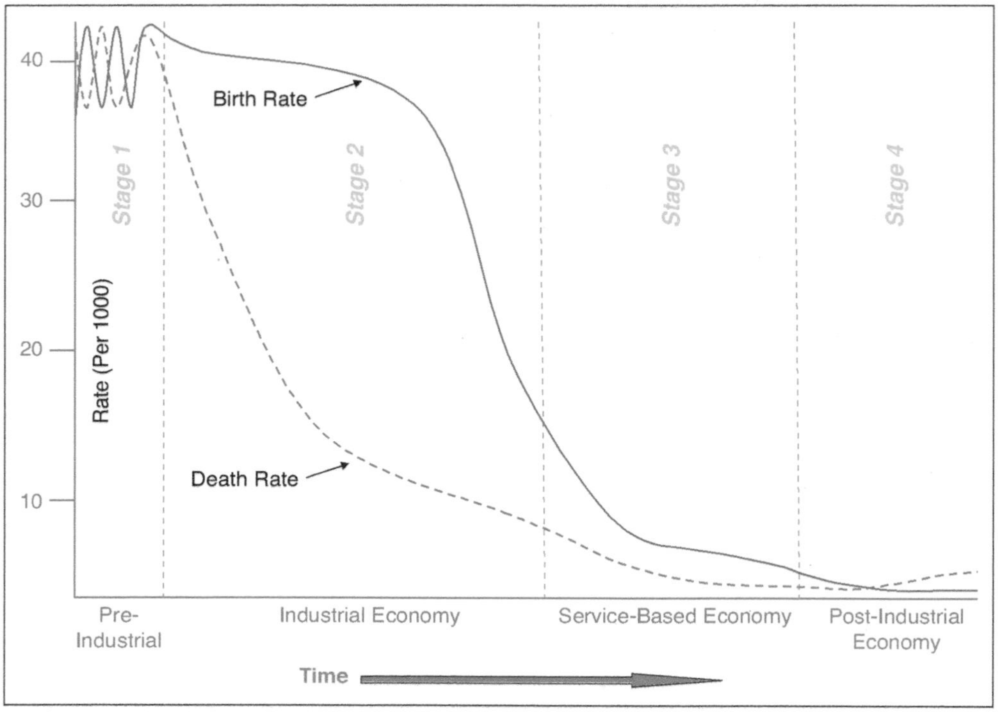

to in the Demographic Transition Model. Today, most countries in the periphery, and some in the semi-periphery, are in Stage 2 of this demographic transition.

In Stage 3 of the Demographic Transition Model, death rates remain quite low and birth and fertility rates begin to drop dramatically, resulting in decreasing rates of natural increase. Most of Europe and North America went through this stage during the late 19th and early 20th as these regions developed mature industrial economies. Today, most of the semi-periphery of the world-economy is currently in Stage 3. By Stage 4, which began sometime in the late 20th century in most of the core, birth rates had fallen so much that some countries have approached **zero population growth**. Indeed, in a handful of core countries today (mainly in northern and eastern Europe) populations are actually declining. The populations in the core regions of the world-economy have passed through each of these stages, and today they are the only regions in Stage 4 of the model.

Why is there such a strong correlation between economic "development" and fertility, and what factors account for these global patterns? These are extremely complex questions, for which there are few easy answers. It is possible, however, to identify some of the most probable explanations. To be sure, traditional values and customs concerning reproduction and conceptualizations of femininity and masculinity, as well as traditional religious customs, in the folk cultures of the periphery are part of the explanation. Access to modern forms of birth control (expanded in the core, more limited in the periphery) may also help to explain these patterns. But it can be argued that both of these explanations fail to take into account the power that women have in most societies around the world with respect to reproductive choice. They also fail to address differences in economies and lifestyles between the core and periphery. The most likely and most plausible explanation for the correlation between fertility and economic development is that the role of women in the societies of the core and periphery are quite different. In the subsistence agricultural societies of the periphery the role of a woman is often what we might call "traditional"—they are not only mothers, but also farmers. In such societies, traditional conceptualizations of women and children predominate and in these traditional economies children are an economic asset—the more hands for the fields, the better. On the other hand, in the core regions of the world-economy, post-modern ideas have

FIGURE 3.2 WORLD REGIONAL MAP WITH POPULATION TOTALS

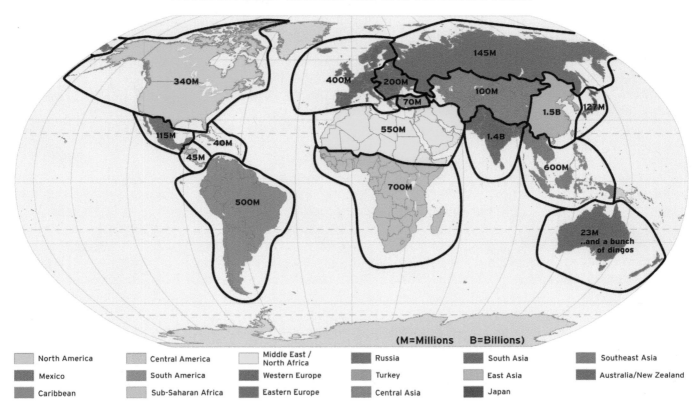

(M=Millions   B=Billions)

North America	Central America	Middle East / North Africa	Russia	South Asia	Southeast Asia
Mexico	South America	Western Europe	Turkey	East Asia	Australia/New Zealand
Caribbean	Sub-Saharan Africa	Eastern Europe	Central Asia	Japan	

led to radical critiques of such traditional roles for women. In most of the core societies, the role of a woman is not seen as just a mother, but also a breadwinner. So too, in these post-industrial economies in which very few people farm for a living and where the costs of living are substantial, children are in fact an economic liability. In the core, then, women have embraced other roles and put off having children until later in life. This has resulted in drastically lower fertility rates, since waiting to have children until later in life statistically reduces the number that a woman could have.

In summary, population "problems" in the periphery and parts of the semi-periphery involve those of an ecological nature. These populations are in Stage 2 of the demographic transition, with very high rates of natural increase. But at the same time, these are precisely the places that are least able to cope with young and growing populations, mostly due to weakly developed political and economic infrastructures.

LATIN AMERICA*

A SINGLE REGION?

Latin America is a region with a question mark after it. The Plaid Avenger has never really considered it a region in the past, because as you know, creating a "region" involves identifying some sort of homogeneous singular trait that you can apply to the geographic space in question. I've said for years this place is too big. You can't apply singular homogeneous traits across all of Latin America because it encompasses everything from the south of the US-Mexican border all the way to Tierra del Fuego, the tip of South America, which is almost in the Antarctic. It's just too darned big, and too darned complex! A

What's the Deal with . . . the Latins in Mexico?

What happened was this freaky-freak French king Napoleon III (not Napoleon the short, dead dude we all know, but one of his later kin) sent over this dude named Maximilian (actually, an Austrian) to sit on the Mexican throne in a feeble effort to reestablish France's presence in the New World in 1862. The French were actually invited to do this by an elite rightist core of Mexican aristocracy who wanted to revive the Mexican monarchy and simultaneously dismantle the leftist movements occurring in what was independent, sovereign Mexico at the time. So this French dude Maximilian was in charge, and when he was out scouting around he said, "Hmmm. . . . We're French, we're not even Spanish. What rallying point can we possibly use to get the locals to think that we should be in charge here?"

Maximilian being executed.

What our brilliant French brethren came up with was that the Spanish, Portuguese, and the French languages are all linguistically linked. They are all part of the Romance, or Latin-derived languages. In a fit of what must have been desperation at the time, Napoleon III told ol' Maxi to say, "Look, I may be just the French guy here in Mexico, but we are all brothers under the Latin language, so we're all family here in Latin America." Again, a totally bogus, politically made-up term, but somehow it stuck. The Mexicans deposed Maximilian's derriere fairly quickly, but the term stayed—in fact, kicking out the French is what the Cinco de Mayo holiday is all about; it's not Mexican Independence Day! That is a different holiday altogether! May 5 is all about forcing out the French! Which is always a good time for all.

single same-ness that applies to everything from the Rio Grande to Rio de Janeiro, or the Atacama to the Andes? Not hardly. Except maybe the term 'Latin America' itself . . . and BTW, where did that come from? A: . . .

After thinking about the French fool Maximilian's antics for a while, the Plaid Avenger has realized that, indeed, there are a lot of traits that can be recognized across the whole of Latin America. We're going to look at some of those traits in this introduction pre-chapter to the more unique subregions of Latin America.

WHAT IS LATIN AMERICA?

So . . . this "Latin America" is a term that everybody recognizes, everybody gets, you know where it is, but it's based on a word that is completely and utterly meaningless: Latin. Do all of these people speak Latin south of our border? Heck no. Latin is a dead language. Nobody speaks that Caesar-ized stuff anymore. In fact, there are really only two predominate languages spoken in the region: Spanish and Portuguese. That, and a smattering of small states that speak English or French in the Caribbean.

How did we get this term Latin America: are they all of Latin or Greek ancestry? Obviously not. As with many things on the planet, we can blame the French. In that pretty little piece of propaganda, the French called for a Latin brotherhood based on all their languages (French, Portuguese, Spanish) being of Latin roots. Pretty shaky foundation for a regional identity, but there it is.

Geographically, the term means everything south of the US/Mexican border. That's common knowledge across the planet. Mexico, which we teased out for obvious reasons earlier, is not part of the US/Canadian region. There are too many differences, including its Latin-based language, levels of development, poverty, and lots of other issues which distinguish Mexico from the United States. So even though it is a NAFTA member and southwestern US is increasingly becoming part of a "greater Tex-Mex" region, for now Mexico still shares more commonalities with states south of that border. . . . Before we get to those similarities, let's identify frequently referenced regions in this part of the planet.

LATIN LOCATIONS

There are a few terms that the Plaid Avenger wants you to know before we get into these subsequent chapters. These definitions will help us understand a lot of terminology used on the global stage. We already talked about Latin America encompassing every single country south of the US/Mexican border, in the entire western hemisphere. Now let's break down some more specific regions within Latin America.

SOUTH AMERICA

The first one is easy enough. South America is both a continent and a region. The South American continent is that other major chunk of land in the western hemisphere. It starts with Colombia and then heads east to Venezuela, Guyana, Suriname, French Guyana, and then turns down south to Brazil. South of Colombia is Ecuador, Peru, Chile, Bolivia, Paraguay, Uruguay, and Argentina. It is important to note that most all of the countries of South America are fairly good sized; Brazil itself is a monster country, being the fifth largest in the world. This makes South American states quite territorially, physically, and economically distinct from the slew of much smaller micro-states to their immediate north. Food for thought, and I'm not talking about salsa either.

MIDDLE AMERICA

The term **Middle America** is everything between the southern border of the United States and the South American continent. It's a fairly generic definition that usually only gets play in high school and college level textbooks anymore, but I suppose it doesn't hurt to know the reference when you see it pop up. For this magnificent treatise of learning, I will further break this area down to more manageable and meaningful subregions, described below. Study the Middle America map equation above to be cartographically clever.

MEXICO

The first subdivision is Mexico itself, solo. Mexico is a country, a state in its own right, but it's also radically different from all the states around it, including the US, and it's also quite different from the Central American and Caribbean states. Mexico is its own subregion that we're going to define in the next chapter. In respect to size, resources, economy, population and level of development, Mexico stands apart from all its neighbors in Middle America: it is a giant of population, economy, and resources when compared to the rest of the Middle America 'hood. And its shared border with the USA make it quite distinct from all the rest of the Latin American states as well . . . as does it's NAFTA membership. Combine that huge economic interaction with the US with the huge cultural interaction with the US, and you have a sub-region that really straddles two different worlds.

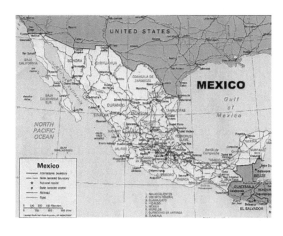

THE CARIBBEAN

The Caribbean is a group of island states comprised of the Greater Antilles and the Lesser Antilles. The Caribbean, when we think about it, calls to mind a distinct culture in terms of . . . everything. Cuisine, language, stuff they drink, how they party—it's all different. Caribbean means something to us, and it means something that's different than Mexico. We all understand that, right? We go to a Caribbean restaurant or a Mexican restaurant, but we never really see the two put together. (Note to self: that might be a good idea. Instead of Tex-Mex cuisine, how about Car-Mex cuisine—or is that a lip balm?) The island nations in the Caribbean Sea, south of the United States, are a distinct subdivision.

CENTRAL AMERICA

The last sub-region of Middle America is **Central America**. This is the one that causes the most confusion for younger students of the world. It is everything between Mexico and South America, noninclusive. Mexico and South America are the bookends. There are seven distinct countries in what is widely accepted as Central America: Belize, Guatemala, El Salvador, Honduras, Nicaragua, Costa Rica, and Panama. You may hear reference to Central American civil wars or Central American gangs; this group of countries is the origin for that descriptor. Yeah, good times.

We've got a bunch of Americas here, and Central America is quite distinct in that it has little in common with Mexico, the Caribbean, or even South America. It has a lot going on, particularly in terms of the violence that we will talk about in much more detail when we get to the Central America chapter. It's a distinct place; a bunch of small states that bridge both North and South America, but they are all quite unique, even from each other, in all many aspects.

South America plus the three regions of Middle America—Mexico, Central America, and the Caribbean—that's Latin America all the way around. Now that we know where it is and where the definition comes from, let's talk a little bit more about what Latin America is. What homogeneous traits can we use to define this vast region?

WHAT IS LATIN ABOUT LATIN AMERICA?

COMMON CULTURE

First, we can point out some distinct elements so we don't have to keep repeating ourselves in subsequent chapters. Number one is culture. There is kind of a common culture south of the US/Mexican border, be it in the Caribbean, Central America, Chile, Argentina, or Brazil; there are some things that do remain the same. One of the kind-of-the-same things that we already pointed out is language. While Latin America is a bizarre term in and of itself, Latin Americans do primarily speak Spanish, and the only other really big language is Portuguese.

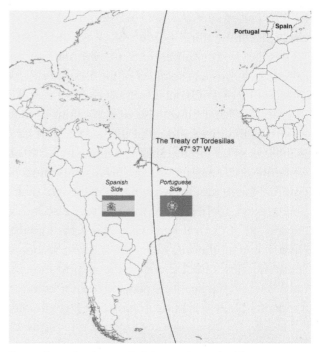

How did this come about? Easy enough. It came about because of colonial endeavors in this region that date as far back as 1492 when Columbus sailed the ocean blue, came over here and bumped into the Caribbean . . . and he was sailing under the Spanish flag. After his "great discovery," the big naval powers at the time, Spain and Portugal, started floating over here in fairly short order as well. (Great Britain didn't come along until later. It wasn't as big of a naval power as the Spanish at the time, so it was a latecomer. It therefore had to head north for colonial expansion which is why it gets the leftovers—North America.)

Portugal got the short end of the continental stick . . .

Let's get back to our story here. Why do we have these two main languages, Spanish and Portuguese? Columbus came over in 1492. Two years later, a fairly important event occurred which I want you to understand and know about: The Treaty of Tordesillas. The **Treaty of Tordesillas** occurred in 1494 when these European countries were bumping into

Pope responsible for Tordesillas . . . and Catholicism in Latin America.

the New World, colonizing, taking over, and staking claim to pieces of it. Of course, there was friction between the countries for hundreds of years at home anyway; now they were just taking the fight abroad. Much of the friction was between Portugal and Spain, two of the main colonizing powers in this part of the world. Who was going to own what? Are we going to fight over it? The scramble for stuff was fast turning into a full-scale fracas. Both these countries, being predominantly Catholic—which is another common cultural tie we'll see more about in just a second—would listen to the papa, the **Pope**, the main man in the Vatican. And therefore to alleviate conflict, the Pope said, "Hey guys, come on now, we're all civilized colonial imperial masters here. Let's all get on the same page, get

around the same table here, let's work together! We don't need to fight! We're going to settle this fair and square. Papa is going to just draw a line on the map and everything on one side we'll give to the Spanish and everything on the other side we'll give to the Portuguese. Now you guys be good!" The line happened to be 45 degrees western longitude. Check out the map above. And the caption.

As you can see from the map, the Portuguese look like they got shafted on this deal. They only got the tip of what is now known as Brazil. Why is this? Did the Pope just not like the Portuguese? Well, maybe a little. But you also have to keep in mind, this was during early exploration in 1494, only a few years after Columbus got there. They didn't even know what was there yet. They were just bumping around the coastline; with no comprehension about the actual size of the continents . . . neither North nor South America, quite frankly. What they thought was a good deal was based on known circumstances at the time, which really wasn't much. As we all now know, this is a pretty big place. South America is the fourth largest continent on the planet. As exploration continued and the true scope of the land's magnitude unfolded, the Portuguese did pick up what is now modern-day Brazil. Its boundaries naturally fell back to the Andes—a nice, easily defined natural border. Anyway, the Treaty of Tordesillas is the cornerstone of why the Spaniards ended up controlling so much of the New World—basically, all the rest of it outside Brazil.

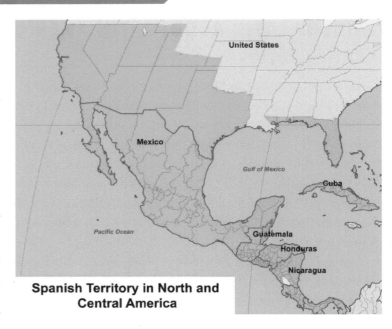

Spanish Territory in North and Central America

And it all started with Mexico. We think of today's Mexico as that place "south of the border," but Mexico in its days as part of the Spanish Empire included all of western North America, as well. It was everything that's now California up to Washington State, to Utah, to New Mexico, to Texas. Spanish territory also included Mexico, all of Central America, most of the Caribbean, Florida, and all the way down the western seaboard of South America. Check it out on the map to the left. That is why today these territories are Spanish-speaking. I guess in today's world the Spanish-speaking territories might even include California, Florida, New Mexico, and Arizona, but that's a separate issue in immigration that we'll get to later.

Latins love the Savior.

This all means that those old colonial ties formed this common culture that's still in place today. It's a common culture primarily based on language, but also on one other thing I mentioned earlier: Catholicism. Two primarily Catholic countries colonized virtually all of Latin America. You can still see the deep, deep-seated Catholicism throughout Latin America today. There are no countries in Latin America that are *not* Catholic. When the Plaid Avenger travels down there, he can see some freaky-freaky stuff that doesn't look like Catholicism. You can see some rites and rituals down there that could make the Pope soil his robes, no doubt. You'd say, "That's not Catholicism! What is this voodoo hoo-ha? What are people doing with voodoo dolls and holding up bloody chickens saying Hail Marys?" Then there's this crazy stuff that's going on down in Brazil; they have these big parties that don't look like traditional Catholicism: the "Carnival." But what you have is indeed the basis of Catholicism mixed in within indigenous culture, local culture, and imported African culture.

So a couple of very strong cultural characteristics, language and Catholicism, do serve as a starting place for some homogeneity of "Latin American-ness." But there are others . . .

URBANIZATION

Urbanization is a primary feature that is consistent across *every place* in Latin America, and I do mean every place. I've never even thought about this until recently, but this region is one of the most urbanized on the planet. Typically, when we think of urbanization, we think it means that most people don't live in the countryside; most people live in cities . . . and we typically associate this with fully developed countries like the US, Canada, Japan, and Western Europe. That's where you *typically* have people who are all jammed in the cities, and hardly anyone out in the country. For some reasons that we will get into in a minute, Latin America is more urbanized than many fully developed places. This means almost everyone in these countries is jammed into something called an urban area, typically a very big city.

On average, upwards of 80–85 percent, sometimes upwards of 90 percent, of the population of a Latin American country is located in just a few major cities. This is highlighted best by Mexico City, which will probably soon be the world's most populated city with over 25 million inhabitants—a full one-fourth of Mexico's total population.

Why do people go to the city? It is suggested that in the developed world people go to the city for all the reasons we already talked about back in chapter 2. That's where jobs are. Jobs are concentrated in urban areas. People want to go to urban areas because there's health care, doctors, clean water, sewer systems, electricity, movie theaters, good restaurants. The reason everybody wants to move to a big city or urban area is because good stuff is there. The standard of living is higher and that's where your jobs are.

Is that true of Latin America? Not all the time. It is a kind of conundrum in Latin America. In most urbanized places in the Latin world, all of the pull factors that we just mentioned—the good things that pull people to the city—are simply not available for the masses that come. There is the perception that all these great things are in the city and the perception that all these jobs are available, but it is simply not true for a lot of folks who end up going to the city.

Latins packed into the cities.

Now those are the pull factors, but there are also some push factors, or the not-so-good things that push people from rural areas to the city. One of the main push factors, a major theme across Latin America that we'll discuss at length, is landlessness. Because most people do not have access to land, or resources on the land, they are pushed away from it. Owning no land in title or having no serious claim to the land is just another reason to go to the city. So there are a lot of pull and push factors that make Latin America one of the most urbanized places on the planet . . . a trend that is growing.

One last note: An interesting phenomenon in Latin American cities that most Americans don't get is that they are set up in a kind of "economically reversed" scenario from the US urban model. Say what? In America, what's found deep in the inner city? In fact, what is the connotation of the term "inner city"? I'll tell you what it connotes: slums, projects, poverty, ghettos. That's not where people want to be. As a result, most people who have money don't live in the inner city in America. They've got money, so what do they do? They move away from the city and into the "burbs," which we talked about back in the North America chapter.

In Latin America, the situation is just the reverse. The city center is still the prime real estate, so the rich people concentrate themselves and their businesses in the true city center. Where do all these impoverished masses go that I

Vertical growth of slums or favelas around Rio.

have been suggesting flood into the cities? They make a ring around it. They get as close to the city as they can, usually in ramshackle, shanty-like town dwellings they put together out of corrugated-cardboard and any other thing they can pull out of a landfill, and build them in undesirable, unused parts of the urban fringe like mountain sides. These shantytowns get very big and grow into almost permanent fixtures in rings around the cities. Whole shantytowns are a phenomenon that's very easy to spot in any major Latin American city. Just drive straight into or out of the city and notice how the poverty line fluctuates one way or the other. It's a distinct characteristic that has something to do with urbanization. It also has something to do with . . .

WEALTH DISPARITY/LANDLESSNESS

Perhaps there is no other region that we will talk about in this book or that you can go visit on the planet that has wealth disparity as extreme as Latin America. What is **wealth disparity**? Disparity is the difference between highest and the lowest, the difference between the greatest and the least. In terms of wealth, no place in the world is like Latin America in that so few people have so much of the wealth. I'm just going to guess-timate some numbers here. These are Figures that vary from country to country, but on average, the extreme amounts of wealth are held by the upper 5 percent of the population. The richest 5 percent typically own something like 80 percent of all the stuff. That's all

Times haven't changed much economically.

the land, all the factories, all the businesses, everything. Maybe you think that's not bad—perhaps that's the way it is in the US as well. Not really. Because the other part of the equation is that 95 percent of people have got to split up that other 20% of the stuff. Of course, there is always a significant, or at least partial, middle class; some people have got to own some stuff, but there's always a significant majority of people in these countries that own nothing. No title to their land, no title to their house, no other economic means except their labor to sell. That is a kind of common feature across Latin America from Mexico down to the tip of Tierra del Fuego, and is particularly nasty in Brazil and lots of other places where people kill each other over land.

This brings us to the issue of **landlessness**. Because of this lopsided scenario of so few people owning so much, including the land, the vast majority of people can't stake a claim to anything. In such societies where people don't have anything, their options are very limited. Mostly they go to cities, as already suggested. They can also try to work on someone else's land without the landowner's consent. This is a very unstable situation because the owner can show up at any time and kick them off. This is a particularly resonate issue in Brazil where landlessness has turned into an occupational hazard (see box on page 89).

The George Washington of America: Simón Bolívar.

This problem of wealth disparity and landlessness is a common theme across Latin America, and was a primary motivation of the independence movements across the region as well (these movements in Latin America began in the 1820s–30s). This dude named **Simón Bolívar** headed up many of these egalitarian independence movements, and is viewed historically as the George Washington of Latin America. Wealth disparity/economic equality was one of his central themes to rally folks to fight. But the issue wasn't resolved at state inception, and it has plagued Latin America ever since. Just one example for now: The Mexican Revolution in 1910 was fought over land: so many people got so frustrated about landlessness, that they had a revolution to remedy it. One of the core parts of their new constitution included equal rights for people and access to land. All politicians of any stripe have had to deal with this issue historically, and still do today.

Pancho Villa and his crew fought in the Mexican revolution for land reform. But even 100 years later, Mexican President Enrique Peña Nieto had to include land reform in his political platform. If you're going to run for office at all in Mexico, no matter what political party you are from, you have to address the land issue. The main political party in Mexico is called the PRI. It's a land reform party—go figure. It was founded to redistribute land and work out ways to give people access to land. We already referred to Dilma in Brazil (see previous page) dealing with land reform, and she has also had to deal with massive citizen protest in 2013–14 that are demanding better government services and opportunities for the masses. Then there are peeps like that late Hugo Chavez in Venezuela

What's the Deal with Landlessness in Brazil?

Unofficial stats in Brazil refer to 1.6 percent of the landowners control roughly half (46.8 percent) of the land on which crops could be grown. Just 3 percent of the population owns two-thirds of all arable lands. The Brazilian constitution requires that land serve a social function. [Article 5, Section XXIII.] As such, the constitution requires the Brazilian government "expropriate for the purpose of agrarian reform, rural property that is not performing its social function." [Article 184.]

This is a big deal for everybody. It concerns politicians, business people, and the landless poor. In Brazil, they have a law on the books that basically says, "If you can successfully squat on a piece of land, cultivate it, and make it produce something for one year, then you have a legal stake to it. You can take it! It's yours! Here's the deed in your name!" You may say, "Hey, that sounds like a pretty good policy," and maybe it is.

Dilma sez: "Land reform is necessary to build a country with justice, food security and peace in rural areas."

However, the scenario develops a lot of squatter settlements: large groups of people who are squatting on owned land that is typically owned by some rich businessperson or some rich urban dweller who is not out there in the countryside. If they find out that people are squatting on their land, they send in henchmen to go clear them out. When the landless poor fight back or try to continue their stay on the land to finish out the year so they can have legal claim to it, the situation can turn violent.

This situation has also led to the formation of the Landless Workers' Movement, or in Portuguese, Movimento dos Trabalhadores Rurais Sem Terra (MST). This is the largest social movement in Latin America with an estimated 1.5 million landless members organized in 23 out of 27 states. And they can get violent as well.

This is still an issue that is alive today in Brazil and all across Latin America where hired gunmen are going and literally cleaning out villages and killing everybody on site. Because it is private property, the owners, ranchers, and other types of folks can just say "This is legally mine and these people wouldn't leave, so I killed them." This is a very hot issue. In her first months in office President Dilma Rouseff pledged millions to resettle landless folks and encourage farm production in rural areas.

Pancho, Peña, Rouseff, Castro and Hugo: All leaders for land reform?

who was all about equal economic rights in a country with huge wealth disparity, and started his socialist "Bolivarian Revolution" program in order to level the playing field. And don't forget historic figures like Fidel Castro, who led the Cuban revolution to basically reclaim land and businesses from the rich and redistribute them to the poor in an old-school communist-style revolution. This issue of wealth disparity and landlessness is a prevalent theme whether you are in Brazil, Mexico, Cuba, Nicargua, or Chile.

IN-"DOCTRINE"-ATION: A HISTORY OF US INVOLVEMENT

Nobody hates any specific person in the United States, but taken as a whole, Latin America's historical relationship with the US in today's world is largely seen as negative. That may seem like a bold statement, and certainly there are those that still support US policies, but I can tell you who those people are. They are the rich people in Latin America. If life is good, you've got no reason to have qualms with the United States. Unfortunately, as we've already pointed out, that's not the majority by any stretch. Most people see the United States with a bit of an imperial taint, or in a bit of a **hegemonic** light. Certainly, given the US's involvement in virtually every country in Latin America, it's not hard to see why they have kind of a bad taste in their mouth when it comes to historical intervention by the US.

"Don't mess in our hemisphere!"

To what is the Plaid Avenger referring? You've got to know this, because it still applies in today's world. The number one thing you've got to remember from the US's history of influence is **The Monroe Doctrine**. It's this antiquated, 200-year-old statement that was made in 1823 by President James Monroe. It was a foreign policy statement that said in essence *Disclaimer: this is a Plaid Avenger Interpretation:* "If any European power messes in any place in this hemisphere, the US will consider it an act of aggression against the United States." In essence, if Spain were to go try to retake Mexico or Chile, if the Portuguese were to try to retake Brazil, if the British try to take back Jamaica—the US would consider it an attack on US soil. Heck, it doesn't even have to be a full-on takeover; any intervention at all would be considered an act of aggression. "If you mess with anyone down there, we will consider it as you screwing with us." It's extremely similar to NATO article 5 in saying, "If anybody here gets messed with, it's an attack on all of us."

Uncle Sam ready to whip some imperialistic Eurotrash back in the day.

Now why on earth would the US say that? It seems kind of silly. I mean, in 1823, what sort of position is James Monroe in? The US was a new country that had only been around about 30 or 40 years. It had only expanded *slightly* over the 13 original colonies, and they were certainly not a world power. They did a great job shooting the British from behind trees, but other than that, they weren't capable of fighting anybody. They could take care of their soil, but weren't up for fighting anybody else on foreign soil. This was largely a toothless threat. Maybe you are thinking, "Why is the Plaid Avenger telling us that this is important?" Here's why: This statement became a cornerstone that remains relevant in US/Latin American policy TO THIS DAY.

It didn't mean anything at the time when Monroe said it, but it has come to mean *everything*. Why did Monroe make it at that particular time in 1823? Mexico declared independence in 1820. The Central Americas seceded and most of the South American countries were declaring independence at or around this time. It was seen largely as a supportive gesture. The US was basically thinking, "Yes, our Latin brothers, kick all Spanish and other colonial powers out. We just did it in America so we will encourage everyone else to do it. We're the good guys and we're helping the other good guys, so it's all good!" Again, it was a show of support more than a credible threat to the Europeans: the US could not really take on the Spanish or the British in a foreign land war at that time. Forget about it. No contest.

The Monroe Doctrine led to a bunch of other things, such as the Roosevelt Corollary. The **Roosevelt Corollary** was issued in 1904, about a hundred years later, and at that point, the US was quite a bit more powerful than it was during Monroe's tenure. The US was also under a very powerful president at the time, President "Rough Rider" Teddy Roosevelt.

What was Teddy known for? What was one of his most popular sayings during that time? "Speak softly and carry a big stick." Indeed, that saying can be applied directly to what became known as the Roosevelt Corollary, which was the foreign policy towards Latin America at the time. Teddy said, "I like the Monroe Doctrine's policy that if anyone messes around in our backyard we'll consider it an act of aggression against us. That's good, but let's take it a step further. If there is any flagrant wrongdoing by a Latin American state *ITSELF*, then the US has the right to intervene."

"Speak softly, and carry a big stick. Just in case a piñata party breaks out."

In other words, if any Latin American countries south of the US border attack each other, then the US gave itself the right to intervene. More than that, if they just screw up internally, the US was giving itself the right to intervene as well. This had serious repercussions for what sovereignty meant at the time. Of course, I can't go back in time to hear their exact thoughts, but there is no doubt that it was not held in high esteem by Latin American states that were considered sovereign then.

In other words, you had the United States saying, "Sure, you guys are sovereign, as long as we agree with your sovereignty. Otherwise, we give ourselves the right to intervene." This became kind of a big deal because Teddy was carrying that big stick, and he was not afraid to smack people, or entire Latin American countries, down with it. Under Roosevelt's corollary, relations deteriorated slightly between these regions. However, a bright spot in US/Latin American relations under the Good Neighbor Policy was just around the corner . . . maybe.

FDR sez: "Sup. We gonna chill over here. If y'all need us, holla."

The **Good Neighbor Policy** was a popular name for foreign policy at the time of the next President Roosevelt—Franklin D., that is—in the 1930s. In a marked departure from the heavy-handed foreign policies up to this point, FDR said, "You know, we're good guys, we're your buddies. We don't need to come down there and beat you with a big stick. My fifth cousin Teddy was a funny guy, but we don't really need to be that heavy-handed. We'll throw out that Roosevelt Corollary and we'll just be here to help if any leader needs us."

That sounded pretty good, and it was certainly an improvement over the Roosevelt Corollary. But under the Good Neighbor Policy, there were multiple scenarios where US troops were sent down at the request of "leaders" that sometimes could also be referred to as, oh, I don't know, let's call them *military dictators*, who just happened to be supporting US foreign and economic policies at the time. Even though it sounded better, there were still slight implications that perhaps things were not completely on the up and up. That brings us to the last part of the US's history of involvement.

COLD WAR EFFECTS

After WWII, there came the War of Coldness. The Cold War has already been referred to several times in this book; you may be thinking, "Ah I'm tired of freakin' history; I don't need to know any of this stuff," but you can't understand the world unless you understand the historical and political movements of at least the last hundred years. Nothing has affected the world more than the Cold War and its politics. Even Latin America was affected.

We might think of Latin America and say, "What? There was no hot war down there, much less a cold war! There are no Commies down there; there's no Cold War frictions in them parts!" Not so! Latin America was actually quite radically impacted by the Cold War because of US anti-communist policies that were applied across the planet. That's the reason the US got involved in Vietnam, Korea, and dang near everything else that was active at the time. The US even supported leaders with questionable character, just as long as they didn't associate with the Soviets. Supporting a brutal dictator who was suppressing his own peoples in Latin America? Sure! No problem, as long as he ain't a commie!

When we think of the Cold War and Latin America, the first thing that pops into the American mind is Cuba and Fidel: that flagrant flaming Commie that the US still hates to this day. There have been lots of repercussions between the US and Cuba (i.e. why Cuba's pretty impoverished today), but all of the other Latin American countries were impacted as well. Some of them had much more violence with a much greater death toll than Cuba ever did.

What I'm talking about is a renewed distrust of the US as a result of its Cold War activities. The American government was so rabidly anti-communist that any movement towards the political left by any Latin American country was viewed by the US as a hostile act. And so it became ingrained in US foreign policy for the last 50–60 years that it was absolutely intolerable for anybody to be left-leaning.

Sole Western Hemisphere Commie.

If you think back to our chapter on global politics and governments, not every single system on the left is Commie—but all forms of socialism were viewed as being a slippery slope that would eventually lead to the Soviets marching into Arizona. The US government believed that any form of socialism, however mild, would lead to communism or would lead to an opening for the USSR to make inroads. It really was battled at all cost. No cost was too high; no moral too low to violate in order to ensure that Latin America stayed firmly in the US's backyard of influence.

Let's get into some specifics. . . .

What did the US do in this all-out barrage to stop communism in Latin America? Be forewarned: This is going to hurt for the uninitiated. It's a little hard for many proud Americans to hear, but the US did some pretty nasty things, quite frankly. While they typically champion democracy on the planet, at the time of the Cold War, the USSR was seen as such a threat that the US said, "Well, we're all about democracy, and it would be great if we had democracies there, but we can't allow anybody to go near the left. So it would be better to support someone on the extreme right as opposed to anybody that might be even the slightest bit Commie."

What this equated to was US support for people who might be considered brutal dictators at worst, and elitist dudes of questionable ethics and character at best. During the Reagan years in the 1980s, the threat of Soviet infiltration by arms sales to places like Nicaragua was interpreted as an immediate hostile threat to the US. In response, support for dictators was sometimes pitched to the US Congress as basically, "We're about to get freakin' *invaded* by the Commies. They're going to get into Central America and they're going to sweep through Mexico and then they'll be knocking down Texas's door!"

In hindsight, this seems a bit preposterous. To be fair to the Reaganites and their ilk, we do have to consider that at the time, the Soviets were as aggressive as the US was, had as many nukes as the US did, and had previously tried to hide missiles in Cuba. So the commie-infiltration precedent was present. There was a very real fear of global domination by the Soviets.

You can listen to speeches by Henry Kissinger and the like that say, "You young people just don't understand that we had to do these horrible things because if we didn't, you'd all be wearing red right now and we would all be slaves of an oppressive giant Soviet Empire." What this equated to in the Reagan era was not only supporting extreme dictatorships, but also hatching plots to overthrow—and sometimes assassinate—democratically elected leaders. This also equated to supporting extremist rightist factions and rebel groups. Right wing death squads in Nicaragua comes to mind, also found in Guatemala and El Salvador too. The US supported *anybody* as long as they weren't left-leaning and weren't socialists and didn't support any of those other ideologies, especially communism.

A lot of these groups ended up slaughtering thousands, and tens of thousands, of their own people. Many just ended up as bands of guerillas running around the countryside causing mayhem to the elected governments. The end result of all these anti-communist policies in Latin America, particularly Central America, was civil war. These excessively destructive civil wars were supported in part by movements of US funds or arms, or funds for arms. One of the more famous ones is the **Iran/Contra scandal**, during the Reagan era, when guns were floated into Nicaragua in support of anti-leftist movements to overthrow the democratically elected leftist regime, which we'll discuss more in the Central America chapter.

All these things together, in terms of US intervention or involvement in Latin American affairs, bring up a term that's often used for Latin America: **the US's backyard**. You'll see this term used even in modern political science magazines and international news. It really summarizes the way the US has felt about Latin America, which is, "It's not really our house, but it's our backyard. We're not really cleaning it up or taking care of it unless someone starts coming around and messing around in it." The US doesn't want anybody messing around in its backyard; that's why it's been heavy-handed at times throughout history.

2014 Update: This era may be coming to a close, as the USA has recently renounced the Monroe Doctrine altogether for the first time ever, and is increasingly losing influence within the region as a whole as more and more states turn to other powers like China and India, or become big powers themselves, like Brazil and Mexico. More on this later.

DRUGS

Oh yeah, I almost forgot about drugs! That is the hot new thing with US intervention in Latin America. The Cold War is over. There is not really any Good Neighbor Policy or Roosevelt Corollary going on. No foreign power is invading—that we know of—and the US is not going to do anything about it if they invade each other. The current and active deal with US intervention in Latin America is all centered on **drugs**. What the Plaid Avenger already knows from personal travels is that Latin America produces the bulk of the world's cocaine . . . and the United States consumes the bulk of the world's cocaine. Talk about a most horrific symbiotic relationship.

This creates a situation where the US government, armed with an anti-drug policy named "The War on Drugs," facilitates or makes it an imperative for us to intervene in other countries to stop drug production all over the world. The Plaid Avenger

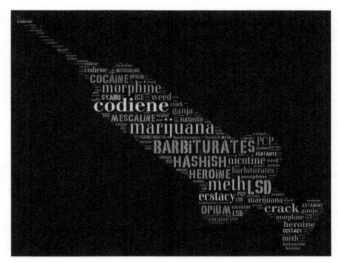

Drugs: the current "injection" of US influence into the region.

won't get into a big debate about the pure insanity of such an endeavor. I'll leave it to you students of the world to figure out if it's a good method to stop drug use or not, but certainly it is the US policy: "We don't really care so much that people are hooked on drugs up here; we just want to make sure they don't produce them down there." That will somehow solve the problem? Yeah, good luck with that one!.

This brings up one particular aspect of US foreign policy today called **Plan Colombia** (read box on page 94).

US direct intervention right now is mostly focused on Colombia, Ecuador, Peru, and Bolivia: the big drug-producing and exporting countries. Plan Colombia, in particular, has equated to around 6 billion US tax dollars in an effort to stop Colombia from producing drugs. Money well spent! (Did you read the inset box yet?) Mexico is the current hot spot for huge drug related violence, and an all-out drug war between the Mexican government and the powerful drug cartelts . . . but thus far, the US has not gotten involved too much, and for the first time ever, maybe they really should! This Mexican drug cartel war is radically affecting the US, and spilling over into the states in very real, and very dangerous, ways. Mexico is seeking active US involvement in their drug war; Colombia has worked with the US for years with drug wars and drug policy . . . but other countries like Bolivia, Venezuela, and Ecuador are not so keen on US policy. However some others, like Bolivia, are increasingly not. That brings us to a related topic . . .

What is the Deal with Plan Colombia?

Plan Colombia is a program supported by the United States to eradicate coca production in Colombia. It may sound good, but Plan Colombia has become extremely controversial for several reasons. For one, although the coca plant is used to manufacture cocaine, it has also been used by indigenous peoples in the area for thousands of years for health reasons. Many of these people depend on coca to make a living. Furthermore, some of the methods that the United States uses to eradicate coca, such as aerial fumigation and the application of deadly fungi, pose severe health problems for people exposed to it.

Plan Colombia is also controversial because Colombia was undergoing a civil war at the time. Many people claim that the goal of Plan Colombia wasn't really to stop drugs, but to help the Colombian government fight the Marxist rebel group FARC, which gains much of its funding through the drug trade. When Plan Colombia was first introduced by the president of Colombia in 1998, its main focus was to make peace with the rebels and revitalize the economy of Colombia. However, policymakers from the United States revised it, and the focus became more about military aid to fight the rebels and the elimination of drug trafficking. Human rights organizations are indignant at Plan Colombia because they see it as a way of strengthening right-wing paramilitary groups in Colombia that are committing atrocities against peasants who are speaking out for equal rights and economic reform.

It's your tax dollars, so you should know: the US government has spent close to six billion dollars on Plan Colombia since the year 2000. In 2006, it was reported that coca production actually increased in the last three years. Hmmmmm . . . I'm no mathematician, but something don't add up here. . . .

Fun Plaid Fact: After Plan Colombia was revised by the Americans, the first formal draft was written in English, and a version in Spanish wasn't created until months after the English copy was available. In related news, many Colombians have accused Colombian pop star Shakira of being a language sellout for releasing an album in English.

LEFTWARD LEANING

This is the most fun, exciting, new, and current part of "What's Latin about Latin America." Through a combination of virtually all the above reasons that Latin America is Latin, this notion is the one that pulls it all together. As we've already cited, particularly during the Cold War era, the US was very troubled by and directly intervened in leftist or left-leaning countries in Latin America. God forbid they embrace some sort of socialism, and certainly not communism! Nonetheless, some people went left anyway. Castro springs to mind. Cuba has been communist and certainly in the "lefty" column for this entire time.

By and large, he has been alone in the hemisphere until very recently. What we've seen in the last decade—perhaps because of the wealth disparity or landlessness or poverty in Latin America—is a general bad taste in Latino mouths about the US's history of intervention. This has really culminated in a lot of countries heading back towards the left column. One that springs to mind is Venezuela in South America. A rabidly left country (some might even say it's already communist), Venezuela boasts a new brand of socialism that strives for social equality and redistribution of wealth to a limited degree through social programs, formerly under the tutelage of their self-avowed socialist President Hugo Chavez. Although Hugo is now hu-gone (he died in 2013), the socialist movement he spawned is very much alive. President Evo Morales of Bolivia has recently joined the ranks of hard-core leftist, and has been busy nationalizing industries and moving his country in a fully socialist direction. Ecuadorian President Rafeal Correa is pretty far on the left side of the tracks now too.

Current US President Barack Obama has been an overwhelmingly popular figure across all of Latin America, so the standard anti-US sniping from Castro and Chavez is now not in vogue. Obama has been actively trying to reach out to restore diplomatic street cred in Latin America, and has spoken openly about thawing relations with Cuba . . . and even shook hands with Chavez at an OAS meeting back in April 2009! Change may be coming, and these historical animosities may be softening!

It's not just the bold and brash loud-mouths in Latin America who are embracing the left. Brazil's Dilma Rousseff is a left-leaning president as well. Argentina has been led by a left-leaning government for years. Ecuador, Nicaragua, Haiti, and Peru have headed that way as well. As you can see from the map on the following page, the future of Latin America does seem to be in the left-leaning categories. Let's explore why that is.

Why is the left progressing and gaining popularity? Many people in the world are starting to look at Latin America as a singular entity—one of the reasons I decided to do this chapter—that perhaps may become a new axis of power on the planet. What am I talking about? Well, as a group of disparate countries that didn't have a lot of common economic or political goals, now they do. As they have this leftward move, we can look at this entire region as representing a more common, singular ideology. There is no other region like that on the planet at this second. We can look at most of the planet which has progressively over the last 50–100 years been going toward something that's more on the right, more strictly capitalist, democratic systems. While certainly these are all democracies down there, they are going more left in terms of social and cultural issues, becoming much more openly liberal. In other words, the overwhelming focus in Latin America is to remedy the very wealth disparity that we talked about earlier.

Why is this happening here? Doesn't everyone want social justice and equality for people across the planet? Well yeah, lots of people do. But as I already suggested, this is the place on the planet that has the greatest wealth disparity. (The Middle East is a close second, and is partly the reason they are undergoing the revolutionary activity of the Arab Spring.) The landless, impoverished masses make this a perfect lab setting for this kind of experiment to evolve. Why is it here? Why is it right now? You have to understand that when you have any state, country, or place where most people are incredibly poor, you are asking for a revolution. When most people have no stake in the land, no claims—it's all fine as long as the minority, who has the power and wealth, can keep them down. When it becomes too lopsided, it becomes tougher and tougher for the elite to keep a lid on it, and things will eventually boil over.

Societies like this are always on the brink of revolution (see Mexican Revolution, Communist Revolution, Bolivarian Revolution, and the very much current 2011 Arab Revolutions!), and Latin America is leaning back to that point today. It is because of this inequality that people are voting for the left, voting for parties whose primary goal is to alleviate wealth disparity. They want to make things more equal. They are striving to improve infrastructure like roads and schools as part of their primary goal. I'm not saying that this isn't a goal of the other parties, but this is the primary goal of leftist parties.

The primary leftist agenda involves things like human rights, investment in education and healthcare, and equal access to land for the impoverished masses. That is the main unspoken priority, and that is why the leftist agenda is so popular. In democratic countries, where nobody has jack to their name, the leftist candidates are ex-tremely appealing because they are telling the people, "Hey, we are trying to make this better for you or more equal for you." Thus, it should not be a radical surprise that there is a big movement towards the left across the region that has the greatest wealth disparity on the planet. That's why people are voting for the leftist candidates.

Why are they not voting anymore for the rightist candidates? It's got a lot to do with the

2014 Political Positions

Far left
Left leaning
Center right

Just look at all that leaning!

US's historic involvement and where we are in today's world. These are all established democracies, and in the 21st century, it is increasingly hard for a military dictatorship or for a government supported by an foreign entity like the US, to hold power because issues are getting clearer and fewer people are being influenced to vote against their interests. The blatant corruption is getting easier to identify, so it's very difficult for extreme right-wingers to hold power anymore.

2014 Leftist Latin Update: the trend described above still holds true generally speaking, but with a subtle shift: the center-right and right-wing political powers of Latin America continue to lose power, mostly because the left-leaning governments/parties are becoming much more moderate, moving to be just center-left. Anti-capitalist, anti-US extremism still exists in places like Venezuela, Argentina, Ecuador, and Bolivia, but is fast falling out of fashion for the region as a whole. The most successful countries currently are the ones on the moderate center-left path. Despite this change, and the fact that the majority of Latin American governments are much more moderate leftist without an extreme agenda, the US still really despises the far-left folks like Castro of Cuba and Morales of Bolivia, and they really, really hated Chavez of Venezuela. . . .

What's the Deal with the US Hatred of Venezuela?

Why did the US despise Hugo Chavez so much, and still hate the Venezuelan government?

Venezuela has the means to produce over 3 billion barrels of oil per day, 60 percent of which is bought by the United States. Using this revenue, Venezuela has started to pick up some serious military hardware in recent years. Former President Hugo Chavez worked hard to secure arms deals with countries such as Brazil, China, Russia, and Spain. Take these newly acquired MiG jets, attack helicopters, AK-47s and Scud missiles, put them into the hands of two million well-trained Venezuelans and you've got yourself a bit of a hemispheric headache. While Venezuela may not have this strength right now, these are the plans set in motion by Chavez, convinced of a coming invasion by the United States.

Why in the world did Chavez believe that the big bad US is going to march into little Venezuela? Maybe it was the name-calling by former Secretary of State Condoleezza Rice when she accused Venezuela of being a "sidekick" to Iran. Perhaps Chavez was offended when former Secretary of Defense Donald Rumsfeld compared him to Hitler. Or maybe Chavez took it personally when evangelist Pat Robertson called for his assassination on national television. In any case, Chavez certainly became suspicious of his neighbors to the north after the US refused to denounce an unconstitutional, and extremely undemocratic, 2002 coup which would overthrow his democratically elected self, even though it only lasted a few days. Virtually all countries denounced the coup within hours, but the US stayed mysteriously quiet. Oops. Probably a bad call on the US's part.

Hugo on the US Hate List.

The US continues to criticize the policies of democratically elected Chavez as being "undemocratic," and in heated response he at times has vowed to cut every drop of oil exports and to mount an all-out guerilla war should the US step foot in Venezuela. And, given the former Bush's administration's doctrine of preemptive war, a simple cry of "terrorist!" could drive hundreds of thousands of US troops to South America. Sound crazy? Yeah, I guess. But that's probably what people in Afghanistan and Iraq thought as well.

Hang on! Don't be misled; I'm not promoting the pronouncements of this proud peacock. He was certainly leftist, certainly socialist, and certainly legally democratically elected, but that doesn't mean he did a particularly good job. While billions were made from oil revenues in his tenure as president, and perhaps millions were pulled out of poverty, Venezuela is far from a socialist utopia . . . or even a functioning society! Crime is rampant, infrastructure is crumbling, and the economy is largely mismanaged at an epic scale. The country has been rocked in 2014 with mass protests, dozens killed, and an increasingly authoritarian government clamping down hard in order to hold the country together. Hugo himself died in 2013, and many of us are wondering if his socialist dream may be slowly fading away as well. . . .

LOSING THE LATINS: US UNDONE

Why the US is *really, really* worried all about this: The US is troubled by this leftist lean because it is largely seen across the planet like this: "The US has screwed up their foreign policy so bad that they have lost control and influence in their own backyard." This may be an extreme statement, but it is a fair statement, nonetheless. Part of the reason they have done that, and part of the reason rightist regimes have lost power, is that they have been perhaps a little too heavy-handed over time. Most of the US involvement and incursions into the region were for reasons previously mentioned, such as anticommunist intervention, but even through the Cold War and to the present day, another reason for US involvement is to protect US economic interests. This is seen as extremely problematic for locals, who are usually on the losing side of what benefits the US economically.

In situations when the US has invaded places like Nicaragua or Belize, or has helped assassinate a democratically elected leader in Chile, it was because of this fear that in lefty/commie countries, a redistribution of land and resources was going to occur—you remember: **nationalization**. That was unacceptable to the US, largely because there were US corporations down there yelling, "Hey, that land/resources you poor people are taking is US property!" This was argued during the Cold War era, about Chile, in fact. A company said, "Hey! US government! You can't let commies take over Chile. They will nationalize our company (which means the commies will take it over and make the profit), and this is US property." A lot of US involvement has been due to protecting these US corporate interests. We'll examine this more in future chapters. Suffice it to say for now, the US's pro-capitalism, pro-free markets, pro-US corporations attitude is hugely distrusted by many Latin Americans . . . and as the region becomes more independent and wealthy, they are pivoting away from US policies and US leadership because there are . . .

NEW KIDS IN THE BLOCK . . .

The perception of US involvement as self-serving, coupled with massive wealth disparity, in part explains why much of Latin America has gone to the left. A lot of leftist candidates, particularly Chavez, Castro, and Evo Morales in Bolivia are saying, "We're not even pretending to do the neo-liberal capitalist policies. To be a free market with free and open trade works for the United States, and it looks good on paper, but it isn't working for us."

A lot of these leaders are saying, "We're not anti-US, we're just anti-free trade." The current president of Bolivia says, "We've done it. We've tried to do free trade, and we're still poor as squat! We won't do it anymore! We're not going to give priority to American corporations, and we're not going to give tax breaks to American anything!" Why is that? Because there's some new kids in town that many countries may give incentives to: China and India and Japan just to name a few. These countries not only are grabbing up tons of natural resources, but are expanding their exports into Latin America as well. Lots of investment flowing in to start new businesses and partnerships with the Latins from all points abroad has been the theme. That's another reason America fears a loss of influence in Latin America.

To be honest, you can't really blame the Latins for taking advantage of the international interest in their region. Specifically, the Chinese are courting countries around the world, making sweet trade deals with them, in order to feed the Chinese economy's ever-increasing hunger for natural resources. Chinese foreign trade/foreign aid deals are even sweeter because they come with no strings attached, unlike deals from the US. On top of that—and this is critical—in the last decade, the Chinese and Indian and even many African leaders have personally visited virtually every single Latin state, and multiple times, on multiple visits, in multiple years. Former US President Bush only managed to head south of the border twice in eight years. Obama has been twice, and one time his Secret Service agents got busted with Colombian hookers. Starting to get the picture here? It gets worse: Russia has shown a renewed interest in floating more military ships into the area and strengthening ties with old and new allies. Iran is even in this game, having opened dozens of cultural centers and working on economic pacts across the region. Meanwhile, the USA does not even have a diplomat in 10 Latin American countries due to confirmations being grid-locked in Congress. 'Nuf said.

And we must also point out that Latin America as a whole is on the upswing, meaning there is a lot more "local" demand for goods and services as the middle class grows, and many of those goods and services are being provided by local companies themselves. Add to the growing powerhouses of Brazil ad Mexico, which are both top 20 world economies and natural leaders within the region . . . and the picture of a much diminished US role starts to take shape.

Now I have spent a lot of time explaining the viewpoint of the Latin Lefties, but I don't want to exclude the ideas of the other side of the political/economic spectrum. About people on the political right: I don't want to suggest that they only want to make money and they don't give a damn about people; that's not the case. Even these people on the opposite side of the spectrum (real anti-socialists) would argue, "No, we want to make these countries richer, too. We want their citizens to have more stuff and not be poor, but we don't think social programs are the way to do it. We think the way to make people richer in Brazil and Colombia and Mexico is to have free trade." Free trade is the typical conservative approach to alleviating poverty. I don't want to deify anyone that is socialist because there are people on the other side who also want good things. They just don't think leftist methods will work: "You can't just give them a welfare check! That's not going to solve anything! Then nobody will be rich! What we need is free trade." Thus . . .

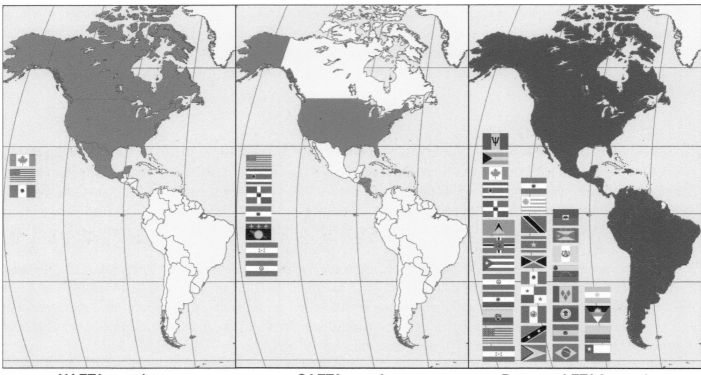

NAFTA members   **CAFTA members**   **Proposed FTAA members**

The **Free Trade Area of the Americas** (FTAA) already is and will be a hot issue for this hemisphere over the next decade. It's going to be a rallying point for anti-globalization as well as globalization powers. There are going to be big revolts over it; there are going to be heavy stones thrown over it. It's already a big, big deal that's just going to get bigger as the adversaries of capitalism (and/or the US) use it as fodder for their fire.

I bring this up because free-traders say, "Oh, we want Brazilians to be happy, but the way to accomplish that is a free trade union." **NAFTA** is a free-trade union between Canada, Mexico, and the United States. **CAFTA** is an ever-growing free trade area between the US and a handful of Caribbean and Central American countries with the inevitable goal of being the FTAA. The **FTAA** is a proposed free-trade area of every single country in the Americas. The United States, of course having an edge in all of this, is a big fan and proponent of the FTAA. It thinks the FTAA will make it richer, for one, but it also thinks it will help it to reestablish influence in the Latin American region as a leader of the FTAA. Anti-US, anti-globalization forces and leftist politicians in Latin America say, "No. We've been playing that game for a hundred years and we're still poor. We don't like it. We don't buy it."

We shall see how the battle for free trade pans out in the coming decades.

Chapter closer: These pervasive themes are not only historical, but they play into today's Latin America. Now let's take a look into some of Latin America's subregions to provide more specific details, so you can understand how each one works into today's and tomorrow's world.

DON'T GET LEFT OUT OF LATIN AMERICA!

HOW FAR LEFT ARE THESE LEADERS?

The Latin American leaders below are arranged in order of increasing dedication to full-on liberal socialist policy; the further to the left, the more fully the incorporation of socialist ideology into their political and economic policies. But can you name all these peeps and the countries they lead? Hint: US President Obama is thrown in as a marker of just barley left-of-center, and the dude to his right is actually center-right, conservative.

_____ _____ _____ _____ _____

_____ _____ _____ _____

Leaders: Rousseff, Obama, Castro, Correa, Santos, Chavez, Ortega, Kirchner, Morales
Countries: US, Bolivia, Brazil, Nicaragua, Colombia, Argentina, Cuba, Venezuela, Ecuador

LATIN AMERICA RUNDOWN & RESOURCES

 **BIG PLUSES**

- → Sizable chunk of real estate on planet earth with a boat load of resources
- → Has never invaded or infuriated any other country or region
- → Not a target of international terrorism at all. Who else can brag about that?
- → Becoming a serious place of interest and investment for China and other rising powers

 BIG PROBLEMS

- → Biggest wealth disparity on the planet
- → Political instability chronic in some areas, possible just about anywhere; Venezuela is in trouble right now
- → History of outside political and economic domination has left many residents with an inferiority complex
- → Heavily dependent on exports of natural resources and basic manufactures
- → Environmental degradation becoming rampant in exchange for economic growth

DEAD DUDES OF NOTE:

Simón Bolívar: The '"George Washington of South America": hero, visionary, revolutionary and liberator . . . he led Bolivia, Colombia, Ecuador, Panama, Peru and Venezuela to independence and instilled democracy as the foundations of all Latin American ideology.

James Monroe: 5th President of the US and important for this chapter for his **Monroe Doctrine** which became the cornerstone for US foreign policy in the entire hemisphere, right on up to the present. The Doctrine pretty much sez: any foreign power which messes with Latin America, will also be messing with the US.

Theodore Roosevelt: 26th President of US, and "Rough Rider" that helped invade Cuba in his spare time prior to becoming a politician. His **Roosevelt Corollary** was an amendment to the Monroe Doctrine which asserted the right of the US to even intervene in the internal affairs of Latin American states to "stabilize" them if necessary.

Franklin Delano Roosevelt: 32nd President of the US and creator of **Good Neighbor Policy** which sought to soften the apparent US hegemony over Latin America by renouncing the US right to intervene unilaterally.

KEY TERMS TO KNOW

Attenuated

Chain Migration

Chiapas

Counter-movement

Creole

Crude Birth Rate

Crude Death Rate

Crude Population Density

Demographic Transition Model

Doubling Rate

Ejidos

Encomienda System

Erosion

Esther Boserup

Hacienda

Hierarchical Diffusion

Human-environmental

Land Form

Leaching

Leeward

Maquiladoras

Middle Passage

Mulatto

Neo-Malthusians

Offshore Banking

Physiologic Population Density

Plantation

Plazas

Population Pyramid

Pulled

Pushed

Rate of Natural Increase

Thomas Malthus

Tierra Caliente

Tierra Fría

Tierra Templada

Total Fertility Rate

Vexillology

Viceroy System

Windward

Zero Population Growth

FURTHER READING

Stephen Castles and Mark J. Miller, *The Age of Migration: International Population Movements in the Modern World*, 4th ed. (New York: Guilford Press, 2009).

Paul R. Ehrlich and Anne H. Ehrlich, *The Population Explosion* (New York: Touchstone, 1991).

W. T. S. Gould, *Population and Development* (New York: Routledge, 2009).

Michael R. Haines and Richard H. Steckel, eds., A *Population History of North America* (Cambridge: Cambridge University Press, 2000).

Richard Jackson and Neil Howe, *The Graying of the Great Powers: Demography and Geopolitics in the 21st Century* (Washington: Center for Strategic and International Studies, 2008).

Massimo Livi-Bacci, *A Concise History of World Population*, 4th ed. (New York: Wiley-Blackwell, 2006).

T. R. Malthus, *An Essay on the Principle of Population* (Dover: Dover Publications, 2007).

Laurie Mazur, ed., *A Pivotal Movement: Population, Justice, and the Environmental Challenge* (Washington: Island Press, 2009).

K. Bruce Newbold, *Six Billion Plus: World Population in the Twenty-First Century*, 2nd ed. (New York: Rowman & Littlefield, 2006).

Fred Pearce, *The Coming Population Crash: And Our Planet's Surprising Future* (London: Beacon Press, 2010).

WEB SITES

U.S. Census Bureau (www.census.gov).

The World Bank (www.worldbank.org).

Population Reference Bureau (www.prb.org).

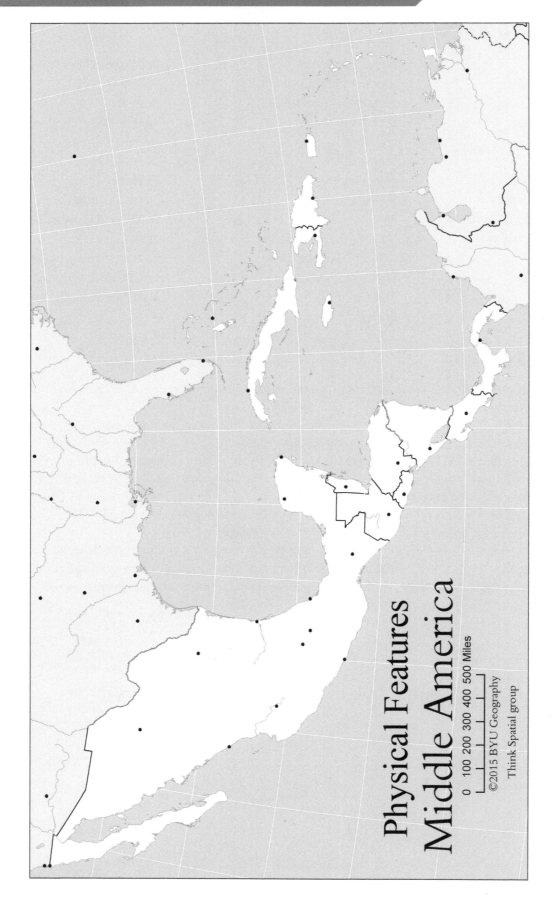

Physical Features
Middle America

0 100 200 300 400 500 Miles

©2015 BYU Geography

Think Spatial group

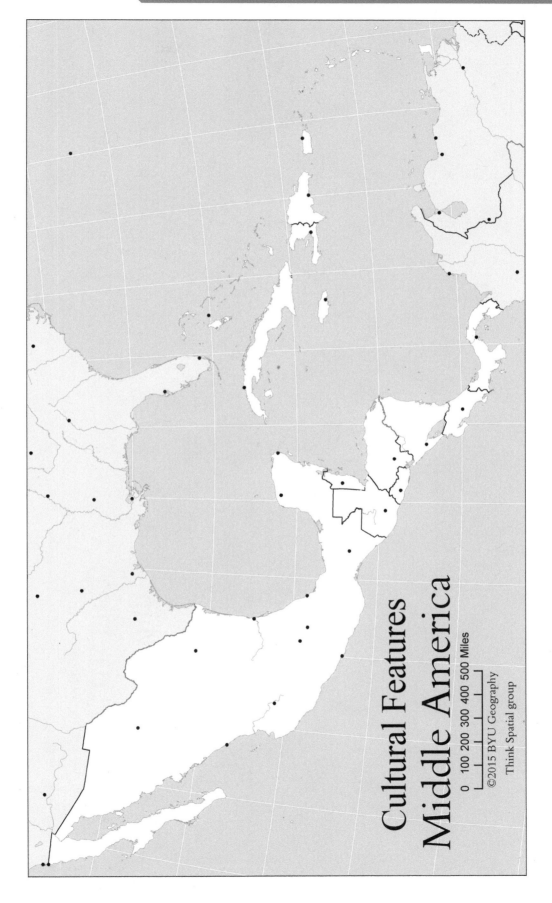

Cultural Features
Middle America

0 100 200 300 400 500 Miles

©2015 BYU Geography

Think Spatial group

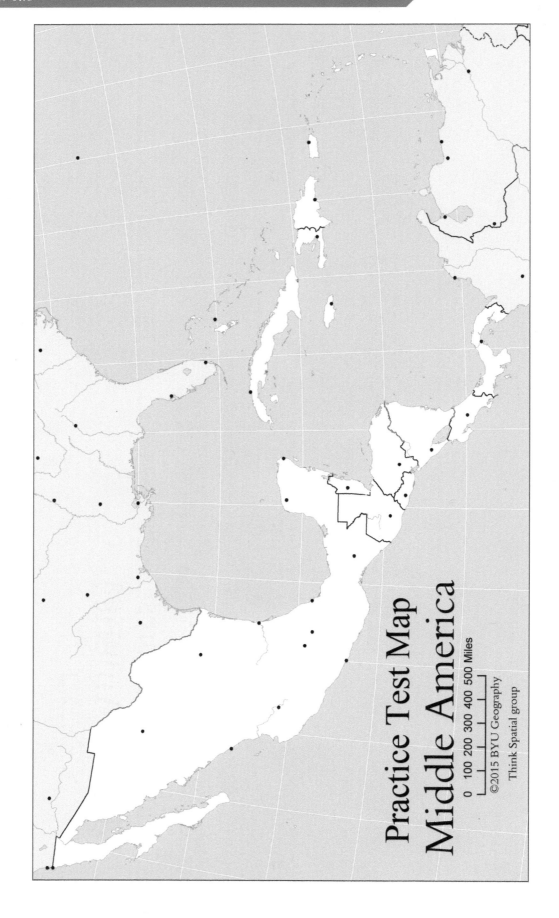

Practice Test Map
Middle America

0 100 200 300 400 500 Miles

©2015 BYU Geography

Think Spatial group

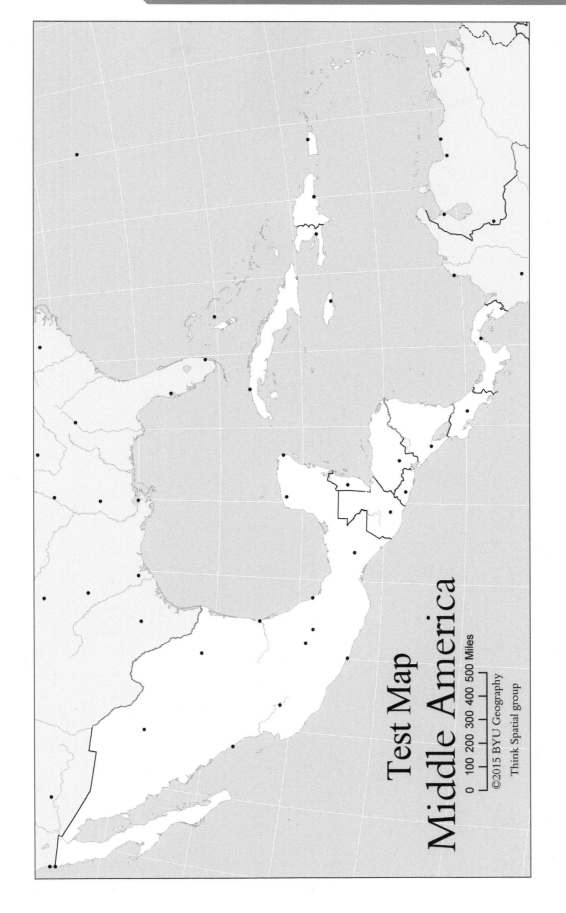

Test Map
Middle America

0 100 200 300 400 500 Miles

©2015 BYU Geography

Think Spatial group

South America and Urban Growth

4

Praise the Lord, all nations! Extol him, all peoples! For great is his steadfast love toward us, and the faithfulness of the Lord endures forever. Praise the Lord! (Psalm 117:1–2)

Now the Lord is the Spirit, and where the Spirit of the Lord is, there is freedom. And we all, with unveiled face, beholding the glory of the Lord, are being transformed into the same image from one degree of glory to another. For this comes from the Lord who is the Spirit. (2 Corinthians 3:17–18)

CHAPTER OUTLINE:

"NO MATTER what I do, I just can't seem to get an A in that class" are words frequently uttered by frustrated students. In fact, sometimes it seems we live on a treadmill, and if we aren't careful we can spend a lifetime staying where we are and never "getting ahead." The scriptures remind us of the hardships of life and their temporal nature. The continent of South America has been described as a challenging one by most geographers, and sometimes it appears the situation there is almost hopeless. Yet, hope never entirely leaves us if we simply give thanks for what we have. Closer in many ways to North America than Europe, South America is an entire continent full of unlimited opportunities for you to find your purpose!

The shape of South America resembles an ice cream cone with the majority of the continent in the tropics and a tapering cone transecting higher latitudes as it approaches the extremes of its southern limits. South America is roughly three-fourths the size of North America making it a large continent with extreme diversity. From the hot humid rainforests

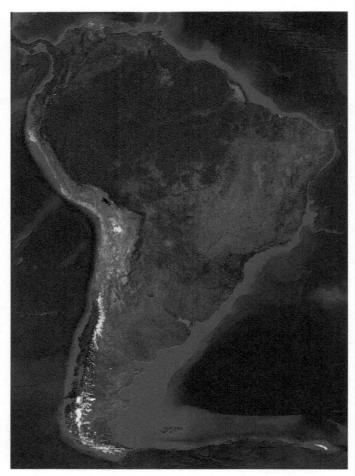

Adapted from *Blue Marble: Next Generation* image by Reto Stockli, NASA Earth
Observatory (NASA Goddard Space Flight Center)

Image © Frontpage, 2013. Used under license from Shutterstock, Inc.

of the interior to the cold dry windswept plains of Patagonia, South America offers an amazing array of animal and plant species—each reflecting a different biome and climate. To understand the climates one must understand the wind currents.

The currents of South America move in directions counterintuitive for the northern hemisphere. Since Earth's southern hemisphere is mostly water, it is quite easy to overlook the impact of the differences in direction of the ocean and atmospheric flows on continental land masses due to the Coriolis effect. These currents are important to understanding the climates of South America. The west-wind drift is responsible for the cold-water current commonly known as the Humboldt Current that proceeds up the coast of South America from the South. This cold water creates turbulence as it meets warmer water and rises up from the ocean floor. This phenomenon is key to understanding the excellent fishing resources and the subsequent effects on culture in places like Peru and Ecuador. El Niño is another natural phenomenon that is unique to South America because of its location on Earth's surface.

The weather event of El Niño is characterized by a shifting of the wind currents toward the west. The moving of the currents across the Pacific Ocean as far as Australia, according to some, may have tended to increase dryness there as it has in South America. As these pressure changes periodically reverse and move the ocean currents to the east of the continent of South America, the Humboldt Current diminishes considerably. On other occasions, the currents of the southern Pacific move farther to the east and this enhances the effects of the cold currents off the coast of South America. An integral aspect of the climate of South America and the history of the area is the relief in the terrain.

OVERCOMING THE "CURSE"

South America contains high ground throughout much of the continent. The Andes Mountains actually extend from Venezuela westward through Colombia and Ecuador and then southward into Peru and Bolivia and even down to Chile and Argentina. The Guinea Highlands are located in the northernmost parts of South America, and to the south of the equator on the eastern side of Brazil are the Brazilian Highlands. These mountains had the effect of making settlement into South America a severe challenge, and this, in turn, has characterized what some geographers call the "Curse" of South America. One of the largest rivers in the world, the Amazon, pours into the ocean and due to its location on the equator makes traveling up the river very challenging. The Amazon is probably the most navigable of the rivers in South America, yet its typically east west flow pattern generally ensures that travel is restricted to this direction; moreover, there are no major cities serviced by the Amazon.

Image © ckchiu, 2013. Used under license from Shutterstock, Inc.

Perhaps the most temperate climate in South America is found in Uruguay and northern Argentina, but even developments in this area were delayed due to the currents of the southern hemisphere and their counter-clockwise motion away from the continent. The interaction of the continent of South America with the rest of the world has had an important influence on the global phenomenon of currents that create the world's climates, but the effect of relief also plays a key role in the rich diversity of the climates displayed and the biomes that exist in South America as well.

The climates of South America are—as you might expect—tropical. One encounters the warm, wet climate and natural landscapes that are typical of the tropical rainforest when traveling from the Caribbean to southern Brazil and Bolivia. Throughout the southern cone-shaped area of South America, the humid subtropical climate is similar to the southeastern United States in terms of moisture but perhaps a bit cooler. As one approaches the snow-covered mountains to the west, the land becomes much drier and grasslands appear. The biome here is similar to the American central plains and at one time, according to Alfred W. Crosby, supported herds of cattle equal in size to those of the great bison herds of North America.[1] It is little wonder that the Europeans eagerly brought their cattle with them to this short, grassland biome. As you would intuit, the mountains of the Andes are very cold and dry and reflect the highland climate mentioned in this book's introductory chapter. As in Middle America, the varying relief patterns produce a diverse agricultural outpouring depending on elevation. One particular type of produce that has changed the world is the potato. The potato originated in the central Andes, possibly in eastern Peru or northern Bolivia. Like the tomato mentioned in a previous chapter, the potato dramatically affected the culture and population of the world by supplying needed carbohydrates.

Image © Vadim Petrakov, 2013. Used under license from Shutterstock, Inc.

Anyone interested in biological diversity will not be disappointed by a study of South America. From the dendrobates frog—one of the most lethal creatures on Earth—to the electric eel, the Amazon basin presents unique wonders of creation. The huge anaconda as well as the vampire bat have been unpleasant neighbors for the denizens of the basin for years, and yet South America's enormous potential has long been overlooked. Located longitudinally midway between the North American continent and Europe, South America is ideally situated for future development within the global economy.

---

[1] *Ecological Imperialism: The Biological Expansion of Europe, 900–1900* by Alfred W. Crosby (New York: Cambridge, 1997), 178.

Bolivia is an interesting case. As one of only two land-locked nations on the continent—the other being Paraguay—it really is two nations in terms of relief as are the other countries of the Andes Mountain sub regions of South America. The **Altiplano** to the west is a high plateau that contains the famous Lake Titicaca. This lake provides much needed moisture to the arid higher elevation of the Altiplano. Extremely high in elevation, this dry, often cold and snow-covered volcanic mountain landscape stands in sharp contrast to the eastern and northern parts of Bolivia. To the east of the Andes Mountains one can see the southern extreme of the Amazon watershed and to the east the land drops from over 20 thousand feet to nearly sea level.

The Altiplano of the Andes Region.
Image © Denis Burdin, 2013. Used under license from Shutterstock, Inc.

The Coloquiri mine near La Paz, the capital city of Bolivia, has has produced silver and tin for years. President Evo Morales, one of the first Native American presidents to be elected in South America, nationalized the mine resulting in discord that brought rioting into the streets of the cities as rival mining workers fought for control. The economic divide is reflective of the ethnic divisions in Bolivia. The ethnicities of the Native American or indigenous peoples of the highlands stand in contrast to the mestizo or people of mixed Spanish-speaking heritage in the Amazon basin or lowlands to the north. The loss of millions of dollars due to struggles between public and private control over utilities, services and the banking sector has done little to help the people economically.[2] Exporting natural resources such as gas to the United States has been delayed due to the landlocked nature of Bolivia and has led to attempting to make exportations through Argentina instead.

© 2007 JupiterImages Corporation

Paraguay is the other landlocked country in South America. Recent interest in **Paraguay** has led numerous mission teams from various churches into the area. The Mormon Church has been very successful in this traditionally Roman Catholic area. The language of **Guarani** has attracted many scholars dedicated to developing scripture into this area. The **Gran Chaco** spreads through the center and west of Paraguay and is a hot plain that is quite dry and borders the **Piranha** and **Paraguay Rivers**. Both rivers flow south into the **Rio de La Plata** off of the coast of Uruguay and the **Pampas** of Argentina. Another interesting area in Paraguay is known as the triple frontier area. This controversial **frontier** is partially claimed by Brazil, Argentina and Paraguay and has even seen the threatened use of military force in an attempt to control this powerful waterway project known as Hidrovia. The Hidrovia waterway project possesses an enormous potential for production of hydroelectric power and for obvious reasons is an area highly sought by the rival nations of the region. Recent economic developments to include the **Mercosur** trading bloc of South American nations has been a problem for Paraguay.

Resentment exists in this landlocked nation since its suspension from the trading bloc for political reasons recently. Particularly galling to the Paraguayans is the recent acceptance of Venezuela, which has been admitted for membership into the trading bloc. The Mercosur trading bloc is one of the largest in the world behind the European Union in terms of **GDP** (Gross Domestic Product), or the total value of goods and services produced within the state. Perhaps another example of the struggles of a landlocked nation is the fact that Paraguay must still utilize the Mercosur system in order to export products abroad as it needs to use transportation routes through surrounding countries to be able to reach the ocean.

---

[2] *Geography: Realms, Regions, and Concepts* by deBlij, Muler and Nijman (USA: Wiley, 2012).

**MERCOSUR
(MERCOSUL)**

Member states
Associate members

Map courtesy of Katie Pritchard

THE APEX

The southern cone or apex of South America consists of Argentina, Chile and Uruguay. With an increasingly temperate and dry climate as one heads south, it is no surprise the cultural landscape of Uruguay has attracted numerous immigrants from Europe over the years. Uruguay possesses enormous potential as a developing economy with the valuable port city of Montevideo on the Rio de la Plata. Although somewhat slowed in settlement by the direction of the currents in the southern hemisphere, nevertheless, Spanish culture continued to grow in this area and spread into Argentina. The famous Pampas region may be the cultural heart of such terms as ranch and lasso and other words reminiscent of the American west. The Argentines have seen their economy challenged in recent years by varying political successes and economic recessions. An inability to harmonize labor interests and the demands of various locals has tended to make foreign investment somewhat problematic for the hopes of developing Argentina. Sovereignty issues surrounding the Falkland Islands generally are concerned with the potential of offshore mineral resources.

The southern cone-like projection of South America.
Blue Marble: Next Generation image by Reto Stockli, NASA Earth Observatory (NASA Goddard Space Flight Center)

IRRIDENTISM AND ATTENUATION

Chile is perhaps the most attenuated nation on Earth. Its strangely elongated shape is a function of the natural geography of the Andes Mountains. In recent years the mines of the Atacama Desert have produced valuable metals to include copper. Perhaps one of the strangest natural landscapes in the Americas can be seen here with ancient monkey puzzle trees replicating our redwoods, small Andean Mountain cats, and tiny deer.

Transportation follows level terrain when possible.
Image © RIRF Stock, 2013. Used under license from Shutterstock, Inc.

A mysterious and wonderful continent, South America offers unparalleled opportunities for serving others through whatever vocation one might choose—ministry, business or agriculture. The Andes Mountains extend north as far as Venezuela and we see the familiar pattern of bifurcated climates as functions of relief. Standing in sharp contrast to the cool western mountain regions are the moist, hot Amazonian basin regions to the east of the countries of Peru, Ecuador and Colombia. **Irredentism** occurs when a government attempts to expand its borders into areas where there are perceived cultural similarities; accusations of this practice can be made when drawn borders do not reflect natural borders. This has not been a particularly large problem in South America, but much success has been made in recent years by the government of Colombia against the FARC insurgency as foreign money has brought in outsiders and gangs seeking to control the drug market.

Strategically important to the United States, these countries off the fertile Pacific Coast have seen insurgencies as varying groups claim authority and potentially dangerous frontier areas such as between Colombia and Venezuela. The most important potential for this area in terms of mobility may be the recent attempts by the Colombian government to provide transportation nodes between the Caribbean Sea and the Pacific Ocean. The Cordillera Central and Cordillera Occidental and Cordillera Oriental mountains transect the industrial areas and Colombia's developing transportation routes. Rainforests also stand in the path of the Colombian government's attempts to reduce shipping costs, which are notably higher for Colombia than Peru or Ecuador on the coast. Using railroads and investments in river dredging may eventually produce much potential for the nation of Colombia.

The sad legacy of drug use from South America through the Caribbean and into Mexico has led to the creation of a trail of misery, which is preventable. The high demand for drugs from the richest nation on Earth has produced armies recruiting boy soldiers and left gang wars in its path. Enormous profits at minimal investment have made drug production very profitable and have made many areas of the northern Andes very dangerous and frontier like. In their attempts to gain power, numerous gangs use the money obtained from their drug deals to achieve parity in power with the elected governments. Eventually the governments and these groups enter into relationships and finally a power struggle exists to clarify who is in charge. Too often it is the mission of the military in this country to maintain or enforce the structural order of the society.

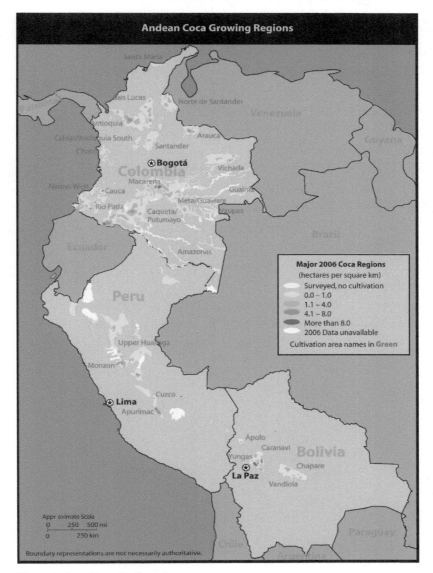

Andean Coca Growing Regions

Santa Maria · San Lucas · Norte de Santander · Venezuela · Guyana
Antioquia · Arauca
Caldas/Antioquia South · Santander
Choco
⊛ **Bogotá** · Vichada
Colombia · Macarena · Guainia
Ninmo West · Cauca · Meta/Guaviare · Vaupes
Rio Patia · Caqueta/Putumayo
Ecuador · Brazil
Amazonas

Major 2006 Coca Regions
(hectares per square km)
Surveyed, no cultivation
0.0 – 1.0
1.1 – 4.0
4.1 – 8.0
More than 8.0
2006 Data unavailable
Cultivation area names in Green

Peru

Upper Huallaga

Monzon

⊛ **Lima** · Cuzco
Apurimac

Apolo · Caranavi · **Bolivia**
Yungas · Chapare
⊛ **La Paz**
Vandiola

Approximate Scale
0   250   500 mi
0   250 km

Paraguay
Chile

Boundary representations are not necessarily authoritative.

The agricultural hearth of cocaine.
Map Source: Office of the National Drug Control Policy

Life is indeed hard, but when one uses these illegal drugs they are not only hurting themselves, but others as well, particularly in South America.

The Caribbean coast of South America is somewhat different ethnically and culturally than other parts of the continent. Oftentimes the descendants of slaves, in the former European colonies of Suriname, French Guiana and Guyana enjoy access to the Atlantic Ocean. It is not surprising to meet people of various European and South Asian backgrounds. Besides amazing diversity, the European Space Port is actually located in French Guiana. The location in relation to the equator allows spacecraft the ability to obtain speeds necessary to depart the atmosphere more easily. In addition to being a prime location for making advances in the space age, traditional resources such as gold have also been discovered there.

NO HABLA ESPAÑOL

The huge nation-state of Brazil illustrates many of the geographic challenges facing South America. According to some sources it has about the fifth highest population of any nation in the world and ranks about fifth in geographic size of the world's nation-states. The language of Portuguese is a legacy of the Treaty of Tordesillas. Possible explanations for the inability of the Portuguese to penetrate the interior of the South American continent would be the currents which tend to blow in unfavorable directions relative to the route of travel for the European discoverers. Also, note the relative lack of specie tended to encourage the settling of the Central American isthmus and the Andean coast where gold and silver is more plentiful. Look at the locations of the major cities such as Rio de Janeiro and Sao Paulo and you will see how they are relatively near the coast. This coastal location of the cities of South America reflects a colonial pattern seen also in Africa and India to take advantage of opportunities of trade but perhaps not permanent settlements.

Recent attempts to settle into the interior of the continent have led the Brazilians to adopt a forward capital strategy similar to Pakistan's. The government of Brazil

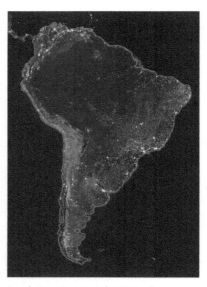

South America at night. Note the majority of lights are coastal!
NASA/Goddard Space Flight Center Scientific Visualization Studio

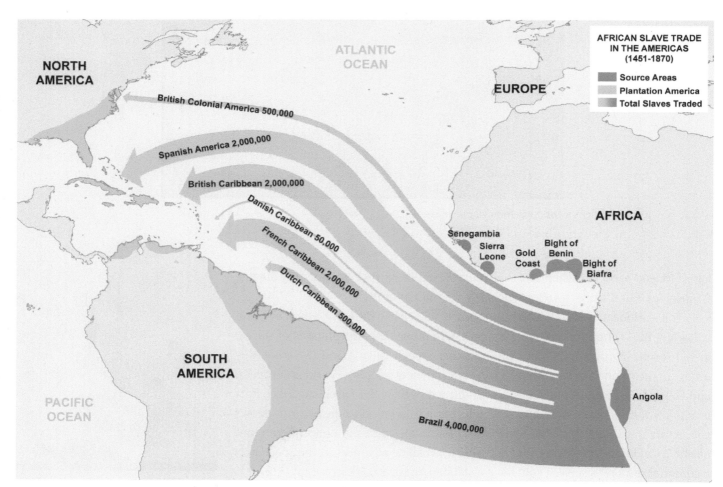

AFRICAN SLAVE TRADE IN THE AMERICAS (1451-1870)

Source Areas
Plantation America
Total Slaves Traded

NORTH AMERICA

ATLANTIC OCEAN

EUROPE

AFRICA

British Colonial America 500,000

Spanish America 2,000,000

British Caribbean 2,000,000

Danish Caribbean 50,000

French Caribbean 2,000,000

Dutch Caribbean 500,000

Senegambia
Sierra Leone
Gold Coast
Bight of Benin
Bight of Biafra

SOUTH AMERICA

PACIFIC OCEAN

Angola

Brazil 4,000,000

Where in the America's did most African's arrive?
Map source: *The World Factbook*

Image © Mark Schwettmann, 2013. Used under license from Shutterstock, Inc.

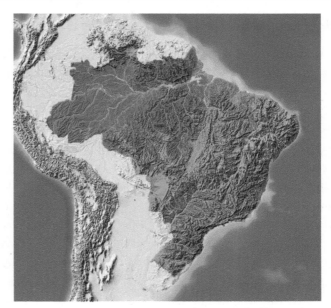

Image © AridOcean, 2013. Used under license from Shutterstock, Inc.

has attempted to reinforce the control of the wild interior region by its placement of Brasilia. By placing the capital city where it is, people are forced to live in the area to work in the government and the hope is this will pull industry into the interior. The city of Manaus has demonstrated the human-environmental effect of settling into the interior. The fragility and diversity of organisms in the rainforest have led to an appreciation of the great sensitivity of many of these specialized species. The southeastern region of Brazil is an industrial core where gold has been discovered in the past. To the far south, more European type agriculture prevails with crops ranging from grapes to rice. This wealthy area stands in contrast to the northeastern part of the country where a large percentage of the population lives but where very little production occurs. The cities of Brazil are somewhat different from what you might expect to find in the United States.

Unlike the United States, many of the more prosperous people in Brazil's huge cities live downtown. More accurately, they live on an urban spine or main road that leads downtown. Again, whereas in the United States the more affluent tend to live in the suburbs, in South America generally and Brazil in particular we find the shantytowns of poor individuals who tend to live in the outskirts of the city where they must travel to get work. The cultural landscape is quite different as well. Regardless, the opportunities for American ministry in Brazil remain despite the growing prosperity of the nation.

SUGAR, SUGAR

In recent years, Brazil has led the way in producing alternative fuels. One fuel in particular is gasohol, an extract from the sugar cane. This is a much more effective energy source than the ethanol used from corn. For several years, Brazil has had enormous potential and probably will continue to present many opportunities for industry and missions in the years to come!

Brazil: Hydro-power house!
Image © AridOcean, 2013. Used under license from Shutterstock, Inc.

URBAN LANDSCAPES OF THE WORLD-ECONOMY*

"Again, the devil took Him up on an exceedingly high mountain, and showed Him all the kingdoms of the world and their glory. And he said to Him, 'All these things I will give You if You will fall down and worship me.'" (Matthew 4:8)

As man settled into cities, we saw the creation of languages in Genesis 11. Likewise, when Satan sought to tempt Jesus, he attempted to use the glory of the cities. Truthfully, most of the human population lives in urban areas now, and we must be able to reach out to them there. To do our Lord's Will and to fulfill the Great Commission stated in Matthew 28:18–20, we must recognize the importance of urban landscapes and spaces to the world economy today.

The previous chapters have outlined the basic structure of the world-economy, especially with respect to relative levels of "development" and "underdevelopment" in the core, semi-periphery, and periphery. This section discusses global patterns of urbanization, especially during the post-industrial era of the late 20th and early 21st centuries, within this world-systems context. Although urbanization has been a phenomenon associated with many cultures for thousands of years (since the Neolithic Revolution), the highest rates of urban growth have materialized during the most recent industrial and post-industrial eras. As a result, there are now more people living in cities than ever before in human history. Where are the most urbanized places in the world? Where are urban populations growing the fastest? What are the potential consequences of high rates of urban growth in different regions of the world-economy? What are some of the distinguishing characteristics of cities and urban landscapes in the post-industrial era? How do urban landscapes differ in the different regions of the world-economy?

GLOBAL PATTERNS OF URBANIZATION

A logical place to begin this discussion is with an overview of some general global patterns of urbanization. The most urbanized populations today are in core and semi-peripheral regions. Indeed, of the regions of the world in which seventy percent or more of the population lives in urban areas, all are either in the core or semi-periphery of the world-economy:

→ North America (The United States and Canada)

→ Mexico

→ Northern and Western Europe

→ South America (with the exception of the Andean highlands)

→ Australia and New Zealand

→ Japan and South Korea

→ Parts of Southwest Asia and North Africa (Libya, Saudi Arabia, Israel, Jordan, United Arab Emirates)

At the same time, however, data compiled over the last ten years indicate that urban populations are growing the fastest in the periphery and parts of the semi-periphery. The overall global trend in the post-industrial, then, has been stagnant or negative rates of urban growth in the core but very high rates of urban growth in the semi-periphery and periphery. The growth of urban populations in the semi-periphery and periphery is cause for concern because it is precisely these

* Pages 118–125 from *Introduction to Human Geography: A World-Systems Approach,* 4th Edition by Timothy G. Anderson. Copyright © 2012 by Timothy G. Anderson. Reprinted by permission of Kendall Hunt Publishing Company.

economies that can ill afford the social and economic pressures resulting from ever-increasing urban populations. Such pressures might include:

→ Housing—to shelter newly arrived migrants (most such migrants in the semi-periphery and periphery have moved from rural areas to urban areas in search of jobs

→ Food—urban dwellers do not produce food, but rather consume food produced in rural areas by fewer and fewer farmers

→ Natural Resources—clean air and water

→ Public Services—water and sewage services, trash collection, communication and transportation infrastructures, security (police), social services

→ Social Problems—crime, ethnic conflict, economic inequalities, unemployment

→ Jobs—to provide a living for newly-arrived migrants from the rural countryside

In order for an urban region to function smoothly, each of these must be addressed or provided. While similar problems plague all cities worldwide, including those in core regions, core economies are much better equipped to handle large urban populations. In many parts of the periphery and in parts of the semi-periphery such services are woefully inadequate at best and nonexistent at worst. Large and growing urban populations, then, present such areas with a myriad of issues and problems, and only add to the many economic and social problems that afflict these areas of the world-economy.

The growth of large cities in the semi-periphery and periphery of the world-economy during the late industrial and early post-industrial eras is illustrated in the following list of the world's largest metropolitan areas in 2005 according to the United Nations:

1. Tokyo-Yokohama, Japan (33.2 million) [*core*]
2. New York, USA (17.8 million) [*core*]
3. São Paulo, Brazil (17.7 million) [*semi-periphery*]
4. Seoul-Inchon, South Korea (17.5 million) [*core*]
5. Ciudad Mexico, Mexico (17.4 million) [*semi-periphery*]
6. Osaka-Kobe-Kyoto, Japan (16.4 million) [*core*]
7. Manila, Philippines (14.8 million) [*semi-periphery*]
8. Mumbai, India (14.4 million) [*periphery*]
9. Jakarta, Indonesia (14.3 million) [*semi-periphery*]
10. Lagos, Nigeria (13.4 million) [*periphery*]
11. Calcutta, India (12.7 million) [*periphery*]
12. Delhi, India (12.3 million) [*periphery*]
13. Cairo, Egypt (12.2 million) [*periphery*]
14. Los Angeles, USA (11.8 million) [*core*]
15. Buenos Aires, Argentina (11.2 million) [*semi-periphery*]
16. Rio de Janeiro, Brazil (10.8 million) [*semi-periphery*]
17. Moscow, Russia (10.5 million) [*semi-periphery*]
18. Shanghai, China (10.0 million) [*semi-periphery*]
19. Karachi, Pakistan (9.8 million) [*periphery*]
20. Paris, France (9.6 million) [*core*]

As late as the mid-20th century, nearly all of the most populous cities in the world were located in Europe or North America. Today, however, of the twenty largest cities in the world only six are located in core regions of the world-economy. Eight are in semi-peripheral locations and six are in the periphery. Thus, while large cities are associated with the wealthiest countries in most people's minds, it is clear that large urban populations are increasingly phenomena of the semi-periphery and periphery as well. Indeed, experts speculate that Mexico City will overtake Tokyo as the world's largest metropolitan area by 2025 and that only two or three cities in core regions (Tokyo-Yokohama, New York, and perhaps Osaka) will remain on the list of the twenty largest cities in the world.

THE NATURE OF CITIES

Why do cities exist in the first place? What advantages are offered by the agglomeration of large populations in a certain place? What functions do cities perform? The answers to these questions are extremely complex and may differ from place to place, not only within the same country, but within the different regions of the world-economy as well. But we can begin to understand the nature of cities by pointing out a few basic caveats concerning urban regions agreed upon by most experts:

1. *Cities perform certain basic economic functions*

   This is the reason that cities exist at all: for the efficient performance of functions that could not be adequately or efficiently carried out if a population were randomly dispersed through space. For example, producers are nearer to the consumers of their products, and workers are nearer to their places of employment. Time, money, and efficiency is saved by the agglomeration of people in space.

2. *Cities function as markets*

   This has been the case since the earliest Neolithic Revolutions. As such, they have close reciprocal relationships with their rural hinterlands. For example, cities consume food that is produced primarily in rural areas. Cities are also the places where raw materials from rural areas (such as agricultural products or natural resources) are processed into consumable goods. Cities dispense goods and services not only for their own urban populations, but rural populations as well.

3. *Cities tend to be efficiently located*

   Cities are most commonly located at certain advantageous sites:

   → Sites that offer security and/or defense such as a hilltop or island (many Neolithic and Medieval cities occupy such sites)

   → Sites that are economically advantageous, such as a **head of navigation site** on a river, a river fording or portage site, or a railhead site

4. *Cities function as "central places"*

   The idea of cities as "central places" stems from the work of the German geographer Walter Christaller, who in 1933 published a theoretical study concerning the distribution of service centers. Christaller wanted to understand the theoretical spatial patterns that would result when rural residents traded with a central market town providing goods and services. The results of this study are referred to as **Central Place Theory** by human geographers. The theory has been applied in many different places around the world in order to more fully understand urban patterns and appears to have stood the test of time in terms of its explanatory value. Christaller was concerned with why cities are located where they are, why some cities grow while others do not, and why there is an apparently non-random pattern with respect to the location of cities with respect to other cities. Central Place Theory holds that the importance of a market city is directly related to its centrality—the relative importance of a place with respect to its surrounding region.

The central concept of Central Place Theory concerns the "range" of a good or service—the distance that people are willing to travel to obtain a certain good or service. Some goods and services, such as bread or food in general, are **low-range goods**, those for which people are not willing to travel far to obtain. Others, such as cars or furniture, are **high-range goods**, for which people are willing to travel long distances to buy. Christaller argued that the "centrality," or relative importance, of a market town or city is directly proportional to the types of goods and services offered there, and that a natural hierarchy of size will arise with respect to market locations based upon what types of goods or services are offered there. This central place hierarchy ranges from low-order places that offer only low-range goods, to high-order places that offer both low- and high-range goods and services. Central Place Theory thus gives us insight into the functions that cities perform, why some cities grow and others do not, and why there are many small central places but only a few very large central places.

URBAN LANDSCAPES OF THE CORE

The following sections list the distinguishing characteristics of urban regions in selected locations in the core, semi-periphery, and periphery of the world-economy:

The core regions of the world-economy contain some of the largest cities in the world. Several of these cities developed into the nodes or "command and control centers" of the world-economy and remain so today. These so-called **world cities** are distinguished by the following characteristics:

→ Financial centers—head offices, stock market locations

→ Command and control centers of world capitalist economy—corporate headquarters of multi-national firms

→ Political centers

→ Cultural centers

→ Nodes of international linkages

WESTERN EUROPEAN CITIES

MEDIEVAL ORIGINS

One of the most striking features of most western European cities today is the juxtaposition of pre-industrial and post-industrial features in the region's urban landscapes. Many western European urban centers can trace their origins to the medieval era (from the 10th to the 15th century), when they emerged as high-order places in the central place hierarchy due to economic and/or political importance. The siting and the pre-industrial landscapes of many western European cities reflect these medieval origins, as well as their early roles as transportation centers and economic and political centers. Many of the largest western European cities, for example, are located either on the coast or on a large, navigable river. Some occupy early head-of-navigation sites (London, for example) or river crossing sites (Frankfurt, Germany). Others occupy "defensive" sites, such as hill-top or island locations, and are often associated with the sites of medieval-era castles and fortifications (Edinburgh, Scotland and Heidelberg, Germany, for example). Another medieval feature of many western European urban landscapes is the seemingly haphazard arrangement of streets, reflective of a lack of centralized city planning and growth by "accretion" over long periods of time. It was only in the 19th and 20th centuries that many western European cities began to enact city planning schemes that began to alter ancient medieval urban city plans. The presence of straight, ceremonial boulevards, such as the Champs Elysées in Paris and Unter den Linden in Berlin, reflect this kind of centralized city planning.

ABUNDANT "GREEN" SPACES

Also reflective of the more modern trend toward centralized urban planning is the presence of relatively large areas set aside for public use. Such spaces include pedestrian zones, public markets and parks. Many western European cities are

surrounded by large tracts of forests and parks located at the urban-rural fringe on the outskirts of urban areas; these tracts are collectively known as **greenbelts**.

DENSE PUBLIC TRANSPORTATION NETWORKS

Modern western European cities are characterized by well-developed, efficient urban transportation networks. These networks usually include bus, tram (streetcar), subway and light rail transportation.

"LOW" PROFILES

Compared to many North American cities, most large western European cities have relatively few tall skyscrapers. With the exception of London, Paris and Frankfurt, all of which have downtown central business districts with prominent skyscrapers, most European cities have a comparatively "low" landscape profile comprised of many square miles of multi-unit housing structures and retail services.

EMERGING POST-INDUSTRIAL MULTIETHNIC CITIES

Although most western European states emerged from the colonial era as very strong nation-states with ethnically homogeneous societies, the late 20th and early 21st centuries have been characterized by significant influxes of ethnic minorities, primarily from southern and eastern Europe, Asia and Africa. From Moroccans and Algerians in Spain and France, to Indians and Pakistanis in England, to Serbs and Africans in Germany and the Scandinavian countries, the larger urban centers of western Europe have received the largest number of these immigrants and their presence has fundamentally altered the ethnic makeup of the region such that most western European societies can now be characterized as fundamentally multiethnic in nature. Increased immigration has altered European societies in many positive ways, but it has also not occurred without some significant social issues. For example, many of these new immigrants are refugees escaping political and economic turmoil in their home countries, and arrive in European cities without jobs or housing. Given the immigrants' need for jobs and housing, some cities have experienced significant strains in dealing with large influxes of unemployed immigrants. In response, many cities have constructed large numbers of immigrant housing developments in specific areas set aside for such housing, often on the outskirts of urban regions. This has created **ethnic enclaves**, neighborhoods that are numerically dominated by specific ethnic groups, with businesses, such as restaurants and retain shops, catering to those groups.

NORTH AMERICAN CITIES

CHANGING FORMS OF TRANSPORTATION

Historically, the most significant factor that has influenced the structure and size of North American cities has been changing dominant forms of transportation over time. In the pre-industrial era—from the colonial period until the late 19th century—most cities were oriented toward pedestrian and/or animal transportation. These pedestrian cities tended to be rather compact, with zones of varying land use forming concentric circles around a **central business district** nucleus dominated by retail services and high-rent residential housing. During the early industrial period, between roughly 1880 and 1940, urban transportation came to be dominated by streetcars, subways and railways. As a result, cities expanded dramatically in size as commuters could now live further from the central business district due to faster and more efficient forms of public transportation. Zones of varying urban land use expanded outward from the central business district along these public transportation arteries. These "streetcar cities" came to resemble a wheel, with the central business district as the hub and railway and streetcar lines as the spokes radiating outward from the center. After World War II, the widespread use of the automobile as the dominant mode of transportation engendered even more radical changes in the size and shape of American cities. With the construction of interstate highway systems, commuters began to live further and further away from the central city, leading to the development of intense **suburbanization** at the urban-rural fringe dominated by upper-income residential housing and services catering to that income group. Cities expanded dramatically in size such that most American urban regions can now be characterized as being comprised of multiple "cities within cities" covering hundreds of square miles and connected via a dense and efficient network of large highways.

SPATIAL DIFFERENTIATION BASED ON ETHNICITY, RACE AND INCOME

One of the most distinctive characteristics of North American cities today is the development of conspicuous sectors of varying residential land use that reflect societal differences in ethnicity, race, class and income. This has led to the development of distinctive ethnic and class-based neighborhoods within American urban regions. This trend began in the 19th and early 20th centuries with the immigration of millions of Europeans, especially those coming from southern and eastern Europe who settled in the large industrial cities of the Northeast, and the migration of hundreds of thousands of African-Americans from rural areas of the South to industrial cities of the North. Many of these immigrants and migrants settled among one another in distinctive ethnic neighborhoods near central business districts, downtown areas that were abandoned by upper-income whites in favor of suburban locales at the urban-rural fringe.

ZONING

Zoning refers to the detailed urban land use planning that city governments in the United States undertake; it is yet another distinguishing characteristic of American cities. Through the use and enforcement of zoning laws, city governments have the power to authorize and enforce what types of economic activities can take place in certain areas and what kind of structures can and cannot be built in certain areas. Such areas are said to be "zoned" for certain activities or kinds of structures. Such zoning laws have had a significant impact on the spatial differentiation of American cities with respect to both residential and business land use.

GENTRIFICATION

Gentrification refers to the revitalization of formerly abandoned properties in the central business district of American cities, a trend that is increasingly characteristic of American cities in the post-industrial era of the late 20th and early 21st centuries. As upper-income whites moved to suburban areas from the central city during the 1950s, 60s and 70s, and as warehousing and light industry activities also moved to urban peripheral regions during the same period, downtown areas in many American cities fell into disrepair. In order to revitalize downtown areas and to entice suburbanites back to the central business district, city governments began to support the efforts of wealthy investors in purchasing and revitalizing formerly abandoned downtown properties. These gentrification schemes often involve the construction of pedestrian malls dominated by expensive restaurants and specialty shops that cater to upper-income customers. While these activities have given many downtown areas a second life and contributed to economic revitalization, gentrification does not occur without some social costs. For example, as downtown areas were abandoned during the era of rapid suburban growth they were often repopulated by lower-income residents and recent immigrants. Because such residents lack the political and economic power of wealthy investors and developers, gentrification schemes often result in such residents being forced to move.

URBAN LANDSCAPES OF THE SEMI-PERIPHERY AND PERIPHERY

LATIN AMERICAN CITIES

AN IBERIAN COLONIAL IMPRINT

The most conspicuous urban landscape features of Latin American cities reflect the Iberian (Spanish and Portuguese) colonial imprint that is common throughout the region, from the southwestern United States in the north to the southern tip of South America. Spanish colonial goals were focused on the expansion of empire, the expansion of Christendom, and the extraction of valuable natural resources such as gold and silver, and distinctive urban landscape features reflect these colonial goals. For example, in order to accomplish these goals the Spanish instituted a centrally-planned and ordered network of urban centers that was built upon pre-existing networks of Native American towns. By law and in practice, all Spanish colonial towns were constructed on a rectilinear grid of streets oriented to the cardinal directions surrounding an open, public *plaza*. Almost all colonial towns were associated with *presidios*, forts of garrisoned military troops that exerted political and military control, and cathedrals staffed by Jesuit priests who were charged with converting Native Americans to Christianity. Other Iberian landscape features common in Latin American cities include Spanish architectural features such as *adobe* construction and red tile roofs.

SPATIAL DIFFERENTIATION REFLECTING STRONG CLASS DIFFERENCES

Like other semiperipheral areas of the world-economy, Latin America is a region characterized by relatively intense social stratification based upon class, race, ethnicity and income. The urban landscapes of the region reflect this stratification. In contrast to American cities, the wealthiest members of Latin American societies often live very near city centers, in elite sectors or neighborhoods. These elite sectors are surrounded by distinctive neighborhood sectors according to race, ethnicity and income. The poorest members of Latin American societies, especially those that are homeless, live in so-called squatter belts in urban-rural fringe areas on the outskirts of cities in very poor conditions devoid of urban services such as running water, electricity and sewage and trash disposal.

SOUTHEAST AND SOUTH ASIAN CITIES

A WESTERN EUROPEAN COLONIAL IMPRINT

In contrast to Iberian colonial goals, the goals of western European colonial powers (such as Holland, England and France) in South Asia (e.g. Pakistan, India and Sri Lanka) and Southeast Asia (e.g. Vietnam, Myanmar and Malaysia) were decidedly merchant capitalist in nature. That is, profit based on the establishment of privately-financed plantations specializing in the production of tropical and subtropical agricultural products (such as tea, coffee, sugar, rubber and spices) was more important than the expansion of empire. After politically securing a colonial area, private companies typically established plantations in interior areas and warehousing and port facilities on the coast, often at the mouth of a major river. These port cities, which existed prior to European colonialism, came to be "remade" into European colonial outposts with distinctive European urban landscape features. Such features included European architectural styles employed in the construction of public buildings and the dwellings of European plantation managers, retail services catering to a European clientele, and European schools and churches.

RESIDENTIAL SEGREGATION BASED ON INCOME AND CLASS

As is the case in most urban centers around the world, South Asian and Southeast Asian urban residential sectors reflect differences in income and class. Colonial port cities (such as Mumbai and Calcutta in India and Hanoi, Vietnam) are usually characterized by three distinct types of residential zones: 1) an elite, European sector surrounding old warehousing facilities near port zones where European colonial managers and civil servants lived, worked, shopped and sent their children to school; 2) a sector of low-income housing also near historical port and warehousing zones numerically dominated by lower and middle class workers; and 3) a sector on the outskirts of cities dominated by low-income landless families. These sectors, called shantytowns, resemble the squatter belts that can be found on the outskirts of many Latin American cities and like squatter belts lack basic services such as running water, public sewage systems and electricity.

Cities are strictly organized into separate zones and neighborhoods. An understanding of the part transportation nodes have played in the historical development of urban areas is a fun study for developing a strategy designed to reach these areas consisting of the majority of the world's population. Understanding the linguistic and ethnic associations of human activity enables us to respect and identify political leadership to further our ultimate aims of sharing our faith through service. Human geography is a subject "made to order" for the purpose of seizing the opportunities available to us to make eternal differences on this enchanted planet. Remember what Luke said in Chapter 12:48b: "For everyone to whom much is given, from him much will be required; and to whom much has been committed, of him they will ask the more." We have indeed been given much. Let us not cease giving thanks for the opportunity to have life and the myriad possibilities available to us to become busy serving God through humanity by cheerfully sharing our talents and faith today!

KEY TERMS TO KNOW

Altiplano

Attenuated

Central Business District

Central Place Theory

Ethnic Enclaves

Frontier

GDP

Gentrification

Gran Chaco

Greenbelts

Guarani

Head of Navigation Site

High-Order Places

High-Range Goods

Irridentism

Low-Order Places

Low-Range Goods

Mercosur

Pampas

Paraguay

Paraguay River

Piranha River

Rio de La Plata

Shantytowns

Squatter Belts

Suburbanization

World Cities

Zoning

FURTHER READING

Mark Abrahamson, *Global Cities* (Oxford: Oxford University Press, 2004).

Robert Bruegmann, *Sprawl: A Compact History* (Chicago: University of Chicago Press, 2006).

Stanley Brunn, Maureen Hays-Mitchell, and Donald Zeigler, eds., *Cities of the World: World Regional Urban Development*, 4th ed. (New York: Rowman & Littlefield, 2008).

Joel Garreau, *Edge City: Life on the New Frontier* (New York: Anchor, 1992).

Alan Gilbert, *The Mega-City in Latin America* (New York: United Nations University Press, 1996).

Susan Hanson and Genevieve Giuliano, eds., *The Geography of Urban Transportation*, 3rd ed. (New York: Guilford Press, 2004).

R. J. Johnston, *City and Society: An Outline for Urban Geography* (New York: Routledge, 2007).

Yeong-Hyun Kim and John Rennie Short, *Cities and Economies* (New York: Routledge, 2008).

Paul Knox and Steven Pinch, *Urban Social Geography: An Introduction*, 6th ed. (New York: Prentice-Hall, 2009).

Joel Kotkin, *The City: A Global History* (New York: Modern Library, 2006).

James Howard Kunstler, *The Geography of Nowhere: The Rise and Decline of America's Man-Made Landscape* (Washington: Free Press, 1994).

Robert Neuwirth, *Shadow Cities: A Billion Squatters, a New Urban World* (New York: Routledge, 2006).

Lewis Mumford, *The City in History: Its Origins, Its Transformations, and Its Prospects* (New York: Mariner Books, 1968).

David Smith, *Third World Cities in Global Perspective: The Political Economy of Uneven Urbanization* (Boulder: Westview Press, 1996).

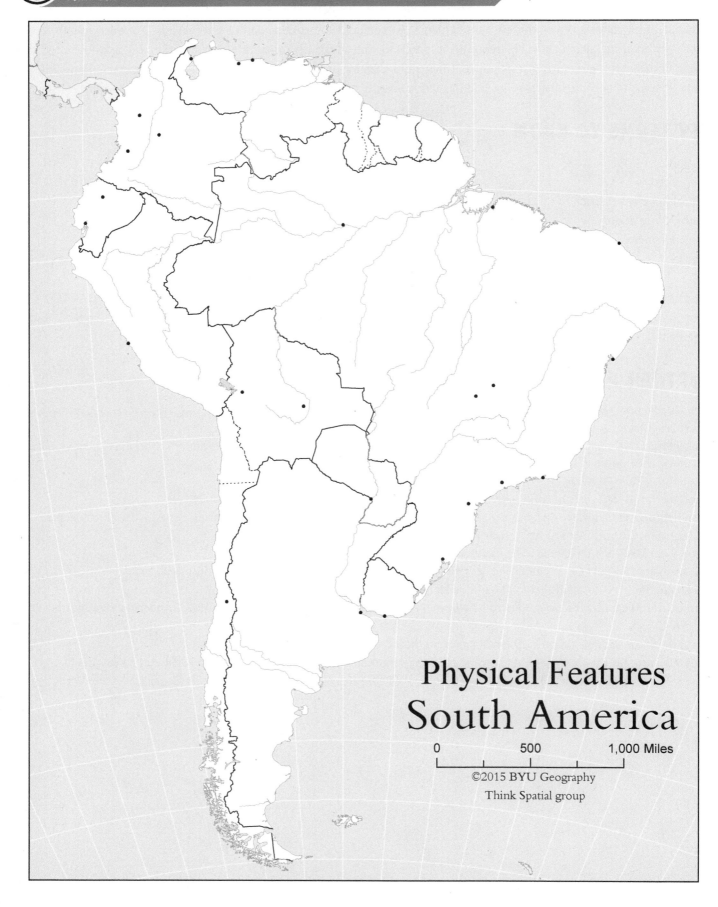

Physical Features
South America

0   500   1,000 Miles

Cultural Features
South America

0   500   1,000 Miles

©2015 BYU Geography

Think Spatial group

Practice Test Map
South America

0   500   1,000 Miles

©2015 BYU Geography

Think Spatial group

Test Map
South America

0   500   1,000 Miles

©2015 BYU Geography

Think Spatial group

North Africa and the Middle East: Changing Borders

5

But the Lord sits enthroned forever; he has established his throne for justice, and he judges the world with righteousness; he judges the peoples with uprightness. The Lord is a stronghold for the oppressed, a stronghold in times of trouble. And those who know your name put their trust in you, for you, O Lord, have not forsaken those who seek you. (Psalm 9:7–10)

CHAPTER OUTLINE:

HAVE you ever felt your best just isn't good enough? This burdensome thought often pervades our thinking and can be a cruel fallacy that knows no boundaries. In the region we will cover in this chapter dwell various peoples, many of whom are doers of their religion rather than just spectators. The religiosity of this people can be treated with nothing less than respect. And yet, a loving relationship with a creator is an increasingly new experience for many of them. The part you might play in the future of this region by demonstrating your faith is unlimited. We can be cheerful in the realization our best is good enough for our Creator, thanks to His Son!

WHY SO DRY?

When Jesus said, "Don't worry about what you will eat or drink . . . " very few of us consider the drink part of the statement. In North Africa and what is referred to as the Middle East, this is a real problem however. In fact, it is not a stretch to identify the culture of this part of the world as a function in large measure of the paucity of fresh water. The rivers and melting snows of this part of the world have been instrumental in producing patterns of human distribution. With these distributions of population have emerged different outlooks and worldviews. From these schools of thought come many of the news headlines we see today.

RIFTS AND RAFTS

The first thing one notices about this region of the world is its aridity. The obvious similarities in the shape and direction of the Red Sea and the Persian Gulf betray a pattern of tectonic activity that may, in part, explain this desiccation. In fact, the processes affecting the Earth's surface are largely responsible for the relief and mountains that both block precious moisture and yet unevenly distribute much of the valuable fresh water this area so desperately needs. Famous for merchantry, Arabs and Persians alike have traveled the Indian Ocean spreading trade and Islam as far as Indonesia. The ocean, or "wet road," in many ways rivaled the famous overland trade route between Europe and China called the "Silk Road" since ancient times. The ocean can carry more supplies with less expense due to the buoyant nature of water and the minimal security risks besides an occasional pirate. The prices of products carried by water and driven by the monsoon winds necessarily offered lower prices to their land borne competition.

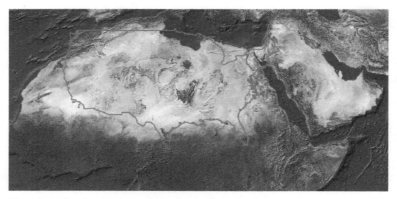

The Sahara (Arabic for Great Desert) is great indeed!

Blue Marble: Next Generation image by Reto Stockli, NASA Earth Observatory (NASA Goddard Space Flight Center)

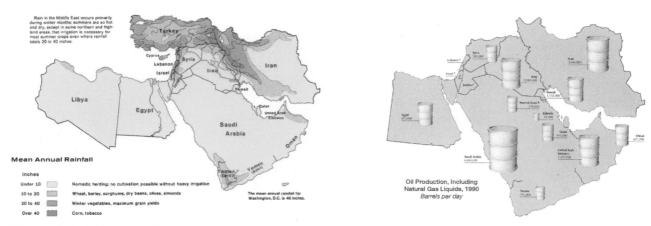

Water and Oil—both essential for survival and trade.

CIA maps courtesy of University of Texas Libraries

The precious fresh water initially in the form of mountain springs and eventually flowing as rivers are in different measures captured and distributed by the Atlas Mountains, the Zagros Mountains in Iran, and the Caucasus Mountains to the north standing guard next to the Taurus Mountains in Turkey that also work in tandem with the Hadley Cells (see the introduction) to block moisture from entering the area. These rivers and their precious water help explain the locations of the earliest civilizations where Noah and his family settled down, and many Bible scholars believe Nimrod settled into this area. The availability of fresh water in this area also explained how civilization occurred. Differentiation of labor—different jobs in a society—existed once a surplus of agriculture could be obtained. The resulting population and subsequent growth of cities brought into being an important dynamic in the human experience that remains with us today.

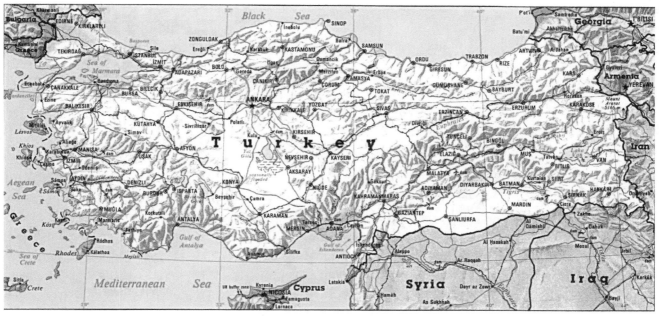

Asia Minor—Note the plateau that makes up most of the peninsula of Turkey—long a frontier between cultures.
Map scanned from *Atlas of the Middle East* (CIA, 1993), courtesy of University of Texas Libraries

Map courtesy of Katie Pritchard and NOAA

Numerous empires have emerged in response to the availability of water, but why did they fall and how does this relate to the tumultuous nature of the North African and Middle Eastern region today? The answer may lie in the concept of nomadism. Nomadism has been a powerful influence in the development of world civilizations generally but especially in this particular region. Nomads engage in what is known as movement, often somewhat cyclical in nature. This means basically they have to travel with the seasons to keep their livestock alive. Regardless, the coveted fresh water and soft life of the cities held a simultaneous appeal and repulsion that may in part explain the inherent totalitarianism that seems to exist throughout the region and the need for the state to remain vigilant against opposing ideas that may threaten it.

Modern day nomadism continues to exist along the coast of northern Africa and is responsible for a culture that is referred to as transhumance. A popular expression, "heading to greener pastures," describes this well. Nomads follow their herds of livestock since the wet, or windward side, of the mountains is moister. The mountains of this region are responsible for the melting snows that flow into the Persian Gulf from the Euphrates and Tigris Rivers. Additionally, the Nile River branches into the Blue Nile, with origins in the highlands of Ethiopia in the horn of Africa, and the White Nile, which originates in Kenya with Lake Albert in Uganda. So, the importance of tectonic plate activity over time and its effect on water supplies can easily be seen. How has this affected the Middle East today?

THIS LITTLE LIGHT OF MINE

Some scholars believe the political systems of this region today reflect disdain for softness correlated with wickedness and proof of irreligious or corrupting influences such as existed in harems and paradise gardens where water flowed and life was easy. To this day, political upheavals occur when people perceive the authorities to be unable to defend their interests, or when it appears they have become dangerously close to the Dar al Harb, or "house or land in rebellion," most often seen as the west. This idea is at odds with the Dar al Islam or idea of a world in submission to the "house or land of Allah." Perhaps a dualistic worldview of good vs. evil is a perspective that can be correlated to the physical geography of the region of North Africa and the Middle East! We must respect the sometimes intense and conservative nature of Islam today. The very religious people who live in this region are without question sincere, and this is proven by their adherence to the five pillars of Islam. It is important for Christians have borne this in mind when dealing with Muslims, amazing inroads of faith have been made in recent years. Religiosity can sometimes be wearying, as one can never be certain of obtaining salvation since one's works are never perfect. We do know through our God-given faith that Jesus is God through His perfect obedience and submission to the Father and His sinless life. This positive message of hope will always have an audience among our friends both in the Middle East and the entire globe!

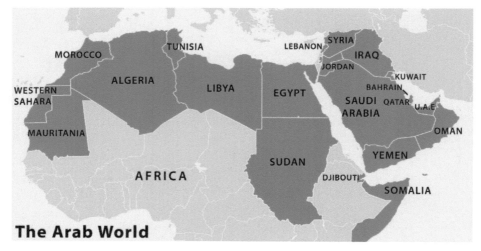

The Arab World

Map courtesy of Katie Pritchard

SUBMIT

We see three patterns of high population in this region. The North African Arabs are generally more sparsely populated over this area with the exception of Egypt. The aforementioned Nile River partly explains why Egypt is the demographic center of the **Arab** World. The Arab peoples share much in common with each other including ethnicity, language, and culture, but perhaps the greatest single cultural factor uniting this part of the world is the religion of **Islam**. The ethnicity of Arabism is in large measure a function of language. From Iraq to Senegal the language of Arabic can be heard. This has given rise to the concept of pan-Arabism, or a sense of pride based on an Arab identity.

Image © ssuaphotos, 2013. Used under license from Shutterstock, Inc.

The wealthy states of the Persian Gulf with the power of oil money have, according to some, seen themselves as the leaders of the Arab World and may serve as opposing forces within modern Islam to the Turks and Persians.

The religion of Islam is the primary unifying force in the North Africa and Middle East world today with their faith in what they call the one God Allah, the belief that his prophet Mohammed's writings have been preserved in the sacred text of the Koran and the various scholarly traditions associated with it. Muslims has successfully endured some of the harshest and most diverse climates on Earth. Still, differences do exist within the faith. It is particularly important not to overstate these differences, but they do exist. The Persians are generally of the **Shiite** form of Islam whereby the problems of the line of the succession from the Prophet extended to the son-in-law Ali. Most of the adherents of this form of Islam believe only God can choose the leader of their religion. On the other hand, **Sunni** Muslims, the predominant persuasion of the Arabs and Turks, believe that the succession from the Prophet should follow from the companion, or father-in-law, of Mohammed. Either way, though we as Christians are united by our faith and belief in the individual relationship we have with our Creator through His Son, we can certainly respect and appreciate the conviction of these people.

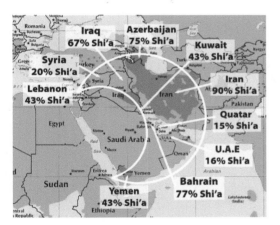

Shia and Sunni Islam in the Middle East.
CIA map adapted by Katie Pritchard, courtesy of University of Texas Libraries

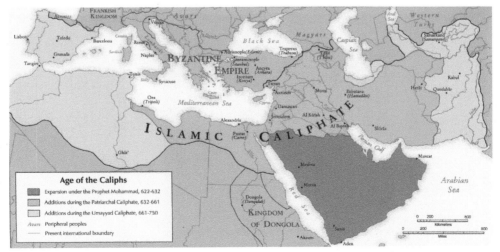

Map scanned from *Atlas of the Middle East* (CIA, 1993), courtesy of University of Texas Libraries

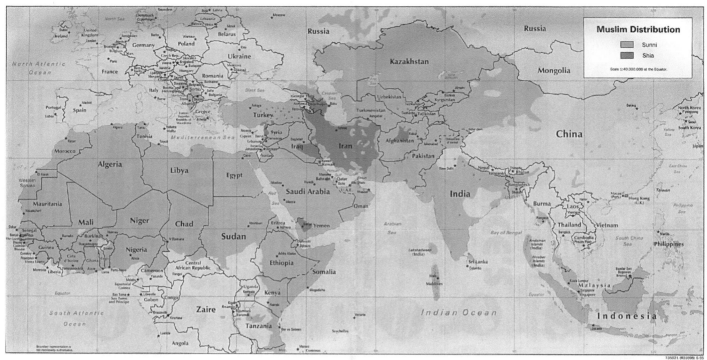

Map source: CIA map courtesy of University of Texas Libraries

735021 (R01098) 8-95

Ethnic and sectarian patterns in Iraq.

Map from *Iraq: A Map Folio* (CIA, 1992), courtesy of University of Texas Libraries

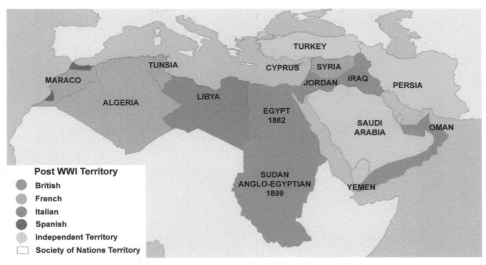

How many borders do you recognize here?

Map source: *The World Factbook*

MODEST TRADE

Another important aspect to the cultures of Islam is recognition of the social structures of gender. Women are to be respectful of men, and a married woman of age covers her head in humility and submission. The men likewise see that the windows of the homes and the corridors leading into the home from the front door are protected to prevent outsiders from being able to see into the home or "inner space." This use of interior spaces represents a historical appreciation for modesty. There is much to be admired in the strength of the Muslim family, and the increase in their population shows that fruitfulness abounds. In fact, the young increasingly inhabit cities in this part of the world. One might ask, "What future changes might the changing demographics of this region portend?" With increasing numbers of young people, the shape of governments might change. Some people believe the recent "Arab Spring" movement is a function of youth and social networks like Facebook.[1]

Trade has always been very important in this part of the world and the souks and bazaars generally are expected located near the old city walls, or medinas, of the cities. Trade through the Middle East existed since Roman times and various empires of Europe have over time sought access to the Euphrates and Tigris Rivers to gain access to the Persian Gulf as an alternative means of movement to China. Today we see a similar mixture of interests from both East and West in the Indian Ocean where China is attempting to gain access to valuable oil supplies. The western gateway to the Indian Ocean will continue to be an important region from a strategic viewpoint and one shared by Iranian and Arab peoples.

The cities in this region of the world tend to be similar to those in Europe in that a city wall often exists reminiscent of ancient and medieval times when

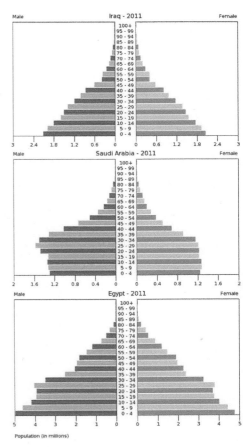

Population pyramids—note the number of young people in these Arab countries.

From *The Plaid Avenger's World #6 Nuclear Insecurity Edition.* Copyright © 2008, 2009, 2010, 2011, 2012 by John Boyer. Copyright © 2006 by Kendall Hunt Publishing Company.

[1] Kaplan, Robert D. *The Revenge of Geography* (New York: Random House, 2012), 122.

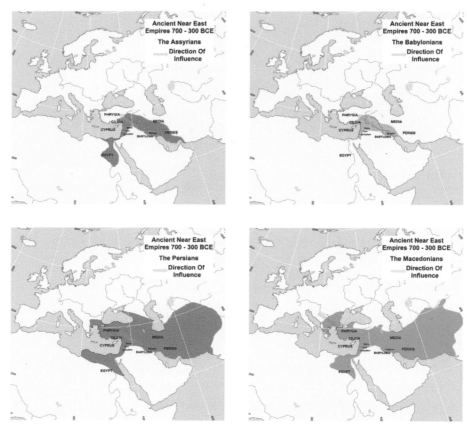

Maps from *The World Factbook*

By Elihy Vedder, 1896, Library of Congress

Stereo by Underwood & Underwood, 1907, Library of Congress

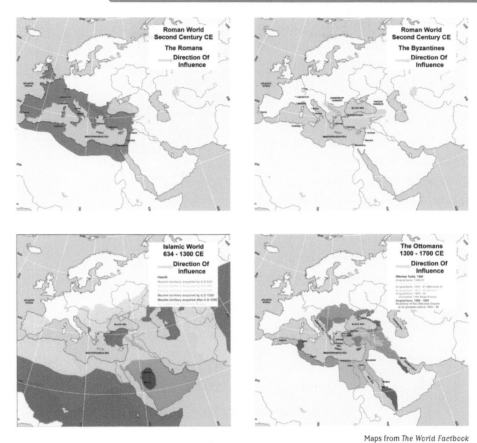

Maps from *The World Factbook*

security remained a paramount aspect of life. The **medina** often contains the old section of the city and can be called a Kasbah. These aged roads are narrow and surrounded by high-walled buildings and are frequented by merchants in covered areas called **souks**. When the merchandisers are permanent and there is an enclosed area for their business dealings, it is referred to by the Persian word *bazaar*.

IT'S HUMANITY'S FAULT (LINE)

Two other important cultural zones of the North African/Middle East area are Turkey and Iran. These peoples have a shared identity like the Arabs and tend to exist in greater numbers as a direct correlation to the amount of water avail-

able. The Turks today live on the Anatolian Plateau and share their nation-state with different peoples such as the Kurds. Water supplies are sufficient due to the elevation and the snow covered mountains in winter that provide a checkerboard of high elevation lakes. The Turkish people also are believed to have originated as nomads in Central Asia and arrived on the Anatolian Plateau in the last millennium, replacing the ancient Greeks and Persians before them. The Turkish language is responsible for the many nations we see with

From *Historical Atlas* (1911) by William Shepherd, courtesy of University of Texas Libraries

the suffix "-stan" meaning "land of." The Turkish peoples have undergone many changes since the First World War when they were a part of the Axis Powers. Increased conservatism in Muslim circles may be hindering the liberalization that took place throughout the 20th century. We will see more about the Turks in the chapter of this book on Central Asia. Another large nation with a high population that occupies the headlines these days is Iran.

Iran, or its historical name Persia, is named for the region of Pars, or Fars, near the Persian Gulf in the Zagros Mountains. The term Iran is generally believed to be taken from the name **Aryan**, mysterious peoples believed by some to be from the cultural hearth of Eastern Europe or central Asia north of the **Caspian Sea**. Like so many other

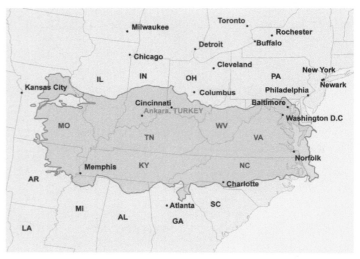

Map courtesy of Katie Pritchard

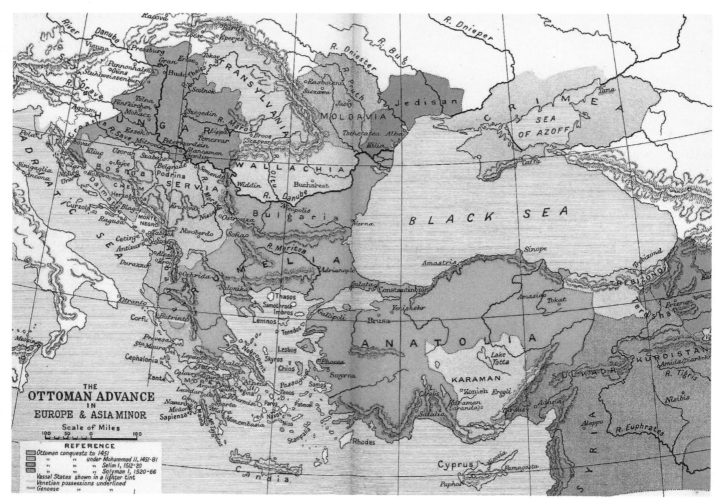

Note the Anatolian Plateau as a frontier region between east and west or Asia and Europe.

Map from *The Cambridge Modern History Atlas*, 1912, courtesy of University of Texas Libraries

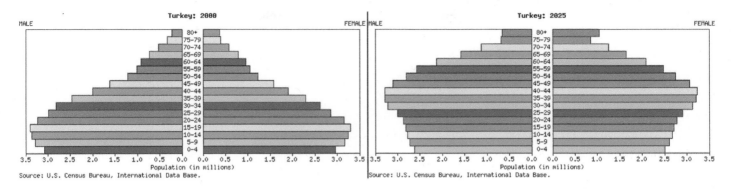

nomadic groups from this region such as the Scythians and Parthians, it is hard to be sure since written records are non-existent or at least very rare. The language of **Farsi**, the language mostly spoken in Iran today, may have originated in and be named for this area. The ancient city of Persepolis is located in Pars and was the capital of the Persians who replaced the Medes and Elamites in controlling vital trade routes through the Zagros Mountains. The famous Persian bureaucracy consisting of expert bookkeepers managed a huge empire. This empire under Darius eventually would advance as far as Europe and India establishing a great civilization that traveled on roads and utilized post offices. Many scholars believe the ideas of heaven and hell were first explained there with the dualistic religion of **Zoroastrianism**. In fact, Darius was

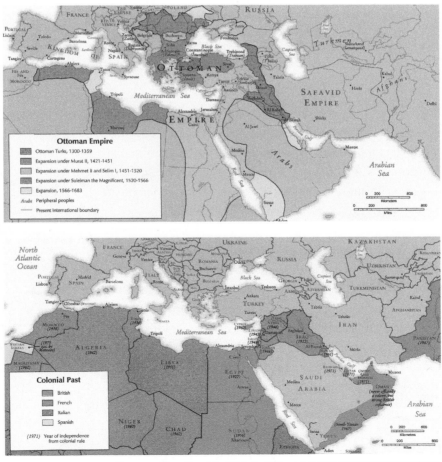

Maps scanned from *Atlas of the Middle East* (CIA, 1993), courtesy of University of Texas Libraries

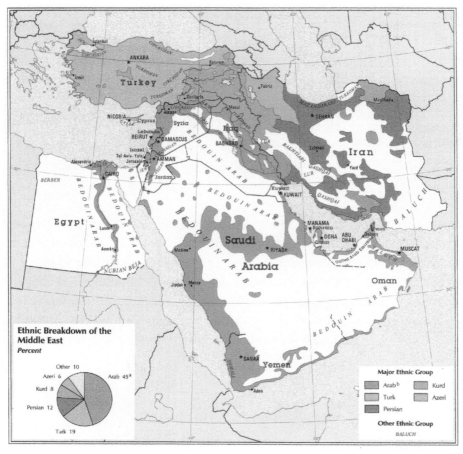

Ethnic Breakdown of the Middle East
Percent

Map scanned from *Atlas of the Middle East* (CIA, 1993), courtesy of University of Texas Libraries

able to convince his followers that foreigners were infidels and duty towards the king was in essence a religious requirement.[2] Regardless, the Iranian people have been a powerful influence in the world extending as far as Afghanistan to the east and as far west into the eastern Mediterranean for thousands of years. It is fascinating to think of the prophet Daniel witnessing his faith to the king of Persia, and to watch the news headlines today as Iran again emerges as a power influencing the areas of its former ancient empire.

Wedged in-between the great states of Turkey and Iran is the area referred to as **Kurdistan**. Kurdistan is an example of a nation, but not a state. To be a **state** the borders should be recognizable to both the members (**citizens**) of that nation and others outside of the nation. **Nationalism**, the sense of shared history, culture and traditions is strong with the Kurds. While mostly Muslim, there are Jewish populations that attribute their ancestry to the deportation of the Jews to Assyria that you may have read about in the book of II Kings in the Bible. The amount of faith required to maintain a culture when immersed in another is a testament to disciplined and faithful scriptural study and the education or transmission of culture to subsequent generations. Iraq's employment of chemical weapons against the Kurdish population at the end of the

Kurdish refugees. All our families have been refugees or strangers at some time in the past.

---

[2] Holland, Tom. *Persian Fire* (New York: Anchor, 2005).

CIA map, courtesy of University of Texas Libraries

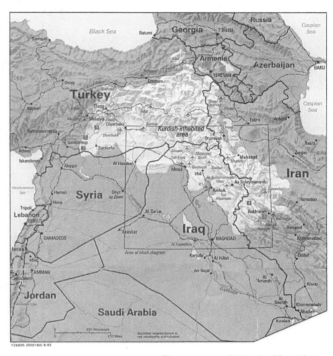

CIA map, courtesy of University of Texas Libraries

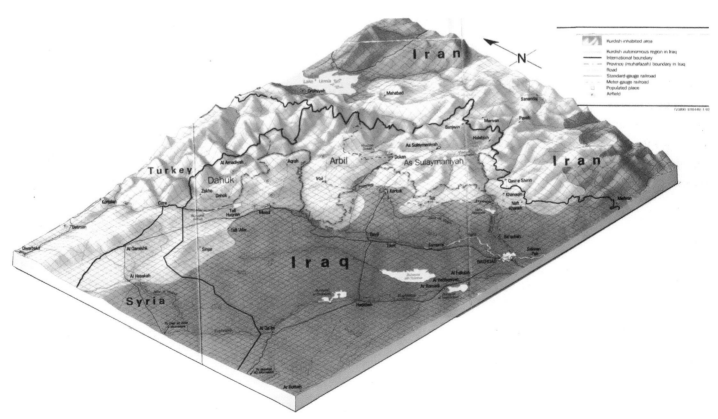

CIA map, courtesy of University of Texas Libraries

Gulf War in 1991 accentuated the plight of these people to the United States. A stateless people, the Kurds to this day are playing a very important role in the affairs of Arabs, Turks and Persians alike.

LONE BUT NEVER ALONE

The nation-state of **Israel** has survived against all odds in the Middle East as it is the only predominately Jewish population in the region. On the eastern border of the Mediterranean Sea, the size of modern Israel is marginally larger than the state of New Jersey. The history of Israel in some ways seems to be a history of the world. Destroyed by the ancient Romans and then spread out during the diaspora, the ethnic religion of the Jews has survived against all odds and its existence is miraculous. After an unprecedented nearly two millennium diaspora or expulsion, the Jews returned to Israel under British authority after World War II while the world for a very brief period agreed to the existence of the state, partially as an outpouring of horror as the evils of the Holocaust became known. The indigenous and now stateless peoples are the Palestinians who had been ruled by the Turks until the end of World War I. After World War II, the United Nations in a rare display of unanimity agreed to recognize the state of Israel, but the age-old challenge of Jewish relations with the recently indigenous populations of Palestinians is reminiscent of the relations the Israelites had with the surrounding tribes at the end of the original Exodus you have read about. The United States remains a close ally with Israel and some believe the Israeli proficiency in technology has

Maps from *The World Factbook*

benefited this mutual relationship in terms of intelligence with the United States. Many Christians in the United States today sincerely hope the fate of the United States will continue to be intertwined with that of the nation-state of Israel, the only representative democracy in the Middle East today.

WORLD POLITICAL GEOGRAPHY*

"And hath made of one blood all nations of men for to dwell on all the face of the earth and hath determined the times before appointed, and the bounds of their habitation . . ."
(Acts 17:26)

* Pages 146–150 from *Introduction to Human Geography: A World-Systems Approach,* 4th Edition by Timothy G. Anderson. Copyright © 2012 by Timothy G. Anderson. Reprinted by permission of Kendall Hunt Publishing Company.

In an era of increasing globalization, we see the significance of the borders of the various nation-states somewhat superseded as various supra and sub-states emerge. A study of world political geography demonstrates the ongoing and dynamic nature of the changes occurring across the globe.

For most people the term *political* brings to mind things like elections, political parties, and vigorous partisan debates about various issues that dominate the airwaves and that are so much a part of our popular culture these days. While these concepts are each certainly political in nature, in an academic sense *political* refers largely to the structure and function of governments, issues related to territoriality, and power structures in various types of societies. **Political geography**, then, is the sub-field of human geography that is concerned with the analysis of the spatial expression of these issues. How do people govern themselves and how has this changed over time? How are governments structured in different parts of the world-economy? What are the spatial characteristics of different types of political organization around the world? This section addresses these and other issues.

TYPES OF HISTORICAL POLITICAL ORGANIZATION

World-systems analysts argue that historically there have been only three types of political, social, and economic organization: mini-systems, world-empires, and the capitalist world-economy. We can more fully understand such historical change by coupling the world-systems model with descriptions of societal structure borrowed from the fields of anthropology and political geography. Anthropologists and political geographers generally identify five different types of political organization that have occurred throughout human history, four of which can still be observed in societies today:

BAND SOCIETIES

A **band society** is one in which the there are no formal positions of power and in which members of the society are united by ethnicity, cultural traditions, and kinship. Bands are usually quite small, perhaps only a few dozen extended families, and order is based in and around these extended nuclear families. Such societies exhibit no formal political claim to territory, with the exception, perhaps, of a claimed hunting territory. There is, however, strong territorial identity that is often associated with cultural identity. Until the Neolithic Revolution, most human societies were organized in such a manner, but few examples survive today. Band societies are limited to a few populations in southwest Africa, parts of tropical Southeast Asia, and some tropical rainforest regions such as the Amazon river basin of South America.

TRIBAL SOCIETIES

A **tribal society** is comprised of a few, perhaps many, bands of people united by common descent, linguistic similarity, and cultural values and traditions. Political leadership in these societies is usually transitory and is usually determined by virtue of perceived courage, bravery, or wisdom. Tribes are largely egalitarian in nature with respect to the communal use of resources and with respect to formalized class structures. Tribes usually claim a home "territory," but the defense of this territory is rarely undertaken with organized military power. Nevertheless, as is the case with band societies, tribal societies exhibit a very strong identity with a specific territory that is often conceived of as a people's "homeland." There are today many societies around the world that can be described as tribal in nature. The vast majority of these societies are located in the periphery and in some parts of the semi-periphery of the world-economy: much of sub-Saharan Africa, parts of tropical Southeast Asia, parts of Southwest Asia, and parts of Middle and South America. In world-systems analysis, both band and tribal societies are considered to be *mini-systems* in which production and exchange (mode of production) is largely egalitarian and reciprocal in nature. Tribal societies today, however, are also part of the capitalist world-economy, whose economies and societies are increasingly influenced by it.

CHIEFDOMS

A **chiefdom** describes a feudal social and economic order in which a powerful royal and aristocratic elite in a centralized control center controls the production and redistribution of agricultural products from different parts of a claimed political territory. World-systems analysis refers to this organization as the *redistributive-tributary* mode of production. Leadership in such societies is hereditary (by blood birth) and they often claim special divine authority to rule. Chiefdom societies are highly stratified by royalty and occupation, and social rank is largely determined by birth. Agricultural surpluses are generated by coercion of a large peasant class through peonage, serfdom, or slavery. These surpluses are then collected, controlled, stored and redistributed to the rest of the society by the royalty and aristocracy living in a central urban control center. Chiefdoms usually claim large territories, from which natural resources and agricultural products are extracted, and raise large, organized militaries to defend these territories by force. The central geographical feature of the societies is the *city-state*. World-systems analysis refers to this type of political and social organization as a *world-empire*. Today there are no surviving examples of chiefdoms. They were at one time, however, found all over the world as emerged after the Neolithic Revolution in the various so-called culture hearths: Mesopotamia, the Nile Valley, lowland Middle America, West and Southeast Africa, northern China, and the Indus Valley. This political and social organization also describes the situation in feudal Europe and Japan.

STATES

A **state** is an independent political unit occupying a defined, well-populated territory, the borders of which are recognized by surrounding states and militarily defended. All of the countries of the world today are in this sense states. This type of political organization represents a significant departure from that of bands, tribes, and chiefdoms for territory rather than cultural or ethnic affiliation is the basis of organization. Most of the world's states are multi-ethnic and multi-cultural in nature, and thus they are not "defined" by a certain culture or language or religion, but rather by place, by a territory. This basis of political organization developed in several areas around the world as feudal orders ended, especially in Europe and East Asia.

All states have a government, within which political institutions of the state function in order to exert control over the state's population and territory. Through such political power, governments are empowered to impose laws, exact taxes, and wage wars. The structure of this empowerment is basically found in two different forms of internal state political structure today. In **unitary states** governmental power and authority is centralized in a very strong central government operating from the state's capital city. The vast majority of the nearly 200 states in the world today are unitary states. A handful of states today, however, are **federal states**, in which governmental power and authority is vested in several different "levels." There is thus a hierarchy of power from the national level (federal governments) to the regional level (state or provincial governments) to the local level (city governments). Authority and control in states are vested in governments, but for those governments to have *political legitimacy* they must have some sort of ideology behind it that unites disparate groups within the society. Brute force, through the use of the military for example, may work in the short run, but without some central integrating philosophy behind it (such as "freedom" or "democracy" in the United States), a state's government can lose legitimacy and find its right to govern questioned by those it governs.

NATION-STATES

The idea of the nation-state emerged in Europe in the 18th and 19th centuries. By definition, a **nation-state** is a state (a territory) that is inhabited by a group of people (a nation) bound together by a general sense of cohesion resulting from a common history, ancestry, language, religion, and political philosophy. A *nation* in this sense refers not to a country, but to a group of people, and a *state* refers to a political territory. The ideal of the nation-state, then, combines these two concepts. As such, this type of political organization involves very strong allegiance to nationality and to territory on the part of the nation-state's citizenry. This political ideal probably first emerged after the Industrial Revolution in Europe as improved communication and transportation technologies enabled more efficient control of large territories. All of the major European colonial powers developed a strong nation-state ideal of political organization in the 18th and 19th centuries and by later exported this ideal all around the world as European power and influence grew very strong during the colonial era.

CORE-PERIPHERY PATTERNS OF POLITICAL COHESIVENESS

Today, all of the countries of the world-economy are politically organized as a *state*, but most aspire to the ideal of the *nation-state*. In practice, however, there are very few true nation-states as the term is defined above: one *nation* of people living in and claiming a defined political territory as its homeland. Why is this? To begin with, even in the strongest nation-states there are always threats to national cohesion. Economic inequality, racial and/or ethnic hostilities and injustices, or perceived disenfranchisement on the part of certain groups in a society may threaten national ideals. But the biggest obstacle to the ideal of the nation-state is the increasingly globalized world that has emerged since the Merchant Capitalist Revolution. Large-scale migrations (some voluntary, some involuntary) during this era, and continue today, have resulted in the creation of states that are fundamentally multi-national in nature, plural societies in which a variety of ethnic and national groups count themselves as citizens.

Some of the strongest (politically speaking) of such **multi-national states** have developed a strong sense of nationality in spite of the plural nature of the society. In some instances this occurred by "happenstance" as a result of a group or groups of people occupying a large territory over a long period (e.g. the European nation-states). In other instances, strong central governments sought to foster nationality overtly through public education systems and the development of a strong sense of patriotism (e.g. the United States, the Soviet Union, and China).

At the other end of the spectrum, multi-national states without a central organizing "principal" can sometimes degenerate into civil war or ethnic conflict in which there is a struggle on the part of the various nations of a state for political power (the former Yugoslavia and present-day conflicts in Africa, for example). At the same time, nations of people without their own state often engage in violence in order to achieve their own state and thus their own political power. On-going examples of such conflicts include the struggle of Palestinians, Kurds, and Basques to forge independent governments and states. Most such struggles, civil war, and political instability in general occur today in the periphery and parts of the semi-periphery of the world-economy. While the core states are not without their conflicts, such states usually have very strong central governments which may seek to quell such conflicts. Ethnic problems and conflicts are also usually worked out in the core through democratic processes or through public debate, both of which are relatively peaceful methods compared to the civil wars and military coups that are common in the periphery.

Political cohesiveness in most states today is influenced by two main types of "forces" working in society. **Centripetal political forces** are those that tend to bring together disparate groups in multi-ethnic, multi-national states. Such forces might include:

→ Nationalism—identification with the state and acceptance of its national goals, ideals, and way of life

→ Iconography—symbols of unification (flags, national heroes, rituals and holidays, patriotic songs, royalty, etc.)

→ Institutions—national education systems, armed forces, state churches, common language

→ Effective state organization and administration—public confidence in the organization of the state, security from aggression, fair allocation of resources, equal opportunity to participate, law and order, efficient transportation and communication networks

On the other hand, **centrifugal political forces** are those which tend to destabilize a society and pull disparate groups apart in multi-ethnic and multi-national societies. Examples of centrifugal forces might include:

→ Internal discord and challenge to the authority of the state which can lead to **political devolution** in which national or ethnic groups seek to form separate political authority

→ Ethnic separatism and regionalism—this is often seen in states where disparate populations have not been fully integrated, (nations without states); this can lead to **Balkanization** in which multi-national states break apart along ethnic lines

→ Trouble integrating peripheral locations—this is especially a problem where disparate rural populations are located far away from the capital

→ Social and economic inequality—this is most often seen in multi-nation states where the dominant group is seen to exploit minority groups in terms of control of wealth and social services, etc.

Borders are of extreme significance to the human geographer. Boundaries were used in the ancient world; in fact, the Bible exhorts us, in Deut. 19:14, not to remove the ancient landmarks. After the Middle Ages and during the Age of Discovery, with its subsequent process of colonization, the technologically advanced Europeans tended to fight among themselves for control of the world's resources and peoples. Drawing borders was a solution to the resulting chaos. Many feel the origins of the nation-state after the treaty of Westphalia in 1648 reflected the process of European borders reflecting local nationalism. French people, for example, began to not see themselves in a medieval way, as vassals to a particular local authority in a nearby castle; rather, the French began to see themselves as members of a larger group of people with whom they felt a commonality. The centripetal forces of sharing languages, religion, history, and money manifested themselves politically with the settling of boundaries. Borders are extremely tenuous concepts, but important ones subject to change. **Natural borders** such as the Pyrenees Mountains between Spain and France tend to divide people into linguistic and ethnic differences naturally and served as natural barriers to diffusion. Many historical geographers point out the relative autonomy of the city-states of Greece in antiquity, and the Cantons of Switzerland as examples of how natural borders tend to separate and can affect independence, identity, and even political processes. Irredentism is the source of much of the world's conflicts today. Irredentism can result when **drawn borders** separate a people who share identities. This frustrated nationalism or persecution can have terrible results, such as has been witnessed in the Caucasus mountains in the last decade, or the horrible genocide in Rwanda. Hopefully, by recognizing the importance of borders and their relationship to centrifugal and centripetal forces within a nation-state, we can serve as peacemakers and prevent violence.

We participate in organizations with various loyalties: states, communities, clubs, and churches. How these different social groupings make decisions politically is a fascinating study and of great importance to the geographer. Understanding the variations in these societal patterns is vital to seizing the opportunity we now have to share our faith. Peter said, in 1 Peter 3:15, "But sanctify the Lord God in your hearts, and always be ready to give a defense to everyone who asks you a reason for the hope that is in you, with meekness and fear. . . ."

TURKEY*

Turkey is a:

(a) Nation

(b) State

(c) Nation-state

(d) Region

(e) Delicious bird, when roasted properly

(f) All of the above

You know it's all of the above, baby! Turkey is like Japan and Russia, in that it is a sovereign state in its own right, with enough distinctions about it to set it apart from all of the places nearby, which makes it a Plaid Avenger world region as well. What is the deal with Turkey? What's going on here? Why is it different enough to be apart from the Middle East— the region with which it has classically been associated? Anybody from the Middle East, and particularly from Turkey itself, would say, "No way! We don't have anything in common with each other!" Here in the West, we just see Turkey and the Middle East as kind of a common whole. There is a lot about Turkey historically, presently, and in the future, that distinguishes it from all other Middle Eastern countries. Does that mean it's more like Europe? Nope, it's quite distinct from Europe as well. A lot of people in Europe want to distance themselves from Turkey, while some want to embrace it. Turkey is in a strange place in the world right now; it has its own thing going on, while at the same time acting as a bridge between two cultures and regions. That's the Plaid Avenger keyword for Turkey: **the Bridge**.

Turkish Delight: aka lokum or rahat, is a 15th century tasty confection made from starch and sugar, with a soft, jelly-like consistency. It is often cut into small cubes and dusted with powdered sugar before serving. It is sometimes flavored with cinnamon, mint, lemon, or rosewater . . . that last one giving the candy its characteristic pink hue. I always wondered what the heck it was since the White Witch fed some to Edmund in *The Lion, the Witch, & the Wardrobe*. Oh Edmund, you little traitorous scumbucket.

As we discussed in the Middle Eastern region, there are lots of different ethnic groups in what is collectively referred to as the Middle East, and one of those is Turkic. Turkic is also a distinct linguistic group. Where are the Turkic people? Many people think, "Turk?, they've always been in Turkey, of course!" Not necessarily. In modern day Turkey, they refer to themselves ethnically and linguistically as Turkic or Turks; however, this group is originally from Central Asia. Turks are relative newcomers to the territory that we now call Turket. The folks who would become the Turks came from Central Asia, across what is now Iran, through Iraq, and into modern day Turkey about 1000 years ago. After several hundred years, they took over the region, built a mighty empire, their culture took hold, and they made it their own. Perhaps we get ahead of ourselves, but that's what makes them distinct: a different ethnicity and linguistic group from all the surrounding countries. Why and how is Turkey so different from every place around it?

Turkey is more 'western' than virtually every other Middle Eastern country in terms of economic development and government. At the same time, they are not as 'western' as European countries. This is the first of many things that puts Turkey in a sort of no-man's land, somewhere in the middle. They have a capitalist system that is fully integrated into world economic dynamics. Nobody would disagree with that. They are the strongest fully established, democratic Muslim country.

That's important enough to pause and restate: democratic, fully Muslim country. Perhaps the main reason why they are typically classified as Middle Eastern is their single commonality, which is Islam. Turkey's population is almost 100 percent Islamic. However, it is not an **Islamic republic**, and that makes all the difference. Turkey has religious ties to its Middle Eastern brothers, but all other aspects of their society, like democracy and capitalism, make them very European. This duality reinforces the idea that Turkey is a bridge. Before we get to a more cultural background, let's talk Turkey physically.

PHYSICALLY . . .

It's pretty easy since Turkey is a single country. If we were to compare Turkey to the US, and overlay it with size held consistent, we see that it stretches from Washington D.C. to Kansas City, Missouri. It's about the same latitude, and covers West Virginia, Virginia, parts of North Carolina, Tennessee, Kentucky, Indiana, Illinois, and Missouri. It's a decently sized country.

Given that its latitude is roughly the same as the states I just mentioned, Turkey has a pretty similar climate. It is humid continental, with some humid subtropical, in the southern parts. However, given that it borders the Mediterranean Sea, we know it must have some Mediterranean climate. Indeed, it does around its coastal fringes. Turkey has four distinguishable seasons, just like the US.

The only other major physical factor to consider is that the terrain is quite different from the eastern United States. Turkey, as a whole, is a pretty mountainous country. It has an uplifted plateau in the middle, with some high mountains, particularly on its eastern borders, where you can

Cappadocia, central Turkey.

find Mt. Ararat (16,000+ ft.) as you go into the Caucasus Mountains. Even over in the western part, the elevated plateau is on average over 6000 feet in elevation.

This is an **escarpment**, as we talked about in South America and will again in Sub-Saharan Africa. There are not a lot of coastal plains or flat areas in Turkey, so as soon as you leave those nice, sunny Mediterranean shores, the terrain and the elevation make it much cooler, much faster. As you get to the far eastern side of Turkey, there can be fairly tough winters. It does have some Mediterranean and Black Sea frontage, which moderates things a bit, but it still is a fairly cooler place than eastern North America at similar latitudes.

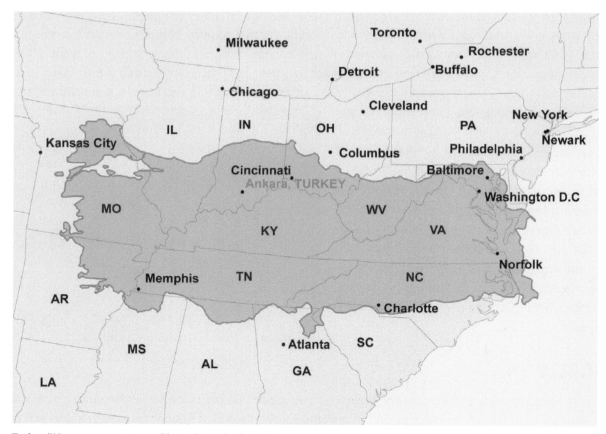

Turkey/US comparison: size and latitude are both held constant.

Turkey is such a mountainous country because it is on the border of several different tectonic plates. It is at the confluence of the Eurasian, African, and Arabian plates, very much like we had several plates coming together in Japan. As with all geologically active mountain-building scenarios, it's an active earthquake region. In the last decade, Turkey experienced some serious earthquakes, having tremendous impact in terms of human lives and property damage. That about wraps up the physical traits of Turkey—let's get back to the cultural. Why is Turkey a crossroads of culture?

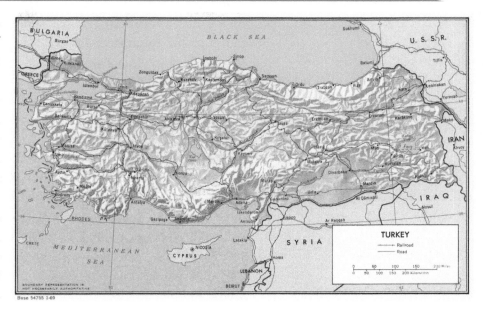

CROSSROADS OF CULTURE

Assyria

Babylonians

Turkey is one of those places, like the classic Middle East, where people have been hanging out a long time. We can go way back to 1000 bce to the Assyrians, a tough band of dudes. There is too much rich history to get into it in detail, but I want you to understand how the **Anatolian peninsula** has been a crossroads of culture throughout time. The Babylonians hung out here in control until around 1000 BCE, and the Assyrians shortly after that.

We've had people hanging out here for a good long time, people who became the core of Western Civilization. The core cultural hearth for Western Civilization is here in the classic Middle East and Turkey. Fast forward to the Persian ethnic group, whose descendants live in modern day Iran, had a mighty empire around 500–600 BCE. These are the guys that Alexander the Great and all of his ilk were fighting against. In this historical situation, the Persians swept from modern day Iran, their core, across the Anatolian peninsula to invade Europe proper, specifically Greece. Turkey was the bridge for the Persians coming from the east to invade Europe in the west. As a result of that, Alexander the Great and his Greek buddies invaded back. You may see reference to the Greeks calling this area **Asia Minor**. To combat the Persians, Turkey became the bridge of invasion for Alexander's forces out of Greece. They swept over, beat up on the Persians, and established a European empire all the way into India, making use of Turkey as a crossroads. And we're not finished yet!

The Greeks, of course, were beaten by the Romans, whose empire eventually included the Anatolian peninsula. To spare you the long blow-by-blow history of the Roman Empire, eventually, the Romans decided they needed to establish a secondary capital in the East, as the Empire proper was expanding. The main capital stayed in Rome, but the eastern, secondary capital was formed in Constantinople, where the Turkish city of Istanbul is today. After the disintegration of the Roman

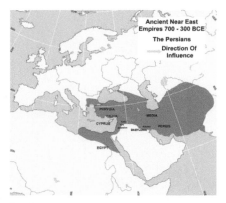

Persians

Macedonians

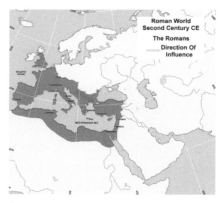

Romans

Byzantines

Empire, the eastern section of the empire became known as the **Byzantine Empire**. You've probably heard of the Anatolian peninsula referred to as "**Byzantium**" in ancient texts. This occurred between 300–500 CE.

Again, the Plaid Avenger is reinforcing that this is a crossroads. We've had the east coming from one direction, and the west coming back from the other. Turkey is always the central pivot point of these movements. Not only was the Anatolian Peninsula the stage for the political division that results in the Byzantine Empire, it was the stage for a division of religion as well. We know it as the **Great Schism**, the split between the Catholic Church and the Eastern Orthodox Church in 1054. The Eastern Orthodox Church set up camp, still in the Byzantine Empire, and spread Eastern Orthodoxy into Russia, the Middle East, and Eastern Europe. The Catholic story, which was centered in Rome and spread to most of the rest of Europe proper. But our story's still not finished!

The Greeks and Romans did their Turkish tours too.

We're taking it up to the time of Muhammad, peace be with him. Born in 570 CE, not in Turkey, but down in the Arabian peninsula in the holy city of Mecca, where Allah used him as the vehicle for the creation of the Koran. This is where the religion that we now know as Islam started.

Where did it go? It diffused outward from the Arabian Peninsula to lots of places: Africa, Central Asia, South Asia, and also Turkey. A reversal took place in the Byzantine Empire. Eastern Orthodox Christianity was displaced on the Anatolian Peninsula because of the expansion of Islam from the east. In addition, the Turkic peoples from Central Asia emigrated to the area during this whole movement of Islam to settle into the Anatolian Peninsula, making this **the bridge** between Eastern and Western cultures

Orthodox Byzantine boys in black getting their chant on.

and religions once again. This is a happening place, man. Everything's been going on in Turkey! But it's not even Turkey yet!

ENTER, THE OTTOMANS

How'd it get to be Turkey? There's one more group we haven't talked about yet: **the Ottomans**. After the Turkic ethnic and linguistic group took over the geographic area, they built a big empire: the Ottoman Empire. Starting in roughly 1300 CE, the little core that was only around Constantinople grew in all directions over the course of the next 400 years, to the point that they took over virtually the entire Mediterranean coast of Africa and much of the classic Middle East, including the Arabian Peninsula and what is modern day Iraq. The Ottomans expanded into the Tigris and Euphrates river valley, Mesopotamia, where the Babylonians once were.

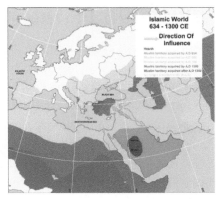

The growth of the Arab-Islamic Empire sets the stage for the Turks . . .

These Ottoman dudes took over the Black Sea area and even invaded Europe; they were on a roll. This event perhaps impacts today's world the most: the intro-duction of Islam deep into European territory via the Ottoman Empire, primarily in Eastern Europe, modern day Yugoslavia, Albania, Romania, Bulgaria, and Greece. The Ottomans, a predominately Islamic empire, controlled all of this territory. At the height of their empire, they were knocking on the doors of Vienna. They even attempted to take Austria a few times. Hey! That's all the way in Western Europe! They nearly conquered it several times, but never quite succeeded. Nevertheless, that gives you an idea of how far they penetrated into the continent, taking their culture, particularly their religious culture, with them.

Ottomans

This ties back to why we talked about southeastern Europe in particular being a trouble spot in today's world because of the infusion of so many different religions, ethnicities, languages, and groups of people. This is part of it, part of the same story. Turkey is the bridge, the platform for the movement of people from Europe to the Middle East—but just as importantly, from the Middle East to Europe. Religion typically seems to be the sticking point for a lot of this, an issue that we see cropping up even in the modern era as Turkey tries to join the European Union. More on that later.

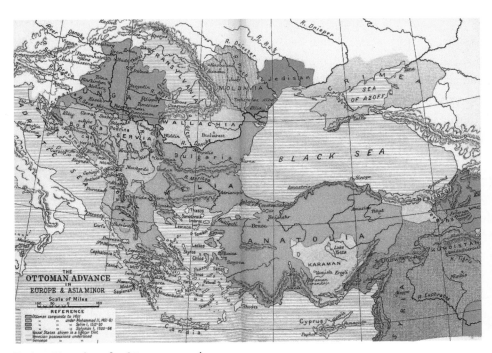

Turkey: Home base for Ottoman spread.

Side note: The Ottomans are the guys that brought coffee to Europe, via their face-off with the Austrians at Vienna. Ever heard of Viennese coffee? The brewing and consumption of coffee were passed across the battle lines during pro-tracted wars over the centuries. We often think of the Austrians and Italians as coffee connois-seurs, but it was introduced via our Turkish friends. Ever had Turkish coffee? Wow! It will knock your socks off, man!

FORMATION OF THE MODERN STATE

As you have guessed by now, the Ottomans didn't last forever. They were large and in charge and had an expanding empire from about 1300 to 1700 ce, but from 1700 onwards, they went into a steady state of decline. Why? It has a lot to do with stuff we've already talked about, but to put it simply, these guys got behind the times. In the Middle Ages, they were superior to the Europeans in terms of technology, army, and firepower. They had the finest universities in all of Europe and the Middle East. All great centers of learning, the oldest colleges, as well as all sci-entific revolution and innovation during the Middle Ages were happening in Ottoman-Turk ter-ritory. They eventually got a little insulated, headstrong, and arrogant, and saw the Europeans as a kind of backward race that they needn't mess around with that much.

Sultans simply scoffed at backwoods Euro-trash.

On top of that, think about how they got so rich. What did they produce? Were they a big colonial empire producing and trading goods, building an economy? Nope. They largely obtained their wealth and sustained themselves by continually pillaging for booty. The Plaid Avenger is all about pillaging the booty. However, in this context it means that through conquest of war, you take valuable stuff, like gold and other riches, from other places. That's okay for a while, but it eventually leads to a lack of economic infrastructure. A booty economy, unfortunately, can't stand on its own two feet in the long run. If the only way you get rich is just by taking stuff, when you get to the point that you can't take stuff anymore, you go broke.

Indeed, this was part of the reason for the collapse of the Ottoman Empire. They became economically stagnant, meaning they weren't producing or innovating anything. This caused the Ottomans to get behind economically and tech-nologically. While they were sitting around stagnating, the Europeans were in full bloom: first the Renaissance, followed by colonialism and industrialization. The Ottomans, on the other hand, were starting to fall down just as the Europeans were starting to stand up, and by 1900 re-calibration of this Europe-Turkish power situation was imminent. Politically, the Ottomans were stagnant as well, being an empire ruled by 600 years of sultans in their sultanates.

A **sultanate** is a monarchy, a royal imperial line of centralized power, a system that was disintegrating through most of Europe at this time but the Ottomans held fast to their old school system for a bit longer, which further increased their stagnation and perceived backwardness. Things were going downhill; there was internal dissent because the economy was not in good shape, and they were losing territories to the expanding Europeans. They lost Greece due to internal revolts and rebellions all across the Balkan Peninsula. By 1909, Abdul Hamid became the last of the unbroken line of Ottoman Sultans; but his line was about to be snapped in half. The people had enough and said, "We've got to change! This sucks!" Again, a very common theme we have talked about in other parts of the world: "We have to get rid of this guy! This system's not working, so let's get rid of HIM!" They deposed the Sultan, but the state were still on shaky ground.

In another part of this tale, which I haven't incorporated yet, the Russians and the Ottomans fought several wars. There was major animosity between the Russians and the Turks, the remnants of which still exist today. The Russians tried to expand out to get more shore frontage on the Black Sea, on the Mediterranean Sea, anywhere they could grab coastline in order to build their naval superiority—and the Ottoman-Turkish territory made for a good target. The Russians tried to interject power into Ottoman territory. The Europeans, at the exact same time, attempted counter Russian influence in Ottoman territory. This is another component of why Turkey became the bridge; it became a battleground for the competing powers of the Europeans and the Russians.

After the Ottomans deposed the sultan, they tried to form something close to a democracy, or a republic as it were, in an attempt to copy the European model, but they were very young and very weak economically. They were looking around and saying, "Hey, who can we ally with? We're getting so weak that Russia could probably beat us. Somebody be our buddy! Somebody sign a pact with us saying you'll come help us if Russia invades. Anybody! How about you guys in the UK, will you sign a pact? France? Spain? Italy? Germany? Come on, somebody help us out here!" At this point, the Ottomans were looked upon as kind of the "old decrepit man" of Europe. They're physically close to Europe. They were a competing power with Europe when they invaded Austria. It's not as if they're unknown or a foreign territory that nobody knows about . . . but nobody wanted to deal with them. It's like they had the plague. Untouchable.

The British and the French thought, "Well, these guys are kind of losers. What are we supposed to do with these guys? We don't want to ally with them. We don't want to go to war with the Russians just for these guys. We don't want the Russians to win, but we don't want to engage them in open battle either, so we're just going to hedge our bets." Nobody wanted to deal with the Ottomans. However, Germany under Wilhelm I, knowing full well that it was about to start World War I, said, "Hey, we'll help you poor suckers out! We'll be your buddy, because we don't like the Russians either. *We'll* be allies with you." In 1914, the Ottomans set up an alliance with Germany. Then Germany went to war with everybody, and ultimately lost. Long story short: Oops! Bad choice! Bad alliance!

The Ottomans, having not chosen wisely, were basically dismantled at the end of World War I. In 1919, at the conclusion of World War I, something happened that the Plaid Avenger refers to as **A Peace to End All Peace**. It's actually a book title you should check out sometime; it's really sweet, even if a bit dry and a lot long (David Fromkin, 1989). It explains a lot about why there's not peace in the Middle East right now and perhaps why there never will be. What is the Plaid Avenger referring to? It's got to do with a bad job hacking up of the Turkey, and it wasn't even Thanksgiving Day.

German leader Wilhelm II helped out the Ottomans—helped them out of an empire!

CARVING UP THE TURKEY

The European powers that were victorious in World War I not only carved up Germany's Prussian Empire, but the Ottoman Empire got the chop too. The way they disemboweled it set up the political, ethnic, linguistic, and religious differences that still perpetuate conflict today. In other words, they took a map of Ottoman territory and said, "Okay, here's this section. You guys go to Greece. Let's call this chunk over here Romania. Oops, I spilled my coffee there. Let's just call that coffee stain Jordan. This over here, Iraq." Almost all the modern day boundaries in this entire area were drawn in part or whole

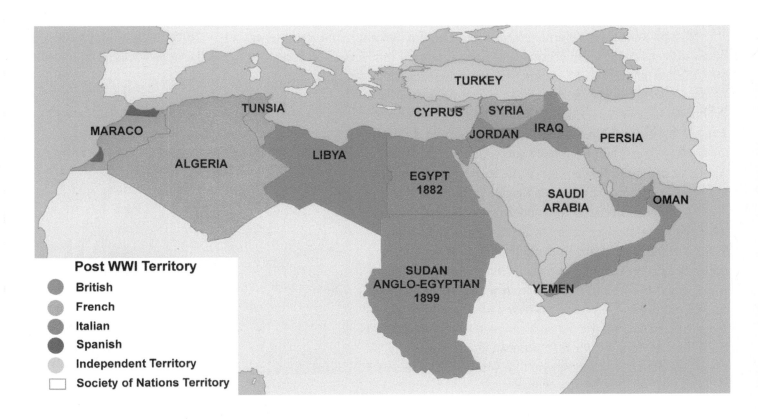

Post WWI Territory
- British
- French
- Italian
- Spanish
- Independent Territory
- Society of Nations Territory

TURKEY
TUNISIA
CYPRUS
SYRIA
MARACO
JORDAN
IRAQ
PERSIA
ALGERIA
LIBYA
EGYPT
1882
SAUDI
ARABIA
OMAN
SUDAN
ANGLO-EGYPTIAN
1899
YEMEN

during this redistribution or reallocation of land after World War I. The leftovers from the hack job became what we call Turkey today. It always was the central power core area of the Ottoman Empire, the piece where the actual ethnic Turks were concentrated. Turkey has remained stable to this day partly because it was primarily Turk, made up of Turkic people. All the other boundaries sucked because, in many cases, the Europeans drew them arbitrarily, with no consideration of the cultures within the newly drawn borders. On top of that, a lot of the new states were given to European powers, which were to supervise them as a "big brother." For example, the French became the "big brother" of Syria. The British became the "big brother" to Egypt and the Sudan. These new "baby brother" states were called **mandates**.

See, all these European powers were still really into the colonialism thing, and the end of the war spelled more territory and influence and resources for them to control. I do need to point out one resource in particular that was just being discovered in parts of the Middle East during this period: oil. Even though oil was not as central to the world economy in 1919 as it is today, these Europeans knew it was worth a lot; therefore, they wanted to control it. Ever wonder why one of the world's biggest oil companies, British Petroleum (BP), is from a country that has no oil? Look again at the mandate map above. The UK had mandates on Oman, the Gulf states, and Iraq? What a shocker.

The mandate system effectively meant that political control still rested with the European powers occupying these Middle Eastern lands, even after the Ottomans went away. This is part of the problem with today's world, but we'll talk about that more when we get to the Middle East. That's "the peace to end all peace" in 1919. Essentially, these countries were carved up in such a way as to ensure perpetual conflict, which of course is precisely what has happened in the greater Middle East. But Turkey has actually held together quite nicely since the war. How?

AWESOME ATATURK

Another politically noteworthy event happened in 1919. A guy named Mustafa Kemal led a resistance to the European allies' plan to carve up Turkey. As the Ottoman Empire was being dissected, a lot of Europeans wanted to divide Turkey up even more. They said, "If we don't completely eliminate them, they'll be a future threat, so let's chop 'em up!" Many Turks in the ever-dwindling territory decided to fight back against this total elimination plan, and one such man was a heroic heavyweight known as Kemal.

Mustafa Kemal was an army officer and an ardent, fervent nationalist. He was a big pro-Turkey guy who loved and believed in his country. He became very popular leading this resistance; and along with his brothers-in-arms in the military, they stymied European efforts to destroy their state fully. He immediately became the figurehead savior of the state, and eventually became a heck of a lot more. He continued with the struggle to regain full control of Turkey away from the allied European powers, and eventually succeeded in 1923. That's a red-letter date to remember; 1923 is the foundation of the modern republic of Turkey by Mustafa Kemal.

After Kemal declared the modern republic, he helps get what the Plaid Avenger refers to as the "Meiji Restoration of Turkey" rolling. What am I talking about? From 1924 to 1934, under the guidance of the main man Kemal, Turkey completely transformed itself, not unlike Japan, from a primarily agriculture-based economy into what is now a more modernized and industrialized country. He pulled it completely into the modern framework of things, using the European model. Politically, he revamped the entire government. The Ottoman Empire was a

What's the Deal with Ataturk?
The power, the prestige, and the honor accorded to Mustafa Kemal Ataturk cannot be undervalued in Turkey of 1923, nor of Turkey in 2014. He is the man. He is the main man. He is the George Washington of Turkey, the father of the country, a hero, a savior, a military and political genius. He transformed an entire country from epic fail to epic win in impressively short order. That's one of the reasons he was beloved. He was then, and he is today. He's the father of Turkey in the sense that he helped hold it together at its very formation, but even more so because he put it on the path of modernization.

monarchy until that government was disbanded after World War I. Mustafa reinvented it, not even as a constitutional monarchy, but as a democratic republic. Period. He was elected President in short order and set about revamping the country.

Under his influence, Turkey established a parliamentary system and laws modeled after those of Europe. The Turks started building roads, opened schools introducing Western-style education across the entire country, and modernized their architecture. They built modern energy grids, telephone lines, communications. You name it, they did it. In ten short years, they really caught up. Now is Turkey a completely modernized and wealthy country like the rest of Europe? No, but it's pretty far down that path of modernization and wealth due to Mustafa and the movements that happened during his leadership almost a century ago.

Kemal embraced the West even in style. What a snappy dresser!

Perhaps what sets Turkey apart from the rest of the Middle East the most is that Kemal made it a staunchly **secular** state. Staunchly! Kemal said, *"In this country, we're of Turkic ethnicity, we are of Islamic religion, but we will not be a religious state."* I can't stress this enough. There is a separation of church and state at the outset. The government is the government, religion is religion, and the two will never mix. The United States may also have the separation of church and state, but it's more staunchly defended in Turkey than it is in the United States.

Kemal went so far as to outlaw religious dress in public buildings, a law that still holds today. In fact, there was a court case in Turkey in 2008 in which an Islamic woman wanted to wear her headscarf. In Turkey, they said, "Wear headscarves! Wear a full burqa if you want! You're just not walking into a government building or university with it on. Period." You cannot wear religious garb into the Grand National Assembly, the Turkish equivalent of Congress. If you have a cross on a chain around your neck, that's got to come off as well. That is the ultimate separation of church and state. We're going to see how that has propelled Turkey along its path and made it extremely different from the Middle Eastern countries in its neighborhood as this story progresses. That is a huge deal. The Plaid Avenger can't stress it enough.

TODAY'S TURKEY

While Turkey was still doing its restoration—and getting along pretty well but not quite as far along as Western European nations—World War II broke out. Turkey stayed mostly neutral, but was actually chomping at the bit to fight: "Yeah, we want to get in! The Germans screwed us over during World War I, so we want to come help you beat them this time!" However, the Allies, including the United States, thought it would be more problematic than helpful for Turkey to join the war. The US, UK, and France said, "Okay, thanks, but just hold on, we don't really need you right now. Just sit tight and don't side up with the Germans. We'll make sure nothing bad happens to you." Most of the war played out like this, but

in February 1945, when it was evident that the Germans had lost World War II, the Turks said, "We declare war on Germany too!" In a funny historical circumstance, they showed up at the victory party after the Germans were defeated. Since Turkey declared itself an ally, on the side of the good and righteous, they got to sit at the peace negotiating table and thus started to become a playa' in the world powers game.

Turkey is a proud NATO member!

They also joined the UN as a founding member in 1945. Something even more interesting is that in 1952, Turkey joined NATO, the *North Atlantic* Treaty Organization. We all know how much Atlantic territory Turkey has, but if you had to pick one single event that really made Turkey different from all its neighbors, joining NATO was it. In one fell swoop, they embedded themselves firmly in the Western camp. From this point forward, on the books, Turkey is a US ally. This becomes of particular consequence as we get into current events because there are crucial NATO bases in Turkey still today.

Why is that important, Plaid Avenger? A lot of US operations into Iraq and Central Asia were made possible by bases in Turkey. When the United States fought the first Gulf War against Saddam Hussein, most of the planes carrying the bombs took off from Turkey. Joining NATO is the critical juncture, as I see it. Turkey said, "We're throwing in our lot, and we're throwing it in with the West. Yes, we're Islamic like our Eastern neighbors, but politically, we're more like our Western neighbors."

MAJOR MILITARY

Being a NATO member requires that you have a strong and updated military in order to be a true member of the club, and Turkey does fulfill its duties in every respect. But NATO is not the only reason that the Turks have a strong military tradition, as well as having strong nationalistic pride of their professional soldiers. There's so much more, and it still plays in today's events.

Don't mess!

During Kemal's reign, he knew the only way to hold the state together and keep it from being dominated by the Europeans was to have an effective military deterrent. He therefore immediately modernized his military up to world standards. In addition, Kemal had other motivations for a strong centralized force: to ensure the separation of church and state in his staunchly secular new republic. Maintaining the secular state is one of the primary roles of the Turkish military, and one that has seen the most action.

In 1960, 1971, and 1980, the Turkish military staged coups and took over the government. From a Western perspective, we'd say, "Ooh, that's not very democratic. That doesn't look good." Most people would consider that a sign of an up-and-coming fascist military dictatorship. In Turkey's case, those people would be wrong. That is why Turkey is incredibly unique. The military has been strong since Mustafa "The Main Man" Kemal Ataturk, and it is seen as the staunch protector of secularism in Turkey. Every time there has been a military coup in Turkey, it has either been because the government has been excessively corrupt, or there's been an impending threat that religious folks are going to take over the government. In both cases, the military interjected and said, "Nope, our allegiance is to the state first." Therefore, you can't even really consider it a military dictatorship because there is no singular dictator in any of these circumstances.

The military said, "Our first role is to protect the state and the state's constitution, and that says we're going to be secular, and that's the way it is." In each of these coups, the military came in, wiped the slate clean, and said, "Okay, do it over." Once new elections are held, the military then get back out of the way. I'm telling you this story because the military in Turkey is still seen by the Turkish people in a largely positive light. Turkey's pro-military stance is one of the sticking point for Turkey's membership into the EU. The EU feels that Turkey's military is too strong, a little too excessive for European taste, which brings us to the next current conundrum for our Turkish friends.

EU IRRITATIONS

A big issue today is the attempted EU entry, which started over fifty years ago, but in real earnest for the last twenty years. In 1987, Turkey started seriously talking about entering the EU. They are a UN member, NATO member, right next door to Europe, and closing in on fully developed economic status. Twenty years ago, Turkey said, "Hey, the European Union looks pretty cool. We'd like to be a part of that. We're your NATO buddy; we've helped fight the Cold War with you. Let us in." This brings up a very important point. They were a US and European ally all during WWII and the Cold War. Turkey very rightfully says, "Hey, we're your boy; we're your ally. Haven't we been helping out for fifty years? Let us in the EU!" However, there has always been a little bit of foot dragging by Western European countries to allow that to happen. Like a foot stuck in the mud that hasn't moved for decades.

First, the EU told Turkey that their economy wasn't liberalized enough, so Turkey opened it up and diversified. Then Turkey was accused of human rights abuses against the Kurds, so they passed laws against discrimination and calmed the Kurd situation down for a decade. Then the EU said that their military was too strong, and that was problematic for a true democracy, so the Turks divested the military of some of its constitutional power. Now the EU is saying talks are stalled until Turkey allows unfettered port access to the Turkic parts of Cyprus, never mind that the Greek parts of Cyprus do not have to reciprocate this situation. Geez! Is this sounding like a runaround or what?

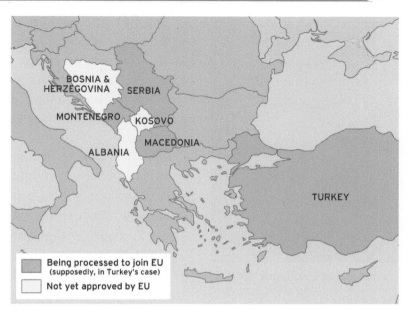

Turkey in the EU? Ah . . . promises, promises . . .

What's the real deal with this EU-inspired nonsense? I'm afraid it's outright cultural racism, for lack of a better phrase. A lot of Europeans are simply not comfortable having an Islamic member of their Christian country club. Remember back in the Western Europe chapter when I told you about the cultural friction developing in Europe? This is apparently part of the same story.

Author as Interpreter:

What the EU Says:
"We're kind of worried that Turkey's not rich enough, not democratic enough, that they have issues persecuting their Kurdish minority, that they don't have enough human rights, that they have some human rights abuses."

What the EU Really Means:
"We're really worried about you guys being Muslims!" The Western European take on this, in the Plaid Avenger's opinion, is that they are worried about admitting an Islamic country. Turkey has a secular government, but it's nearly 100 percent Islamic, and all Westernized states are scared of the .001% of Muslims in the world that are radicalized. They are worried about people moving freely between an Islamic country and the rest of Europe: "If we let Turkey in the EU, then that means that anybody who gets into Turkey, especially radical extremist Muslims that want to blow up European stuff, can come from Turkey and get to anywhere they want in the EU." Valid point. I guess.

What Turkey Says:
"We're working on all that; we've made great strides. We have settled a lot of disputes with the Kurds. We've given them autonomy and given them some rights. We're doing what we can, and yes, we are staunchly secular and we're going to remain that way. We're proud of our military and we're going to try and keep it strong, but we're willing to concede."

What Turkey Really Means:
"Get off your high horse and let us into the Union. It's not the Christian Union, so quit jerking us around. We've been here helping you guys out as a staunch NATO member for fifty years. Don't hold our country not being Christian against us. Please let us into the CU, I mean the EU." It's largely seen by Turkey as a Christian-Muslim issue, so they know the real deal. I agree with the Turks on this one.

Nick was all for a Turkish Club Med.

The Turkish population is starting to get fed up with the entire mess. Public opinion used to be fairly high supporting their EU membership bid, but it has dropped steadily for several years now. Maybe 75% of the Turks were all for EU entry a decade ago, but it is more like less than 25% today. I can't say that I blame them, either. Turkey sat on the sidelines and watched as state after state in what used to be Iron Curtain territory get accepted into the EU, while the Turks continued to get put off, and insulted.

Most recently, former President Nicolas Sarkozy of France suggested that Turkey should become the foundation cornerstone of a **Mediterranean Union**, comprised of states from North Africa, southern Europe, and the Middle East. In other words, Turkey should just start a different club since they are not going to get into the EU club. I think you can probably figure out how that idea was received in Turkey. But honestly, given that the EU has been stagnant economically for years now, maybe the MU is not such a bad idea after all?

KURDISH KONFLICT

To continue on with our current events in Turkey, I've just referred to one of the previous sticking points for EU entry: human rights, particularly for folks in eastern Turkey. Who are we talking about here? The Kurds. The Kurds are not Turks; the Kurds are an ethnic group of people that we've referenced before in the Middle East chapter who have no state. They are a stateless nation of people. They are a people who have been *promised* a state several different times in European and world history, but they never got it. Their ethno-linguitic group is located largely in the mountains of eastern Turkey, in Iran, in northern Iraq, and in Syria. You may have heard of these guys because the US actually had a no-fly zone in northern Iraq to protect the Kurds from Saddam Hussein, who actively mustard gassed them on several occasions.

Turkey's track record with the Kurds is somewhat sketchy, and here's where I'll have to upset some of my Turkish friends in order to give you the full picture. While they are a modern state and there's a lot going on that's really good, the long track record with the Kurds is somewhat negative in that they have not been given equal rights. There's been an active cam-

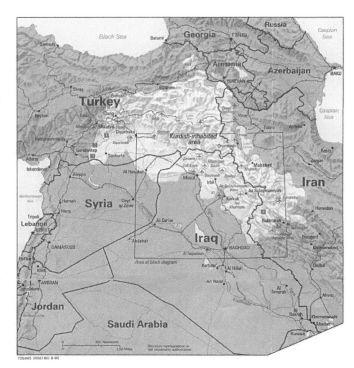

Kurdistan: a nation without a state.

paign of discrimination—it may sound a little strong but it's not too far off—over the last several decades. There have been so many human rights violations that a countermovement has arisen within **Kurdistan**, as the Kurdish areas are sometimes known. The **Kurdistan Workers' Party**, or **PKK**, is a radical group that has spearheaded this countermovement, unfortunately by using violence.

This is an extremist group that is actually considered a terrorist organization by Turkey, the US and the EU. They have undertaken a campaign since the 1970s of blowing up Turkish stuff and Turkish people in order to fight for their rights, with an ultimate goal of gaining independent territory and national autonomy. This has been a main sticking point for Turkey, as I've already referred to earlier, in that the European Union says, "You guys are picking on those

Kurds too much. That's not good." At the same time, Turkey says, "Hey, wait a minute! These guys are terrorists, just like you guys are dealing with terrorists!"

This was a big stain on the Turks' reputation for 50 years, but it actually has settled out quite a bit in the last decade. The Turkish military captured Abdullah Öcalan, the head of the PKK, in 1999 and made plans for his execution, but in an obvious show that they were trying to be conciliatory to the EU, they said, "Okay, we won't execute him. We'll keep him alive in jail and, even though he's a terrorist, we're not going to kill him." He's still in a Turkish jail, and if you know anything about Turkish prisons, you understand that Öcalan is probably begging to be put to death.

Since Öcalan's arrest, the conflict died down, the number of terrorist attacks diminished, and the Turkish parliament passed lots of laws to try to equalize human rights and give these guys a little more power. Things in Turkey had been getting much better for the Kurds, but then the American invasion of Iraq stirred the pot again. The PKK regrouped in northern Iraq and has carried out several attacks on Turkish soil in the last several years. This led to renewed trouble for everyone, and remains a persistence problem even today.

Kurdish refugees.

Remember I told you that Turkey has a strong and proud military? Well, the Turks were none too happy about these PKK attacks, and decided to act. Much to the displeasure of the United States, Turkey conducted several bombing raids and mobilized ground troops to go into Iraq and destroy the PKK forces. The US is not happy, mostly because the northern part of Iraq is the only peaceful part of the country, and no one wants to see more people with guns getting into this mess. But those Turks are not sitting still for the fight, and future movements are probably inevitable. Keep an eye on that one, my friends.

TURKS TRANSITIONED

Another big reason that Turkey deserves to be teased out as an independent region from its Middle Eastern and Eastern European neighbors is the situation of its people and economy.

As you can see from the population pyramid here, Turkey, even just thirty years ago, was considered "developing" because of its explosive population. But look how much has changed since then. Their pyramid has basically rounded out, tapered off, and as you can project ahead by 2025, will become completely stable and even slightly shrinking. It would be exactly what we would consider a fully developed nation, demographically. Its population total is pretty stabilized, but currently maintains slight growth every year, which is the gold standard for the labor force.

And the cash? Economically speaking, they're not the richest place in the world. However, they're doing pretty well, and are a top 20 global economy at this point. A G-20 member, my friends! They have a diversified economy—10 percent of an agricultural sector, and 30 or 40 percent of what we'd call an industrialized sector, and almost 60 percent service sector

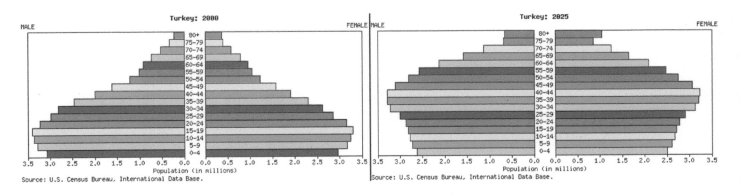

. . . and that's the economic sector structure of a rich, developed country! Or close to it! It's yet another example that they are quite different from all the places around them, just for their modern situation if nothing else. And while they may never get into the EU, at this point they are performing better than every single EU country, with an average 5–8% growth rate every year. Suck it, EU! With no reservations, I say that from just about any measure you want to examine, Turkey will certainly be considered a fully developed country within the next decade or two. The are already a regional power player, and possible contender to be invited to join the BRIC club. Remember that from chapter 6?

Let's end this modern assessment with the strife over secularism. While you now understand that the secular nature of Turkey is what sets it apart from its Middle East neighbors, I should warn you that this status is still heavily debated in their society. The country is almost exactly split on the extreme nature of their secularism. The more western, urban populations favor the staunch separation of Islam from politics, but the rural populations are much more conservative in thought and perhaps want to change things to incorporate religion into the state a little more.

Turkish President Abdullah Gul: perhaps not so secular.

You know what? It's about exactly how the US is right now—think about it. Red states with big rural populations vote Republican, while blue states with big urban centers generally vote Democrat, an almost even split in the last several elections. Republicans would vote for school prayer, Democrats for getting rid of a nativity scene in a government building. That about exactly outlines the Turkish situation as well, only with Islam instead of Christianity. What a great analogy; am I good or what?

Back to Turkey: the current president and prime minister are from a political party that is more conservative and religious (The Justice and Development Party, or AK Party for short.). Many people from the

Prime Minister Recep Erdoğan also accused of not being secular enough.

other big political party are staunch secularists, and distrust the President greatly, fearing that his party is changing laws and possibly amending the constitution to allow for more religious stuff in the government. Some opposition leaders have recently called for the president and prime minister to be fired and their entire political party to be banned on constitutional grounds that they are threatening the republic's very existence. As absurd as that sounds, the case has been cleared to be heard by the Supreme Court. This is serious business, serious enough for the head generals to issue some veiled threats to government leaders to check themselves before they wreck themselves, as a house cleaning may be in order if the pendulum starts to swing too far away from the secular roots of the nation. **2014 Update:** Flash point reached! Check the HOTSPOTS in Chapter 24 for full details, but know this for now: Erdogan and the AK have seen massive protests engulf the country for months, and he is starting to crack down hard, further dividing the society on the very issues described above! The Turkey is really in the oven now, taking full heat!

FINISH UP YOUR PLATE OF TURKEY

If Turkey is ever admitted to the EU, which still seems possible at this point, their economy will grow. They will become even more incorporated as a member of Europe. They will perhaps become even more incorporated as a Western ally in this new war on terrorism in the Middle East. They will be distancing themselves from all of the radical elements that people in the West are frightened of in the Middle East, which will be to their great benefit. The Plaid Avenger is not making fun of anybody in the Middle East. I'm just saying what's going to happen to Turkey if they get into the EU.

If they *don't* get into the EU, things are going to swing the exact opposite way, and that's what's so fascinating about Turkey's position in the world today. They really are in the catbird seat. If they go the EU route, they will become more Western, there's no doubt. If the EU turns their back on Turkey, Turkey is independent enough and economically strong enough, and it's certainly militarily strong and competent enough that it will say, "Fine. The heck with the CU (Christian Union). We'll C U later!" They will turn right around and become a leader within the Middle East. Or the world in general.

That does not mean Turkey's government will become an Islamic Republic, but it will turn around and become a regional leader in the opposite direction. If it turns its back on Europe and embraces the Middle East, it would become, instantaneously, one of the wealthiest countries in the region. It would easily be one of the most diversified economic countries, and probably the one with the most ties to different parts of the world—and one that is not reliant on a single commodity for its wealth, like so many countries nearby are dependent on oil. That gives Turkey a lot more power. **The bridge** of Turkey is actually strengthening. The Turks are gaining more power in today's world because they are becoming what the Plaid Avenger calls a *pivotal historic player.*

Unlike most states in this neighborhood, national pride abounds.

THE PLAYER

How are they a player? Turkey is a player because they've got two different directions they can go. Just like we've talked about with Russia already—they can go China, or they can go Europe. Turkey can go Europe, or Turkey can go Middle East. And they already have aspirations to link up with Central Asian states too, as they share common Turk roots. A possible TU? A Turkic Union? Or a Mediterranean Union? Or both? It's really up to them.

But they are still firmly entrenched as part of Team West. They're a strong NATO member, but they actually refused to help the US movement in Iraq. This actually came as a great surprise to the United States. The US said, "Hey, NATO member since 1952! Turkey, you're our boy! Sweet! We're going to launch our Iraq invasion from your soil!" Turkey said, "Not gonna happen. We don't like Saddam either, but those are Islamic people about to get bombed. We don't support this war. You're not going to bomb people from our soil." What was the direct result of this? The challenge of, dare I say, *smacking down* the world's military and economic superpower's request? What were the repercussions? What happened? Jack nothing is what happened.

Turkey: the other, other, other white meat.

Turkey is *that* strong now; it is a pivotal historic player on the world stage, and we know that because nothing happened to them. The US even threatened to cut off international aid, saying, "You're going to do what we want or we're not going to give you money anymore!" and Turkey said, "Fine. Don't." The threat of pulling aid was all smoke and mirrors; the US was powerless and totally knew it. In fact, the exact same scenario played out again when Turkey sent troops into northern Iraq to get the PKK in 2008. The US begged them not to do it, but Turkey went ahead anyway; the US was forced to support their old ally, even though they weren't happy about it.

Evidence of Turkey's increasing independence and regional leadership abilities continue to grow here in 2014. Not only has Turkish public opinion about EU membership waned, but increasingly they could care less, and moves have been made to secure economic relationships with more Middle Eastern AND Central Asian countries, and the idea of a Mediterranean Union is also starting to look more promising. Why? Well, mostly because the Western European economies are all in the toilet right now thanks to bad management and the recession; many think it's a one-two punch that Europe is not likely to recover from as fast as the rest of the world will. Why should Turkey throw in their lot with a geriatric basket-case at this point? A lot of Turks think the EU option is not smelling so sweet.

Erdŏgan: in a powerful position.

Politically, Turkey is increasingly shaping up to be a major player in Middle Eastern affairs, perhaps the only party with enough street cred to pull it off. Turkey is stable, a democracy, Muslim, moderate, and part of the neighborhood . . . but with no inherent conflicts of interest in the Middle East messes. That makes them quite unique, and uniquely positioned to help. Under Erdogan's leadership, Turkey has now become the most assertive country to try and stop the violence in Syria, by condemning the Syrian regime and offering safe haven for fleeing citizens. On top of that, Erdogan is seen as something of a hero during this time of the 'Arab Spring' as many Middle Easterners view him as an assertive, nationalistic, popularly elected leader . . . which is what all the Egyptians, Tunisians, and Syrians want for their own countries too!

Add to that a strong sense of nationalistic pride, a strong and independent military, a strong and functioning democracy, and strong political leaders, and you have one heck of a world region. We'll have to watch them quite closely to see in which other regions they project their power into. It could be one, or it could be several, because the Turkish position continues to strengthen, just as all of it's neighbors are shrinking. I consider Turkey already to be a second-tier world power; that is, a major regional power on the world stage, just underneath the big boys like the US, China, et al.

TURKEY RUNDOWN & RESOURCES

 ## BIG PLUSES

→ 100% Muslim, secular, fully functioning democracy: the only one of its kind

→ NATO member and huge political/military ally of the US

→ Strong and modern military in its own right

→ Diversified economy that is growing

→ Slow-growing and therefore stable population

→ Modernized state that is fast becoming a most significant regional power; I predict it will soon join the BRICS. It is already in the G-20.

→ Muslim culture AND pro-Western relationships make Turkey the most awesome mediator of conflict for Middle Eastern affairs; trusted ally of the US, Israel, Syria, Palestine, Iran . . . . heck, that's everyone!

 ## BIG PROBLEMS

→ Due to its "Muslim-ness" and current cultural/economic attitude of Europe they are not likely to get into the EU

→ While gains have been made in the last decade, the Kurdish situation and problems with the PKK still remain and could get hot again

→ The secular vs. religious "crisis of character" that the Turkish state is dealing with is far from over, is very divisive to the general population, and may result in political gridlock of epic proportions at some point; I sure hope it doesn't come to that, but it might. 2014 Update: totally called this one years ago, as Erdogan and his administration are now battling non-stop protest from the secularists, which is widening the gulf between the two groups in Turkey.

DEAD DUDES OF NOTE:

Mustafa Keaml aka Atatürk: "The George Washington of Turkey." Turkish army officer, revolutionary statesman, writer, and founder of the Republic of Turkey as well as its first president. Staved off attacks from the Allied powers and held together the core of the Turkish homeland after the defeat and collapse of the Ottoman Empire in WWI. Once in office, embarked upon a program of political, economic, and cultural reforms and entrenched Turkey as a modern, secular state.

LIVE LEADERS YOU SHOULD KNOW:

Recep Erdŏgan: Popular Prime Minister of Turkey and chairman of the Justice and Development Party (AK Party). Despite internal friction due to his association with the religious-based AK Party, he has overseen an era of economic prosperity and democratization, and Turkey's rise as a serious regional power and trusted mediator of Middle Eastern conflicts.

Abdullah Gül: President of Turkey and first former-Islamist party President in the modern history of Turkey. His AK political party is accused of being too religiously inspired by staunch secularists. He mostly deals with domestic affairs while Erdogan is really the head cheese in charge.

ADDENDUM: THE COMPETITION FOR CONTROL

Now for the last, and most complicated, component of understanding the Middle East mess in the modern context. Many, if not all, of the conflicts, trials, and tribulations that have been described so far in this chapter can be related to a much bigger competition that is occurring across this region, and that is a competition among big powers for influence, if not outright control, of the destiny of this troubled chunk of land.

This is a total Plaid Avenger assessment, and would be labeled by experts as completely unorthodox and overly-simplistic in its approach, which means it's probably exactly right! I admit right from the get-go that this approach does not do a thorough analysis of all the details and all of the complex inter-relationships that exist between these groups; if you want to understand them better, then by all means use this text as a launching point for further research into this tangled web of geopolitical skullduggery. Wow! How often do you see the word skullduggery in a textbook? But I digress.

Here is what you need to know to know your Middle Eastern mess: there are many ethnic and religious and political differences among people in this region which cause fractures and frictions, and almost all of these things are totally unrelated to input from the outside world. In other words, many players are vying for influence and control from within the Middle East itself. Moves from outside forces like the US, the EU, Russia, or even al-Qaeda simply accentuate, benefit or stymie the game being played internally. You dig what I'm saying so far?

Before you go any further, go back to the addendum opening page and examine the map for major ethnic distinctions within the Middle East, as they are the primary basis for the Team divisions.

The following is a list of distinct teams that the Plaid Avenger feels are currently in this competition. Again, some of these teams are vying for influence, some are vying to control resources, some are vying to control major events, some are trying to outright control territory, and some are just trying to disrupt the other teams from winning. I will point out what makes each team distinct from the others, their motives, who they like, and who they hate. I will start with the easiest ones and continue to get further complicated as we go. You ready for this reality? Then let's do this!

TEAM TURK

We start with the easiest and least complicated: the Turks. First and foremost, this team's distinctive characteristic is its ethnicity. Turkey is full of Turks. Go figure. But Turks are not Arabs, nor Persians, nor Middle Easterners, whatever that means. As you will find out soon enough in the following pages, I do not count Turkey as a part of the Middle East region at all. In a sense, it's a bit of an outside entity itself.

Having said that, Turkey is in a very unique position to influence affairs in the Middle East in multiple ways. First, unlike any other outside entities, Turkey is overwhelmingly Muslim, which makes it more culturally sensitive and sympathetic to all its Muslim Middle Eastern neighbors. But that is about where the similarities end.

Unlike its regional neighbors to the south, Turkey is a staunchly secular society that maintains a strict separation between Islam and the state. Turkey is also a well-established democracy, a founding member of the UN, one of the earliest inductees into NATO, a possible future member of the EU, and has a well-established and strong, modern military. All those factors together make it much more Western in outlook, and indeed,

Turks takin' it to the streets!

Turkey is an ingrained part of most Western institutions, especially that NATO one; all operations in the Middle East utilize the Turkish corridor as an operating platform. Most NATO and UN operations in the region would be screwed without Turkish support. That gives Turkey a loud voice on all Middle Eastern matters.

That's why Team Turk is in a strong position to influence some Middle Eastern policies, especially ones that are advantageous to itself. Since the Turks are US/NATO allies, they typically work toward the goals agreed upon by those entities—to a point. Turkey openly refused to support the current US invasion of Iraq on the grounds that it could not be party to an illegal invasion of a fellow Muslim state. Recently, Turkey actually invaded Iraq itself to go after Kurdish terrorists, much to the chagrin of the US. That is important to note: Turkey has a large population of Kurds in its own territory, and thus has a vested interest in any/all Kurdish developments in the region. Stability of Iran and Iraq play prominently into the Kurd situation, and thus the Turkish response.

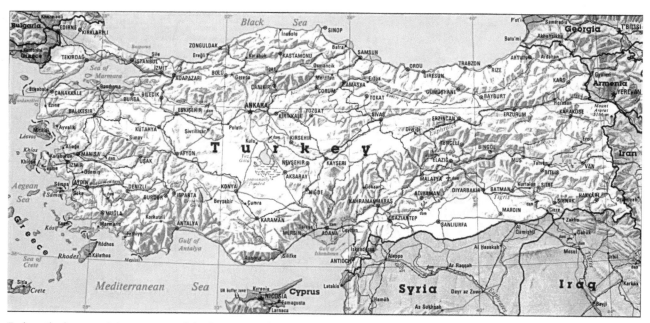

Turkey: the last remaining remnant of the Ottoman Turk Empire.

Other important considerations: Turkey controls the all-important headwaters to the Tigris and Euphrates river systems, which supply the critical resource to many countries to the south, meaning they have a very effective control measure should they ever decide to use it. In addition, Turkey has quietly and methodically been increasing its role as a mediator and host between sparring Middle Eastern parties. Turkey has hosted talks between Team West and Iran, and is really the only true mediating force between these two embittered enemies. In addition, Turkey has led the charge to condemn the violence in Syria, has called for the ousting of the Assad regime, has offered its own country as a safe haven for fleeing Syrian refugees, and hosted international coalitions of forces to deal with the whole Syrian mess. And finally, Turkey has been held up as a model for other aspiring democracies in the Middle East to mimic, as the Arab Spring sees Arab state after Arab state throw out their dictatorial rulers in favor of a more representative system. As a stable, successful, Muslim state in the neighborhood, Turkey is uniquely fit as such a role model.

TEAM HEBREW

Onto the next team, and what a doozy this one has become! Team Hebrew represents one state and one state only, the state that may be a central issue for tons of regional friction: Israel. Now I know what you may be thinking already: "What? How can this be the next easiest team to comprehend when it is the source of so much animosity and conflict?" I'm glad you asked. While many folks believe the Israeli-Palestinian issue may be the source of all conflict across this region, Team Hebrew's motives and motivations are actually quite easy to identify and understand.

Before you read any further, you must understand the basics of how this state came to be and what that has meant for the modern world. Check out the "What's the Deal with Israel" box on the next page.

Okay, now that you are back, we can make some generalizations. All the neighboring Arab states have attacked Israel on multiple occasions ever since its inception, and Israel has pretty much won each of these wars, resulting in increased territory for

The Kosher Club. On unleavened bread.

Israel and control of other areas that are not even part of their state. It also has resulted in the displacement of most of the ethnically Arab Palestinians, who are without a state of their own. Many of these Palestinian refugees are located in the surrounding Arab states. For these reasons and perhaps many others, Team Arab really hates Team Hebrew. Team Persia hates Team Hebrew as well, even though they have never openly engaged each other, militarily or otherwise.

Let's call out some other characteristics of Team Hebrew. Obviously, they are Jewish, and as such, the only non-Muslim actors within this region vying for influence and/or control. Team Hebrew is also a strong, secular, well-established democracy as well as a fully developed country with extremely high standards of living. All of those factors make it quite unique within the Middle East region. Team Hebrew is indeed a fully Western country in every sense of the word when it comes to development, economic and political systems, and even religious and cultural values.

Most importantly, Israel also has high levels of technology, especially military technology, and is an undeclared nuclear power. That is crucial. Team Hebrew is the only entity in the Middle East that has nuclear weapons. They are an undeclared nuclear power because the US and others insist that they do not openly declare this fact, otherwise everyone in the neighborhood will want a nuclear weapon, thus launching a regional arms race. Nobody wants that right now, or ever.

Speaking of the US, it should also be evident to you that Team Hebrew is a staunch ally of the US, the EU, and all western institutions in general, with the possible exception of the UN, since UN members often give Israel the smackdown in that forum. The inverse holds true as well: the US has supported Israel openly and covertly for decades. Israel is regularly the top recipient of US foreign aid, even though they are a rich country in their own right, and have a powerful and well-trained military. The US has regularly shared/sold weapons, weapons technology, and military intelligence with Israel. You can go ahead and make the assumption that this relationship infuriates all the other teams in this regional competition.

What's the Deal with Israel?

Israel is the loaded topic of all loaded topics. There are so many emotions surrounding the formation and history of Israel, it is hard to be completely unbiased about the issue. If you lean too far in either direction, you can be quickly charged with anti-Semitism or, conversely, Zionism. But, by the mercy of Elijah, here is the deal with Israel.

I won't get into Biblical history too much, but will make mention of this: This

BACK IN THE BIBLICAL DAY: CIRCA 1000 BCE

David's Kingdom (c.970 B.C.)
Solomon's Kingdom (c.930 B.C.)

PALESTINE British Mandate (1920-1948)

Jewish settlements, 1947

PALESTINE U.N. Partition Plan (1947)

Jewish state
Arab state

area was the historic homeland of the Jews, "The Promised Land" that Moses brought them to many millennia ago. Fast forward to the Roman Empire, which persecuted the Jews (one in particular you may have heard of: Jesus Christ) and eventually totally kicked the vast majority of them out of the area in the second century ad. Since then, locals have been living there (let's call them Palestinians), hanging out for a couple thousand years.

The modern history of Israel begins around 1900, when Theodor Herzl, among others, began pressing for the creation of a national Jewish state—this movement became known as the Zionist movement. Jews began buying property in Palestine and forming small Jewish communities. In 1917, the British government issued the Balfour Declaration, which endorsed the establishment of a Jewish homeland in Palestine. Shortly after, Palestine became a protectorate of the British government because of dismemberment of Ottoman Empire after WWI (see next chapter).

This caused a surge of Jewish immigration that continued until 1939, when the British government issued the **White Paper**, which limited, but did not stop, Jewish immigration to Palestine. This angered the Jewish community, who viewed the White Paper as a rejection of Zionism, and also pissed off the Arabs in Palestine, who wanted Jewish immigration to stop completely.

After World War II, when the horrors of the Holocaust were revealed, there was a giant surge of sympathy for Jews and for the Zionist movement in general. Simultaneously, Jewish armed forces in Palestine became more aggressively defensive of Jewish territory, attacking both British and Arab forces. The British decided that keeping Palestine as a protectorate was no longer in their interest, so they ceded control to the UN. The UN accelerated the development of a plan

Why would I suggest that Team Hebrew motives are easy to explain? Because they are fairly well defined and there aren't that many of them. Their primary objective? Maintain their territorial integrity. Keep the boundaries of their 67-year-old state intact at all costs. Second objective: have national security—that is, to be secure in their state by not having terrorist attacks or open declarations of war against them. Third objective: get all other neighboring states to recognize their right to exist, which, of course, helps out with the first two objectives.

That's it. That's all. Pretty easy to digest. Now I am not saying its right or wrong, or good or bad, or that it will ever even happen. I'm just saying their objectives are pretty easy to identify and understand. All of Team Hebrew's foreign policies, international alliances, and strategic actions serve to accomplish one or more of those very simple objectives.

Israel's occupation of the West Bank, of the Gaza Strip, and even areas within Syria referred to as the Golan Heights, is maintained to achieve their national security goal. Israeli development of a strong military and nuclear weapons was done to ensure its existence by discouraging any more open attacks across its borders. Israel usually sides up with Team West (particularly the US) when it serves their purposes to achieve those objectives. Usually, but not always. Israel sometimes goes against US foreign policy by negotiating with neighboring states that the US hates in order to achieve the "recognition" goal.

to partition Palestine into separate Jewish and Arab states. This formed the Jewish State of Israel, which was quickly recognized by the United States and eventually by most everyone else, but not any of the surrounding Arab countries. This established the 1947 UN partition borders (see map insert).

Violence erupted soon after the partition. Armies from Egypt, Syria, Jordan, Iraq, and Lebanon all clashed with Israeli forces.

Jordan gained a foothold in the West Bank and Egypt gained a similar foothold in the Gaza Strip. This is important to note because these are still disputed areas today. A ceasefire was finally brokered in 1949, creating a new set of de-facto borders now known as the green line (see map insert).

Israel frequently sparred with its Arab neighbors throughout the 1950s and 1960s. The culmination of this aggression was the 1967 Six-Day War, where Israel, using advanced weaponry from the US, went gangbusters after being attacked by Egypt and Syria, and conquered the West Bank (including East Jerusalem), Gaza Strip, Sinai Peninsula, and Golan Heights (see map insert). Egypt and Syria were enraged about this, and attempted to regain their lost territory in the 1973 Yom Kippur War. This war was more of a stalemate—little territory changed hands—but the Yom Kippur War was significant because the United States sided with Israel, irritating EVERY Arab nation and inciting the 1973 oil embargo.

The last major shift in Israel's borders was peaceful. In 1979, President Jimmy Carter invited Egyptian leader Anwar Sadat and Israeli Prime Minister Menachem Begin to the United States and helped negotiate the Camp David Accords. This treaty returned the Sinai Peninsula to Egypt and granted more autonomy to Palestinians residing in occupied territory. Since 1979, most violence associated with Israel has been within its borders. The Palestinians living within Israeli-controlled territory have garnered significant international support for statehood. Recently, Israel has begun taking steps that seem to suggest that the formation of a sovereign Palestine is imminent. However, Israel is not prepared to cede the amount of territory that the Palestinians believe they deserve, and has even started construction of a wall encircling much of the disputed area in East Jerusalem.

TEAM ARAB

Now we are onto the real cultural anchor of the Middle East, and here's where it starts to get really messy. Team Arab refers to an ethnic group, just like Turks or Persians. However, this ethnic group happens to be the majority within the Middle East as an entire region, and many of the states/groups of this region are overwhelmingly Arab in descent. Maybe you've heard of Saudi **Arab**ia or the **Arab**ian Peninsula or perhaps even Aladdin and other tales of "The **Arab**ian Nights"?

Before the advent of Islam, the Arabic ethnic group was centered in the Arabian peninsula, natch. Although the Arabs were originally primarily a nomadic people with an itinerant lifestyle centered around the use of the camel and an extensive network of desert oases, there were a number of urban centers. These were mostly located in the western portion of the peninsula, al-Hejaz. Cities like Makka (Mecca) and Yathrib were important centers of commerce and trade;

Is that you, Aladdin?

The Arab World

Arab world stretches far and wide, but their Arab League is, so far, shallow and useless.

the former was also a crucial center for the pre-Islamic polytheistic religions that abounded. Muhammad himself was an important merchant in Makka and was well known in the region even before his religious undertakings began.

Within a century of Muhammad's death, the Arabs had expanded well into Sassanid Persia, pushed the **Byzantine** Eastern Empire out of the **Levant**, and were trading blows with Charles Martel in France. The expansion of Islam, except in the instances in which it came into contact with a resident imperial power, was overwhelmingly peaceful.

Local populations of Christians, Jews, and other random monotheistic religions willingly accepted their new rulers, as, especially in the case of the Byzantine Empire, Christian sects were persecuted terribly. Conversion was not enforced, but had concomitant economic benefits, such as the relief from a poll-tax on non-Muslims, but this was not encouraged as it would rob the Arab polities of cold hard cash.

Politically, this "Arab Empire" was not terribly cohesive, especially with regard to the territories in Spain and northwest Africa. Eventually, after a couple of centuries, the core at Baghdad held only nominal sway over anything significantly placed away from it, but the religion itself had spread from France to the Oxus River. From there you, can tie the story back into the end of the Golden Age of Islam, when Genghis Khan totally destroyed Baghdad, thus effectively ending the Arab Empire reign, then onto the rise of the Ottoman Empire, which we have discussed above and will again in the next chapter. Team Arab essentially became just another component of the multi-ethnic empire that consolidated this region from about 1300 to 1923, when the last vestiges of the Ottomans were swept away and the modern map of the Middle East was drawn.

Of course, this new map of the Middle East was not a singular Arab entity—far from it. Of the new political boundaries drawn up after WWI, there were roughly twenty two countries which were predominately Arab in ethnicity. How do I know that? Because in 1945, six of those countries formed an entity known as the Arab League, and membership has jumped to twenty two in the intervening years. This League was formed to ". . . draw closer the relations between member States and co-ordinate collaboration between them, to safeguard their independence and sovereignty, and to consider in a general way the affairs and interests of the Arab countries."

In other words, the Arab states want to work together to achieve common Arab goals. In point of fact, at various points in the last sixty years, there have been several attempts to reunify their states under a pan-Arab banner. However, it has never worked. The **Arab League** itself has to be considered a fairly dismal failure, as the member countries more often fight with each other than actually agree on anything at all. They still get together once a year or so to discuss solutions to common issues like the Palestinian debacle, the Sudanese civil war, or the invasion of Iraq, but they never decide on anything nor do anything about it. I guess they have a nice all-you-can-eat hummus bar that keeps people coming back.

Fellow monarchs discussing oil revenues.

That gets us up to now. Team Arab is still a potent force of influence in the region for various reasons:

1. Many Arab states are excessively rich off of oil production. With wealth comes power.

2. Arab culture is the incubator of Muslim culture. Remember, Islam started in Saudi Arabia, where the two holiest cities of the religion are located. That still counts for something on a planet where there are roughly a billion Muslims. Please note: Arabs are overwhelmingly of the Sunni division of Islam, in contrast to the Shi'a brand which is primarily found in Iran. More on that below.

3. Arab is the most populous ethnic group across the region, including such groups as the Saudis, the Egyptians, the Palestinians, the northern Sudanese, a minority of the Iraqis, the Lebanese, and most of the Algerians, just to name a few.

A few final points of analysis: Team Arab totally hates Team Hebrew, since the Palestinian debacle is now an Arab debacle. That hate has been softening for some Arab leaders, though, as Egypt officially recognized Israel and made peace with them in 1979, and Jordan did it in 1994. Ponder this: Egypt officially recognized Israel in 1979, and Egypt has been the #3 top recipient of US foreign aid every year since 1979. Coincidence? Not.

The bigger issue that you might not know about is this: Team Arab also really hates Team Persia (Iran), and it is the competition between these two teams that has become the hottest current showdown. Team Arab is getting very wary of the growing power and influence of Iran within the region. Dig this: Team Arab is exceptionally worried about the possibility of Iran getting nuclear weapons, perhaps even more worried than Israel or the US! Team Arab will be working with the US and others to make sure that Team Persia does not get such weapons.

Speaking of which, the core of Team Arab lies in Saudi Arabia and Egypt. As fate would have it, both countries are strong allies of the US—well, at least the leaders of the countries are. You should know this fact well: the US generally supports Team Arab for various political (US hates Iran) and economic reasons (they have tons of oil). Got all that? Then let's get to that Iranian team now to see why they are so despised by the Arabs.

ANOTHER PLAID AVENGER INSIDE TIP: Virtually all Arabs are Muslims, but most Muslims in the world are not Arab. Does that make sense? Islam is a worldwide phenomenon now, and places like Indonesia actually have more followers of the faith than any single country in the Middle East. The reason this is confusing for outsiders is because Islam started in Arab lands, and the original and only true Koran is written in Arabic. No matter where you go in the world, from Morocco to Malaysia, Muslims will be reading the Arabic Koran and praying in Arabic tongue, but that don't make them ethnically Arab. Of course, I'm so savvy that I can read Arabic numerals . . .

TEAM PERSIA

Onto the next, and final, of the major teams in our Middle East round-up. For this one, we have to go back even further in time to tell their tale. Further back than the Ottoman Empire, further back than the Arab Empire, and, indeed, even further back than the advent of Islam itself. Team Persia's story goes back several thousand years, but we will just hit the highlights here so that you can understand how very different and distinct this team is from all of its neighbors. Never heard of Persia? In today's world, we call it Iran. While Iran is the core of this Team, there are other players on their side as well. But let's start at the beginning.

Persia can track its origins back to 1500 bce when Aryan tribes—namely, the Medes and the Persians—settled the Iranian Plateau. Power went back and forth between these groups until the Persians came out on top in 558 BCE when Cyrus the Great beat out the competition and then went on to consolidate the peoples of the plateau and expand the empire outward into Lydia and Babylonia and eventually Egypt and beyond, making it the most powerful political entity of the time.

Later, Darius the Great tried and failed to incorporate the Greeks into the Persian fold. Look up the epic Battle of Marathon as to how that went down. Also, watch **The 300** to see an excellent, but not terribly accurate, depiction of this Greek/Persian face-off period—Iran actually protested the release of the film as being racist and inflammatory. Next name of note: Alexander the Great united the Greek city-states and led his famous campaign across

Never should have took on the Spartans . . .

Asia primarily to squash the power of the Persians for good; he was fairly successful in this endeavor, as Darius III was killed (and the empire with him) in 330 bc. While the entity lasted a bit longer off and on, it never regained its past power or glory. Persians have a long, proud history that is still looked back upon with reverence and distinction. Thus, their ire at the negative depiction of their forefathers in **The 300**.

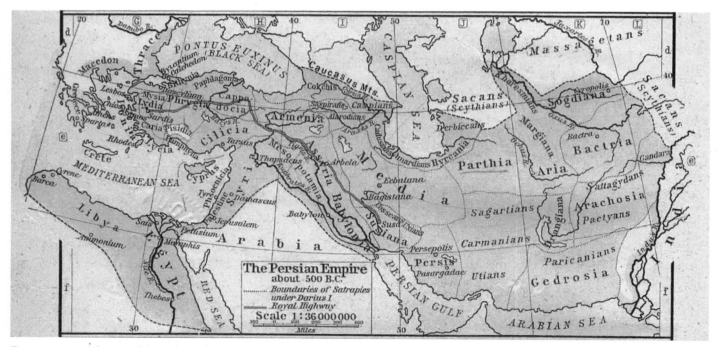

The 300 stopped them, and then Alexander whipped them.

Let's now fast forward. The Arab/Muslim conquest of Persia in the 640s BCE was to forever change their destiny. Not only did this convert the Persians to Islam, but it set the stage for the current Islamic sect differences that are still very much alive today. As referenced earlier in this text, after the death of Muhammad, the religion gets split into two main branches: Sunni and Shi'a.

Now, the liturgical and ritual differences between Sunnis and Shi'as (or Shi'ites) are minimal and mostly unimportant. The branching off of the Shi'a (literally, partisans) of Ali occurred when certain members of the early Islamic community disputed the line of succession following Muhammad's death. Most folks supported one of Muhammad's top men Abu Bakr, but others backed Ali, who was Muhammad's son-in-law. In this contest for power, Ali and then his son Husayn ibn Ali were killed and these early partisans were dispersed throughout the Islamic realms. The death of Husayn ibn Ali particularly became a focal point for the community, especially in modern times, with open manifestations of "passion plays" mourning him and his followers' deaths in the desert near present-day Karbala.

Iran: the core of the Persian power, Shi'a power, and a whole lot more!

What's all that ancient history have to do with modern day Iran? Just this: Iran became the haven for the supporters of Ali, and over time evolved into the core of Shi'a Islam. While Shi'a is the minority sect worldwide (world = 85% Sunni, 15% = Shi'a), it is the majority of followers in modern day Iran (Iran = 90% Shi'a, 10% Sunni) and that makes all the difference for our team competition.

Let's make another jump forward. Persia was picked on and beat up on by Russia and Britain during the European powers' imperial battles for world control. However, it always managed to maintain its independence and was never colonized. By 1906, they instituted a constitutional monarchy and by 1941 a dude named **Reza Shah** was on the throne and increasingly ruling with an autocratic iron fist. It should be noted that this dude came to power with the assistance of the US and the UK who had forced Reza's father to abdicate the throne for his supposed Nazi-sympathizing behavior. Later in 1953, the US and UK coordinated a secret plot responsible for deposing the popular Iranian Prime Minister Dr. **Mohammad Mossadeq** because he had nationalized the Iranian oil industry, much to the chagrin of western countries. That left the bumbling puppet King Reza Shah in charge of Iran via his western allies. It was not to last.

Foreign powers had to get involved in the Middle East to "protect" their oil industries. (Hmmm, what is that old saying about history repeating itself?) This Shah guy was a little too over the top and ticked off too many Iranians, and the Shi'ite soon hits the fan. Time for the Iranian Revolution! Check the box on the next page for details.

Now we arrive at the present. The **Iranian Revolution** has been called the last great revolution of the 20th century, and for good reason. It is a grand experiment of Islamic theocracy with limited democratic elements, but it is, in essence, a blending of church and state into one big ruling system. It's really the first of its kind, and one of the reasons that other Middle Eastern countries are terrified of Iran is the fear that this type of revolution could spread, meaning that Iranian-style theocracy could sweep into Egypt or Jordan or Algeria and displace the systems/leaders that currently have the power. And no one wants to lose the power!

Let's get to the meat and potatoes: Team Persia is doing quite a bit to assert its power and influence across the Middle East, which severely worries Team Arab. These two teams HATE each other, and as I have alluded to earlier, are the biggest

What's the Deal with the Iranian Revolution?

Up until 1979, Shah Mohammad Reza Pahlavi was the constitutional monarch of Iran; he was happy exporting oil, quashing dissent, and resisting Soviet influence. The US and the Shah were the best of friends because he was basically a US puppet. He liked playing with big guns, and the US liked supplying him with just such toys, just to make sure the Soviets wouldn't. Unfortunately for the United States, most people in Iran were very unhappy with the Shah. It seems that Iranians did not like living under a dictatorship.

The Shah and the Ayatollah.

Open protests of the Shah's government began to occur in January 1978, and the Shah ordered the military to shut them down. Hundreds of demonstrators, including students, were killed. This happened several times, and popular support against the Shah really took hold. A populist coalition of liberals, students, leftists, and Islamists escalated the protests to the point of revolution and ousted the Shah in January of 1979. We call this the 1979 **Iranian Revolution** or Islamic Revolution.

The primary political figure during the revolution was the Ayatollah Khomeini, a Muslim cleric who encouraged demonstrations against the US and Israel, whom he declared "enemies of Islam." During the final chaotic days in November of 1979, a group of young radical Muslim students took over the US Embassy and held 52 Americans hostage for 444 days . . . aka **The Iranian Hostage Crisis**.

The Ayatollah Khomeini was the most popular figure in Iran because of its strong Muslim background, but there was also a powerful secular movement. A power sharing system of government was attempted, but eventually Iran became a true Islamic Republic, with all of the power vested in the religious elite led by Ayatollah Khomeini. The United States became even more enraged, especially since it was unable to rescue the hostages, and thus cut all diplomatic ties with Iran.

To this day, there is no official US presence in Iran—no diplomats, no troops, nothing. This is one of the reasons there is currently so much animosity between the United States and Iran. Many US leaders see Iran as a threat to Western stability in general, *especially* if they become a nuclear power. This is a critical issue in today's world, as Iran continues to be in the spotlight by developing nuclear technology—which they say will be used for peaceful purposes, but the West is convinced will be used to build a bomb. The rhetoric is running full steam on both sides right now.

Just know this: the US is all about shutting Iran down, either diplomatically or ultimately by force. The EU wants to shut them down, but only diplomatically. Russia is kind of an ally, and wants to help them develop their nuclear program for energy production, and China wants to do nothing about the Iranian situation because they are all about the sovereignty. Whatever goes down, don't look for it to be dealt with at the UN Security Council, which is already deadlocked before the process even begins.

forces facing off within the region. This hate started in earnest when Saddam Hussein (with backing from other Arab states, the US, Russia, and Europe) invaded Iran, thus starting the decade long **Iran-Iraq War** which ended with about a million casualties. The hate between Arab and Persia exists to this day. Why so much hate? Dig this:

HATE number 1: As just mentioned, Team Persia has no problem exporting the ideals of its revolution to other places. Iran increasingly sees itself as the only true Islamic country of the region, the only real and true purveyor of the faith. It mocks the Arab countries for their morally corrupt leadership as well as for being stooges of US foreign policy. Iran is trying to assume a moral and ethical leadership role on issues like the Palestinian debacle, the existence of Israel, and any foreign intervention into Middle Eastern affairs.

Since Team Persia is the only fully Shi'a state, we have come to define Team Persia in this religious sense as well, but they are not totally alone. I want you to be aware of a new entity that is named the **Shi'a Crescent**. It is based on a projection of Iranian influence and power to other Shi'a groups across the wider region, but please don't mistake this Persia/Arab animosity as a religious war of any sort. The Shi'a Crescent is simply a nice way to highlight areas in which Team Persia has increasing influence. Take a look at the map.

The map shows you the percentages of Shi'a folks in selective states, and we should note a few choice groups that make daily news and effect events. Iran is the core of this crescent, but its effective reach spans from the Persian Gulf countries all the way through Iraq, which has a majority of Shi'a and is why Iran is helping fund and arm Shi'a militant groups in the current Iraq civil war. The Shi'a vein also runs thru Syria, a country whose leadership belongs to a sub-sect of Shi'a, and therefore is malleable to Iranian Shi'a influences. To continue on the road, you get to southern Lebanon, where Shi'as also have a slight majority, and also where a radical Shi'a terrorist group named Hezbollah hangs out. Iran funnels money and guns through Syria to Hezbollah to help them fight against Team Hebrew. You starting to get a sense of the Iranian deal here?

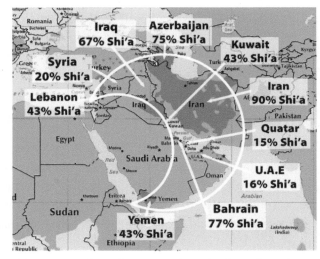

Shi'a Crescent: this ain't no Lucky Charm for the Arabs or the US.

"Israel? I do not recognize. Does not compute."

HATE number 2: Team Persia still openly utterly and completely hates Team Hebrew, and the former, ultra-conservative Iranian President Ahmadinejad many times caused controversy by making statements to the effect that the Israeli regime should be wiped off the face of the map . . . or something to that effect, as the translation itself is a matter of debate. We've already pointed out that Team Arab hates Israel in general, but some Arab leaders have recognized Israel, while still others have been softening their rhetoric about the Jewish state over the years. Meanwhile, Team Persia leaders have held the hardliner attitude of refusal to recognize, but also calling for its destruction. Hardcore! That's why Iran supports Hezbollah, but it also supports Hamas down in the Gaza Strip, just to keep jabbing a sharp stick into the Israel flank at every available opportunity,

This also puts Iran in the position of appearing to be a true helper and defender of the Palestinians, reinforcing the idea that the Arab leaders are just a bunch of losers that can't look out after their own. It makes them appear to be the only country in the region able and willing to stand up not only to Israel, but also to other outside forces.

HATE number 3: Team Persia is developing a nuclear industry. Hold on! I didn't say nuclear weapons! But that is what all the other Teams think is going on. Iran has been thumbing its nose at its neighbors and world powers by its persistence in developing fuel to be able to develop a nuclear power industry, and this is the issue that is really causing the most consternation across the region, and outside the region as well. As of this writing, Iran has not broken any international law in its nuclear pursuit—it is not illegal to produce nuclear energy.

However, virtually all Western nations, including the US and the EU, all Arab nations, and the UN as a whole, have condemned the move harshly and are busy setting up international sanctions to stop the Iranians. For their part, Iran has claimed they have a God-given right for nuclear energy, and nothing will stop them in their quest. Iran's open defiance sets them apart from all the other states in the region, and is seemingly a point of pride with Iranian leadership. Great. Like we needed another problem in this region.

Here's why it is a problem: Iran getting the tools to create a nuclear bomb tips the balance of power in the region radically. No one else has the bomb—except Israel, who won't admit it. While most folks assume that Israel would be the country most frightened of a nuclear-armed Iran, in fact that's probably not quite true. Israel has the bomb as well, making for an effective deterrent to an Iranian attack. Why would Iran nuke Israel, knowing that 100 nukes would be sent back in response? The folks who are really worried are Team Arab, because they don't have a bomb! A nuclear Iran would be stronger than any Arab country, and would contain the most powerful deterrent to prevent any attacks against their Persian motherland, including any attacks potentially from the US or others!

Sum up: Team Persia hates Team Arab. Team Persia hates Team Hebrew. Team Persia hates Team Foreigner. In return, all those teams hate Team Persia right back, and may be working with each other in order to stop any further rise of Persian power. You dig? Arab states might work with the US, and maybe even Israel, to make sure the Persians get sidelined. Interesting stuff. Or should I say, interesting Shi'ite?

TEAM FOREIGNER

Now our job starts getting way easier. The major forces have been outlined in perhaps too much detail, but we still have some other characters in this Middle Eastern drama that are powerful players affecting events. I don't have to spend as much time with these, because they are in the news on an hourly basis and if you don't know these cats yet, then you must be living under a rock. Team Foreigner consists of those states not actually from the region, but somehow find themselves here anyway.

With no reservation, the US is the primary outside entity that has dramatically altered events of the past decades, if not the entire last century. The US has supported kings, monarchs, dictators and dirtbags a'plenty in its quest to influence events and resources in the Middle East that are of strategic interest. Well, I guess all countries do that, but the US is the most powerful at this game. The Saudi royal family, the Shah of Iran, Saddam Hussein, and Hosni Mubarak all quickly spring to mind as entities the US has supported in the past.

It also should be noted that Israel and Egypt are the number 1 and number 2 recipients of US foreign aid, and have been for decades. Simultaneously, US companies have sold tens of billions of dollars worth of arms to Saudi Arabia. But the US is also a staunch ally and weapons

Foreigners love to party in the Middle East!

supplier to Israel. Cha-ching! The weapons business is good! Uncle Sam has now invaded Iraq not once but twice, and the US maintains a troop presence in multiple states of the region; a visible sign of its presence can be seen almost everywhere. The US-led War on Terror is actively engaged across the entire region, both overtly and covertly, to root out rebel/terrorist groups whose ideology conflicts with the US world view. Of course, there are also diplomatic interjections: the US is typically the sponsor of most Israeli/Palestinian peace talks; the US is leading the trade embargo against Iran; the US is a builder of democracy in Iraq, or at least they are trying. The list could go on.

Of course, they are not totally alone. The US does have its "Coalition of the Willing" in Iraq which includes its typical lapdogs, the UK and Australia, along with 49 other countries—well, it had 49 originally, but that number has been dwindling fast as of late. The UN is, of course, intimately involved in trying to help stem the tide of conflict in the region, albeit with not much results. Now there is something called the **Quartet on the Middle East**: a foursome of entities collectively working together to mediate the peace process in the Israeli-Palestinian conflict. The Quartet consists of the US, Russia, the EU, and the UN, and former UK Prime Minister Tony Blair is the group's current Special Envoy. Go get 'em, Tony!

Let's cut to the chase on this Team Foreigner. Team Persia hates 'em, especially the US. In return, Team Foreigner really hates Team Persia, especially the US. Did you know that the US and Iran don't like each other? Well, now you do. The US typically supports and sides up with Team Arab on a whole host of issues, but mostly on the hating Iran one. Unfortunately, the US position with Arab states gets extremely complicated because the US is best friends with Team Hebrew too. Let's rank the Teams in order of how much the US likes them:

Special Envoy Blair. Oh, Tony, you are always special in my book!

1. Team Hebrew comes out on top.

2. Team Turk is close behind. Remember, they are a NATO member!

3. Team Arab is supported by US on most issues, but not all.

4. Team Persia comes dead last. Perhaps that's exactly how the US wants it right now. Dead. Last.

Starting to make sense? I doubt it. Let me add in one last complicating clause of this foreign influence: While many Arab state leaders' support the US, and vice-versa, most people in those Arab states are increasingly angry about US presence and their leaders' support of US policies. This is making the water even hotter for a lot of these folks, as they become increasingly unpopular with their own people. One need only point out that Osama bin Laden's original mission was to drive US forces out of the Saudi holy land. Not accomplishing that, he then turned to start attacking the Saudi government themselves for being corrupt stooges of the US. And the al-Qaeaders aren't alone, as there are other groups within the region expressing their displeasure in any number of violent ways.

ADDITIONAL NON-STATE ACTORS

Now for the last. I won't go into great detail on these groups, but I did want you to be aware that there are other entities that are affecting the flow of events within the region—much smaller entities that are not states nor governments. Therefore, these groups do not behave like states or governments, cannot be coerced the same way as states or governments, and cannot be effectively attacked or punished like states or governments can. That makes them very sticky entities indeed, and they are often thorns in the sides of the bigger Teams we have already talked about. But, they can also be used as tools by the big Teams to attack or influence one another as well. Hmmmm, tricky, tricky, sticky, sticky.

I know that terrorist groups are probably the first things that come to mind, but there is actually a much bigger and more important one to tackle first: the Kurds. The Kurds are an ethnic group (check back to map on opening page again) that are located in the mountainous areas of Iran, Iraq, Turkey and Syria. They are a nation of people without a state, and they very much would like to have one. For over a century, the Kurds have dreamt of, and petitioned big powers to have, a Kurdistan country of their own. They have been used, abused and betrayed by virtually every regional and world power in this quest. The Kurds are seemingly a damned bunch of people, and I mean damned in terms of doomed.

They have been sporadically beaten down by the governments of all the states that they occupy, mostly because no government wants to see a free Kurdistan. That would entail the loss of some of their own territory, and therefore no one is going for it. Team Turkey, Team Persia as well as components of Team Arab, mainly Syria and Iraq, actively work against any Kurd movement toward independence. You may remember that Saddam Hussein used biological weapons on these people multiple times to keep them in their place. You may not have known that Turkey has also been fairly oppressive in keeping the Turkish Kurds down.

I bring up Team Turk in particular because they invaded Iraq back in February 2008, much to the ire of the US, who is trying desperately to keep a lid on the Iraq mess. Why did the Turks invade? To root out a radical pro-independence Kurdish terrorist group named the PKK that has been responsible for multiple attacks within Turkey. Just so you can see how convoluted alliances can become in the Middle East, Team Turk is actually getting help from Team Persia in this effort, while the normally pro-Turk US is trying to stymie the effort. Geez! How confusing!

Even smaller groups of note are political parties, rebel groups, and terrorists groups. Yeah, that's a mouthful, and I don't mean political party or rebel group—I mean political party AND rebel group. Entities like the Shi'a group **Hezbollah** in southern Lebanon, **Hamas** in the Gaza Strip, or various groups across the Middle East referred to as the **Muslim Brotherhood** all fit this description. What do I mean? They are all political parties that in some countries under some circumstances actually put up candidates to run for office in all the countries in which they are located. However, most but not all of them also have militant wings of their party that actively go out and commit violent acts like suicide bombings, rocket launches into populated areas, or open armed warfare against other entities or states.

Why would anybody vote for a political party that does things like that? Mostly because in the areas where they are located, they are seen as actually fighting for the people, as opposed to the ineffective and corrupt governments that are

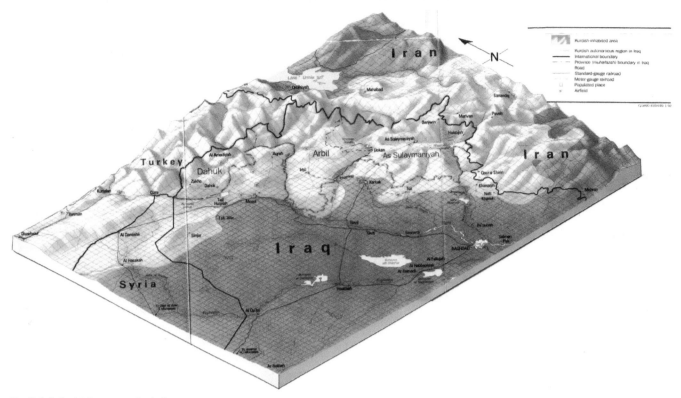

Kurdish folks hiding up in the hills.

in charge. In both Gaza and southern Lebanon, the political party sides of Hamas and Hezbollah build schools, run hospitals and soup kitchens, and generally protect the citizens, which seems like a pretty good thing. That seems like the types of things that the government is supposed to do, but at which they are failing. In such circumstances, the locals vote for those political parties, and why wouldn't they? An example: Hamas actually swept the Palestinian elections a few years ago because the alternative party, **Fatah**, was largely seen as corrupt, ineffective, and lackeys of western powers.

But wait a minute! We have a problem here! These same parties have militant wings which do violent things, and as such, are labeled as terrorist groups by Israel, the US, and the EU. You know what happens to terrorists groups: they get embargoed, stone-walled, and cut down. Western powers don't like these groups and try to shut them out of the political process, causing ever more conflict. Even though Hamas was democratically elected, they were not recognized by the West, and all of the Gaza Strip is still being punished for their support of the group.

Some of these groups are simple to understand, though. Al-Qaeda just is a terrorist group, with no aspirations for

Part of the Non-State Actor Action Faction.

political participation. Hezbollah is close to being in that category as well, as they are mostly a tool of Iran just to cause trouble with Israel. Varying groups named Muslim Brotherhood can go either way: there is an Egyptian Brotherhood, a Jordanian Brotherhood, a Syrian Brotherhood, et al. Many of the Brotherhoods have been banned from political participation in total (Egypt and Syria), mostly because they are proponents of bringing Islam into the political process, meaning that they might want to eventually change the government structure to be more in line with the Iranian model or with Islamic law. None of those Arab governments want that to happen, so they just label them as terrorist groups and ban them from politics, if not make their existence completely illegal altogether. Some of these groups have in turn resorted to the terrorist track because what else are they gonna do after they have been banned from participating? Other individuals from these groups just run as "Independents" on the ballot, but their existence is made no easier once their political beliefs are espoused. Fun times either way.

2014 Update: The Muslim Brotherhood political party candidate in Egypt won the presidential election there in 2013! Since the Arab Spring toppled Hosni Mubarak and brought a representative democracy to the country, the Brotherhood now finds itself as a legitimate power within the system. Oh, but wait! Before the year was done, a military coup deposed that democratically elected dude (his name was Mohammed Morsi), jailed him, and once again re-outlawed the Brotherhood. Same as it ever was, once again.

All these non-state players have the potential to cause varying degrees of trouble for the major powers. It all depends on who they decide to side up with to achieve their own objectives, and it appears that that can change on a daily basis.

CONCLUSION

What you see from this Team breakdown is that we have had a series of empires which controlled the Middle East at different times; first the Persians, then the Arabs/Islamic, and finally the Ottoman Turks. While all of them are now Islamic, each of these three empires has resulted in a distinctly unique strong team—that are radically different from each other, with distinct ethnic character and political affiliations—that is vying for influence across the wider region.

The three main Islamic historic entities still present in today's world are the Persians (now Iran), the Ottoman Turks (now Turkey), and the Arabs/Islamic (now twenty individual Arab states, with Saudi Arabia and Egypt as the core). These main three have been recently been joined by Team Hebrew (Israel), Team Foreigner (the US, EU, et al), and various other non-state actors, like Team Kurd and smaller political party/terrorist groups. In the end, they are all fighting for their own agendas to either control territories within the region, control governments in the region, control resources, or the direction of the region as a whole.

2014 Holy Hot-spots of Activity Update: Well, this whole chapter has been radically impacted in 2011 by the Arab Revolutions across the entire region. The Arab world is in total upheaval right now, with Tunisia, Libya, Yemen, and Egypt (twice) already chucking out their rulers, and Syria in a total Arab civil war scenario that can't last much longer. Oh, and that Osama bin Laden guy got killed too! How will that affect the power struggles in Middle East?

Well, Iran has certainly benefitted from the Arab turmoil, and is pressing on in haste to get its nuclear option going and spread its influence further before the Arabs can regroup. In addition, the US and other outside influences are certainly going to have to re-do all their relationships with the Arab powers as new leaders, possibly of true democracies, come to power with different attitudes than their predecessors about Israel, Iran, and the US itself. And while al-Qaeda has been slightly rattled by the loss of their lascivious leader, don't look for their little death franchise to be that greatly impacted worldwide . . . although we can always remain hopeful that they will crash and burn, into themselves that is.

All in all, it's still too early to call in how all this Middle Eastern revolutionary activity will be playing out even in the short term . . . long-term projections would be a total joke right now.

Whew. That's it. What a freakin' mess. Let's get out of this chapter!

ADDENDUM! IT'S EXTRA! SO HAVE SOME MORE STUFF!

DEAD DUDES OF NOTE:

Mohammad Mosaddegh: Democratically elected Prime Minister of Iran from 1951 to 1953, until he was overthrown in a coup d'état backed by the US CIA. Was an ardent nationalist with a passionate opposition to foreign intervention in Iran. He nationalized Iranian oil, which so infuriated the UK and US that they labeled him as a dangerous commie and got him jacked out of power, which is one of the reason Iranians hate the US and UK.

Mohammad Rezā Shāh Pahlavi: The Shah of Iran from 1941 to his overthrow in 1979; only actually held all the real power from 1953 onward, after the US CIA had chucked out Mosaddegh (see above). The Shah did industri- alize and modernize Iran, but unfortunately he was a typical corrupt and clueless monarch, as well as an US/Western stooge, who increasingly pissed off his people until they had a revolution to throw him out.

Ayatollah Khomeini: Fiery Shi'a orator and holy man, he inspired and led the Iranian Revolution of 1979 that deposed the Shah and created the first Islamic Republic of Iran, of which he then became the Supreme Leader. Has god-like status to millions of Iranians, and a demon-like status to millions more Westerners due to his con- doning the Iranian Hostage Crisis of 1979–1980 and his extrem- ist view of the US as "the Great Satan."

Saddam Hussein: Him again? Yep. In this context, know that Saddam was supported, funded and armed by the US, Western powers, and even other Arab states in the 1980's for his willing- ness to declare war on Iran, the country that Team West and Team Arab started despising after the Iranian Revolution. Saddam was seen as an Arab fist that would smash this Shi'a/Persian theocracy threat. Good call on that one, everybody.

LIVE LEADERS YOU SHOULD KNOW:

TEAM ARAB:

King Abdullah: Monarch/leader of Saudi Arabia, the birthplace and holy center of Islam. Arab, Sunni, Saudi, center-right, conser- vative, massively wealthy, and a huge US ally.

Hosni Mubarak: ex-President of Egypt (a one-party state), a posi- tion he held for 30 years. Arab, Sunni, Egyptian, center-right, conservative, now a prisoner.

TEAM HEBREW:

Benjamin Netanyahu: Current Prime Minister of Israel, and a conservative, center-right, hawk- ish fellow. Don't look for any Palestinian "peace process" to go forward on his watch. Is a US ally, even if relations are strained with the Obama administration right now.

TEAM TURK:

Recep Erdŏgan: The kick-ass Prime Minister of Turkey and chairman of the Justice and Development Party (AK Party). Despite internal friction due to his association with the religious-based AK Party, he has overseen an era of eco- nomic prosperity and democratization, and Turkey's rise as a serious regional power and trusted mediator of Middle Eastern conflicts.

Abdullah Gül: President of Turkey and first former-Islamist party President in the mod- ern history of Turkey. His AK political party is accused of being too religiously inspired by staunch secular- ists. He mostly deals with domestic affairs while Erdogan is really the head cheese in charge.

PLAID CINEMA SELECTION:

Take a Middle Eastern Team Tour via Cinema! In the interest of saving space, I won't describe all these in detail, and many are repeats from the previous chapter, but check out these team lists of films to watch to get a fuller sense of the cultures and backgrounds of each team. Total TV team party!

TEAM ARAB:

The Battle of Algiers (1966)
Gaza Strip (2002)
The Kingdom (2007)
Lawrence of Arabia (1962)
Paradise Now (2005)
The Yacoubian Building (2006)

TEAM PERSIA:

About Elly (2009)
Bashu, the Little Stranger (1986)
Dayereh aka The Circle (2000)
Ekhrajiha (2007)
Marmoulak aka The Lizard (2004)
Persepolis (2007)
300 (2006)
Two Women (1999)

TEAM TURK:

Eskiya (1996)
Günese yolculuk (1999)
Propaganda (1999)
Uzak aka Distant (2002)
Yol (1982)

TEAM HEBREW:

Ajami (2009)
Bab el shams (2004)
Sallah Shabati (1964)
The Ten Commandments (1956)
Walk on Water (2004)
Waltz with Bashir (2008)

TEAM KURD:

Günese yolculuk (1999)
Marooned in Iraq (2002)
A Time for Drunken Horses (2000)
Turtles Can Fly (2004)

TEAM PERSIA:

Ayatollah Khamenei: Religious/Spiritual leader and ultimate wielder of political power in Iran, since it is a theocracy. Persian, Shia, ultra-right, conservative, and totally hates the US. A lot.

Mahmoud Ahmadinejad: Former President of Iran, and great agitator to make the entire world hate his country by continuing to flaunt uranium enrichment. Persian, Shia, ultra-right, conservative, and totally hates the US. A lot. Like a lot, lot.

Hassan Rouhani: Current and much more sane President of Iran. Wants Iran to re-join the world from its insane self-imposed isolation, and perhaps modernize the system a bit, despite being a conservative cleric himself.

KEY TERMS TO KNOW

Arab	Drawn Borders	Political Devolution
Aryan	Farsi	Political Geography
Balkanization	Federal State	Shiite
Band Society	Islam	Souks
Caspian Sea	Kurdistan	State
Centrifugal Political Forces	Medina	Sunni
Centripetal Political Forces	Multi-National State	Transhumance
Chiefdom	Nationalism	Tribal Society
Citizens	Nation-State	Unitary State
Dar al Harb	Natural Borders	Zoroastrianism
Dar al Islam	Nomadism	

FURTHER READING

John Agnew, *Hegemony: The New Shape of Global Power* (Philadelphia: Temple University Press, 2005).

John Agnew, Katharyne Mitchell, and Gerard Toal, eds., *A Companion to Political Geography* (New York: Wiley-Blackwell, 2007).

John Agnew, *Globalization and Sovereignty* (New York: Rowman & Littlefield, 2009).

Saul B. Cohen, *Geopolitics of the World System*, 2nd ed. (New York: Rowman & Littlefield, 2008).

Kevin Cox, *Political Geography: Territory, State, and Society* (New York: Wiley-Blackwell, 2002).

Jason Dittmer, *Popular Culture, Geopolitics and Identity* (New York: Rowman & Littlefield, 2010).

Wilma Dunaway, Colin Flint and Peter Taylor, *Political Geography: World-Economy, Nation-State and Locality*, 5th ed. (New York: Prentice-Hall, 2007).

Francis Fukuyama, *State Building: Governance and World Order in the 21st Century* (Ithaca: Cornell University Press, 2004).

Derek Gregory, *The Colonial Present: Afghanistan, Palestine, Iraq* (New York: Wiley-Blackwell, 2004).

Gerry Kearns, *Geopolitics and Empire: The Legacy of Halford Mackinder* (Oxford: Oxford University Press, 2009).

Baldev Raj Nayar, *The Geopolitics of Globalization: The Consequences for Development* (Oxford: Oxford University Press, 2007).

Edward Said, *Orientalism* (New York: Vintage, 1979).

Joanne Sharp, *Geographies of Postcolonialism* (London: Sage Publications, 2008).

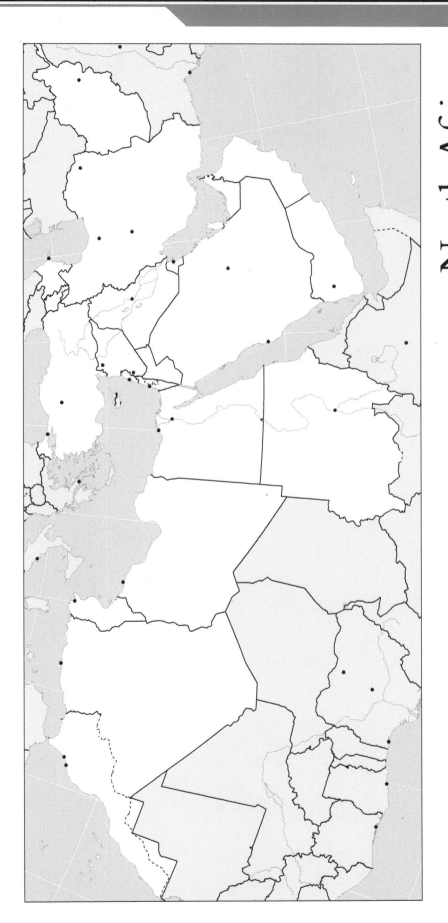

North Africa
Southwest Asia

©2015 BYU Geography
Think Spatial group

0   500   1,000 Miles

Test Map

North Africa
Southwest Asia

Practice Test Map

0   500   1,000 Miles

©2015 BYU Geography
Think Spatial group

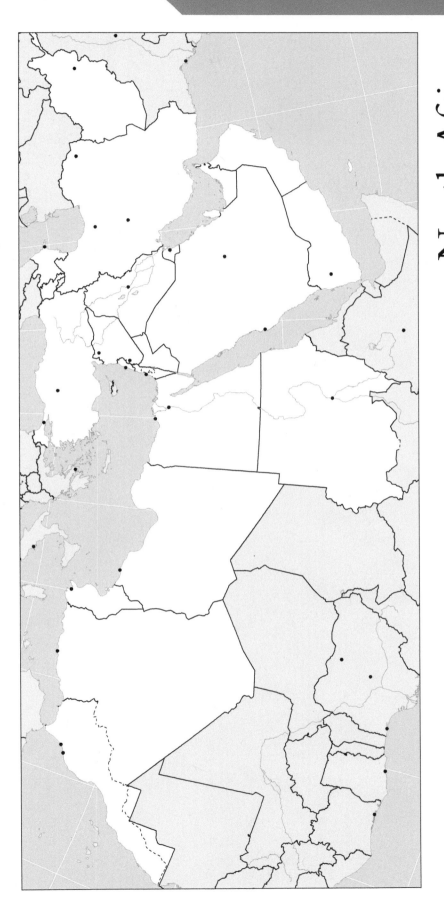

North Africa
Southwest Asia

Cultural Features

0   500   1,000 Miles

©2015 BYU Geography
Think Spatial group

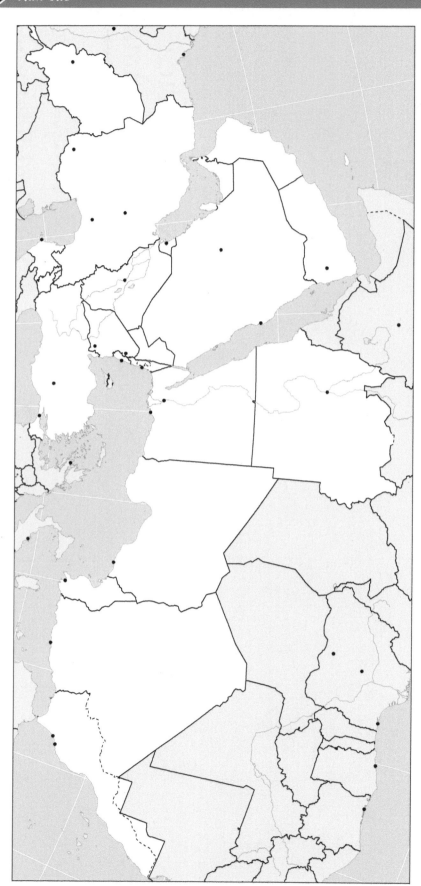

North Africa
Southwest Asia

Physical Features

0   500   1,000 Miles

©2015 BYU Geography
Think Spatial group

Sub-Saharan Africa and Culture

6

Then children were brought to him that he might lay his hands on them and pray. The disciples rebuked the people, but Jesus said, "Let the little children come to me and do not hinder them, for to such belongs the kingdom of heaven." (Matthew 19:13–14)

Then the righteous will answer him, saying, 'Lord, when did we see you hungry and feed you, or thirsty and give you drink? And when did we see you a stranger and welcome you, or naked and clothe you? And when did we see you sick or in prison and visit you?' And the King will answer them, 'Truly, I say to you, as you did it to one of the least of these my brothers, you did it to me.' (Matthew 25:37–40)

CHAPTER OUTLINE:

SOMETIMES it seems we are drowning in wealth. A challenge for the author is simply to remain healthy by not eating too much! When sadness or depression strikes, as it inevitably does, we must immediately give thanks for what we have. Not everyone, however, has difficulty remembering the blessings they receive, and consequently they can easily see the challenges that others around them face. The continent of Africa is rife with problems like AIDS and malnutrition. Many children can become orphans or at least, not get the care they need. But God cares even for the children. When we begin to see the world through the eyes of Him who loves us, we can shake off our minor worries and concerns and be refreshed by serving those He also loves and who need some of our wealth and blessings. Where are the areas of need in Sub-Saharan Africa today?

Africa needs help just like the rest of the world! Its challenges, like all of ours, are primarily spiritual. To reach the continent with hope and to make a real difference, we must notice the uniqueness of the continent. Since the "great desert," or

Sahara as it's called in Arabic, is such a geographic barrier, we are dividing our discussion into two different regions—the area of the north that is predominantly dry and typically Islamic in culture and the southern part of the continent.

Notice the location of the continent of Africa on your globe. In addition to being one of the largest, it is also the most tropical continent on Earth. The equator runs midway through the continent and, as one might expect, this accounts for the warm moist climate of the interior. As one proceeds both north and south, the Hadley cells discussed in the first section begin to reveal the phenomenon of wet-dry climates. These enormous areas of grasslands made famous in so many animal documentaries spread far and wide, but there are deserts as well. The deserts of the Kalahari and the Namib are the result of maritime influences below the equator where the cold waters of the Southern Ocean begin to spread into the Atlantic. It is the Atlantic Ocean that brought European explorers into contact with the Indian Ocean.

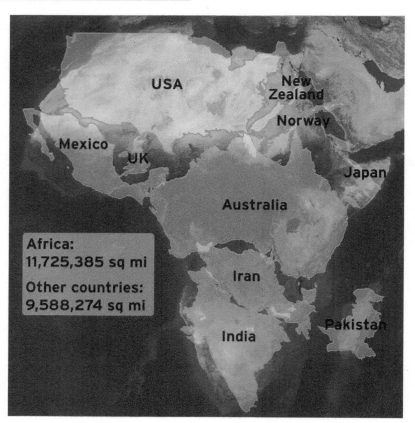

Adapted from *Blue Marble: Next Generation* image produced by Reto Stockli, NASA Earth Observatory (NASA Goddard Space Flight Center)

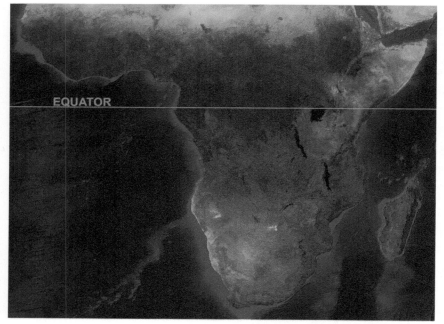

Adapted from *Blue Marble: Next Generation* image produced by Reto Stockli, NASA Earth Observatory (NASA Goddard Space Flight Center)

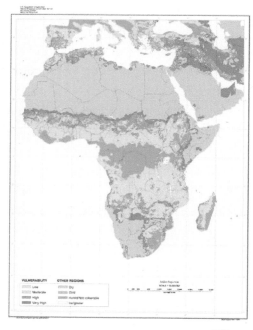

USDA map

SMOOTH AND GROOVE

One of the challenges facing the student of geography is locating the numerous nations in Africa. One of the reasons there are so many small countries, particularly in West Africa, has to do with the tendency of the explorers to trade from the coast. The explanation for this is simple; look at the coastline of Africa. It is one of the smoothest continents in the world. In fact, according to Robert D. Kaplan in the book, *The Revenge of Geography*: "Though Africa is the second largest continent, with an area five times that of Europe, its coastline south of the Sahara is little more than a quarter as long."[1] The additional problem is that the continent possesses few good harbors and very few navigable rivers from the ocean, unlike Europe. The result was that European and Arab traders were forced to build temporary forts and deal with local chieftains. The arrival of the industrial revolution only intensified this hunger for resources and the manipulation of cultures to obtain them. The question then becomes: Did Europe progress technologically faster because of its incredibly complex coastline totally different from the continent of Africa with its smooth coastlines and because of the paucity of ports, navigable rivers and smooth coastlines?

A deterministic perspective, or one that excludes a faith in God's power, would seek a geographical explanation. As Christians we should bear in mind, however, that God is the ultimate Author of the earth and all that is in it, and that occasionally we get glimpses of patterns in His Creation. One interesting suggestion from the famous author Jared Diamond is that a lack of domesticated beasts of burden resulted in lower agricultural yields, which in turn prevented industry from prospering in Africa.[2] Another perspective explaining the challenges facing the continent historically and today is the landform. Essentially Africa is surrounded by escarpments from the coast and in the rift valley—a huge tectonic separation running from the horn of Africa in the northeast to Zambia. This difficulty penetrating the continent is reminiscent of South America and has been instrumental in the shaping of internal movements of various peoples, particularly from the west heading into the south. The European experience in Africa has been particularly important in shaping the political geography of Africa today.

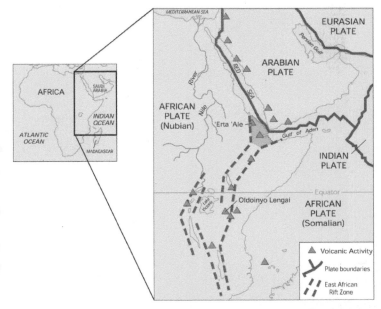

From *This Dynamic Earth: The Story of Plate Tectonics* (online edition) by W. Jacqueline Kious and Robert I. Tilling, prepared by U.S. Geological Survey

Internally, we can see the effect on culture of the Great Rift Valley too. As you may remember from our discussion of North America, most movement of peoples and culture is often latitudinal in direction. When the Bantu-speaking peoples began to move, for example, from the western and central parts of the continent to the east, they encountered this famous rift valley. Note on the map the highlands and giant lakes such as Lake Victoria, Malawi and Tanganika. These lakes in large measure have the characteristic shape such as Lake Baikal in Siberia because they have also been formed by tectonic activity. So, the direction of water and high rifts or chasms tended to funnel travel towards the south where many of the peoples of western Africa eventually emerged. As late as the 17th century the Lundas continued this trend and the result has been a sort of geographic cul-de-sac on the southern end of the African continent! Once again physical geography has shaped the cultures and movements of people upon the earth.[3]

---

[1] Robert D. Kaplan, "The Revenge of Geography" (Random House, 2012), 31.

[2] Diamond, Jared. *Guns, Germs, and Steel: The Fates of Human Societies*, 91.

[3] *The Times Atlas of World History* (United Kingdom: Hammond, 1984).

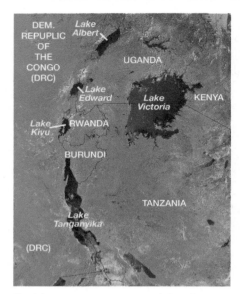

The lakes are representative of the rift caused by tectonic plates.

Jacques Descloitres, MODIS Land Rapid Response Team, NASA/ GSFC and Katie Pritchard

Africa at night. The lights represent coastal-populated areas important for trade like in South America.

NASA/Goddard Space Flight Center Scientific Visualization Studio

The religious "fault line" of Africa.

From *Literary & Historical Atlas of Africa and Australasia* by J.G. Bartholomew (1913), courtesy of University of Texas

DRAW, MISTER!

Drawn borders in the south of Africa really represented European interests more than a genuine reflection of African political realities. Indigenous peoples travel the borders relatively freely, while those who are obviously of foreign ancestry will be carefully scrutinized. The legacy of colonial administration is that the border often is drawn irrelevantly to the direction of most people's movement. The European powers for the geographic reasons mentioned above tended to share and balance interests in Africa at various times. The French were particularly strong in western Africa while the British tended to

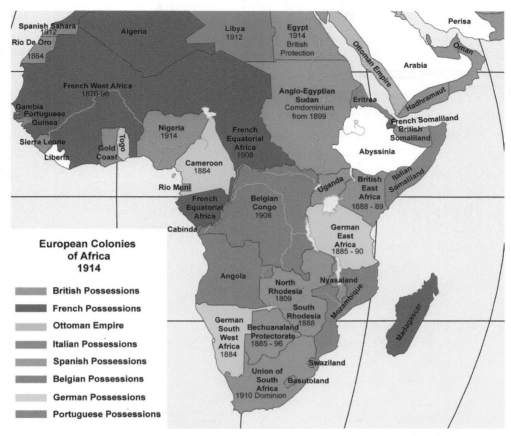

European Colonies of Africa 1914

- British Possessions
- French Possessions
- Ottoman Empire
- Italian Possessions
- Spanish Possessions
- Belgian Possessions
- German Possessions
- Portuguese Possessions

Notice the "French Lake" in West Africa and the British attempt at a cape to Cairo railroad blocked by the Germans and Belgians.

Map source: *The World Factbook*

be interested in areas adjacent to bodies of water important for naval purposes such as Egypt and South Africa. Even little Belgium claimed part of the action in the 19th century in the central part of Africa while the Germans and Italians nibbled at the eastern and Horn regions of the continent. These generalizations aside, the fact is today we are seeing a strong counter movement from the continent back to Europe.

COUNTER MOVEMENTS

Many of those teeming and sometimes desperate folks attempting to arrive in Europe are attempting to escape terrible conditions and are seeking asylum. Many of these young people are heading to the countries that initially colonized the African countries in a type of counter movement. The colonial experience is controversial; some say hospitals, roads, and faith are good things that were brought from Europe, others point out that the exploitation of resources and the pitting of one group against another have more than compensated for

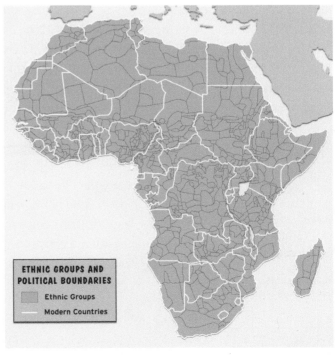

ETHNIC GROUPS AND POLITICAL BOUNDARIES
- Ethnic Groups
- Modern Countries

Map source: *The World Factbook*

the good deeds done by the colonial powers. Perhaps one of the most dire legacies of colonialism is the European tradition of drawn borders. These drawn borders do not correspond in most cases with ethnic groups, and this has had fatal and horrific consequences. One relatively recent example of some of the problems resulting from colonization is the genocides that have plagued the continent. In particular the tendency to favor one tribe over another has ignited tensions in places like Rwanda in the 1990's where the Belgium government had favored the Tutsi tribe over the Hutu tribe. A tribe consists of people who claim relationships with one another on the basis of ancestry or ethnicity. When the Hutu tribes took power, retribution began. The outside intervention of Tutsi's from neighboring Uganda exacerbated the tensions and led to a fierce civil war claiming up to a million people. Violence by the Hutu majority may have resulted in up to one million deaths.[4] This tragedy has been followed by genocides in Sudan where countless numbers of Christians have dedicated their lives and money towards helping.

HELP! HOLD THE LINE

In fact, Christians have held the line so to speak against Islam from Senegal to Ethiopia. It is of great interest to the geographer how this line representing the differing religions and cultures generally follows the Sahel—the dry area bordering the Sahara. Additionally, Christians have flocked to Uganda and Kenya over the years in an effort to help these countries obtain water and other vital resources. Because ministries are welcome here, the western nations have been able to donate time and money to these areas. Areas of Islam generally have been less friendly to Christian missions and simultaneously have intact family structures with high rates of childbirth. One area of need will continue to be AIDS. In many regions of Sub-Saharan Africa this disease has reached gigantic proportions. The average age for males is in the 40s in some of the countries affected by AIDS. Christians know Jesus loves the children and there is no better way to demonstrate His Love than to help His children in Africa today. But all is not bleak on the continent. Let us examine one of the richest areas—South Africa.

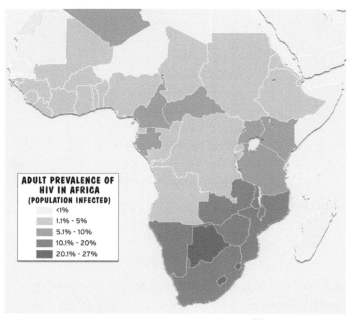

ADULT PREVALENCE OF HIV IN AFRICA (POPULATION INFECTED)

<1%
1.1% - 5%
5.1% - 10%
10.1% - 20%
20.1% - 27%

A sad map of suffering in Africa.
Map courtesy of Katie Pritchard

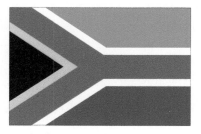

South Africa—one of only a few nation-states that have six colors in their flag!
Image © Globe Turner, 2013. Used under license from Shutterstock, Inc.

Despite the relatively recent plague of AIDS, South Africa's GNP places it around roughly half that of Sub-Saharan Africa.[5] Blessed with numerous minerals and sitting in a strategic location on the southern point of Africa, this area of more moderate climates and mixed cultures ranging from European to indigenous peoples funnels south along the Rift Valley. South Africa is sitting on diamonds and gold as well as other less appreciated but nonetheless important resources like coal. Despite Africa's geographic handicaps and frustrating post-Colonial period, with faithful missionaries on the ground providing a genuine effort against diseases and loving a generation of children, the way could easily be set for the rise of an African revival of spirit and economy.

---

[4] Prunier, Gérard (1995). *The Rwanda Crisis, 1959–1994: History of a Genocide* (2nd ed.). (London: C. Hurst & Co. Publishers), 4.

[5] http://www.tulane.edu/internut/Countries/South%20Africa/southafricaxx.html

THE CONCEPT OF HUMAN "CULTURE"*

"Go therefore and make disciples of all the nations, baptizing them in the name of the Father and of the Son and of the Holy Spirit, teaching them to observe all things that I have commanded you; and lo, I am with you always, even to the end of the age." (Matthew 28:19–20)

To be obedient to Jesus' command, we must understand principles of human culture. Culture is a dynamic activity that changes over time and spreads over space. By defining culture and understanding the global variations of culture, we can better create a "sense of place." An increasingly detached world seeks security and authenticity; it is our job to provide it!

VARYING DEFINITIONS AND CRITIQUES OF "CULTURE"
TRADITIONAL DEFINITIONS

Culture is one of those words that many of us often use without thinking very deeply about what it exactly means. In the post-industrial era, punctuated by the so-called "culture wars" and post-modern dialogue, it is often used as a catch-all term to describe attributes relating to such things as race, ethnicity, and gender. Today, "culture" is a highly-elusive term that means a lot to some social scientists but nothing at all to others. That is, its meaning and importance is highly debated, and in academia today it is one of those ideas that is being rigorously critiqued and deconstructed. But however one might approach or define human culture, there is no denying that the study of how it varies geographically and how it shapes and influences cultural landscapes is central to the field of human geography. It is altogether necessary, then, to have a basic understanding of what culture is, how different academic traditions define it, and how it is expressed in the landscape in different parts of the world-economy.

Although the concept of human culture in its various forms is a central focus of most of the humanities and social sciences, it tends to be defined in different ways by various academic disciplines. In sociology, for example, a stress is placed upon the codes and values of a group of people. Sociologists would argue that if you really want to know the "culture" of a group, whether it be an ethnic group or a class of people or an entire society, one must understand the rules of conduct that have been agreed upon by members of that group. These rules of conduct might include laws, social mores and traditions, and codes relating to family and societal structure. Sociologists tend to look at how order is achieved and how society is organized in order to understand the values and traditions of that society. Anthropology, a discipline defined as the study of human cultures, tends to focus on the everyday ways of life of a group or society. Such ways of life might include linguistic norms, religious ideals, food, dress, music, and political structures. In order to uncover and understand the cultural values and traditions of a group or society, anthropologists study these ways of life.

While there is a long tradition of debate about the meaning and nature of culture in the academic literature of many of the social sciences, cultural geographers, until recently, have traditionally spent little time defining what culture is. Instead, traditional human and cultural geography in the United States focused on how culture was expressed in the landscapes of places—the landscape, especially its physical manifestations such as houses, fields, settlement patterns, neighborhoods, etc., were "read" and analyzed in order to uncover clues as to the cultural values and traditions of the people that built them. Until post-modern thought began to influence the field in the 1970s, most cultural geographers were content to rely on definitions of culture that had been developed by sociologists and cultural anthropologists in the 20th century. For these cultural geographers, then, human culture consists of many aspects of a group or society that, when combined together, results in a distinctive way of life that distinguishes that group or society from others:

* Pages 197–200 from *Introduction to Human Geography: A World-Systems Approach,* 4th Edition by Timothy G. Anderson. Copyright © 2012 by Timothy G. Anderson. Reprinted by permission of Kendall Hunt Publishing Company.

→ Beliefs (religious beliefs and political ideals, for example)

→ Speech (language and linguistic norms and ideals)

→ Institutions (such as governmental and legal institutions)

→ Technology (skills, tools, use of natural resources)

→ Values and Traditions (art, architecture, food, dress, music, etc.)

As such, cultural values and traditions are not biological in nature. Such traditions are learned, not genetically inherited (that is, we are not born with these values), and are passed on from generation to generation through a mutually intelligible language and a common symbol pool or iconography. For cultural geographers, who seek to identify the spatial expression of culture, a **culture region** is an area in which a distinctive way of life (as defined above) is dominant.

THE NEW CULTURAL GEOGRAPHY

Post-modern thought has had a dramatic effect on the field of cultural geography over the past twenty years, especially in Great Britain and the United States. In what has come to be known as **The New Cultural Geography**, a new generation of scholars is turning traditional notions of culture and its expression on the landscape on its head. Heavily influenced by post-modern literary and philosophical traditions and by neo-Marxist thought, one of the leading voices in this new movement has gone so far as to argue that culture does not even exist and that we learn little about the nature of the world and of the societies in it by approaching culture in the traditional ways outlined above. Rather, it is argued by the new cultural geographers that what we might call "culture" for lack of a better term is not a thing, but rather a process that shapes values, traditions, and ideals. These processes and their accompanying values and traditions differ significantly not from society to society or country to country, but from person to person and are influenced by such things as an individual's class, gender, race, and sexuality. In this line of thought it follows that our perception of the world is influenced by the same factors, and it is argued that cultural landscapes hold clues to such factors working in society. They can be read and deconstructed and analyzed in the same way that a literary text can. The cultural landscape, then, is not seen as simply the built environment or the human imprint on the physical landscape. Instead, the New Cultural Geography conceptualizes it as a place or a "stage" upon which, and within which, societal problems and processes are worked out, especially with respect to struggles relating to class, race, ethnicity, gender, and politics.

CORE-PERIPHERY RELATIONSHIPS

FOLK CULTURES

Both traditional and post-modern conceptualizations are valuable for a broad understanding of how "culture" varies around the world. Employing these ideas alongside the core, semi-periphery, periphery model of the world-economy from world-systems analysis, we can understand basic, general differences in ways of life around the world. It should be noted that these basic differences do not translate well down to the local or individual level. To understand the cultural processes at work at such scales one must analyze cultural processes and patterns at those scales. Here, we are concerned with broad global patterns.

At the global scale, and in a broad sense, we can distinguish between two primary "types" of culture operating today. At one end of the spectrum are so-called folk cultures. This term describes human societies and cultures that existed in most parts of the world until the Industrial Revolution. At this point, as the core, semi-peripheral, and peripheral areas of the modern world-economy became better defined, a major divergence occurred. Folk cultures remained the norm in the periphery and parts of the semi-periphery, but in the core, and today in some parts of the semi-periphery, cultural values and traditions came to be increasingly modified by "popular" tropes, fads and ideas (this is discussed in more detail below).

A **folk culture** refers to a way of life practiced by a group that is usually, rural, cohesive, and relatively homogeneous in nature with respect to traditions, lifestyles, and customs. Such groups and societies are characterized by relatively weak social stratification, goods and tools are handmade according to tradition that is passed on by word of mouth through tales, stories and songs, and non-material cultural traits (e.g. stories, lore, religious ideals) are more important than material traits (e.g. structures, technologies). The economies of folk societies are most characteristically subsistence in nature—farming or artisan activities are undertaken not to necessarily make a profit, but rather to simply survive—and the markets for such products are usually local or regional in nature. Finally, order in folk cultures is based around the structure of the nuclear family, ancient traditions, and religious ideals. If we define a folk culture by these characteristics (and this is a very "conservative" definition), then such cultures and societies are practically nonexistent in the core of the world-economy. Instead, they describe most societies that are tribal or "traditional" in nature in the periphery and in some remote, rural parts of the semi-periphery (the Amazon Basin of Brazil, for example, which is part of the semi-periphery). Even so, some folk culture traits almost always persevere even in the post-industrial societies of the core; they are "holdovers" of our folk cultural roots from hundreds or even thousands of years ago. Some examples would include the popularity of astrology or tarot card reading, the fairy tales that each of us learns as children, and folk songs from long ago that are still passed on today ("Auld Lange Syne" or "Yankee Doodle").

POPULAR CULTURES

While folk cultural traditions dominate most societies in the periphery, the societies of the core and parts of the semi-periphery today are best described by the term popular culture. Although some ethnic groups in core regions of the world-economy attempt to live in a "traditional" or folk manner, or practice a "traditional" lifestyle, it is nearly impossible for such groups do so in the core because popular culture is so pervasive and far-reaching. For example, even though many Americans see the Amish as a distinctively "folk" society, closer inspection reveals that compared to true folk cultures in the periphery Amish society today is not truly folk in nature. Although most Amish do not use electricity and employ rather simple machinery in their agricultural systems, that machinery is mass-produced, material for barns and houses is purchased from retail stores, and their agricultural endeavors are capitalistic, profit-making undertakings.

Compared with folk cultures, **popular cultures** are based in large, heterogeneous societies that are most often ethnically plural, with a concomitant plurality of values, traditions, and ideals. While folk cultures are by definition conservative (that is, resistant to change), popular cultures are constantly changing. This is due to the power and influence of fads and trends that change rapidly and often in core societies, as well as to the dominance and influence of mass communication in the core. While ideas and trends are slow to move from place to place in folk cultures (usually through hierarchical diffusion), they can move around the planet instantaneously by means of mass communication technology (satellites, the internet, television, radio, etc.) in a popular culture (by means of contagious diffusion). This, in fact, is the central defining characteristic of popular cultures—such fast change and quick diffusion is what makes a culture subject to "popular" (read trends and fads) ideas. In the post-modern era, such trends and fads have significantly shaped how people in core societies receive news, what music they listen to, what books they read, what movies they see, what food they eat, and what clothes they wear. In a folk culture, such things are dictated by tradition that has been passed on by word of mouth over many generations.

Other characteristics of popular cultures include the use of material goods that are invariably mass-produced, and societies in which secular institutions (government, the film industry, MTV, multi-national corporations employing advertisements to entice people to buy their products) are of increasing importance in shaping the "look", the landscapes and ways of life in core societies. The power of such popular fads, trends, and ideas is expressed in the standardized landscapes that are a hallmark of the core and parts of the semi-periphery. That is, popular culture tends to produce standardization that is reproduced *everywhere* in such societies. This can be seen in styles of architecture, music, clothing, dialects, etc., that are the same throughout large, populous societies and over large distances. Currently, the strongest popular culture in the world stems from the United States. Things "American" (music, food, films, styles and the like) affect nearly every place on the planet, including even traditional folk societies in the periphery. Because the diffusion of popular ideas and fads occurs via mass communication, even traditional societies in far corners of the "developing" world are not immune to the influences of popular culture from the core.

Delimiting and defining the geography of popular culture presents a challenge to human geographers because such cultures tend to produce "placelessness" that challenges unique regional expression. For example, ranch style homes became popular in the United States in the 1950s. Although the style probably originated in the eastern part of the country, such houses became so popular so fast that they soon could be found everywhere around the country, including Alaska and Hawaii. Another example would be popular music. When a song goes out over the radio or on television it is heard by millions of people at once, all over the country, or even the world. That song, then, becomes known by millions of people of varying ethnicities, cultures, and nationalities—it has become a song known to millions, not just a few members of a specific tribe or ethnic group as is the case with a folk song. In this way, popular culture fads and trends are extremely powerful. Popular culture supersedes ethnic and national boundaries and spreads rapidly across large distances, often at the expense of local or regional folk cultures. Even so, regional expression often still exists in the form of such things as regional dialects, accents, and food preferences, even in societies such as the United States where strong popular cultures predominate. For example, many people in the South today continue to speak English with a strong regional accent in spite of the fact that most people there are exposed to standardized, accent-neutral English in schools and on television news programs and the like.

The post-industrial world-economy of today is punctuated by stark divisions within it. In the previous chapters we have seen how this plays with respect to vast differences between the core, semi-periphery, and periphery in such things as standards of living, population structure, modes of production, and social relations of production. This chapter has demonstrated that these differences are also seen in the types of "cultures" operating in the world-economy: popular cultures in the core, folk cultures in the periphery, and a mix of each in the semi-periphery. The following chapters will address a variety of other aspects of culture and argue that these too vary with respect to "location" in the world-economy.

Perhaps the greatest contribution any of us can make is to create or contribute to "a sense of place." This can begin in our own homes and on campus as we attempt to create and reinforce a cultural hearth. A loving and safe place is analogous to a military base of operations and is vital for the security of those we see sent out into the world. A **cultural hearth** is the source of a unique way of life and is the area serving as a source of cultural diffusion. **Cultural diffusion** is the process of spreading the aspects of culture from the cultural hearth. Cultural diffusion can spread in unexpected ways and at different speeds. It is absolutely essential that the cultural hearth we build be like the home in Matthew 7:24: built on solid rock.

What services and encouragement are you providing your family and school, thereby creating a genuine cultural hearth capable of changing the world? Remember Jesus' admonition in Matthew 5:13, "You are the salt of the earth; but if the salt loses its flavor, how shall it be seasoned? It is then good for nothing but to be thrown out and trampled underfoot by men." As we turn to the study of language and religion, we see concrete examples of how a cultural hearth can be instrumental in cultural diffusion.

KEY TERMS TO KNOW

Cultural Diffusion	Culture Region	Sahel
Cultural Hearth	Folk Culture	The New Cultural Geography
Culture	Popular Culture	

FURTHER READING

Jeff Chang and D. J. Kool Herc, *Can't Stop Won't Stop: A History of the Hip-Hop Generation* (New York: Picador, 2005).

Henry Glassie, *Pattern in the Material Folk Culture of the Eastern United States* (Philadelphia: University of Pennsylvania Press, 1968).

Marvin Harris, *Cows, Pigs, Wars, and Witches: The Riddles of Culture* (New York: Vintage Books, 1989).

Marvin Harris, *Theories of Culture in Postmodern Times* (Lanham: AltaMira Press, 1998)

Alan Light, ed., *The Vibe History of Hip-Hop* (New York: Plexus, 1999).

D. W. Meinig, ed., *The Interpretation of Ordinary Landscapes: Geographical Essays* (New York: Oxford University Press, 1979).

Donald Mitchell, *Cultural Geography: A Critical Introduction* (Oxford: Blackwell, 2000).

S. Craig Watkins, *Hip Hop Matters: Politics, Pop Culture, and the Struggle for the Soul of a Movement* (New York: Beacon Press, 2006).

Wilbur Zelinsky, *Exploring the Beloved Country: Geographic Forays into American Society and Culture* (Iowa City: University of Iowa Press, 1994).

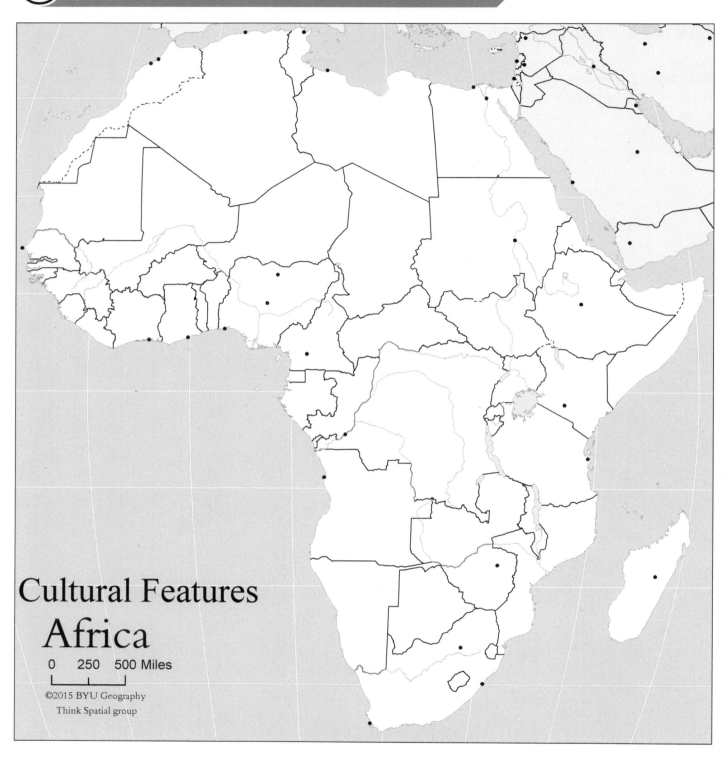

Cultural Features
Africa

0   250   500 Miles

©2015 BYU Geography
Think Spatial group

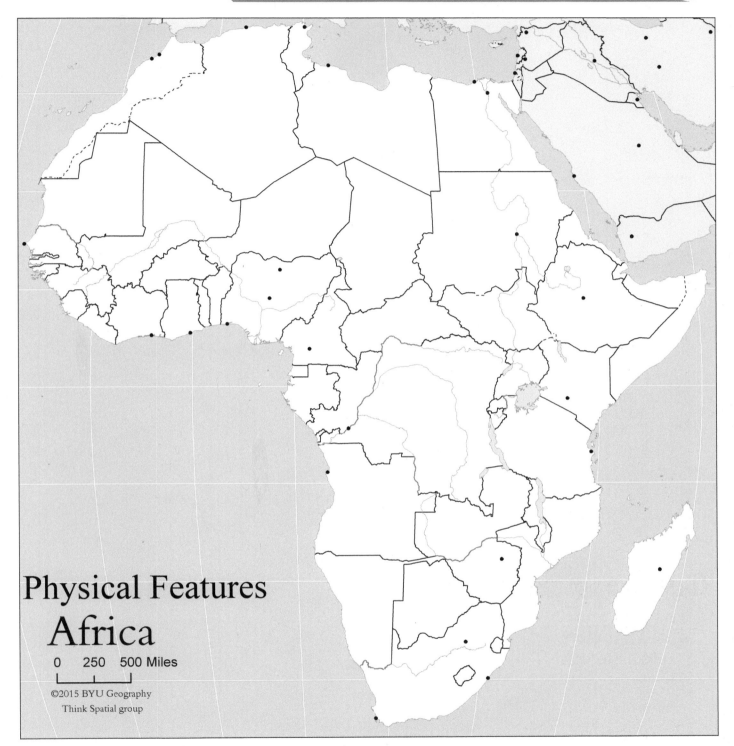

Physical Features
Africa

0   250   500 Miles

©2015 BYU Geography
Think Spatial group

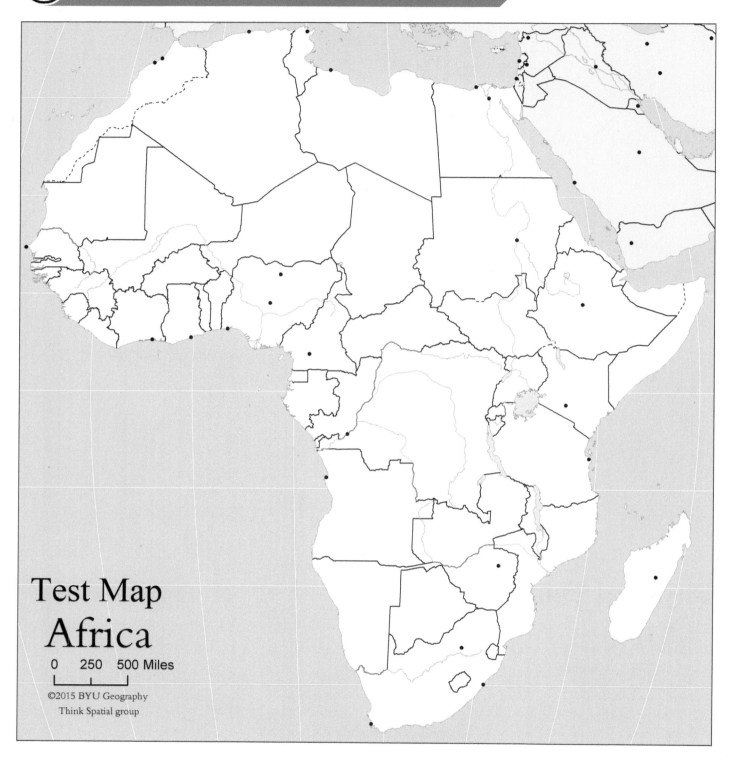

Test Map

Africa

0   250   500 Miles

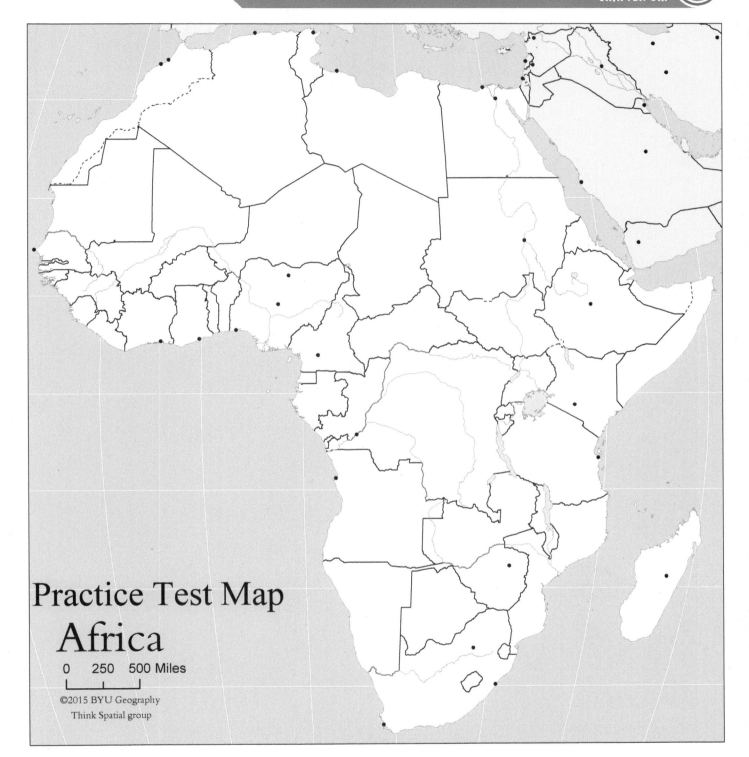

Practice Test Map
Africa

0   250   500 Miles

©2015 BYU Geography
   Think Spatial group

Europe: From Industrial Revolution to Globalization

7

But our citizenship is in heaven, and from it we await a Savior, the Lord Jesus Christ, who will transform our lowly body to be like his glorious body, by the power that enables him even to subject all things to himself. (Philippians 3:20)

CHAPTER OUTLINE:

FOR centuries people migrated from Europe in record numbers. A counter-migration is occurring today with people from the former colonies of the British Empire emigrating instead to the United Kingdom. Regardless of where people are, or where they are going, we must anticipate the opportunity to serve others where needs exist. The greatest need is—hope!

PENINSULA OF PENINSULAS

Europe is an exciting peninsula! In fact, it is a series of peninsulas growing like tree branches out of a bigger peninsula. This series of landforms combines with the continent's relative location on the planet to reveal an amazing display of variety in cultures. These cultures have placed Europe on the cutting edge of changes from the Industrial Revolution to the growth of a supra-national government today. Europe is an exciting proof of the impact of physical geography on global cultures!

If you look at a globe and observe Europe, you will probably notice it is located to the extreme north of the world. In fact, New England in North America roughly corresponds with the region of southern France in terms of latitude. How is it that we see such differences in climate? The answer is simple, the Gulf Stream of currents travels from the warm, sunny tropics and carries with it latent energy in the form of moisture. As this moisture precipitates, latent energy is released into the atmosphere and the result is warmth. This moisture is obviously an aid to the agriculture of the continent, but perhaps as important to the agriculture is the location of the seas and oceans that mark the borders of Europe.

GOING THE WRONG WAY—NATURALLY!

If you look at the map of Europe you will first notice the relatively east-west pattern of the high mountains in the center—the Alps. As you look to the west, you will see the Carpathian Mountains that run in a semi-circular pattern through Romania, and as you follow the line of these mountains to the south, you notice the Dinaric Alps on the side of the longitudinally shaped Adriatic Sea. These mountains proceed south through the Balkan Peninsula and mark the different plates, which come together to form Europe. In fact, if you look at all of the seas surrounding Europe, they all seem to be flowing in roughly parallel and distinct north-east-west patterns like the Black Sea and the Mediterranean. Besides evidence of plate activity, the connection between these two seas has caused an interesting phenomenon of currents, which generally flow in a counter-clockwise direction. The Mediterranean actually moves in a

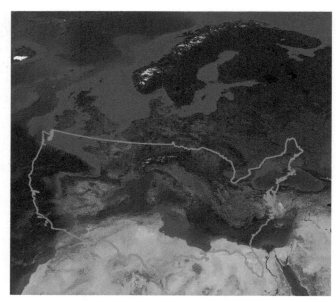

From *Blue Marble: Next Generation* image produced by Reto Stockli, NASA Earth Observatory (NASA Goddard Space Flight Center)

direction opposite what one would expect from the Coriolis effect in the northern hemisphere. Due to the high evaporation rate of water in the Mediterranean Sea, the water entering in from the Black Sea tends to create a type of vacuum according to David Abulafia in his book, *The Sea*. The result of this back flow of water has been the diffusion of culture and people from the Eastern Mediterranean to the west over time. Because there are relatively few rivers to replenish the sea, the majority of the water comes flowing into it from the Black Sea.[1] This inflow and the subsequent phenomenon of a counterclockwise pattern in part explain how the Apostle Paul and Christianity diffused west into the European continent from the Holy Land instead of the normal eastward pattern of the clockwise Coriolis effect. It also explains some of the climatic differences we see within Europe, which also may, in an indirect way, help explain some of the challenges this continent confronts economically today.

Image © JeniFoto, 2013. Used under license from Shutterstock, Inc.

WET SKIES/DRY SKIES

The Mediterranean climate we described in the beginning of this book is certainly evident here! The warm, dry, sunny slopes of the mountainous regions of southern Europe no doubt contributed to the rise of a western

---

[1] Abulafia, David. *The Great Sea* (New York: Oxford, 2011), xxvii.

civilization with a distinct culture. Many geographers believe the hills and mountains of the southeastern part of Europe lent themselves to independence and autonomy. These various city-states each had convenient climates for agriculture and enabled outdoor events such as the Olympic races or events at the **acropolis** to take place. Because the physical barrier of mountains tended to prevent any empire from taking control (such as ancient Persia), this tradition of the independent state probably was a key to many of the freedoms Europeans were able to eventually develop.

Climatic differences in many ways present a **centrifugal** effect on the attempts to create a new European Union. Centrifugal effects are those factors that tend to pull cultures apart. As you may know, **centripetal** forces pull us together. Examples of centripetal forces would be language, religion, ethnicity, etc.

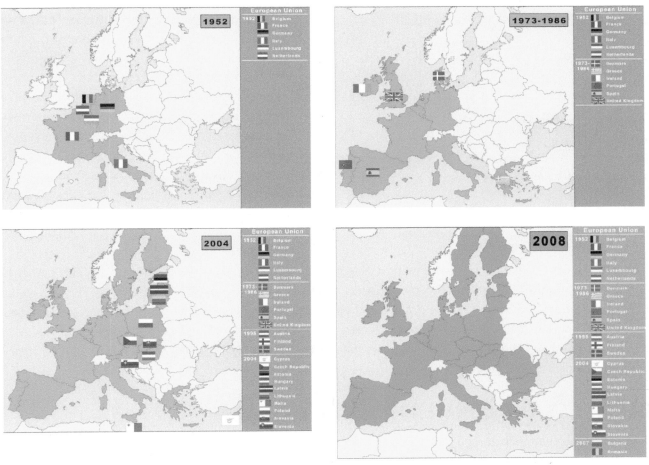

Maps courtesy of Katie Pritchard

SPREAD OF STEAM

A good starting point for discussing the effects of physical geography on European culture and how Europe is alternately pulled together and pushed apart is to note the spread, or **diffusion**, of the **Industrial Revolution**. The Industrial Revolution was a result of the agricultural advances which will be described in the third section of this book. With advances in population and with a surplus of food, more manpower was available to be used in the burgeoning factories. The Industrial Revolution first appeared in Great Britain, which was spatially favored with an abundance of coal, iron, and harbors for shipping. The rail lines, which quickly connected these resources, were instrumental in bringing people to the mines and factories where more machines were created. The development of steam engines continued to enable miners to dig deeper while the use of new iron smelting techniques utilizing purified carbon also known as coke produced better grades of iron that were capable of withstanding higher stresses and temperatures. Cities began to grow near the coalfields and the mines, and harbors enabled the goods to be shipped to a global market. This process would diffuse to the continent of Europe.

The Belgians and Dutch were the next nations to receive the Industrial Revolution probably due to the proximity of the harbors to Great Britain, which delivered the machines and people to work on them. From the "low countries" the Industrial Revolution moved east and south into Germany and France. It is interesting to note that many of the debtor nations in the European Union today were some of the last to receive the revolution. It is quite possible that some of the cultural challenges we are seeing today are a result of the uneven spread of the Industrial Revolution. The urban landscape of Europe is also unique and reflects the history of Europe.

Generally the **CBD** or central business district of Europe's great cities presents relatively unimpressive skylines, at least compared to those of the bright shining cities of East Asia and even North America. European cities display their past

The unique cultural landscape of Paris on the Seine River.
Image © Crobard, 2013. Used under license from Shutterstock, Inc.

with narrow roads designed for travel by horseback or on foot. With plentiful land, European cities have obviously not needed to grow taller buildings. The result is urban skylines of relative uniformity and only moderately tall buildings. These large, low skylined cities are generally surrounded by **greenbelts** or park areas where nature can be seen in contrast to the often denuded cities. The urban pattern of European cities often is centered on a river such as Paris with the Seine or London on the Thames. These **gateway cities** essentially represented portals to the world at various times throughout the Age of Exploration until the scramble for colonization. So how do gateway cities and the harnessing of the power of steam relate to European unity and dissension today?

With the European Union's emergence we have begun to see what is known as **supranationalism**. This is when people of various **nation states** begin to work together to form a sort of supra state. Interestingly, because of Europe's history and because of the strong influence of Middle Eastern and central Asian invaders through the valleys and plains of southern Europe, those in northern Europe seem to hold a different view of the individual's relationship to the state. Today, we see Germany attempting to persuade the Mediterranean nations of Portugal, Spain, Greece, and Italy to control their social spending programs. These attempts at unifying a **Eurozone** have led to some problems in recent years.

WHERE DOES THE CORE END?

A major area of population in Europe is referred to today as the core. This **core** region consists of huge populations in the Low Countries of **Benelux**. Here we see Belgium with its famous port of Antwerp—so highly sought in World War II to supply invading allied forces—to Amsterdam in the Netherlands, the most populous city in the region. This area is the densely populated core of Europe located relatively closely to the great gateway city of London. Moving west from the Low Countries on the continent, we come to the Ruhr Valley with over eight million people. It is a massive urban agglomeration connected by a huge river. Some historians consider the diffusion of the Industrial Revolution to the mainland after the British and the Dutch had embarked on the conquest of colonies for their resources to be the reason why Germany gained technological advancements and the social changes associated with the revolution. This particular model of diffusion explains that the Central Powers during World War I might have attempted to contest established borders partly because of an aggressive worldview and frustration with a new era. Regardless, the rate of population growth necessary to sustain any civilization is considered to be around 2.3 children per couple. However, Europe is not sustaining this rate of childbirth evenly across the continent. Poland currently has one of the highest rates of population on the continent followed by some of the other more peripheral countries. We see that the increased employment opportunities in Europe that are the effect of the diffusion of the Industrial Revolution from the Island of Great Britain to the mainland have beckoned immigrants and resulted in areas of higher population.

PUSH ME/PULL ME

One of the major challenges facing Europe as it attempts to unite in a spirit of supranationalism is the **demographics** of the continent. The population is increasingly seen as aging and this brings with it many consequences. One of the consequences of a population's aging is the paucity or lack of available labor. This causes friction because many of the immigrants are entering Europe through the **periphery**. By periphery we refer to an area outside of the **core**, or more industrial area. In the case of Europe, this periphery is generally Eastern Europe, and this area, with its relative poverty and high rate of

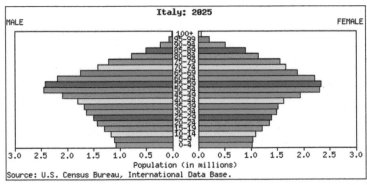

Map by Katie Pritchard

immigration from lower latitudes, threatens to put a halt to the Schengen Zone of Europe.

Europe has what can be described as various sub regions within the continent. The historical differences in physical geography and climate are apparent today in economic matters and in areas of borders. For example, in an attempt to control illegal immigration and to protect indigenous or local labor, the Europeans have created a zone that allows border controls to be lifted called the **Schengen Zone**. Romania and Bulgaria are currently unable to enter the Eurozone and are sensing a second-class status by not being able to enter this Schengen Zone. The states of northern and western Europe are increasingly separated into a regional and cultural block from those of southeastern Europe because of their desires to pro-

I owe, you owe—Euro!
Image © Gena96, 2013. Used under license from Shutterstock, Inc.

tect the various social programs that protect their peoples. These sensitivities to economic concerns are not merely limited to the borders, however. One change that has superseded borders has been recent attempts to lift air border controls.[2] Easier travel amongst these nation-states in Europe will no doubt have cultural ramifications and may serve to continue building transnational unity (centripetally), pulling the continent closer together.

The creation of a European Union has also created restrictions on which nations can enter this supranational organization, particularly in the economic downturn that occurred after 2008. As a result, Germany has pressured various southern European

---

countries to control their spending in order to keep the **euro** strong. Germany can stand to lose if southern European countries such as Italy, Spain, Portugal, and Greece do not repay their debts to it since, in many ways, it is the largest economy in Europe. These Mediterranean nations sometimes feel as if such austere measures are designed to exploit them by forcing them to pay out to what they perceive to be conquering-type powers. There are other regional considerations besides interregional problems between the creditor states of northern Europe and the emerging economies of the south.

MEMBERS ONLY

Turkey has attempted to enter the **European Union** unsuccessfully for many years. The exclusion of one of the more liberal Middle Eastern countries has caused some backlash as a result of perceived discrimination on the basis of religion. European exclusivity has not been confined to Turkey alone, however. The Baltic States recently were accepted into the European Union, but states like Moldova, Belarus, and Ukraine have been excluded, and there is very little prospect of these eastern European states or Russia being allowed to join the union

The vital ports of "Benelux" are easily seen in this map.
Image © Rainer Lesniewski, 2013. Used under license from Shutterstock, Inc.

any time soon. It will be interesting to see if the supranationalism of Europe will be a centripetal force pulling the various **nation-states** together or if the region will instead begin to come apart under the pressures arising between the creditor regions of northern Europe and the debtor regions of southern Europe. Only time will tell whether this continent that at one time squabbled in the shadows of the greater empires of the Middle East and Far East will maintain the advantage it gained in the Industrial Revolution and the spread of capitalism and Christianity, or if it will decline into a multitude of different regions and peninsulas.

Remember Proverbs 14:4 —"Righteousness exalts a nation, but sin is a reproach to any people." (KJV)

The limit of supranationalism?
Map courtesy of Katie Pritchard

THE GEOGRAPHY OF THE WORLD-ECONOMY*

"Now the great city was divided into three parts, and the cities of the nations fell. And great Babylon was remembered before God, to give her the cup of the wine of the fierceness of His wrath." (Revelations 16:19)

Although the scriptures are replete with warnings against idolatry, the Bible states that at the tower of Babel, mankind attempted to become greater than God. Interestingly, we see in the above verse that the generally accepted interpretation for the world-system is Babylon, and it will be brought down when the Lord returns. Today, we see the globalization of the world's economy unlike anything ever seen in history. It is imperative that we study the economy for the purpose of obedience. In Psalm 41:1–3, we read, "The Lord cares for the poor." We can help people by carefully investing our talents and treasure in locations that will return profit and create wealth enabling us to fulfill these instructions.

We have discussed the structure of the world-economy and defined the core, semi-periphery, and periphery. This chapter examines the nature of economic activities in the three regions of the world-economy in greater detail. How do people make a living in different parts of the world? How has globalization influenced the structure of the world-economy? What factors determine what kind of economic activities are undertaken in certain places?

THE CLASSIFICATION OF ECONOMIC ACTIVITIES

Economists identify five different types of economic activities, usually referred to as "sectors" of an economy. This classification of the various sectors of an economy will be employed in this chapter to compare and contrast economic activities, modes of production, and social relations of production in the various regions of the capitalist world-economy of today. The **primary sector** of an economy refers to activities related to the extraction of natural resources. This includes fishing, hunting, lumbering, mining, and agriculture. The **secondary sector** describes so-called "heavy" or "blue-collar" manufacturing industries that involve the processing of raw materials (usually natural resources) into finished products. This includes such activities as steel production, automobile assembly, the production of chemicals, food processing industries, and paper production. The following three sectors are defined as service industries, those businesses that provide services to individuals and the community at large. The **tertiary sector** refers to financial, business, professional, and clerical services, including retail and wholesale trade. The **quaternary sector** of an economy describes jobs and industries that involve the processing and dissemination of information, as well as administration and control of various enterprises. These jobs are often described by the term "white collar," and consist of professionals working in a variety of industries such as education, government, research, health care, and information management. Finally, the **quinary sector** refers to high-level management and decision-making in large organizations and corporations.

FACTORS IN INDUSTRIAL LOCATION

The location of various types of industrial activities is influenced by a variety of geographical factors. These factors are at work not only at the regional scale, but at the national and global scale as well. Among these factors are the following:

→ The Costs of Production
  * **Geographically fixed costs**—costs relatively unaffected by location of the enterprise (e.g. capital, interest)
  * **Geographically variable costs**—costs that vary spatially (e.g. labor, land, power, transportation)

Introduction to Geography

- → Capitalist Ideology and Logic
  - ✦ Since the goal of almost all industries is the minimization of costs and the maximization of profit, the location of an industry is most likely to be where the total costs of production are minimized
- → Complexity of the Manufacturing Process
  - ✦ The more interdependent a manufacturing process is, the more its costs of production are affected by location (e.g. steel production)
- → Type of Raw Materials Involved in the Manufacturing Process
  - ✦ Raw materials that are bulky and heavy, perishable, or undergo great weight loss or gain in processing have the greatest effect on siting
  - ✦ Examples: pulp, paper and sawmills; fruit and vegetable canning; meat processing; soft drink canning
- → Source of Power
  - ✦ Important when a source of power is immovable (e.g. the aluminum industry)
- → Costs of Labor (Wages)
- → The Market for the Product
  - ✦ **Market orientation**—placing the last stage of a manufacturing process as close to the market for that product as possible; products that undergo much weight gain during the manufacturing process (e.g. soft drink canning and bottling)
  - ✦ **Raw material orientation**—locating the manufacturing plant as close to the raw material that is used as possible; usually applies in industries that use very heavy or bulky raw materials (e.g. paper production)
- → Transportation costs
  - ✦ Water—the least expensive for of transportation for bulky and heavy goods
  - ✦ Rail—also relatively inexpensive for bulky and heavy goods but with less flexible routes
  - ✦ Trucking—relatively expensive but carries the advantage of very flexible routes
  - ✦ Air—the most expensive form of transportation; employed usually for transporting very valuable goods or those that are time-sensitive (e.g. overnight mail service)

ECONOMIES OF THE SEMI-PERIPHERY AND PERIPHERY

The economies of the periphery and parts of the semi-periphery are dominated by primary economic activities. The vast majority of people in the periphery live in rural areas and make their living from subsistence farming. Subsistence farming systems involve agricultural activities that are undertaken not necessarily for money profit, but rather for daily sustenance. Three main types of subsistence agricultural systems are dominant in the peripheral regions of the world-economy today:

SHIFTING CULTIVATION

Shifting cultivation, sometimes called *slash and burn agriculture*, involves a complex set of farming practices employed in tropical wet regions where environmental conditions (heavy annual rainfall and very poor soils) are delicately balanced. It is an ancient practice that probably dates back to the earliest stages of the Neolithic Revolution and represents a rather sound ecological solution to the vagaries of life in such an environment. Shifting cultivation cannot support very large populations because tropical soils do not support the kinds of crops that can feed very large populations (such as grain crops). Rather, it involves the use of low technology tools such as machetes, hoes, digging sticks, and fire to harvest crops such as bananas, taro, cassava, and manioc. It is practiced by very small groups (bands and tribes) practicing a nomadic or semi-nomadic lifestyle in which a small area is cleared of brush with a machete, covered to allow it to dry, and burned, which fixes nitrogen into the soil. Plants that reproduce vegetatively (like those listed above) are then planted in the ashes. After one or two growing cycles the soils are exhausted and the group moves on to another site. This lifestyle is practiced mainly in tropical rainforest regions today by a relatively small number of people in locations such as the Amazon and Orinoco basins of South America, parts of Java and other Indonesian islands, parts of tropical central Africa, and parts of Middle America and the Caribbean islands.

PASTORAL NOMADISM

Pastoral Nomadism, sometimes referred to as *extensive subsistence agriculture*, describes the practice of following or hunting herds of game or herding domesticated animals. It is practiced by large numbers of people in tropical grassland environments (savannahs) in central and east-central Africa and some mid-latitude grasslands regions in south-central Asia. Pastoral nomadism involves an almost wholly nomadic lifestyle that is extensive in its use of land since livestock like cattle, sheep, and goats require much land per animal in order to thrive. Contrary to popular belief, most pastoral nomads each meat very infrequently because the animals they heard are the main source of wealth and income. Such societies are usually tribal in nature and the lifestyle involves seasonal movements to "greener" grazing lands. As such, there are few permanent settlements in the areas where pastoral nomadism is the dominant economic activity.

EASTERN EUROPE*
THE BOOKENDED REGION

We just wrapped up Western Europe. We now know of its extremely rich and multifaceted world influence. And we are aware of how it destroyed itself in the 20th century. Let's move on to Eastern Europe. It has slightly different physical geography, slightly different cultural geography, and it's a place that has been radically different from Western Europe in the past. But things are changing . . .

Eastern Europe is the *bookended region,* as the Plaid Avenger refers to it, and bookended in a variety of ways. The bookend reference means there is something in the middle being held up, propped up, or pinched between two bigger sides. In this respect, physically, Eastern Europe is between the giant Russia to its east, and those Western European states to its west. (Western Europe—the region we just covered.) Ideologically, it's also in the pinch in the middle. Eastern Europe has been a buffer zone between these two same giants—Russian culture to the East, Western European ideas to its west. Because of this, it has essentially been the battleground between these two teams, and therefore a fuse to the powderkeg of big conflicts. Where did the major transgressions of the 20ᵗʰ century start? World War I, World War II, the Cold War? Yep . . . all got launched in Eastern Europe.

This region had been marked for a very long time by **devolution**, or shattering apart; big political entities breaking down into smaller units. This region is also distinct because it has been a **buffer zone** and a battleground of ideologies throughout the 20th century. It was a buffer between Russia and Western Europe, and as such, was a battleground for the Cold War. Commies in Russia thought one way, dude-ocracies in Western Europe thought another way, and everyone in Eastern Europe was caught in the middle. The book getting squished by the bookends.

On top of that, we're going to see that there are a lot of historical influences that have permeated Eastern Europe proper, making it extremely diverse in every aspect you can imagine: linguistically, ideologically, religiously, ethnically, culturally . . . It's all here. This diversity is one of the reasons Eastern Europe has historically yanked itself in lots of different directions and shattered apart. That's why it's a region, and just one of the many reasons why it's distinct from surrounding regions. Russia is Russia, Western Europe is Western Europe, and Eastern Europe is something in-between, for now.

But that was then and this is now. If you want a key word, a single word that keeps coming up in the discussion of Eastern Europe, the Plaid Avenger has one: transition. This is a region in transition. Of course, every place on the planet is changing and moving around, but this is an entire region in which just about all facets of life are on the shift: with an obvious place they came from, and with an obvious place they're going to. All the states in this part of the world are going through this process of transition. What sort of transition, Plaid man? A transition from what to what? We'll talk about that in more detail as we go along.

THE PHYSICAL MIDDLE GROUND

When we think of Eastern Europe as I just suggested, we think of that region that's sandwiched between two much bigger and more powerful other world regions. But if we just look at it physically, just physically for a minute, it's also kind of a buffer zone in a very natural sense, in a variety of different ways.

CLIMATE

The temperature and precipitation patterns of Eastern Europe are more **continental** in character than areas to its west. Being further and further away from the Atlantic Ocean, and particularly its most moderating Gulf Stream effect, makes these states more prone to increased temperature disparity: that is, it can get much colder in the winters and perhaps even more hotter in the summers. As we progress further into the continent and away from large bodies of water, this continentality effect becomes even more pronounced. We'll see this played out to the extreme when we get into Russia, deep in the continent's interior, where it's way colder still. Nothing is quite like Russia, where the only thing that's not frozen is the vodka. Even the Plaid Avenger's beard froze up in that place multiple times on secret subversive missions. When we think of the inverse, we think about the coastlines of Western Europe. We think of a more moderated climate. It's not too bad there; it rarely freezes, even in places as far north as Great Britain or parts of Norway.

Eastern Europe is in between, in that it's not quite as bad, not quite as extreme as Russia but not quite as moderated as the West. As you progress from Western Europe inward, into the continent, you get more continentality—meaning less moderation, more temperature extremes during the year, and certainly cooler, if not downright colder, in the wintertime. Climatically, they're in a zone of transition, as well, from the West to Russia. But don't let me mislead you: this place does have some nice climates, and in fact some areas are exceptionally rich in natural resources and soils. In fact, Eastern Europe actually produces a lot of food. Ukraine has classically been one of the breadbaskets of all of Europe, and it continues to be. There is a lot going on here physically, but it's just not quite as pleasant, not quite as moderated as its Western European counterpart.

LAND SITUATION

Another big physical feature to consider with Eastern Europe is that big sections of these countries are landlocked. When we think of Western Europe, we think of maritime powers—all of them have sea access, all of them have coastlines, and, therefore, they all have big navies and armadas, and they're all big traders and were all big colonizers. Not so much so for the Eastern Europeans, if at all. Yes, there is some coastline along the Black Sea for some countries like Bulgaria and Romania. Yes, some of them have Mediterranean Sea access like Albania and Croatia. And yes, some have Baltic Sea frontage like Poland and Estonia. But none of these states in what we're considering Eastern Europe were ever big maritime powers, and they're still not. Unlike their Western European counterparts, none of these states were big colonial forces in the world. They didn't have the advantages we talked about in Western Europe of sapping off the resources and riches of the planet by controlling vast areas outside of Europe proper. Those Eastern European empires and states were never major maritime powers, nor colonizers, nor absorbers of wealth from other points abroad. And that sets them quite apart from their Western European counterparts, even up to this day.

CULTURE CLUB SANDWICHED

The last big physical thing to consider is that they were physically in the middle of some very big powers, that have very different cultures. What do I mean by this? Well, Eastern Europeans have their own culture, obviously: Polish folk have their Polish culture, Ukrainians have their Ukrainian culture, etc, etc. But all these folks in all the countries in this region are sandwiched between entities that have bigger histories, bigger cultures, and are bigger powers. By bigger, don't misinterpret, I'm not talking about "bigger is better." In some cases that's true, but I'm just suggesting here that they're *bigger* in terms of having more of an impact within their regions, but also beyond their regions as well. A main one to consider is Mother Russia: a very distinct culture, language, religion, and ideology that by its very size and strength has dominated areas close to its Moscow core. And what's close to the Moscow core? That would be all the Eastern European countries.

Order up! Who wanted the Turkey, Polish sausage with Russian dressing on rye?

But Russia is not the only big power heavily influencing this regions. Team West and its cultures influenced the region, as has North Africa. You also have the former Ottoman Turk Empire down south. The Ottoman Empire is long since gone but Turkey took its place, and so there's a Turkish culture that abuts them. An important part of that Turkish culture is the fact that they are Muslim, and the Ottomans brought Islam into the region in general. That's important to note, by the way, because perhaps unlike any other world region or other part of the world, you have a variety of world religions that come together in one place, albeit, not always on friendly terms. You do have Islam which did penetrate into Christian Europe via Eastern Europe. But even before you get to that, you have to think about the divisions of Christianity. Of course, there's Old School Christianity that got divided into Eastern Orthodox; hey, that's in Eastern Europe. That then got divided into Catholicism; hey, that's in Eastern Europe, too. And then Catholicism was further branched into Protestantism, and that's in Eastern Europe as well.

A quick summary here: Eastern Europe in the middle of major world cultures—Russian versus Western versus, say, Arab/Islamic. Also, it's the middle ground of major world religions—Christianity, Judaism, Islam, and all the divisions of Christianity itself, as well. These divisions all coming together—different ethnicities, different people, different religions, different cultures, different histories—has served to play out within Eastern Europe, more often than not, as a battlefield of all these differences. This leads us to the concept of Eastern Europe as, not just a battleground, but a **shatter belt** . . . a zone of breaking down, breaking apart bigger entities to smaller and smaller ones. Perhaps I'm getting ahead of myself. Before we get to this idea of a shatter belt and a breaking down, a.k.a. **devolution** (another great word), let's back up the history boat for just a second and take these things one at a time as they proceeded throughout history.

IDEOLOGICAL MIDDLE GROUND

As pointed out above, Eastern Europe is in the middle of lots of different cultures, different religions, and different ethnicities who have battled it out across the plains of Eastern Europe over the centuries. I'm not going to bore you with all the details; I'll just set the stage so you understand the modern era, how some of these outside influences came to cohabitate within this region over time, and why perhaps it's still causing some conflicts today.

HISTORICALLY

TEAM WEST VS. RUSSIA VS. THE OTTOMANS

Historically speaking, and I'm only going back a few hundred years, Eastern Europe was kind of a battleground of big empires. We have some of the Team Western players like the Austrian empire, the Prussian empire (which evolved into Germany), the Russians, and the Ottoman Empire. Four big entities which virtually controlled all of

1815: Four main players.

what we consider now Eastern Europe. Just so you know the Russians and the Ottomans fought it out for long periods of time vying for control of these territories. The Russians and the Austrians did the same thing, as did the Prussians and the Russians.

These empires have radically different cultures as well, in things like language, ethnicity, economies, politics, and religions. Differences which sow the seeds of conflict for generations. Consider only the religious differences for a moment more. The Germans and Austrians were mostly Protestants and Catholics; the Russians were staunchly Christian Orthodox; the Ottomans were Islamic. They are all vying for cultural and political influence in this region. Are you starting to get a sense of the confluence of conflicting ideas in this area? Empires trying to expand, fighting each other, in this fringe zone we're going to call Eastern Europe, as we get a little further along.

Now that's the background set-up. We need to get into the modern era, and in the modern era there are some major devolutions that occur. Evolution has the connotation of growth, of building into something bigger and more complex. **Devolution** is just the opposite: a term that simply means breaking down—devolving into smaller, simpler pieces. Devolution in a political sense involves a big empire or a big country breaking down into smaller countries; that's a real big theme for Eastern Europe historically and perhaps even into today's world.

1915: Goodbye, Ottoman Empire!

WORLD WAR I

The starting phase of devolution in the modern era was with the Ottoman Empire. The Ottoman Empire, which was a Turkish-Islamic empire, introduced Islam into southeastern Europe. Places like Bosnia, Albania, and Kosovo are still predominately Muslim today as a result of the cultural influences that the Ottomans brought in. You have to remember, the Ottomans were actually knocking on the door of Austria for a very long time in a bid to expand their empire into Europe, but they were weakened considerably and were on the brink of collapse by the time the 20th century rolled around. Let's fast forward: this declining empire most unwisely allied themselves with the Germans during the lead up to World War I (you can read more about that in the chapter on Turkey). But the Germans lost that war, which means their Turkish allies were also losers, which resulted in the carving up of the Ottoman territory. By 1915, there was no such thing as the Ottoman Empire. It's subdivided, as you see in the map into a variety of new states already—places like Romania, Yugoslavia, and Bulgaria.

The Austria-Hungarian Empire, which kind of started World War I—or at least the assassination of their leader Franz Ferdinand triggered it—they lost the war as well, and thus their Empire became the next shatter zone. Ottomans shattered first, then the Austria-Hungarian Empire shattered next, which created several more countries—individual entities which declared independence.

We also have to point out that Russia was unwittingly pulled into World War I, and it got the crap kicked out of it by the Germans. World War I was just so bizarre! All of the major, and minor, powers in Europe had a complex web of military alliances with each other that—once the war started—obligated everyone to jump into the fray to protect their buddies. Pretty much everyone declared war on somebody within days of the Austro-Hungarian assassination spark. Thus, the Russians got sucked into a land war to help their Serb allies, just at a time when dissent and revolution were internally brewing back at home too. The results of which were that Russia, under the command of the inept Tsar Nicholas, sucked so bad on the battlefield that they lost big chunks of their western fringe to the

1919: Goodbye Austro-Hungarians!

Germans . . . and of course this did not sit well with already-aggravated Ruskies back home. And that was when our main Commie friend, Vladimir Lenin came to power in Russia in 1917, he essentially said, "Hey, we want out! We're out of World War I. We surrender. Germany can have all the territory that we have ceded thus far. They can have it. Take it! We're done!" This was actually a very popular policy back home in Russia, which was on its way to becoming the USSR. We will pick up the importance of this WWI Russian territorial loss in a couple of paragraphs from now . . .

Even after the Russian withdrawal, Germany didn't really win the war either, and therefore, they lost that recently-gained Russian territory, as well as some of their own. The result of which was a bunch of new countries that popped up around 1920—places like Estonia, Latvia, Lithuania, Poland, even parts of Ukraine. They all declared independence. They were in this middle zone, again, this Eastern European middle zone between these major battling powers. A whole bunch of new countries popped up then, but they ain't gonna last . . .

Goodbye, German Empire!

WORLD WAR II

Because, of course, Russia then became the Soviet Union under Lenin's tutelage and, as the USSR grew in power, it wanted its Eastern European territories back. That brings us up to World War II, because the Germans were still over there hopping mad under Hitler and they wanted their lost territories and prestige back as well. Under "der Fuehrer," they decided to have a re-conquest of Eastern Europe and, what the heck, they'll take over Western Europe too, just for good measure. To ensure they could pull this off, Hitler got together with Stalin—what a fun party that must have been—and together these two signed what is called the **Pact of Non-Aggression**. Basically, Hitler said, "Uhh, okay, I'm gonna take over the world, but hey Stalin, you seem cool enough to me. Yeah, so we won't attack you as long as you don't get in the way of us wiping out Western Europe. And we know you want some of these Eastern European territories back, so you can have those and we'll take everything else." Stalin, in all his wisdom, said, "Word; that sounds cool to me. Game on!"

However, about halfway through the war, Hitler reneged and decided to attack the Soviet Union, pulling them into the war. Why would Hitler have done that? Oh . . . that's right: because he was an insane megalomaniac. You know the end of the story: Since Hitler attacked the USSR, the Soviet Union sided up with the US and Western Europe to beat the Nazis. During this Nazi smack-down, the Soviet Union swept in from the East, the US and allies swept in from the West, and the Nazi smack-down finished up with everyone high-five-ing right square in the middle of Eastern Europe.

Amazingly, if you look at the line where the Allies met, it's right in the middle of what we call Europe today. The troops met and high-fived each other when they finished killing all the Germans, or at least the completely crazy ones who continued to fight, then that was it. World War II was over. Peace! Peace out!

Well, kind of. It was great, for all of five minutes, before it became a face-off, again, but with new players . . .

THE COLD WAR

After the success of the allied powers with the Soviet Union at the end of World War II, you essentially have what happens for the

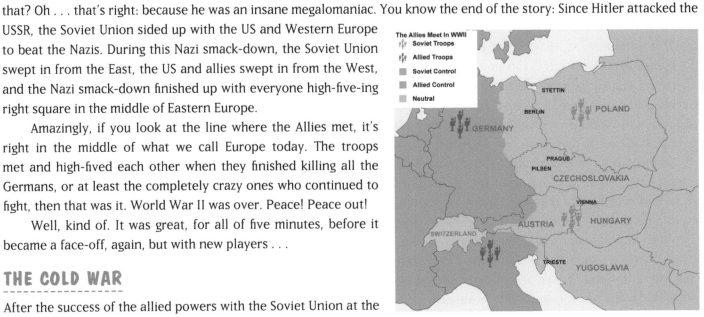

As Borat would say: "Aaah-High-Five" in Europe.

next 60 years—this ideological face-off. You've got the Soviets controlling Eastern Europe, where they swept in and cleaned out the Nazis, and the West who swept in from the West and cleaned out the rest of the Nazis. Although they were playing for the same side during World War II, where they met became the new face-off of major powers. This is more familiar ground we're getting into—the modern era. You, of course, know that this is the Commie world squaring off against Team West: the freedom-loving democratic capitalists in Western Europe. But how did this go down? How did this happen since they were all allies during the War?

SOVIETS TAKE OVER

Well, as I suggested, we had the Soviets, who were buddies of Team West at the time and helped wipe out the German threat. When it was over the Soviets said, "Well, you know what, you guys over in Western Europe are kind of crazy. We keep getting invaded by you. This Hitler guy was just the most recent one; the German Kaiser Wilhelm before that, and even Napoleon did it before him! You Western Europeans are nuts!"

The Soviets said, "Hey, you people over in Western Europe keep invading us, so we're going to stay here in Eastern Europe, for a couple of different reasons." One: they wanted to reclaim a lot of these territories they lost in World War I. That is Estonia, Latvia, Lithuania, parts of Poland and Ukraine were pulled back into Mother Russia—now the Soviet Union. Two: Russia very openly said, "We're going to stay in Eastern Europe to create a buffer zone so that there are no further intrusions from the West." Now the way this turned into essentially a Soviet takeover to make this Soviet buffer zone, but it was done in such a way that it wasn't as imperialistic-looking as it truly was.

Here's how they did it. Just as the Americans, French and British hung around Western Europe to help with reconstruction, the Soviets claimed they were following the same playbook by chilling out in Eastern Europe to mop up and clean up. The Soviet official line was, "Okay, we're here helping out. We'll stay in Poland. We'll stay in Estonia. We're going to stay here in Hungary. We're going to help clean things up and help rebuild just like you guys are going to do in the West. And ya know what? We'll even help them run elections. This'll be great!" Lo and behold, every Soviet-overseen election that occurred in Eastern Europe resulted in a landslide victory for Communist candidates. What a surprise! Go figure. The Communist party just went through the roof in popularity!

Now was any of this real? I don't know, perhaps a little. I wasn't there personally. In the east, the Russians really were liberators of many of these places during the war, so perhaps there was some sympathy, some empathy, and some popularity of communism at the time, but certainly not as much as would have swept a tide of elections across all of these countries. Basically what I'm saying is, this was all farce. Soviets held up these elections and said, "Well, they all elected Soviet leaders, they all elected Communist leaders, so they're just under our umbrella now. We're just here helping out. Isn't that great?" It was, indeed, a Soviet takeover to create this Soviet buffer zone.

People in the Western European realm, and the US as well, were exceptionally unhappy about these developments. Maybe you younger generations today might think, "Why didn't they just go in and get the Commies then?" You have to remember that they were the allies during World War II; they were on our team, and we could not have won the war without them. And World War II had just finished. Nobody wanted to fight a new war. Nobody wanted to then declare war on the USSR and have to do all that fighting all over again.

So this Soviet 'takeover' of Eastern Europe happened as Team West was busy rebuilding the west—rebuilding and restructuring: laying the foundations for the EU to rebuild Western Europe. It happened when the eastern side was being rebuilt and occupied by the Soviet Union. Where these two teams met during the war, of course, became the division between these two forces . . . and is where something called the Iron Curtain then fell. See map box below . . .

When all this action was going down, all of those countries in Eastern Europe fell into one of several different categories of which I want you to be familiar. First, some of this territory of Eastern Europe was simply absorbed back into the Soviet Union and became part of other entities and simply just went away. Parts of Poland were worked back into the Lithuania SSR and the Ukrainian SSR. So **absorption** was one option. But what is an SSR?

What's the Deal with . . . the "Iron Curtain"?

In 1946, British Prime Minister Winston Churchill delivered his "Sinews of Peace" address in Fulton, Missouri. The most famous excerpt:

> From Stettin in the Baltic to Trieste in the Adriatic an "iron curtain" has descended across the Continent. Behind that line lie all the capitals of the ancient states of Central and Eastern Europe. Warsaw, Berlin, Prague, Vienna, Budapest, Belgrade, Bucharest and Sofia; all these famous cities and the populations around them lie in what I must call the Soviet sphere, and all are subject, in one form or another, not only to Soviet influence but to a very high and in some cases increasing measure of control from Moscow.

Iron Curtain. But which side was Iron Maiden on?

This speech introduced the term "iron curtain" to refer to the border between democratic and communist (soviet controlled) states. Because of this—and because most Americans have no clue what a "sinew" is—Churchill's address is commonly called the "Iron Curtain Speech."

The "Iron Curtain Speech" was received extremely well by President Harry Truman, but much of the American public was skeptical. Throughout the speech Churchill's tone was very aggressive towards the Soviet Union. Many Americans and Europeans felt that this was unnecessary and that peaceful coexistence could be achieved. The United States had recently considered the USSR an ally. In fact, Stalin, whom the American press had dubbed "Uncle Joe" to boost his popularity during WWII, was probably the most pissed about the speech, feeling he was betrayed by his allies. The "Iron Curtain Speech," besides coining an important Cold War term, set the tone for the next 50 years of US-Russian relations—which is ironic, considering it was given by a British dude.

Ah! That's the second option: some of these briefly independent states became **republics** of the USSR. See, USSR stands for the Union of Soviet Socialist *Republics*. Kind of like states of the United States of America, like Wisconsin is a political component of the USA. Some of these entities in Eastern Europe actually went from being an independent sovereign state, to becoming a republic of the USSR. I'm thinking specifically of Moldova, Estonia, Latvia, and Lithuania. They ceased to be sovereign states, and became a part of the USSR.

The third and final possibility, and perhaps the most important option, was that some of these sovereign states actually retained their sovereign state status but became **Soviet satellites**. Mainly, I'm thinking of Poland, Romania, Bulgaria, Hungary, Czechoslovakia, and East Germany. These were countries. They were sovereign states. They had a seat at the UN. However, they didn't really have control over their own countries, and everybody knew it. Everybody knew the Soviets were pulling the strings of the Polish government, of the Romanian government, of the Czech government. Everybody knew it. They did have seats at the UN, they were supposedly sovereign states, but the Soviet Union truly controlled them, because the USSR had their patsy commie officials running the show in the governments of these places. Does that make sense?

That's the scenario as it was for about fifty years during the Cold War. The Soviets either controlled directly or indirectly (due to their massive influence) every place that is now called Eastern Europe. Team West was on the other side of the curtain: the democracies, the free market capitalist economies that were supported by the US. Those are your two teams during the Cold War. Eastern Europe was close enough to see all the economic growth and political freedoms stuff going on in the democratic countries, all while they were getting paid visits by the KGB to make sure they still loved being Commies. Now do you see how Eastern Europe is a battleground in the middle of this most recent ideological game? Stuck dead in the middle of the commie vs. capitalism showdown!

So long sovereignty, hello SSR!

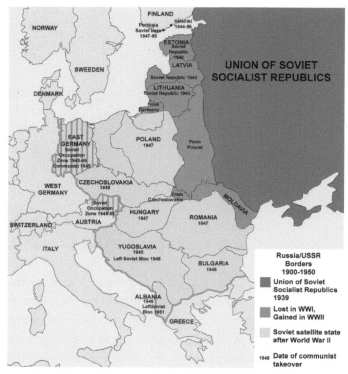

Soviet territory regained, and soviet satellites acquired after WW II.

MARSHALL REJECTED

The other term I want you to think about as well, or at least understand and know, is that the Soviet satellite states did not accept the **Marshall Plan**: a US-sponsored program of aid, loans, and material support to rebuild Europe. Just after WW2, the United States, in its infinite wisdom, said, "You guys are screwed over there! Your whole region has been leveled!" Some smart folks in the United States government said, "We should help them out." In particular, US Secretary of State George Marshall said, "Look, we gotta help these dudes out. We need to get them back on their feet, because if we don't, their economies will get worse, collapse and then turn into complete chaos, and if we don't help Western Europe out, the Soviets—who we are already worried about—are right next door. The Soviets will sweep all the way through if we don't do something!" Probably a fair statement. That Marshall was a smart cookie.

More on the M-Plan: this aid package was offered to all of Europe. The US went to every single one of these countries and said, "Here, we'll give you a ton of money to help you out with industrial capacity and infrastructure to get you back on your feet economically." It was even offered to the Soviet Union. The countries in yellow text on the map accepted the Marshall Plan and took the aid, the sweet cash. Lo and behold, these are the richest countries in Europe right now. The countries labeled in white refused the Marshall Plan, and those are the not as rich countries of Europe right now. While we certainly can't pin everything in the modern era on a singular aid package from the US, it certainly had its impacts back in the day. And it reinforced the division between east and west, as the teams seemed to be now fully set.

If you look at this map of who accepted the Marshall Plan and who didn't, you'll see our same Iron Curtain division between Eastern and Western Europe again demonstrated perfectly. We know that one reason why Eastern Europe is poorer is because the Soviet Union eventually lost the Cold War, mostly because their economy sucked so bad. The second reason is they were stymied out of the gate because they were influenced by Russia to decline the Marshall Plan.

We've already talked about what a Soviet satellite was, and the Soviet influence in Eastern Europe. Make no bones about it; Poland would have probably loved the Marshall Plan. The citizens of the Czech Republic would have loved the Marshall Plan, but they were under Soviet influence, and they were talked into declining it. Maybe there was stricter coercion than just talk going on. Who knows? That's why Eastern Europe is, generally speaking, not quite as developed and rich as Western Europe even today, but they are trying to catch up quickly.

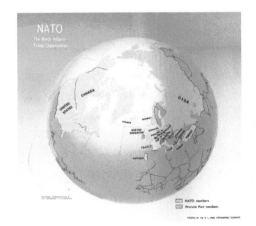

This is a sweet Cold War relic!

NATO VS. THE WARSAW PACT

Back to the story. Once those Soviet Union and Soviet puppet states did not accept the Marshall Plan, the writing is really on the wall at this point. The lines are drawn. The Iron Curtain has fallen, the rift is exposed, and this is most evident by this nice Cold War relic map shown below of the formation of NATO versus the Warsaw Pact. **NATO** was formed by the US and all of those Western European states who were looking over at the USSR and Eastern Europe saying, "Oh, we feel threatened by them. They might invade us. They might come and kill us. Therefore, we're going to form up NATO, which says if any of you Soviet-type people attack any Western European state, the US is going to come to their aid—in fact, all the NATO countries will come to their aid." That's why NATO was formed, on one side of that Iron Curtain.

The USSR, not to be outdone, said, "Oh yeah? You guys have your own NATO club? Fine! We'll start our own club. We're calling it the **Warsaw Pact**." The Warsaw Pact was essentially an anti-NATO device based on the same premise that an attack on one constituted an attack on all. Of course, everyone knows it's bunk, since the participating countries were simply puppets of the USSR anyway. But the Soviets said, "Oh, no, no, no. Everybody's free around here. So we asked Poland if they wanted to join, and they said yes. We asked Romania, and they said yes too." The goofball Soviets even went out of their way to name it the Warsaw Pact to try and demonstrate that it was the Poles who had thought it up. Yeah, right.

What you had was an entrenchment of ideas—NATO on one side, Warsaw Pact on the other. Free western democracies on one side, the Communist Commies on the other. Why am I talking about this in detail? Because that's what Eastern Europe has been through for the last seventy years. Now you know the end of the story, though. It won't last for too much longer as we approach the modern era . . .

THE THAWING

What's happening today? What we've had since around 1988 is that the Soviet Union began to **devolve**. There's that word again! The Soviet Union fully collapsed in 1990. They finally figured out, "We can't do it anymore. We give up." During this entire Cold War, the Soviets kept insisting that the Eastern European states were voluntarily part of their sphere of influence, and specifically told the states themselves, "If you guys don't want to be part of the Soviet Union, just have a vote and we'll let you out." Basically, that was a bluff. Occasionally, the peoples of some of these countries would take them up on it, and have street demonstrations and stuff like that. Every time that happened, the Soviet Union sent in the tanks and shut it all down. Everyone knew the Soviets were lying. They'd say, "No, really, you guys are sovereign states. Really!" Wink wink, nudge nudge. Not really.

As 1989 approached, when the Soviet Union started to collapse, this notion crept back up again. You may have heard of this **Mikhail Gorbachev** guy. He was floating around saying, "The Soviet Union's in trouble for lots of different reasons. However, if you guys over in Poland want self-determination, maybe we'll *think* about letting you guys go again. If you really want self-determination, go ahead and have yourselves a vote." This caused an explosion. Once a little bit of the leash was let out, everybody ran. What you had was an instantaneous devolution, where Eastern Europe pulled away as rapidly as possible from the Soviet Union. You see here in this independence dates map is that the USSR, a single political entity, had turned into fifteen different countries virtually overnight.

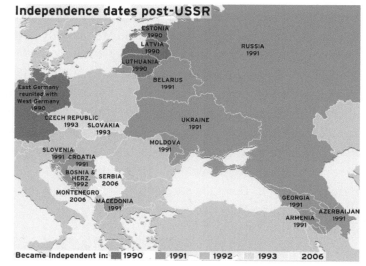

Independence dates post-USSR

Became Independent in: ■ 1990 ■ 1991 ■ 1992 ■ 1993 ■ 2006

Free at last!

Czechoslovakia, Poland, Hungary and lots of other places said, "Yeah! The Soviet Union is going to let us out! Good, we vote to be out, we're out! We're truly sovereign again, we're free!" Places like Estonia, Latvia, and Lithuania said, "We want out too!" Romania, Bulgaria, Ukraine, "We're out!" Everyone wanted out. But it didn't stop there! Even after all these states are out, places like Czechoslovakia said, "Uhh, we're already out, and we want to devolve further into the Czech Republic and Slovakia for lots of different reasons." (See **What's the Deal with . . . the Velvet Revolution?** box on page 228.) There is another, separate wave of devolution in what was Yugoslavia, which all occurs at roughly the same time, but for different reasons, but we will get to that later.

Hotspots of Uprising

FLIGHT OF FANCY

This flight from Soviet influence didn't end with declarations of independence. Oh no! That was just the beginning of their run for the border! Let's take a closer look at three states in particular that set the trend for what was to happen next: Poland, Czechoslovakia, and Hungary. These three states crystallize all of the stuff we've been talking about. During the Soviet occupation and the era of Soviet influence, every time the Soviets bluffed about possible independence, people in these three countries took to the streets to challenge it. As you can see from the map, the biggest street protests and riots in the last 45 years of the Cold War era took place in Poland, the Czech Republic, and Hungary. These countries were never entirely happy with being under Soviet influence. In fact, after World War II, they initially were like, "Yay! We're free! We can be just like Western Europe now!" It's hard to be free when the Soviets set up a puppet government for you.

Now, guess which three countries pushed to get out immediately after the devolution process began in the early 1990's? Guess which three countries wanted in NATO immediately? Guess which three countries petitioned to be in the EU immediately? That's right, kids! It's the same three countries: Poland, Czech Republic, and Hungary. As soon as it was possible, they immediately voted themselves out of Soviet influence. They were the first out of the blocks. But that wasn't enough for them. The peoples within these countries had been so unhappy being under the Soviet yoke for over 50 years that they wanted to ensure that it didn't happen again. All three countries immediately petitioned to join NATO.

We talked about NATO in chapter 6, so you know about the impacts of being a NATO member. These three Eastern European states wanted to become NATO members immediately to ensure no further Russian influence. I say again, *ensuring no further Russian influence.* These guys did not want to be anywhere near the Soviets; they wanted to be a part of Western Europe as fast as they could run to it! You know that NATO Article 5 says, "Hey, anybody that's in our club gets attacked, then we take that as an act of war against all of us." Poland, Czech Republic, and Hungary said, "Sweet! We're in the NATO club! Russia can't touch us. We're like the UK, France . . . we might as well be Canada! Russia can't do anything to us!" That was exactly the case. All these countries also immediately petitioned to get into the EU. Eastern Europe, being under Soviet sway, was just as broke as the Soviets. They had not done well during the Cold War. They started with a blank slate in the early '90s—broke, ideologically bankrupt, embracing the west, embracing NATO, wanting out of Soviet influence, and particularly wanting into the EU as quickly as possible for the sake of their economies.

RE-ALIGNMENT WHILE RUSSIA IS WEAK

As you probably already picked up from some of the details in this chapter, we are in an era of transition for this Eastern Europe region. And now it becomes quite obvious what they are transitioning from and to—from that Soviet era of occupation and control, they are mostly realigning and transitioning to become adopted into the capitalist democracies of the west. This happened at a brisk pace after 1991 when the Soviet Union officially voted itself out of existence, and the entity broke up from one huge power into fifteen sovereign states, as already pointed out.

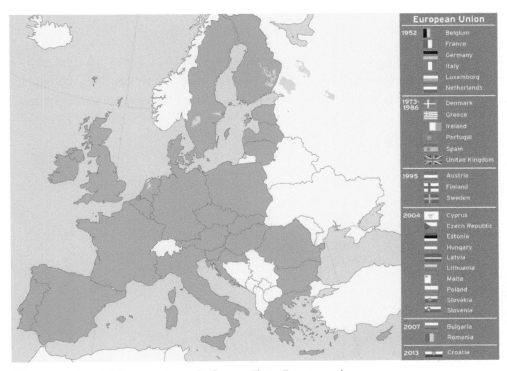

EU creep into Russia's former sphere of influence. Those Euro-creeps!

We also already pointed out, many of the states, Czech Republic, Hungary, and Poland, were quick to embrace the West; Estonia, Latvia, and Lithuania were right on their heels, by the way. Because, as you now have learned in this chapter, they used to be independent sovereign states in between World War I and the end of World War II, but then were reabsorbed by the Soviets. Those guys were always chafing under Soviet rule, so Estonia, Latvia, and Lithuania were the second round of states that jumped ship into the arms of the West. Many others have since followed. All of this realignment and transitioning of these countries of Eastern Europe was occurring right after the Soviet crash . . . quite frankly, when Russia (and Russia became a new country, as well, at this time) was excessively weak. I mean, they just lost the Cold War, and they mostly lost the Cold War because they were freaking broke. You'll read more about that in the next chapter. Their economy was in shambles, their power structure was shattered, and their government was in chaos. It was precisely in this period of Russian weakness that so many of these Eastern European countries just *ran* to the west.

This transition, of course, happened in a couple of distinct ways by some distinct entities we've talked about many, many, many times already in this great text. Number one was the EU—the European Union. Most, if not all, of these countries *immediately* applied for EU entry status. Many of them were soon granted it. As you can see by the maps to the right, there has been a progressive wave of Eastern European countries entering the EU since the very beginning of their "freedom" from Soviet domination.

What's the Deal with . . . the Velvet Revolution?

What's up with Czechoslovakia? Why did it split into two, the Czech Republic and Slovakia? Were there two different ethnic groups that wanted their own self-determination and their own countries? Not really. While there are ethnic Slovaks and ethnic Czechs, neither group had much of an idea that a partition was even occurring. Most couldn't have cared less. There was no animosity between these ethnic groups. So what's the deal? Why did they do it if there wasn't a problem? Why did they split?

They split because the western side of Czechoslovakia, where the Czech Republic is now, was much more like the West. It bordered Germany; it had the industrialized sector, and was richer, with a higher GDP. Slovakia, the eastern side of the state, was much more Russian/Soviet influenced. It was poorer and more agricultural. Folks in the Czech Republic, as the Soviet Union was collapsing, said, "Hey, we want to get into the EU. Help us out, Plaid Avenger! What should we do?"

Where did it go?

I said, "Well, let's look at this strategically. To get into the EU, your economy has to be decent and it has to be stable. As a whole, your economy is neither. But if you were to lop off the poor suckers on the eastern side of your state, your GDP per capita would go up! All your averages would go way up!" Indeed, that's exactly what happened. People in what is now the Czech Republic said, "Hey, we'd be better off on our own without Slovakia dragging us down!" Czechoslovakia underwent a transition called **The Velvet Revolution** in 1993. This revolution was like velvet, nice and smooth. There were no shots fired. From the stroke of a pen, one country became two.

As already suggested, Hungary, Poland, and the Czech Republic were the first three that jumped in. They were followed by a whole host of countries that jumped in by 2004, and the fun's not over yet. Romania and Bulgaria entered in 2007, and there are still multiple potential candidates. Most of the countries that constituted the prior Yugoslavia have asked for entry. It's quite important to note that Ukraine, a huge country and a former part of Russia historically, has applied multiple times and debates within the country are determining just how hard and fast they should push for EU entry. We'll come back to Ukraine in a bit.

Perhaps the most telling tale to tell about the push for the EU in Eastern Europe is Czechoslovakia, a country that used to be one, that's now two—another shattering within Eastern Europe—simply to get into the European Union. That sounds bizarre—you've got to bust up your country just to join a supranationalist organization? Indeed, it is a bit bizarre, and you should know why, when, and where that happened. It was called the **Velvet Revolution**.

That brings us to the other avenue for the Team West love embrace by these newly independent Eastern European countries in the 1990s, and that was NATO. No surprises here. In fact, you can essentially look at the story told by those maps of EU entry and see that the NATO entry tells the same tale. As already suggested, Czech Republic, Hungary, and Poland were quick to escape the Soviets and jump under the NATO security blanket. Estonia, Latvia, and Lithuania right on their heels—just like with the EU. Many of the other countries who joined the EU are also now NATO members.

This has been miffing Russia a bit as they have watched the continued eastward expansion of these western clubs to their Russian borders. And that, my friends, is worthy of an entire new section. . . .

RE-INVIGORATED RUSSIA CAUSING NEW CONSTERNATIONS

What's all that about? Well, this eastward expansion of western institutions, like EU and NATO, is now being slowed, stalled, or stopped, depending upon your point of view. What am I talking about? Hey, I'm talking about Russia, man. These guys are back. As I already pointed out, all of these Eastern European countries jumped the Soviet ship and headed for the west when Russia was weak—right after the Soviet Union crashed, when Russia was down and out, when they were poor, when they were politically bankrupt. But that time has passed. They're back and they're flexing their muscles in more ways than one. And by that, I am thinking specifically of Vladimir Putin. Have you seen this dude topless? Watch out, the man is a menace! But seriously . . . okay, seriously, he is ripped . . . Russia as a great power is back on the world stage, and is no longer allowing encroachment into their arena of influence. We can look at some current events to underscore this Russian resurgence.

ISSUE 1: UKRAINE YEARNING: THE ORANGE REVOLUTION

Like the Velvet in Czechoslovakia, the Orange Revolution in Ukraine was defined by the east/west differences of folks within the state, although unlike the Velvet, it did not end with the permanent division of the state into two new states. Yet. While the Velvet Revolution occurred over two decades ago, the Orange one just happened in 2004–2005. Russia has exerted influence and/or control over Ukraine off and on throughout most of its history. Russian influence is very strong in areas physically closer to it, like eastern Ukraine with whom it shares a border. Back in 2004, Ukraine had a big move towards true democracy. The scenario included two dudes. The first was Victor Yushchenko: the pro-democracy, pro-EU, pro-NATO candidate. A real poster child for the west. The EU supported him, and the United States thought he was awesome. They loved this dude!

His opponent was Victor Yanukovych: an old-schooler, very conservative, more Russian influenced, and in fact, he's the candidate that Vladimir Putin from Russia came over and campaigned for. To try to help lock up the election before it even was held, former elements of the KGB even unsuccessfully poisoned Yushchenko! Talk about a street fight!

How hilarious is this? I can almost hear the boxing announcer: "In the red, white and blue trunks, fighting for Team West, hailing from western Ukraine, it's Victor Yush! And his opponent, in the red trunks, representing the red Russians, from eastern Ukraine, its Victor Yanu!" Hahahaha dudes, could you make this stuff up? Victor vs. Victor—I wonder who will be the victor?

Bush

Yushchenko

Yanukovych

Putin

Bush supported Victor Y. while Victor Y. was backed by Putin. Y ask Y?

Why's the Plaid Avenger bringing up this election in Ukraine of all places? Because it exemplifies what's still happening in Eastern Europe, which is this battle between the West and Russia; it's a battle that's not so much for outright control as it is for influence within this region. Russia still has serious economic and security interests in this area, and still likes to think of it as within Russia's sphere of influence. When this election went down, the pro-Russian candidate won, and everyone in the world said it was a fraudulent election. There were massive street protests, and this ultimately turned into the 2004/05 **Orange Revolution**. So much heat got put on the Ukrainian government, they threw out the election results, re-ran it, and then the pro-western Yushchenko won. We can see this as part of an ongoing battle for influence in Ukraine, which is representative of the broader battle for influence in the region as whole.

The Russian candidate was downed because the pro-western candidate won—downed, but not out. In the Post-Orange Revolution Ukraine, there are still a lot of pro-Russian people. There's a pro-Russian political party, and in March of 2006, the same party gained enough seats to win back control of the Congress. The Plaid Avenger knows that those in the US think democracy's great, that the good guys won, and that's that. Not so! This is an ongoing battle for control. It's not over! Most other countries have gone the way of the west, but it's definitely not over yet for the Ukrainians.

And the country is about evenly split between the western pro-West and the eastern pro-Russia, as you can clearly see from the results of the 2006

election cycle in the map to the right. Could the divisions in this country be any geographically clearer?

Because of the pro-Russian political party gaining the majority of seats in their Congress, they got to choose the Prime Minister position . . . which promptly went to Yanukovych! (The President is voted in by direct election, Prime Minister appointed by majority of Congress.) So after the 2006 election, you had this bizarre situation where President Yushchenko was pro-West, and Prime Minister Yanukovych was pro-East! Talk about a country with a split personality!

This internal division EXACTLY symbolized the fight for influence in Eastern Europe between Team West and Russia. Yushchenko continued to push for EU and NATO entry; Yanukovych actually announced that no way in hell his country would join NATO. The US sent aid to Yuschchenko and his allies; the Russians sent aid to Yanukovych and his party. The Russians and pro-west Ukrainians even had several economic battles over Russian supplies of oil and natural gas to the country.

Meanwhile, the entire country's economy has tanked and they are more reliant on outside aid than ever to keep themselves afloat, which means this pitched battle for influence has gotten even more important. Are you Yanuk-ed out yet? I hope not because the story gets crazier as we play it forward to the present, as the truth is always stranger than fiction. Dig this . . .

The Ukrainians went back to the presidential polls in 2010, and take a wild stab who won. Victor Yanukovych, of course! Seriously? Yep, seriously, and apparently legitimately this time. So is the Orange Revolution dead? Umm . . . kind of, I guess. I mean, Ukraine is still a democracy, and they did have a free and fair election, but unfortunately for those revolutionaries, the same pro-Russian guy they had the revolution against is now the guy who won and is in power.

Second time's the charm for now-President Yanukovych.

The first order of business for Yanukovych was to solidly state there would be no NATO in Ukraine's future, which coincided with Russia giving a huge price break to Ukraine on oil/gas contracts and promising them a bunch of foreign aid as well. Isn't it funny how coincidences like that happen? Yanukovych also immediately extended leases on critical Russian naval bases located on the Crimean Peninsula, a move which infuriated pro-Western politicians who retaliated by throwing eggs at the Speaker of the House when the bill was passed. No. Really. Google it.

In particular, there is a serious fight brewing over the future of Ukraine's southernmost province otherwise known as the Crimean Peninsula, but we will save that story for world hotspots in chapter 25. By all means, flip to it now if you can't wait to find out more! **2012 Update:** tensions within Ukrainian continue to simmer, as European leaders are currently boycotting the Euro 2012 football finals held in Ukraine, in protest against Yanukovych's jailing of former Prime Minister Yulia Tymoshenko. Also, a massive fist-fight broke out in their Parliment over a law put forward by Yanukovych to recognize Russian an official language of their state. I told you this fight was far from over!

Ummmmmmm . . . BOOM! Called It!

2014 Insane on the Brain, Ukrainian Unraveling Update! So glad I pointed out the Ukrainian rift for all these years to all the plaid students who used this text! Because that prediction was dead on the mark . . . literally. Unless you have been comatose since your bobsledding accident in the 2014 Winter Olympics, you have at least heard that Ukraine melted down politically once more. Massive street protests culminated in the overthrow of President Viktor Yanukovych, which prompted the Crimean Peninsula to declare independence and ask for absorption into Russia! Which then happened! The Ruskies have had tens of thousands of troops massed on the Russian/Ukrainian border for months, and now there are similar "join Russia" protests across eastern Ukraine that have turned bloody and violent. Will Russia invade and convert other parts of the state into Russian territory? Who knows? It is all such convoluted chaos right now, that anything could happen! Team West and NATO are furious, the Ruskies are indignant and flexing their muscles further, and some are predicting that this is the start of a new Cold War!

I think that may be a bit of an overstatement, but this is certainly a titanic regional game-changing event.

Ukraine aflame!!!

ISSUE 2: RUSSIAN PETRO POWER

By this, I mean that Russia possesses vast amounts of natural gas and oil, and they are increasingly using this commodity as a political pressure point, or an outright political weapon of choice! For now, know this: Russia provides something on the order of about one-third of all European demands for energy. That's a lot, dudes. By Europe, I mean Eastern and Western Europe depend on Russia for about one-third of its energy needs. What's that got to do with current events? What's that got to do with a reinvigorated Russia? Just this: they have petro power now. To reassert their strength and influence in Eastern Europe, they are playing the fuel card. Meaning, when those Eastern European states get a little too uppity, when they do things Russia doesn't like—for example, if the Ukraine says they want to join NATO—Russia says, "Ho, ho, ho . . . hold on there comrades. You get your oil from us. You make us angry—and the price of your heating fuel might double overnight. You make us more angry—and you get no oil at all. You dig that? It gets cold there in the winter too, don't it? How do you like them apples?"

Russia keeps laying the pipe.

Does this sound like fiction? It shouldn't, my friends; it's already been occurring. Russia has raised prices at various points in the last couple of years to essentially punish states that are doing stuff that they don't want them to do. They did it to Ukraine and Belarus in 1994 and again in 2006 when they doubled the prices to their Belarusian brothers. They completely dried up shipments to Latvia in 2004–06 to pressure the Latvians into a port deal. Most recently, the Russians shut down a section of pipeline that supplied a Lithuanian refinery under the guise that the infrastructure was just too old and needed to be replaced. I'm sure that was totally unrelated to the fact that Lithuania was threatening to veto an EU-Russian economic pact. Total coincidence, I'm sure. Oh, and if you beleive that, then you will also believe that Russia tripling the price of oil to the pro-West government of Ukraine in 2014, and demanding full payment of all outstanding bills, is also a huge coincidence. But of course you don't believe that, because you now know how the rascally Ruskies operate!

Even though the price of oil has dropped drastically since its historic highs of $150/barrel back in 2008, Russia has made trillions over the last decade, with more rubles to come . . . but it's not just about that. Yes, they get rich from it, but more importantly, it has given them political power due to the economic leverage that controlling that commodity brings. Russia is like the crack dealer to Eastern Europe in particular: they've got the stuff that those Europeans need. If you anger the dealer, you might not get your energy fix. That's not a situation the Eastern Europeans are happy about. The Russian influence is back. That's not all that Russia is doing to flex its muscles in this region.

Early version of European Missile Defense Shield.

ISSUE 3: NYET! TO NUKE SHIELD

Another thing that's extremely topical—and we haven't seen the end of this story yet, either—is the proposed missile defense shield system that is being pushed hard by the United States to be implemented specifically in the Eastern European countries that are NATO members. Specifically right this second, we are talking about the Czech Republic and Poland. This is specifically incensing the Russians.

Maybe I should back up—what is this *missile defense shield* nonsense about? This is a concept the United States has been toying with for 60 odd years. It has never really come to fruition. But the idea is, you set up a bunch of missiles and a radar system which can detect if any missiles are coming into Europe. Then, you launch your missile to blow up the incoming missile before they land. Again, it's never worked; out of perhaps ten million tests maybe they've done it once successfully, but that's not the point of this rant right this second.

It's the fact that these new NATO members in Eastern Europe have said, "Uhh, okay, yeah, we'll do it since we still support the US." At a NATO summit which occurred in February of 2008, all the NATO countries—all of them—agreed to go forward with the US plan for the missile defense shield. And just to kind of wrap this up, Russia is seriously aggravated about it. They have been pounding their fists on the table, saying, "Nyet, this is not going down. This is seriously threatening us and we are going to use economic leverage and political leverage to intimidate the Eastern European countries that are going along with this."

What do they mean by that? Well, they've already made the fuel threat. When the Ukraine said they wanted to join NATO, Russia said, "Yeah, that's fine. You join NATO and we will re-target nuclear missiles at you." This is something that Russia has also promised to do to any of the states that jump into the missile defense shield program. This is serious stuff, man. Again, this story is still ongoing. The United States says, "Hey, look Russia. Come on, dudes. The Cold War's over. We're not enemies anymore. This is mostly a defense system against . . . Iran, ooh, Iran. Yeah. Iran's going to send a missile somewhere, so this is about that. Or other terrorists might bomb our NATO allies."

But Russia says (and you've got to empathize with them a little bit here), "Uh, yeah right. Come on. NATO's been encroaching to Russian borders for the last twenty years. Now you're putting up this big missile defense system. It's an obvious attempt to neutralize or marginalize Russian power in this region and in the world."

The United States is probably telling the truth, but Russia sees it in a very different light . . . and, again, I always try to empathize. Not sympathize, not agree with, but empathize. The Russians do have a long history of being screwed with by folks over in Western Europe, from Napoleon, to Hitler, right on up through the 1980 US hockey team's miracle victory over the Soviets at the Winter Olympics. So in no way, shape, or form, or any time soon, are they going to be okay with an encroaching military technology that is set up on their borders. And they're really ticked off about it. Vlad "the man" Putin has been saying, "No, we're going to do everything we can in our power to make sure that does not happen, including veiled threats about missiles and open threats about energy issues." As always, sucks to

The Russians are still riled up about that game.

be in the middle of the mess, but Eastern Europe seems to have a historical niche for it. **2012 Update!!!** At the big NATO summit in 2012, the NATO and the US officially announced that the first phase of the shield network in now up and operating! Two more phases in the future will add more radar stations and missile interceptor launch sites! Russia no like! Yikes! Are we entering a new Cold War phase for Eastern Europe? Let's watch and find out! **2014 Update!!!** Given the new tensions between Russia and the US over Ukraine, look for NATO to increase exercises and troops deployment to Eastern Europe, and look for the US to invest billions more into the missile defense system . . . even if the system don't work, it will infuriate the Russians, so money well spent either way.

ISSUE 4: YUGO-SLOB-IAN MESS: KOSOVO CHAOS

Finally, for the current events underlying what's happening in Eastern Europe with this reinvigorated Russia, comes the latest chapter in the disintegration of what was Yugoslavia. I've intentionally kept the mess out of the chapter up until this point because it's a debacle that muddies the Eastern European waters a little too much to tackle it up front. But now that you understand the rest of the region, I am confident that you can now figure out this confounding coalition of convoluted crap with the greatest of ease. Well, we at least have to give it the old college try, since events here are once more crystallizing the fight between the West and the East. Let's do it.

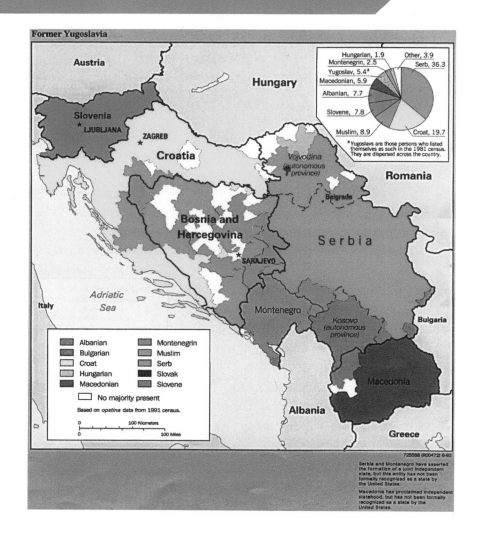

Former Yugoslavia

Yugoslavia was different than the Soviet Union. Let's revisit them so you'll understand what's happening in today's former-Yugoslavia. First off, they were never part of the Soviet umbrella. They were commies, with a commie-run government after WW2, but not Soviet. It was a communist state under the tutelage of **Marshall Joseph Tito**. "Marshall" is really just a nicer sounding title than "oppressive dictator extraordinaire." The Plaid Avenger suggested earlier, as we were talking about the historical background of this region, that perhaps no other singular country has more different ethnicities and religions and cultures as did the former Yugoslavia. Emphasis on *did*.

They had Slovenes there, Croats, Bosnians, Herzegovenians, Serbians, Montenegrins; I don't even know what half these terms mean! On top of that, you have Christian Serbs, Muslim Serbs, also Christian Bosnians, a group that encompasses Orthodox and Catholic Bosnians, Muslim Bosnians . . . It's confusing! It was a multi-ethnic, multi-religious, multi-media mess.

Why has there only been serious conflict and devolution there in the past twenty years? It's all because of Joseph Tito; not the Tito of the Jackson family, the other Tito, the dictator. Tito was a strong-armed dictator, like Stalin and lots of other folks in the communist realm. When ethnic or religious tension reared its ugly head in what was Yugoslavia, he sent in the army and crushed any uprising or disturbance with the iron boot. Through force, Tito held together this state full of all these different ethnicities and religions. Tito was

Tito double vision: Josip or Jackson?

the only reason it didn't devolve years ago. However, he died in 1980. Since Yugoslavia was a military state, it was able to perpetuate itself for another decade, but eventually imploded from the stress of diversity.

Look at the time frame and do the math. Tito died in 1980; the state held on for about ten years. That brings us to about 1990, the same time that all the other parts of Eastern Europe were devolving, splitting, splintering, fracturing, shattering into separate countries because of the implosion of the USSR. It all hit the fan in Yugoslavia as well. Without the strong-armed dictator Tito there to hold the different groups together by force, the whole country collapsed on ethnic lines, followed by wars fraught with ethnic discrimination, vicious bloodshed, and human rights violations. Long story short, Eastern Europe in the last 20 years saw the USSR devolve from one entity to fifteen sovereign states, Czechoslovakia devolve from one country to two, and Yugoslavia devolve from one country to five in the 1990's. In 2006, Montenegro became #6. And the fat lady has not sung yet, my friends, which brings us up to current events. . . .

As you can see, the Yugoslavians have had kind of their own tale that's a bit separate from the USSR, but of course related, and that brings us full circle to today's world. And the shattering may not be over yet! Slippery #7 may be on its way, but this time it won't be without a fight. Kosovo may be the next to go. You by now know the story of Slobodan, genocide in Kosovo, and the US/NATO invasion of Serbia to stop it. The end of the tale goes like this: In February of 2008, Kosovo declared independence. Immediately, a bunch of Western European states and the

What's the Deal with Slobodan?

We all know this guy! He was arrested for war crimes, and died in The Hague before he even got through his trial. What did Slobodan do that was so important that the Plaid Avenger has to tell you about it? The ultimate test of sovereignty is that you can kill citizens in your own country with no international repercussions. The government can do anything it wants to in its country if they are a sovereign state; it's their right.

Slobodan Milosevic

So what's the deal with Slobodan? He's another strong-armed guy like his predecessor pal Tito. Except when he came to power, his state was disintegrating. He helped perpetuate the conflict that led to the breakdown of Yugoslavia. During that period, he essentially allowed genocide to occur. He was an Orthodox Serb, which was the majority ethnic and religious group. He allowed the Serbian people to start picking on and beating the crap out of the ethnic Albanian-Muslim minorities in the Kosovo region, which were perhaps petitioning for independence. He not only allowed violence; he promoted it.

This started to spiral out of control in the early 1990s to the point that people in the outside world—particularly Clinton from the United States and lots of folks from the UN—began debating the limits of sovereignty. After the US/NATO intervention, Slobodan was eventually arrested for war crimes and died while on trial in 2006. Oddly enough, when his body was brought back to Yugoslavia, lots of people were cheering him. He's considered a war criminal by most of the world, yet lots of Serbs have him held up as a hero.

US recognized their independence, which, of course, *infuriated* both Serbia and Russia. Russia is not very happy about this. NOTE: Not all Western European states recognized Kosovo, and therefore, the EU as a singular entity did not recognize it either. The big boys like the UK and France did, but many other minor European powers did not, for reasons we shall get to in a minute.

We have this situation that, again, crystallizes this new power struggle within Eastern Europe that's related to the old power struggle. Russian influence versus Western influence. The Russians, their historic ally Serbia (that is, what's left of Yugoslavia), and other entities around the world said, "You can't recognize Kosovo. You guys were promising all along you wouldn't." Because, of course, the United States and NATO intervened in Kosovo, kicked out Slobodan, and protected the Kosovars. But they always claimed that they were merely preventing genocide, and not carving out a new state. Russia argued, "Hey, all along when you guys did that intervention thing, you said that you weren't setting up a new state, that you weren't going to let them have independence, that you were just going to stop the civil war, and basically you're liars."

Russia is not alone on this issue either. China, as well as many European countries, have refused to support Kosovar independence because they think it sets a bad precedent. If there's one small, little group of people that are pissed off at a country and you're going to allow them to declare independence, then there's going to be a whole lot more shattering going on, and not just in Eastern Europe. You're setting a precedent for other folks to do it around the world. Russia don't want the Chechans declaring independence; China don't want Tibetans declaring independence either. It's interesting to note that places even in Europe like Spain did not support Kosovo independence because they said, "Hey look, we have a small group of radicals who want an independent Basque country, so no way. There's no way we're going to support an independent Kosovo, because then the Basque people will say, 'We're going to be independent too.'" You see how complicated this issue has become.

Kosovo still causing consternation.

Back to the point: this is all about Russian influence here. The Russians are supporting their old Serb allies, saying, "No, this is nonsense. We ain't recognizing Kosovo and nobody else should either." And the West is saying, "Oh, but, you know, they want to be a democracy, and peace-loving and all that stuff. We are going to support 'em.'" Yep. Eastern Europe right back in the middle again. I guess old habits die hard in the end.

The point of this last section is that Russia is back. They are re-flexing their muscles. They are reasserting influence, not control, but influence over their old Eastern European sphere of influence, and the West is still pushing in as well. **2014 UPDATE**—Ok, I guess we can now report that the Russians are actually also reasserting real control, since they officially re-absorbed the Crimean Peninsula of Ukraine back into the Russian motherland! And perhaps they are going to "re-absorb" more parts of Ukraine, as the state teeters on the brink of civil war! Maybe the Ruskies will take back Moldova too! Who knows? Honestly, not even the Russians know quite yet what they are going to do, and all of us will just have to watch as events unfold. But Eastern European states are once again back in the middle ground battle-zone between competing powers . . . that much we can say for sure. Same as it ever was.

CONCLUSION: THE TRANSITION IS ALMOST FINISHED!?

A new era, a new time, and it seems like the Eastern European transition is almost done. Most of these states have joined the western block institutions, with perhaps just a few more to go. Most countries are in the EU. Most are in NATO. They are Europe now. Most are part of Team West. Most. Some may be leaning back towards Mother Russia. Some.

But Eastern Europe has historically been in a vice and perhaps it may be ready to explode again. This region does have a strange history of being under the influence of these greater powers to its east and its west, which is why I started the chapter by calling it "bookended." Somehow, Eastern Europe is the one that sparks all the trouble between the sides. Seriously, think about it. Go back to World War I: It was started in what is now Yugoslavia, when Archduke Franz Ferdinand (it's not just a band) was assassinated in Sarajevo by a young Bosnian Serb from the Serbian radical group named the **Black Hand**. World War II started when Hitler invaded Poland. Then the Cold War battle lines were drawn in Eastern Europe, with most of the missiles either deployed on it, or pointed at it.

Why list all these transgressions again? To stress a not-so-obvious point: all major conflicts of the last century were sparked in this bookended region. Hmmm . . . that certainly is food for thought, isn't it? Major players seem to get sucked into confrontation over this swath of land between Russia and Western Europe. Sucked in, and then life really sucks for all involved.

The disintegration of both Ukraine and Yugoslavia also started here in Eastern Europe—go figure, because they are located there—and it's not done yet. That's why I bring up this Kosovo situation, which is once again pitting an east versus west, a Russia versus Team West. Both the Ukrainian and Kosovo situations are promising to polarize these sides unlike really anything else that's going on here at the dawn of the 21st century. I know what you're thinking: "Hahahahaha what a load of bull, like a major war is going to start over some insignificant little hole in the wall like Kosovo, or Crimea, or **Transnistria**?" Yeah, you're probably right. My bad. How stupid am I being? I mean, the odds of that happening are about as likely as a global war that killed millions being started over the assassination of an unknown Austrian Archduke while motor-cading through Sarajevo. Ummm . . . oops. Bad example.

The question is, when will these people quit? How many more subdivisions can occur? How much more shattering before it's over? Quite frankly, the answer is perhaps never, because it's outside forces which seem to promote, antagonize, and continue to push in this battleground buffer zone that we call Eastern

Eastern Europe today . . . at least for now.

Europe. Ukraine is now the country to watch as the next chapter of Eastern European devolution, and it promises to be the most exciting, action-packed episode yet! And never, ever discount the possibility of Bosnia or Serbia blowing back up again; tensions continue to simmer under the lid of the former Yugoslavia. But that's all for now. We keep talking about Russia being involved as one of the players in this battleground of ideologies in Eastern Europe, so perhaps we should now turn our attention to the Bear. Oh Vlad . . . are you out there . . .

EASTERN EUROPE RUNDOWN & RESOURCES

 BIG PLUSES

→ Due to lower wages and overhead, Eastern Europe is attracting lots of foreign investment and jobs from Western Europe

→ Eastern Europe produces a lot of agricultural products and is becoming a manufacturing hub for Western Europe

→ Incorporation into the EU promises to help raise standards of living and the economy for most Eastern European states to eventually reach Western European standards

→ Incorporation into NATO promises to ensure political and military stability for its Eastern European members

 BIG PROBLEMS

→ Many states of the region have a long way to go to achieve western-style standards of living

→ Possibility of ethnic/religious/nationalist conflict rearing its head again is highly likely, particularly in the Balkans

→ Non-NATO states in the region (Ukraine, Belarus) and even some NATO members (Poland, the Baltic states) will continue to be torn between their western allies and having to suck up to Russia

→ Many Eastern European states are heavily reliant on energy supplies from Russia

→ Ukraine. New Cold War warming up? 'Nuf said for now.

KEY TERMS TO KNOW

Acropolis	Euro	Pastoral Nomadism
Benelux	European Union	Peninsula
CBD	Eurozone	Periphery
Centrifugal	Gateway City	Schengen Zone
Centripetal	Greenbelts	Subsistence Farming
Core	Industrial Revolution	Supranationalism
Demographics	Maritime	
Diffused	Nation States	

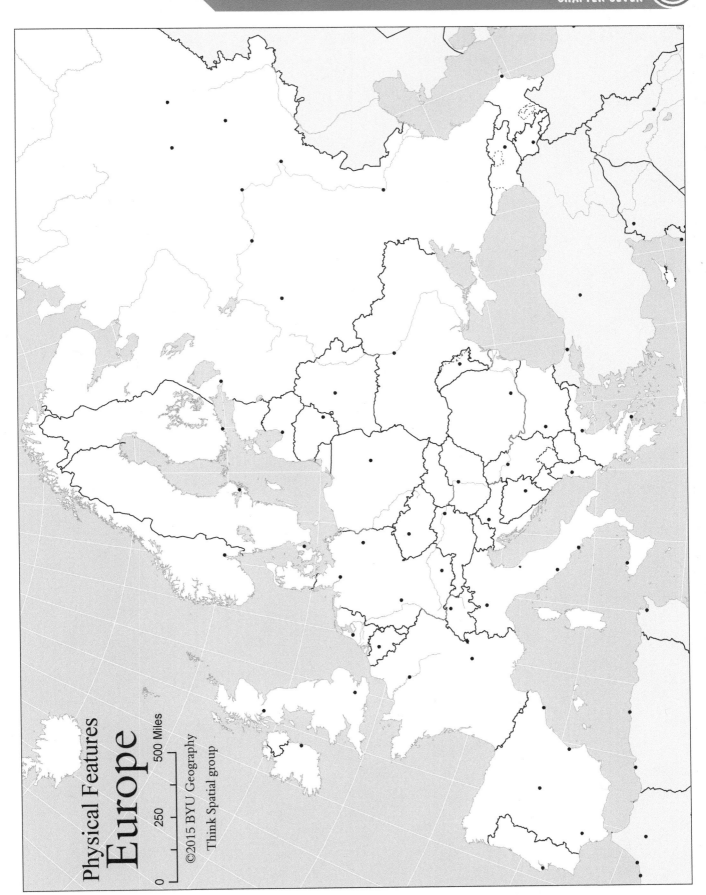

Physical Features
Europe

0   250   500 Miles

©2015 BYU Geography
Think Spatial group

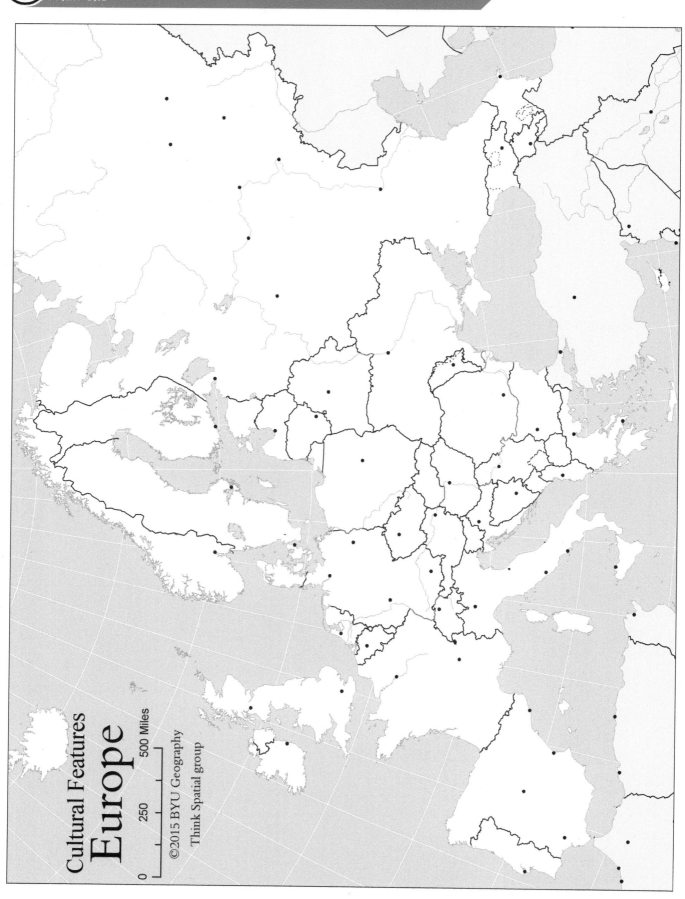

Cultural Features
Europe

500 Miles

250

©2015 BYU Geography
Think Spatial group

0

Russia

8

> Therefore I tell you, do not be anxious about your life, what you will eat or what you will drink, nor about your body, what you will put on. Is not life more than food, and the body more than clothing? Look at the birds of the air: they neither sow nor reap nor gather into barns, and yet your heavenly Father feeds them. Are you not of more value than they? But seek first the kingdom of God and his righteousness, and all these things will be added to you. Therefore do not be anxious about tomorrow, for tomorrow will be anxious for itself. Sufficient for the day is its own trouble. (Matthew 6:25–26; 33–34)

CHAPTER OUTLINE:

THE Russians are a testament to endurance. They have persevered through challenges. As Russia continues to transform today, many geographers ask questions like, "What makes the Russian people so resilient?" and, "Will this resiliency continue? In Romans 5:3 Paul tells us tribulation produces perseverance. Are you able to discern applications for your own life based on the geographic challenges the Russians faced and lessons the they learned.

FROZEN CHOSEN

Russia is huge! The vastness and topography of Russia have implications both for the cultures and the greatness of the largest landmass on the planet.

Adapted from *Blue Marble: Next Generation* image produced by Reto Stockli, NASA Earth Observatory (NASA Goddard Space Flight Center)

Russia is, in many ways, a harsh environment. The sun literally never sets on the nation-state. This huge landmass spreads longitudinally about 170 degrees. Essentially flat in the northern parts of the nation, the ground is in a state of **permafrost** for much of the year. Unlike Scandinavia in Europe, which receives a great deal of snow, there is no maritime or ocean current to provide the requisite moisture for snow. The result is a very cold and dry continental climate.

DISTANCE AND LONGITUDE

When one considers that the southern areas of Russia generally correspond with those along the **ecumene** of Canada, one has a greater appreciation for the coldness and hugeness of Russia. Still, it is important for one to remember that the effects of **topography** play a very important part in creating climate. High mountains both block and trap water as the clouds cool off through a process known as the adiabatic lapse rate that results in orographic uplifting. In the case of Russia, the only real topography that blocks moisture producing a windward and leeward effect are the Caucasus Mountains. These mountains play an extremely important role psychologically in the minds of the Russian peoples as they provide a natural border for the state. The Ural Mountains, go generally north to south and are the dividing range between what is considered European Russia and Asian Russia. The other important areas of relief include the ranges from the Pamirs through the Tien Shan and from the Altay Mountains to the Sayan Mountains to Lake Baikal. These mountains are important not just in how they provide natural borders for Russia to the south but how they affect the hydrography of the region.

Russia is mineral rich!
Map from the U.S. Dept. of Energy

If you look at the map and find Lake Baikal, you will see that it is in the shape of the southern mountain chains too. Like the lakes in East Africa that follow the rift valley, this huge lake also is a function of the folding of the earth's surface. Some experts believe about one-fifth of the freshwater on Earth is in this deep lake. Unfortunately, as we shall see, there are some environmental risks inherent in this. Anyway, the largest lake in the world is the Caspian Sea. Why is this large body of water not a sea and yet another smaller body in central Asia, the Aral Sea, is? This is debatable, but remember this, the definition of a sea is that its salinity levels correspond to the world's oceans. Both the Aral Sea and Lake Baikal do have salt in them, but it is much less than that of the world's oceans because the input of rivers flowing down from both the Pamir's and the Tien Shan Mountains tends to reduce the salinity. Since the definition of a sea is that it has salt water, and since, according to Mikhail S. Blinnikov, the Caspian Sea only has one-third the salt water of the world's oceans, it is therefore a lake![1]

Most of the rivers of Russia flow from south to north because of the southern topography previously described. The area of Russia today referred to as Siberia is neatly divided by two major rivers flowing north. Between the Yenisey and the Ob Rivers is the West Siberian Plain. This area is very swampy when the permafrost begins to melt in spring. The next major river to the east is the Lena River, and this begins what is referred to as the Russian Far East.

GOT GAS?

An aspect of major importance to the Russian economy now is the decision to depend upon the export of natural resources. Energy of course is a major component of these exports, in particular, natural gas from the region of western Siberia bound for energy starved Europe. Coal is another resource the Russians have historically utilized. Next to the United States, Russia is the premier producer of coal in the world. Taken as a whole, the Russians have a huge area they call home that is cold and crossed by rivers and bordered by mountains. Rich in mineral resources, they lack one thing: harbors.

It's not that Russia doesn't have harbors, but it is an unfortunate aspect of their geography that only one major ice-free port near the Atlantic Ocean is adjacent to the Barent's Sea and lay in Murmansk where it receives the benefits of the Atlantic currents; the same currents that bring rain and warmth to Scandinavia and northern Europe. Nevertheless, with temperatures barely above zero degrees Fahrenheit during the winter, Murmansk is no place for a picnic. Note also the distance in relative location from the Norwegian Sea, which, in addition to being largely above the Arctic Circle, is greatly separated from the ice-free port in Vladivostok on the Pacific Ocean. This enormous distance leads population geographers to wonder about the future of Russia and its ability to control its far east possessions.

BABIES ANYONE?

Of increasing interest to **demographers**—or students of population and changes in it—are the statistics and distribution in the population of Russia. Like Europe, the **population pyramids** are increasingly inverted. This inverted population graph indicates an aging population. Additionally, women outlive men in many areas by enormous amounts. Of historical concern to Russians is their ability to maintain control of the Russian Far

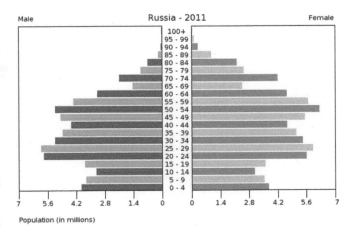

Male   Russia - 2011   Female

100+
95 - 99
90 - 94
85 - 89
80 - 84
75 - 79
70 - 74
65 - 69
60 - 64
55 - 59
50 - 54
45 - 49
40 - 44
35 - 39
30 - 34
25 - 29
20 - 24
15 - 19
10 - 14
5 - 9
0 - 4

7   5.6   4.2   2.8   1.4   0   0   1.4   2.8   4.2   5.6   7

Population (in millions)

An inverted pyramid is the opposite of the arab countries shown in Chapter 5 page 139.

[1] Mikhail S. Blinnikov, "A Geography of Russia and Its Neighbors" (New York: Guilford), 17.

East, an area increasingly being replaced by fast growing populations from Muslim Central Asia. Many experts say a population must increase by at least 2.1% per year merely to continue to exist. If this is the case, the Russians with their 1.6% annual growth rate are in a similar situation to Europe. The government of Russia now pays $10,000 to parents if they have a second child to increase birthrates.[2]

HOW COLD I AM

Unfortunately, like in the United States, many Russians have turned to alcohol because it is cheap and available, and it helps them withstand the toughness of life. While hope can seem far away at times when one studies the Russian peoples and their histories, the truth is God has a special plan for all of us. He is in control of Russia just as He is in control of the United States. Even though life can be hard and very discouraging and frightening at times, one thing we can be sure of is that His love for us remains intact. In the Bible we see the verse, "For I know the thoughts that I think toward you," says the Lord, "thoughts of peace and not of evil, to give you a future and a hope." (Jeremiah 29:10–12 NKJV)

RUSSIA*
THE BEAR IS BACK

Russia is the largest country on the planet. They were the Cold War adversary of the US; you know, the bad guys. We refer to Russia as the Bear, and this wild woolly bear seems to change focus and direction every semester that passes, just as a wild bear hunts in the woods. I'm scared! Not really, but it is a region that is in transition, much like Eastern Europe; in this regard, it is changing every day. However, unlike Eastern Europe which has a very distinct direction of change, Russia is what the Plaid Avenger calls a fence-straddler—that is, it is playing the field in terms of with whom it does business, with whom it has a political alliance, and with whom it will throw in its future.

Time to bear all!

Russia has been a distinct culture, a world power player, a part of the Soviet Union, and a major shaper of world history and events of the 20th century. It has experienced tremendous political and economic upheavals and can almost be considered behind the times here in the 21st century . . . but hang on, because the Putin-inspired takeover of the Crimean Peninsula here in 2014 proves that Russia is back on course for a possible huge power grab! How did this region go from a global power to a globally broken basket-case? More importantly for our understanding of today's world, how is it successfully getting back into the game? All of these things and more, we will explore as we tell the story of the Bear.

One big bear. US/Russia size comparison.

---

[2] Blinnikov, Michael S., "A Geography of Russia and its Neighbors" (New York: Guilford Press, 2011), 140.

HAIR OF THE BEAR

We always start our regional tour with a breakdown of what's happening in the physical world. When we think of Russia, several words typically come to mind: big, cold, and vodka. There's some truth in this. Russia is a cold place, a big place, and those Russians do love their vodka. But kids, just say "No" to the vodka . . . unless you are in Russia, of course, and then you say, "Nyet" . . . although apparently few of them do.

Russia is the largest state on the planet, twice as big as Canada, the second largest. Does size matter? Do I even need to ask? Of course it does! Size matters in most aspects of life. This huge country crosses eleven time zones, spanning almost half of the globe. When the sun is setting on one side of Russia, it is rising on the other side! Just think of all the problems associated with infrastructure and communications in a state this big. Think of all the problems associated with just trying to keep all of your people within your country aware of what's going on at any given second, much less having to move things around in it, like military equipment.

Don't even try to think about having to defend its vast borders. Your head will explode. Russia has a top slot in terms of how many different countries it borders, which I believe is now fourteen in all. But it's not all bad. This massive territory contains loads of stuff—what stuff? All kinds of stuff: oil and gold, coal and trees, water and uranium and lots and lots of land. What do lots of people do with lots of land? They typically can grow lots of food, but that's not always the case in Russia because of its second physical feature—it's cold.

Maybe that's why the bear has such a thick coat. Virtually all of Russia lies north of the latitude of the US-Canadian border. This high latitude equates to the cooler climates found on our planet; indeed, some of Russia lies north of the Arctic Circle, which means most of its northern coast remains frozen year round. To compound matters, Russia lies at the heart of the Eurasian continent; because of this fact, it has perhaps the greatest extremes of **continentality** expressed by its extreme temperature ranges, both season to season and also day to day. There are some parts of Russia in which the temperature within a 24-hour period may change 100 degrees: that is, it could be 60 degrees Fahrenheit at 3:00 in the afternoon and 12 hours later it might be −40 degrees Fahrenheit. That's extreme. Maybe this tough climate has something to do with Russian attitudes and their dour outlook on life in general. A tough place with a tough history which brings us to our last physical descriptor: vodka.

Russia is the greatest consumer of vodka on the planet. Both in terms of total quantity and per capita consumption, no one can touch the Russians when it comes to drinking vodka; alcoholism rates are high, and this mind-numbing stuff continues to impact society and health in a variety of sobering ways. Perhaps it's the bleak and challenging climate that makes the Russian life so hard and vodka consumption a necessity to deal with the challenge. Perhaps this is why vodka is such an integral part of Russian history. Did someone say Russian history?

Siberia! The Spring Break destination of the Soviets!

FROM CUB TO GRIZZLY

How did the Russian state get to be the largest country on the planet? How did it get so much stuff? Where did these guys come from? What does it mean to be Russian? The "original Russians" were actually Swedish Vikings who moved into the area around modern day Moscow to Kiev, probably around 1000 to 1200 C.E. Thus, Russians look European and share many of their cultural traits.

HISTORIC GROWTH

Looking at the maps across the bottom of the page, we can see that Russia, in 1300, was more like a kingdom than an empire; indeed, it was not a huge territory. It was a small enclave of folks that called themselves ethnically Russian but were politi-

From the land of Volvo, we come to conquer Kiev.

Ivan the Great: Mangled the Mongolians.

cally subservient to the real power brokers of that era: the Mongols. **The Golden Horde** was a political subunit of the Mongol Empire that controlled parts of Russia and Eastern Europe at the time. It was not until 1480, under the leadership of Ivan the 3rd, a.k.a. **Ivan the Great**, that Russia stood up to the Mongols, threw off the yoke of Mongol oppression, and began its growth into the land juggernaut we know today.

Ivan united his people, kicked some Mongol butt, and expanded the empire. His son, Ivan the 4th, a.k.a. **Ivan the Terrible**, continued this trend by feeding the Bear and expanding the empire further. Ivan the Terrible earned his name by having an incredibly nasty temper, eventually murdering his son and heir apparent with an iron rod. Nice guy. Good family man.

Soon after the Ivans created this empire called Russia, an imperial family line was put into place that lasted uninterrupted up until the 20th century. They were the Romanovs, to whom you've probably heard references before. I'm not going to go into great detail on all of them, but just to put a few into perspective, let's list some of the more famous names. **Peter the Great** popped up around 1700 and was known for many things, notably creating and building the town of St. Petersburg. Hmmm . . . I wonder where he got the name?

Why is this significant? Because it shows what Peter was all about, and that was embracing Europe. Even back then, Russia lagged significantly behind its European counterparts in terms

Ivan the Terrible: Skewered his son.

Russia in 1300   Russia in 1462   Russia in 1584

of industrialization, technology, military, and economy. Peter was all about catching the Bear up. St. Petersburg was often called "a window to the West" because the city embraced Western technology and architecture and there was a real sense that it was opening Russia to interaction with the rest of Europe. Here's another example of how Russia is tied historically to the West: the term "Czar" in Russia was taken from the Roman Empire's "Caesar." In addition, Moscow was often referred to as "the third Rome" throughout history, as it viewed itself as the protector of true Christianity, in its distinctive Eastern Orthodox tradition. (Rome #1 was Rome; after the Roman Empire collapsed, then Rome #2 was Constantinople; after the Byzantine Empire crashed, then Rome #3 was Moscow). Russia has embraced and associated itself with the West in many ways over the years, although in most cases, it lagged behind.

A Tit Bit for a Russian Ambassador!!!!

Russian ambassador dealing with Napoleon.

Peter expanded the empire in all directions. This brings us to an interesting component of the Russian Empire's growth: Why expand? Many historians have speculated that there has been an inherent drive to expand the state in an effort to gain coastline. All major European nations with world dominance during this era had a lot of coastline and were maritime powers. Because of Russia's position on the continent, it is conceivable that most of its growth occurred in part to reach the coast in order to become a world power as well. This is speculative to be sure, but it does make some common sense.

Peter the Great: Wanted to be Western.

One resource that's certainly driven Russian continental expansion was fur. Given the climate of this part of the world and the way people dressed, it certainly is the case that the trading of fur—that is, stripping the exterior of live animals and making coats out of it—was the main economic engine behind the continued growth of the land empire. Cold climate = animals with lots of fur = animals we want to kill for their fur = fur coats worth lots of money = more land = more furry animals to continue driving the economy. Speaking of continued land growth, we come to another familiar name in history, **Catherine the Great**. She radically expanded the empire across the continent, was a familiar face in European politics, and helped propel the Russian state into serious power player status on the continent during the 1760s and 1770s. Catherine continued to modernize in the Western European style, and her rule re-vitalized Russia, which grew stronger than ever and became recognized as one of the great powers of the continent for the first time ever.

Catherine the Great: Made Russia into a European power playa'.

The last face to consider during our old school Russian tour is that of **Napoleon**. Wait a minute! He's not Russian! True, but he did invade Russia in 1812 in one of the many great blunders

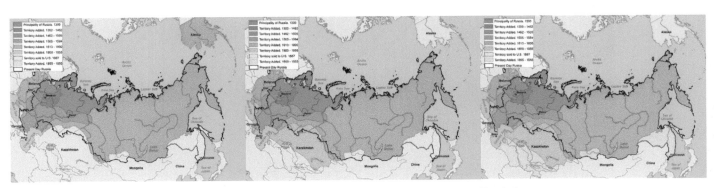

Russia in 1800

Russia in 1867

Russia in 1955

in history. Napoleon was doing pretty well conquering Europe until he invaded Russia. Here's a little Plaid Avenger tip from me to you: don't invade Russia. Don't ever invade Russia. No one has ever successfully invaded Russia. As Napoleon—and later, Adolf Hitler—found out, it's easy to get *into* Russia, but it's impossible to get out. As the French forces advanced to Moscow and eventually took over the city, they did so in the middle of a horrifically cold Russian winter, and because of their dwindling supplies, were forced to retreat. They were attacked continuously on their way out. Bad call, Napoleon. You silly short dead dude! Russian Trix are for kids!

I bring this up to reiterate a common theme in Russian history: Russia always seems to come out stronger at the end of extreme turmoil. After Napoleon's invasion, it comes out sitting pretty as the world's largest territorial empire. We will see this theme again after World War II.

SERF'S UP, DUDES!

I've painted a pretty picture with lots of faces and names that you've heard before, but don't let me suggest that life and times in Russian history were good. Mostly they really sucked, for the lower classes in particular. During the imperial reign I've described thus far, life was fine for the royal court, but life as a commoner was brutal. We have a sort of a feudal system in Russia for most of its history. Serfdom in this country really turns into something more equivalent to full-on slavery. What we're talking about here is a typical feudal structure where there is a lord who owns the land and under him, the mass of people in the country are workers tied to that land. As such, they have no rights to hold landed notes, no human rights, and are basically slaves of the landholder. In combination with the fantastic climate of Russia, one can see how living here for most folks was not a fun time. Chronic food shortages and mass starvations were common occurrences as particularly brutal winters, bad crop harvests, and repressive taxation systems kept the peasants in a perpetual state of misery.

Why talk about the plight of Russian peasantry? Because the dissatisfaction among a vast majority is soon going to culminate in a revolution; a revolution the likes of which the world has never seen. It wasn't like the aristocracy didn't see it coming, either.

Dimitri! Where's the party, dude? Looks like the fun never stopped for the Russian peasantry!

In an attempt to placate the masses, Tsar Alexander II passed the sweeping Emancipation Reform of 1861, which was supposed to end this miserable peasant situation. Say what? Emancipation didn't occur until 1861? That's late in the game! Consider: that was the same year that the US started their Civil War in order to emancipate slaves! Unfortunately for Russia, this proclamation on paper didn't amount to much in real life, and the crappy peasant existence continued. By the turn of the century, freed slaves in the US had more rights than supposedly freed peasants in Russia, not to mention that the vast majority of Europe as a whole had long since abolished both slavery and feudalism.

As we progressed toward the year 1900, many Russian folks were still essentially slaves of the land. This was a time when the rest of Europe was industrializing and getting more connected. Europe had political revolutions, which created states that emphasized equality for individuals, even the lowest classes. People in Russia knew what was going on in the rest of Europe. They heard about the French Revolution and Britain's change to a constitutional monarchy. If you look at a political map of Europe's changes over time, you'll see that things started in the West, progressed eastward, and these changes took a long time to make their way to Russia. Indeed, economic and political revolution reached there last, but when it did come, it came big.

As changes were enacted in Europe, as common folk in Europe gained more and more rights and perhaps even more and more wealth, Russia was stagnant. Russia was falling behind yet again. Dissatisfaction was also on the rise, and things in the 20th century made it quite a bumpy ride for the Russian Bear, particularly for its royal line.

THE BEAR TAKES A BEATING

The 20th century, taken as a whole, was not kind to Russia. It started the century on a downward slide. As Russia fell farther and farther behind the times and as popular dissent increased, the imperial government did the worst thing possible: it lost a war. One of the things that kept peasants in line was the concept of a strong central government, even though it may not be popular or even good. Citizens wanted a government to protect them from foreign powers, invasion or destruction from invaders. Even though the monarchy was not popular, at least they maintained a strong military presence to guard their citizens and territory. When they failed to do that, full failure of the state was not far behind, and Japan provided the first kink in the armor of the **Romanov** line.

GODZILLA ATTACKS

In 1905, the Japanese declared war on Russia. The Japanese, having risen in power for the last fifty years up to that point, began encroaching on Russian lands in the east. This came to a head when the Japanese took over part of mainland Asia claimed by Russia. **Tsar Nicholas II** deemed it necessary to go to war to reclaim this land and to put the Japanese in their place. Easier said than done. The Russians sent their entire fleet from Europe to take on the Japanese in the Sea of Japan. After this extensive voyage, the Russian fleet arrived to be beaten down by the Japanese fleet in a matter of minutes. It was the most one-sided naval battle in history. The entire Russian fleet was obliterated. Back home in Moscow, popular dissent turned into popular hatred after this stunning defeat. The **Russo-Japanese War** was costly and unpopular and served to really get people thinking about replacing the monarchy with a whole new system altogether. But the fun had only just begun!

Back home in Russia, the peasants and workers alike had about enough. Later in 1905, a general strike was enacted empire-wide to protest the slow pace of reform and general discontent with the aristocracy. The system was

Nicholas II rides to the front lines of WWI: "The Monk made me do it!"

broken, and the people knew it. This protest was met with an iron fist by the government, and things turned nasty, fast. The 1905 Russian Revolution is best remembered for the legendary Bloody Sunday massacre; cross-toting, hymn singing protesters were slaughtered by government troops as they marched to the Tsar's Winter Palace. Brutal! The Romanovs had restored order, but not for much longer.

WORLD WAR I

The Russo-Japanese War may have been damaging and unpopular, but at least it was far away. However, the next phase of fun happened closer to home, when Archduke Franz Ferdinand of Austria was assassinated in what later became Yugoslavia. This was the spark that ignited World War I and everybody declared war on each other. Russia had diplomatic ties with Serbia, who was pulled into the war immediately, so the Ruskies had to jump in as well. As we pointed out, this came on the heels of a loss to the Japanese on the other side of the Russian empire. From the onset, World War I was extremely unpopular because people at home were saying, "Hey, you already lost to the Japanese! Things are going down hill fast and now we are in another war? Nobody even knows what we are in it for! Serbia? Serbia who?"

World War I was fought on Russia's front doorstep; for three long years, there was a catastrophic death toll on the Russian side. The Germans made huge gains into Russian territory and snapped up large parts of the front. Meanwhile, back at the bear cave, popular dissent was growing wildly, fueled in part by the shenanigans of yet another famous Russian: **Rasputin**. "The Mad Monk" as he was referred to by many Russians, was a shaman/con-man who worked himself and his magic into the inner circles of aristocratic power, including being a direct advisor to the royal family themselves. In point of fact, Tsar Nicholas II was strongly advised by Rasputin to personally lead the charge on the front lines of WWI, even though everybody knew good old Nick was not up for the challenge. Chaos then ensued both on the war front, but more importantly, back home as well. (check out inset box on Rasputin, next page).

I don't need to get too much more into World War I history after this, because the Russians didn't have a lot to do with it. About halfway through the war, internal dissent within Russia reached an all time high. The Bear reeked of revolt.

THAT'S REVOLTING! COMMIES TAKE COMMAND

The revolt occurred in February of 1917 and Tsar Nicholas II abdicated, meaning he quit before he got fired. When the Tsar, who had been out fighting in the battlefield during World War I, returned to Russia, he discovered his people were in open revolt. He abdicated, and put his brother on the throne. Faster than you can say, "I hereby resign from the throne," his brother abdicated as well. A temporary government was set up, but in terms of who is really going to take power, nobody could tell.

Archduke Franz Ferdinand

My death started WWI. Sorry!

Tsar Nicholas II

My death made the Revolution irreversible. Sorry!

Vladimir Lenin

My death launched Stalin into power. Sorry!

The Mad Monk, Rasputin

I'm still not dead. Someone get me the hell out of this coffin!

On October 25, 1917, this culminated in an event that people all over the world know, **the Bolshevik or Communist revolution**, led by our good friend, **Vladimir Lenin**.

"Let's redistribute some wealth, people!"

Communism was not something that everybody in Russia were just jumping up and down about . . . it is likely that most people had never even heard of it. Karl Marx and Friedrich Engels wrote about the concept in Germany and Austria; they weren't Russians. Lenin was a revolutionary even in his youth, and came in contact with a lot of these ideals during his college years. His older brother, who was executed because of his socialist activism, also influenced him. Lenin became a lawyer, but was eventually exiled to Siberia because of his radical revolutionary activities.

In Siberia, he wrote and published a lot of socialist literature, and became a prominent figure in the revolutionary movement. He eventually fled to Finland, and later Switzerland. While there, Lenin made appearances and speeches to other socialist groups in Europe. When the open revolt in Russia started to happen, Vladimir and other revolutionaries boarded a train and headed back into Russia after a long absence from their homeland. Once Lenin returned, he became a prominent figure in the Bolshevik party and leader of this "Soviet Revolution."

One of his ideas that gained popular support was not necessarily that communism was the greatest thing ever, but instead, "We need to get the heck out of World War I!" He received popular support for his Soviet ideas by default, because everyone pretty much agreed with his other political stances that monarchy, peasantry, and Russian life in general sucked, and change was necessary. The power structure was completely unfair. What also gained popularity was the promise to get Russia out of the war in exchange for the support of his cause. Long story short, the Communist revolution became an internal power struggle, and the Bolsheviks came out on top. "**Bolshevik**" means "majority," but in reality, the movement was anything but a majority. The Bolsheviks had many political rivals in the struggle to control Russia. The commies were actually a very small group of people who enacted this revolution and took over the whole nation.

What's the Deal with Rasputin?

The biography of Grigori Rasputin, a.k.a. the Mad Monk, is often fortified with folklore, which is fine by the Plaid Avenger because the folklore that surrounds Rasputin is hilarious. What is clear is that Rasputin practiced some sort of Christian-like religion and gained favor with Tsar Nicholas II by helping to medically treat his son through prayer. The Russian Orthodox Church didn't like Rasputin—mainly because churches never like competing ideology, but also because Rasputin loved sinning. Much of this sinning involved prostitutes.

Big City Russian Pimpin'.

During World War I, Rasputin advised Tsar Nicholas II to seize command of the Russian military. This turned out poorly for two reasons: (1) Tsar Nicholas II wasn't a good army commander and (2) while Tsar was away, Rasputin gained considerable control of the Russian government and helped screw up the Russian economy. Needless to say, a lot of people were not happy with him. In December of 1916, a group set out to assassinate the Mad Monk. First, they attempted to poison him. The would-be assassins loaded two bottles of wine with poison, which Rasputin drank in their entirety. The assassins waited and waited, but Rasputin continued to display lively behavior. In a panic, they shot him point-blank in the back. When they came back later to deal with the body, Rasputin jumped up, briefly strangled his attempted killer, and then took off running. The group of assassins gave chase, shot Rasputin AGAIN—this time in the head with a large caliber bullet, and then beat him with both blunt and sharp objects. Finally, they wrapped his body in a carpet and threw it off a bridge into the Neva River, but, the river was frozen over, so Rasputin's body just smashed into the layer of ice. The assassins climbed down to the frozen river and broke the ice under Rasputin so his body would sink into the water. They eventually succeeded and Rasputin disappeared into the freezing depths. Three days later, authorities found Rasputin's body and performed an autopsy. The autopsy revealed that Rasputin had died from drowning and, in fact, his arms were frozen in a position that suggested he died while trying to claw his way through the ice and out of the river.

Me-oww! That cat was a freaky-freak. And his lives ran out. Maybe . . .

True to their word, as soon as the Bolsheviks took over the government at the end of 1917, they immediately bailed out of World War I. They agreed to let the Germans have any invaded territory and signed a peace treaty. As you can see from this map, this resulted in huge territorial losses. The Russians lost territories that later became Estonia, Latvia, Lithuania, Moldova, and large parts of Poland. I am telling you this specifically because it is going to become Russian territory again after World War II. The Ruskies never truly intended to give away those lands permanently. In their minds, it was always a temporary fix.

Now as I just suggested, not everybody in Russia immediately embraced this communism gig. In fact, there were many holdouts. As you might expect, there were conservative people in the aristocracy and the military who thought that the monarchy, the old established way, should return and things should revert back to the way they were.

As soon as the Bolsheviks took power, one of the first things Lenin did was put the entire royal family under house arrest. Shortly thereafter, a civil war broke out; from 1918 to about 1920, there was open fighting throughout the country. The **Russian Civil War** was fought between the new party in charge, the Communists, and the old conservative holdouts and much of the military. The parties who fought the Russian Civil War referred to it as "the Reds versus the Whites." Obviously, the Reds were the communists and the Whites, the loyalists. The Red Russians versus the White Russians!

Check out the Russian losses in WWI.

What's the Deal with Anastasia?

Anastasia was born the Grand Duchess of Russia in 1901. When the Bolsheviks took power in Russia, Anastasia's entire family was executed because of their links to the Romanov dynasty. However, legend is that Anastasia and her brother Alexei might have survived the execution. This legend is supported by the fact that Anastasia and Alexei were not buried in the family grave.

Got Dead? Yep.

Several women came forward in the 20th Century claiming to be Anastasia, but guess what: she's dead. Regardless of the intrigue and the animated kid's movie, The Plaid Avenger is certain that Anastasia died of lead poisoning with the rest of her family. Lead poisoning induced by bullets. Lots of bullets.

Again, our story here is that the 20th century was not very kind to Russia. It had a disastrous World War I, and then an internal revolution which turned into a civil war. The civil war is eventually won by the Reds. BTW: once the civil war broke out, Lenin ordered the assassination of the entire royal family. Why did Lenin do this? I thought he was a nice commie. The reason Lenin did this was because he believed that in order to succeed, Communist Russia had to sever all ties with the past: "You people are fighting for a monarchy? You want to restore the imperial royal line? Well, I will fix that! They can't be restored to the throne if they're all dead!"

This made the statement that no matter who won the war, there would be no return to monarchy because there were no more monarchs. The Romanov line ended with the assassination of Nicholas II, his wife, and five children including Anastasia; rumors suggesting Anastasia escaped execution circulated for decades (see inset box on left).

The civil war was finished by 1920, and the Communists had power over the government and the country. Unfortunately, fate was still not smiling on Mother Russia, as the country saw back to back famines in 1921–23 as the result of horrific winter weather—like the weather is even good in the good years? However, by the mid-1920's, the worst seemed to be over and it was time to get that commie transition on!

THE BEAR BATTLES BACK

The next event of great significance happened in 1924, when Lenin died. For vehement anticommunists, that was a day of celebration; for Russians, it was a day of mourning, because Lenin was the founder of the party, a rallying post, and a popular leader of the communist movement in Russia. He had the grand designs about where society was going, with an ideal of equality where everyone shared the wealth, and the whole utopian society thing. Why was his death titanic? Well, because after he died, who would succeed him? The answer was **Joseph Stalin**.

STYLIN' WITH STALIN

From 1924 to 1953, Joseph Stalin, one of the biggest psychos history has ever known, was in power of what was now called the Union of Soviet Socialist Republics (USSR), the post–civil war title for Russia.

Stalin's ascension to power was problematic for several reasons. For starters: he was crazy. After he gained power, he was psychotic and became more psychotic as every year of his reign passed. He consolidated all power to the center under himself, basically as a dictator, and that's what he was. He set up the infamous secret police force, the **KGB**, who sent out spies amongst the people, and conducted assassinations of all political rivals, their families, their cats, their dogs, and their neighbors. Anyone who spoke ill of Stalin ended up in a grave with a tombstone over his head. No, I take that back. They ended up in a mass grave with dirt over their heads. He was a nasty guy in a tough place in an extremely gut-kicking century for Russian history. But our story is not over yet, kids—the fun is still not done!

Before I go any further, I should say there is always a bright side to every individual. Smokin' Joe Stalin's silver lining was that part of the commie plan was to consolidate agricultural lands and speed industrialization. The Reds wanted to catch the country up with the rest of the modern world economically, technologically, and militarily. As much as I hate to admit it, that is exactly what occurred during the Stalin era.

Joseph Stalin: Scariest Psycho of the 20th Century
Unfortunately, the award for "Scariest Psycho" is highly competitive. There's Hitler, Pol Pot, Mussolini, Idi Amin, Pinochet, and many more. But the Plaid Avenger is fairly certain that Joe Stalin is the nastiest of them all. Stalin was one of the Bolshevik leaders who brought communism to Russia in 1918. When Vladimir Lenin, who was the original leader of the Bolsheviks, unexpectedly died in 1924, Stalin slowly assumed power. In 1936, Stalin initiated **The Great Purge** which lasted for two years. During this time, "dissenters" were shipped off to Gulag labor camps and often executed. Most "dissenters" were actually just normal folks who had been wrongly accused by other normal folks who were being tortured at a Gulag labor camp. As you can see, this was a self-perpetuating cycle. If everyone had to sell out eight friends to stop getting tortured, and each friend had to sell out eight friends, pretty soon you would have a ton of political "dissenters." It's like an incredibly violent email pyramid scam. Anyway, The Great Purge resulted in the death of perhaps 10 to 20 million Russians.

While The Great Purge was probably Stalin's sentinel work of psycho-sis, it was far from his last. Stalin organized a giant farm collectivization which left millions homeless and millions more hungry. He also continued purging dissenters until he died in 1953. Historians estimate over 45 million Russians died directly from Stalin's actions (this figure does not include the estimated 20 million Russians that died during WWII). Norman Bates got nothing on Stalin.

Essentially Stalin oversaw a "crash course" of modernization, industrialization, and collectivization. These hugely ambitious goals could not have been achieved just by planning, though; the Russian people themselves have to be given a large share of the credit to have pulled this off. Perhaps Stalin's biggest contribution to these gains was not just organization and planning, but he somehow instilled a die-hard sense of uber-patriotic nationalism in the people that made for explosive societal gains. Specifically, the Russian people were convinced to work harder, work longer, make more out of less, consume less, and totally work hard for the country, not for themselves. And they did!

Of course, Stalin used all sorts of coercion and open force to achieve these aims when necessary. And again, he was nuts. But if we are going to credit Stalin with anything other than intense insanity, we can credit him with the Soviet infrastructure moving forward at breakneck speed, and they indeed caught up in a matter of a few decades. They created the atomic bomb shortly after the US did, and even beat the US into space with the launch of the **Sputnik** satellite. Fifty years earlier, they were all beet farmers. That is some fast progress!

THE BEAR ENTERS WORLD WAR II

By World War II, the USSR was starting to be a major world force. Before the war began in 1939, Joseph Stalin and Adolf Hitler met and signed a nonaggression pact. The meeting went down like this, brought to you Plaid Avenger style: Germany said, "We are going to take over the world, but here's the deal: we won't attack you as long as you don't attack us." Smokin' Joe Stalin, in all of his wisdom, said, "Okay, go ahead and take over Western Europe and the rest of the world. That sounds good to us. They were always kind of bothersome anyway. Also, we won't attack you as long as you don't attack us. Deal!" The war officially kicked off in 1941, when Germany attacked everybody. Hitler later made the fantastically idiotic mistake, after taking over most of Europe, of reneging on the nonaggression pact. In other words, he attacked the USSR anyway.

Russian soldiers line up for their daily vodka ration—of three liters.

As discussed before about Napoleon, one of the greatest historical blunders for anybody is to invade Russia. No one's ever done it successfully, and Hitler was one in a long line of idiots who tried. After he attacked the USSR, the USSR was then obliged to defend itself and was pulled into World War II. This is of great importance: the USSR was the real winner of the war in Europe. Once the Germans declare war on the United States, who joined the rest of the Allies, it became the Germans versus the world. The Germans were really in a pinch as they were attacked on all flanks, the Allies on one side and the Russians on the other. While I am certainly not going to devalue the role of the Allies in the west, Russia really took out a lot of Germans and caused them to divert resources and men to the Russian front, which is the reason why World War II was won by the good guys.

Had the Russo-German nonaggression pact been upheld by Germany, it would be hard to tell what the map of Europe would look like today. It certainly wouldn't look like it does now. And the Russian death toll was astronomical. There were almost 9 million Russian military deaths, and at least 20 million civilian casualties—29 million people total! Take 29 and multiply it by a thousand, then take that and multiply it by another thousand! Not trying to insult your intelligence. I just want you to take a second and think about how many folks died in this thing. Just on the Russian side!

This was some of the most brutal fighting on the entire continent. The Russian front was bloody, nasty, and deadly, and when all was said and done, the Russians beat the crap out of Germany. It was part of the main reason the Germans had to surrender in the end. I really want to stress the Russian losses. 29 million killed. We think about just the US's role in World War II (which was great by the way), but the US suffered maybe half a million to Russia's nearly 30 million deaths! As a result of WWII, Russia experienced an epic population loss and made a major impact on the European theater of war. The war ended in 1945, the end of which set up the scenario for the next fifty years.

IS IT GETTING CHILLY IN HERE? THE COLD WAR

At the conclusion of the war, the Allies came in from the west and the Russians were mopping up in the east and they met each other in the middle of Europe. They all did high fives and then they split Germany in half. Hitler was dead in the bunker, and there was a line where the Allies met after the war. The Allies were the United States, Britain, France, and the USSR. That's right! Smokin' Joe Stalin was on our team! Where the two forces met, as you see on this map, is halfway through Europe. That line quickly became known as **The Iron Curtain**.

What does this mean? Well, the Russians occupied Eastern Europe; after the war was over and everyone finished celebrating, the Soviets said, "Maybe we will just stay here for a little while. We'll make sure everything has settled down. We'll help rebuild this side of the continent." The United States, under the **Marshall Plan**, was building up Western Europe. Both parties occupied Germany to make sure it wouldn't start up trouble again—thus we had an East Germany and a West Germany, which reflected events happening in Europe as a whole.

*Austria remained occupied by Allied powers until 1955, at which time it became independent upon condition of neutrality.

The metallic drapes were drawn.

In their effort to do this, the Soviets were pretty sinister. Under Soviet tutelage, "elections" were held all over, in places like Poland, Czechoslovakia, Hungary, Yugoslavia, and parts of Austria. We all know the outcome: lo and behold, the Communist party won resounding victories in every single place they held a vote! Everyone in the West knew this was a farce, but nobody wanted to start another war by calling out the Russians publicly about this scam. Remember, they were allies at the time, and they helped defeat the Germans together. The United States just wanted to get the heck out of there. They were not about to start a new war with Russia. The Western governments hesitantly went along with these newly "converted" communist states.

This set up a scenario where puppet governments replaced the previous governments. There's a term for this you should already know called **soviet satellites**. Some examples were sovereign states like Poland, Czech Republic, and Hungary that actually had seats at the UN—but everybody knew that the Soviets had the real control. These were puppet governments, to give the illusion of being expressive governments of their people. The situation simply evolved into Soviet occupation and control at this point. One of the reasons Western Europe and the United States allowed this to happen is that the Soviets very adeptly pointed out that they were staying in these Eastern European nations to make them a buffer zone because countries in Western Europe historically kept attacking them. That was true (e.g., Napoleon's France, Hitler's Germany).

These Eastern European states behind the curtain, that ended up being Soviet satellites, were actually parts of the territory that the Soviets had lost at the end of WWI, so the Soviets regained that loss. I like to point this out because it plays into what is happening in today's world. Whole countries that were sovereign states for a while between World War I and World War II (like Estonia, Latvia, Lithuania, Belarus, and the Ukraine), were totally reabsorbed by the Soviet Empire. These places declared independence after World War I, and then suddenly found themselves no longer sovereign states but soviet republics or sub-states of the Soviet Union. That's why the map of Eastern Europe changed so rapidly. Places came and went as sovereign states were reabsorbed.

COMMIE ECON 101

The first half of the 20th century was fairly repressive, brutal, chaotic, and completely violent for Russia, but things stabilized for them internally during the Cold War. We have already set the stage in Europe for the East versus West ideological battle for control of Europe, which is going to turn into a global phenomenon that has shaped virtually every aspect of every other region that we talk about in this book. What's the deal with the USSR as full-fledged commies after World War II?

As I have suggested previously, the Soviet Union experienced massive growth during Stalin's regime. He died shortly after World War II, but a lot of the industrialization which was radically successful under him continued on. Many of the programs, while oppressive and brutal, made the state a stronger world power. Indeed, it was the rise of the USSR which created the bipolar world that defined the Cold War. Stalin did a lot to get them up in the big leagues, and then he checked out. He brought about massive industrialization, particularly in the weapons sector, which continued after his death.

What happened under the Soviet experiment post–World War II? What did they do? What did it look like there? What did they make? What did they produce? How would that be part of their eventual undoing? During the Soviet era one of the primary goals, even under Lenin, was to catch up industrially. They did this by centralizing everything. That's what communism is known for. The state runs everything, all aspects of politics—of course that's easy, it's what all governments do—but also all aspects of the economy—and that's typically not what the governments do. That is the communist way: full political and economical control of the entire country.

This is called **central economic planning**, and it entails the government's approval of where every 7-Eleven is located, where every ounce of grain is grown, where railroad tracks are laid down, where coal will be mined, where whole cities will be built to support something like an automotive industry. Every single aspect of life that we take for granted in a capitalist system was controlled in the Soviet sphere. This was done very early in the Soviet Empire to accomplish specific goals, one of which was called **forced collectivization**. Russia had been a peasant society, land based, rural, farmers digging around in the dirt, for hundreds of years. It couldn't do that in the 20th century if it was going to catch up and become a world power; it needed to get people off the land and into the cities. Why the cities? To work in the big factories that were industrializing and creating stuff.

Tanks for the memories, commies!

Goal number one was forced collectivization. Pool all the land together, because it now belonged to the state: "You used to have rights to it and a deed on it, Dimitri, and maybe you used to grow cabbage and beets on it, but those days are gone now, comrade! You've got two options: go to a city or get in a collective, government regulated, growing commune." The state controlled all the land and agricultural production, because the state had one huge tractor to do the work of 1,000 men, and it had pesticides and fertilizers, which made huge monoculture crops. This new system produced tons more food. Whereas 1,000 dudes had to farm a piece of land before, now just ten can get the job done. 990 dudes headed to the city to accomplish the next goal: **rapid industrialization**.

To achieve rapid industrialization, the Soviet government needed to get everybody in and around the city into big factories. What were the factories going to produce? All of the things the West has that the Soviets needed to catch up on. Things like tanks, petrochemicals, big guns, rifles, maybe some big fur coats every now and again because it's cold there, aircraft carriers, missiles, bulldozers, heavy machinery, etc. The question you need to ask yourself at the same time is: what were they not making? The big difference between the communist expression and the capitalist expression in the 20th century comes down to this question. The answer is that the Soviets didn't make the items that normal people buy. Instead, they made huge stuff that only countries used. Like I told you earlier, the Soviet system stressed more work and *less consumption* by the individual, so that the state can become stronger.

You have to remember, the Soviet way was not focused on individual citizens or individual rights. The Soviet government didn't care about personal expression or fashion statements or if you liked building model airplanes. For them, the Soviet Union was an awesome state and an awesome idea and its citizens should be ardent nationalists and do everything

for their state. By and large, people were pretty cool with that; at least most of them were, because to not be cool with that meant the KGB would run you down in the middle of the night and shoot you. People accepted it. Lots of them even liked it.

What wasn't produced in the Soviet economy? Consumer goods like cool clothes, sunglasses, lawn chairs, microwave ovens, refrigerators, independent-use vehicles. Anything that you'd go to a mall to buy today did not exist in the Soviet world. Knick-knacks did not serve the greater goal of the government, which was to catch up with the capitalists. That's just the way it was.

As much as the Plaid Avenger likes to make fun of knick-knacks and other crap you don't need, it played a role in why the Soviet Union unraveled—but that's for the next section. They did indeed catch up and became a world power under this system. One of the other things the Soviets did with the big things like tractors and bulldozers and weapons was to export them abroad. These are items that the Soviet

The AK-47; Russian export extraordinaire.

Union is still known for. Russia is, to this day, a huge manufacturer and exporter of arms. Example: the Russian made AK-47 is easily the most popular and most widely distributed rifle on the planet.

Russia also exported a lot of this stuff at cut rates to other countries around the world. Why would you export a bunch of missiles to Cuba? That's right, for the Cuban Missile Crisis! You also want to send around machinery and petrochemicals and similar items that you produce to make money and recoup costs, but more importantly, to promote your influence in the world. That is what the Cold War was all about: coaxing other countries to join your side by providing them with things they really need, building bridges, selling them weapons and chemicals, and maybe even lending money. Anything to get people on your side. The United States and the Soviet Union both did this all over the place.

Only in a few places does this battle of ideology turn hot and people start shooting at each other. Two places of note where this occurred were in Korea, resulting in the Korean War and in Vietnam, resulting in the Vietnam War. We might even go as far to say as this also happened in Central America for all the Central American wars, during which both these two giants funneled in weapons so that the locals could fight it out between their opposing ideologies. The commies versus the capitalists in all these places was really what the wars were all about, and the two giants funded their respective sides.

CATASTROPHIC COMMIE IMPLOSION

How did the Cold War end for the Soviets? Why did I suggest that knick-knacks on the shelf may have been part of the winning strategy of the United States, when all is said and done?

RUBLE RUPTURE

The Soviet Union, from the 1960s through the 1980s, simply overextended itself. The United States and Western Europe's economies were based on producing a little bit of everything: tanks, aircraft, and petro-chemicals, but also consumer goods like cars, washing machines, toys, and Slap-Chops.

When you have stuff that individual citizens can buy and sell, you don't have to solely rely on big ticket items to fuel the economy. How many bulldozers can you sell in one season? I mean, the US system ended up championing in the end, because people buy consumer goods all the time. People like to

Hurry! Buy now! These things are flying off the shelves!

buy refrigerators and new cars and blow dryers and pet rocks; but the US also made and sold bulldozers and missiles too. Long story short; US/Team West got richer, Soviets stagnated.

Russia's economy stank because it was based on flawed principles and was focused solely on items that eventually made it unsustainable. But there were other reasons these guys were going broke. Namely, the USSR liked to give lots of money and support to their commie allies worldwide. During this expansion of Soviet influence across the planet, they really started to overextend themselves. They were giving stuff to Angola and the Congo in Africa, buying sugar from Cuba that they didn't even need, just so Cuba would be their ally, and lending money to places in Southeast Asia. By the 1980s, they were going broke while their economy was also stumbling.

Another ruble-draining effort was being stupid enough to get involved in a war in Afghanistan. Geez! Who would be dumb enough to get bogged down in an unwinnable war in Afghanistan? Oh, ummm, sorry US and NATO, I forgot about current events for a second. You guys should have really learned more from the Soviet experience, which was an . . .

AFGHAN BODY SLAM

Let's go into more detail on this mess, because it does play into current events more than most want to admit. Ever since the USSR came into existence, it crept further and further into control of Central Asia. eventually absorbing all of those -stan countries and making them into Soviet republics. All except one: Afghanistan. The USSR stayed very cozy with the leadership of the country, though, and gave them lots of foreign aid to ensure that they stayed under the commie sphere of influence.

However, there were other forces within Afghanistan that didn't like the Soviets and, for that matter, their own leaders. A group of Muslim fighters we'll call the **mujahideen** decided to wage a war to overthrow the Afghan government. The government feared a greater coup was at hand, so it invited the USSR to invade them to help out. Sounds good so far, huh? Yeah, right. This was a huge freakin' mistake for the Soviets.

In 1979, the Soviets invaded Afghanistan, and started a decade-long war which will serve to demoralize and humiliate the USSR. Remember, this was during the Cold War, so the US CIA—under orders from the government—funneled tons of weapons and training to the mujahideen to keep the Soviets pinned down. It worked like a charm. In fact, it worked too well! After the mujahideen eventually repelled the Soviets, they then proceeded to have a civil war, and the winners of that civil war became the Taliban. Crap. Any of this starting to ring a bell yet? The Taliban are buddies with and shelter al-Qaeda, and after 9/11, the US declared war on both groups. Now the US finds itself fighting against forces it helped arm and train back in the 1970s. Oops.

But that is a tale for another time; back to our soviet story. The Soviet/Afghan campaign of 1979–89 was a devastating war for Russia, comparable to the Vietnam War for America. The active conflict in Afghanistan started to siphon away resources at a rapid rate. At the end of ten years of fighting, many Russian lives were lost, many Russian rubles were spent, and all for nothing; they were forced to walk away from the whole mess with their tail between their legs. Check out an awesome flick which shows how this went down: *Charlie Wilson's War* starring Tom Hanks. As a bonus, it has male rear nudity—of Hanks, himself! Actually, ew.

THE GREAT COMMUNICATOR SAYS: "I HATE COMMIES"

Once the 1980s rolled around, a new component is introduced: the Reagan Factor. Ronald Reagan was the most hilarious US president ever, in the Plaid Avenger's opinion, and was certainly one of the reasons for the demise of the USSR. He is often credited as being the entire reason, which is preposterous, but his administration did a lot to accelerate the processes in play regarding Russia's economic decline. The drive for military buildup during the Cold War was all fine and dandy when the Soviet Union was expanding and growing

"I hate commies."

industrially and economically, but this started to take its toll on them in the stagnant 1970s and 1980s. When the Americans developed a new type of missile, so the Russians also developed a similar one in order to stay competitive. Then the Americans built a new type of tank, so the Russians built one as well. It was a "keeping up with the Joneses" sort of situation. You have to be as cool as your neighbors by having the same stuff, the newest and coolest stuff. That had been happening since the 1950s. **The Space Race** was a classic example of that competitiveness. Then in 1980, Ronald Reagan was elected president of the United States, and hilarity ensued.

Reagan hated commies. If you learn one thing about Reagan, it's that Reagan hated commies. He hated commies when he was a Hollywood actor and ratted out a bunch of other actors as communists during the McCarthy hearings. He hated commies when he was the governor of California, and he really hated commies when he became the President of the United States. He hated them so much that he was willing to send money and arms to anybody on the planet under the guise that they were fighting commies. In conclusion: Ronald Reagan hated commies. Have I made that clear yet? When he came in power in 1980, he had to deal with domestic issues because the US economy sucked. One of the primary things that he and his administration did very early on was accelerate military spending. This was a very clever move because it did a few things at once: It helped the American economy by creating jobs building more bombs, and also, Reagan knew that the Soviets were trying to keep up militarily, so he was making them spend more money as well. Money they didn't have.

20TH CENTURY COMMIE COUNTDOWN

Lenin: 20th century premier commie and snappy dresser.

Stalin: 20th century premier psycho with a serious 'stache.

Krushchev: Placed missiles in Cuba. Nice job, laughing boy.

Brezhnev: Patented the Unibrow, kin to Herman Munster.

Andropov: He "dropped ovv" after only 16 months in office.

Chernenko: Not to be outdone, he croaked after only 13 months.

Gorbachev: "Oh crap, this place is falling apart.

This spending craze happened during the entire decade of the 1980s, just as internal dissent became more prevalent within the Soviet Union. Rioting broke out in places like Czechoslovakia, Poland, and Hungary, and as the economy continued to worsen in the Soviet Union, people began to starve. The US media showed scenes of people in long bread lines in the Soviet Union. These guys weren't just broke; they were going hungry at the same time they were spending millions, if not billions, on weaponry to keep up with the US. It was a crafty move that paid off huge dividends for the US. When the final Soviet leader came to power, Communist Party Secretary General **Mikhail Gorbachev**, he inherited a state that was broken and riddled with holes. Gorbachev knew his administration was screwed. The comrades continued to lie to themselves and used the KGB to scare everyone so they wouldn't utter it out loud, but the people at the top knew that the party was over. The Communist Party, that is.

HOLY GLASNOST, THIS IS DRIVING ME MAD! PASS THE SALT, PERESTROIKA

What Gorbachev did was enact three things. Two of them are words you should know. One is **glasnost**, which means "openness." The second is **perestroika**, which means "restructuring." The third thing he did was try and limit military funding. Glasnost meant Gorbachev was tired of the secrecy of the Soviet government. He didn't want to lie to people any more, and he wanted to stop the KGB from terrorizing people. He wanted citizens to be more open with the government so it could improve. Places in Eastern Europe such as Poland, who wanted their own votes and no longer wanted to be a part of the Soviet Union, were let out. Under perestroika, Gorbachev knew they had to restructure the economy and reel in military spending. They also needed to cut off subsidies to countries they could no longer afford to prop up. This all leads to number three: not being able to spend any more money on the military. Or at least not as much.

This third enactment led to the **SALT**, first in the early 1970s and then again late in the decade. The **Strategic Arms Limitation Talks** were about curbing weapons production—not stopping weapons production altogether, not getting rid of weapons—just slowing down the speed at which they were being made. What you have to understand about the Cold War is that, if humans survive for another 1000 years, it will be looked on with hilarity and humor; the whole reason that we haven't killed each other yet is because there are so many bombs on the planet between the Soviets and the USA. There was a principle that basically sustained our life on earth, known as **MAD**. **MAD** stood for **Mutually Assured Destruction**, which essentially meant if anybody lobs a bomb, then the other side will throw a bomb, in which case both sides will throw all their bombs and everyone will die. Who is going to do that? The insane logic of MAD is what stopped us from attacking each other, but it made us continue to make more bombs. Both sides already had enough bombs to annihilate the other side, but they just kept making more. The US would say, "We have 1000 bombs that can each destroy five countries at once!" Then the Soviet Union would build 1001. That's where you get into these astronomical numbers of bombs, a bomb for every darn person on the planet. Why? I can't really answer that, because it makes no sense. But I digress. I get a little agitated thinking about MAD; how fitting.

Back to our story: this equated to Gorbachev realizing that they had to stop doing this. During the SALT talks, they agreed, "Let's not build this type or class of bomb anymore, or build half as many as you were going to and we will build half as many. Also, how about we eliminate this type of submarine, and you guys eliminate that type of submarine." When these talks started, there was some capitulation, but for the really big thing, Gorbachev would say, "Let's not build these big intercontinental ballistic missiles," but Reagan (because he hated commies so much) built more of those missiles because he knew the Soviets were going broke. Indeed, not only did he not agree to weapons elimination, but accelerated production and built more. In the same year, Reagan appeared in front of the US Congress during the State of the Union address and said that the USSR was an "evil empire" and that the USA must do everything in their power to combat evil. He was really throwing all the cards down on the table—what a huge bluff in the global game of poker.

Here is the bluff that won the game: Somebody made mention to Reagan of how cool it would be if the US put missiles in space; thus when the Soviets shot nuclear weapons at anyone, the US could shoot their bombs out of the sky with these space missiles. This missile system evolved further to become a space laser to shoot out bombs and was called **SDI**, the Strategic Defense Initiative, given the nickname **Star Wars**.

When this idea came out it was merely that: an idea. Then some people started throwing around some funds for this idea, which was most likely sketched on a cocktail napkin by a science fiction writer. Somehow Reagan got his hands on this cocktail napkin and started to circulate it around. Anybody heard about this Star Wars thing? It sounds pretty good. When news of this reached Gorbachev in the Soviet Union, the comrades said, "Oh, no! They are going to have an orbiting missile defense system! In space! We are going to have to spend millions and billions of research dollars to figure out how to get one of these too and we are already broke. We can't do it!"

SO LONG TO THE SOVIET: THE END

Gorbachev went to Reagan and said, "We will stop making these bombs, guns, and tanks, just don't do this Star Wars thing!" Reagan, in all of his wisdom, said, "Sorry, we can't; it is already built. We are too far along and can't stop now." This was a complete line of bull. There was nothing in space. Nothing was even tested. This huge bluff paid off because Gorbachev decided to throw up his hands and said, "Well, that's it then; we can't keep up. If we can't have a missile defense, that means you guys win. You could nuke us, so we quit. MAD is over and so are we!" This was when the USSR really fell down, around 1988 1989.

"We're not MAD anymore; we're broke!"

Several things collide here. The Soviet economy was in trouble, they had overextended themselves with their propaganda around the world and with Soviet aid in other countries. They had a war in Afghanistan with active fighting for 10 years, from 1979 to 1989, which proved to be disastrous. It took a huge toll in not only millions of dollars of equipment, but also the death of 25,000 Russian soldiers. When all was said and done, they had not done anything there for 10 years. They failed to control Afghanistan, so they had to walk out of it. Glasnost and Perestroika are starting to be interpreted literally by Eastern Europeans and the Soviet people themselves, who were now calling for restructuring and/or independence. In 1989, the Soviet Union had to throw in the towel: "Poland, you want out? Go ahead. Estonia, Latvia, Lithuania, you want out too? I guess you guys are out as well. Central Asian republics, we're done with you too." It all disintegrated in very short order, and all those countries declared independence.

Fifteen new sovereign states emerged from the ashes of the USSR. The strings of the Soviet satellites' puppet governments were cut. The map was redrawn in one fell swoop. In 1990, they adopted economic and democratic reforms into what is left of the USSR, which became the United Federation of Russia. In 1991, Boris Yeltsin was elected president of Russia. He recognized the independence of all the countries that had declared it. In that same year, the newly formed Congress voted for the official and permanent dissolution of the USSR. It was no more. From 1991 forward, it was called Russia.

RADICAL RUSSIAN TRANSITION

By the 1990s, the commie threat was gone and the USSR was dismembered, so everything in Russia should be all champagne and caviar, right? Unfortunately, everything still sucked for Russia. It has gotten a lot better in the last few years, but we are going to carry it forward from 1991, when all this becomes official in today's world. That first decade of post-Soviet independence for the Ruskies sucked as bad as the decade before it. The pain of a radical transition had to be suffered through as they approached the 21st century. What kind of transition, and what kind of pain?

ECONOMY TAKES A DUMP

As we pointed out during the final decades of the Soviet Union, their economy was in bad shape, and changing their name doesn't change the facts. Now it's the "Russian" economy that stank. Their industries were still in Cold War mindset and only produced big stuff like tanks and missiles that didn't have much use in today's world. You could still sell such items, but not nearly enough to rebuild an economy. They had tremendous problems with shifting their entire industrial sector to more capitalist-centered stuff, normal consumer goods. Capitalism was still a new concept for them, and they were still breaking in their capitalist cowboy boots. The transition from commie to capitalist was full of pitfalls and road bumps and full-on collisions, which made for an extremely rocky economic road to recovery.

On top of that, Russia lost its international status. They used to be "the other world power" but in the 1990s nobody would give them the time of day, much less a bank loan. This has changed in the last few years, but right after 1991, no one was willing to help Russia. Russia used to be a somebody, but now it was a nobody. It's also important to note that the USSR went from a singular sovereign entity to fifteen separate ones. It lost tremendous amounts of territory. Those other fourteen countries that weren't Russia were plots of land it no longer owned and there was the agriculture produced on that land, and mineral and energy resources under it, and the people who live on it were a workforce. Those are the basic foundations of economies. This territorial/resource loss was a tremendous blow to the Russian economy, an economy that was already hurting.

The Bear gets skinned.

On top of this, you have to think about the Soviet era, during which the Soviets used a lot of resources from these countries. For example, most of the food production for the Soviet Union came from the Ukraine and Kazakhstan, which became separate countries. A lot of oil and natural gas came from Central Asia, and not only was Russia no longer profiting from these resources, it now had to buy those resources with cash money . . . and they used to own the stuff! It was a double-edged sword with both edges toward Russia. They were getting sliced and diced. Finally, most of the nuclear and weapons stockpiles were in what was Ukraine and Belarus, and there were a lot of expensive complications involved in destroying and/or relocating those weapons.

The second problem with the economy was the influence of organized crime: the Russian Mafia. Crime is everywhere, so organized crime is everywhere as well . . . but it is really strong in Russia. In fact, if you have ever seen adventure movies or international crime dramas, there is always a Russian mafia somewhere. They have always had a presence, even back in the days of communism when they created the Russian black market, which moved untaxed alcohol, weapons, and hardcore drugs. Indeed, they are still doing a tremendous job trafficking drugs and weapons internationally. They're also trafficking people. Yes, I did say people. A lot of the Russian mafia deals with the movement of women and children for sexual exploitation purposes. It's a fairly nasty business, but I guess that's what mafias do.

During the confusion and transformation of the Russian economy in the 1990's, these illegal criminal elements had a complete field day. The Soviet system was fairly corrupt internally to begin with, and that was perpetuated under free market capitalism where there are not a lot of rules. In addition, the government was broke and couldn't pay federal employees, such as the military and police. In such circumstances, corruption becomes endemic and easy to do at all levels. People do what they have to do to stay alive. The influence of organized crime is a major function of what is happening, even in today's Russia. Vladimir Putin himself came to power on a platform of being tough on crime, and in fact, nobody would be elected who is not.

This powerful criminal element in Russia not only stymies local business, but has negatively influenced international investment as well. But the open and outright blue-collar criminals were not alone in the 1990s. A new dimension of white-collar crime arose to give the underworld a run for their money (ha! Pun intended!): **the Oligarchs**.

OLIGARCHS OVER-DO IT

The **oligarchs** were white-collar criminals who, during the transition period of communism to democracy and capitalism in the 1990s, worked for the government or had insider ties with the government and bought whole industries and businesses under the table before they would go to auction. In communism, everything is owned by the state—all the oil, land, timber, energy production, everything. In capitalism, individuals own everything and the state owns virtually nothing. When a government goes from communism to capitalism, it starts selling everything. Ideally, this process would allow any individual the opportunity to buy former state industries at fair market prices. However, in Russia, this did not happen; most purchases were made by government insiders at rock bottom prices.

Here's an example: Yukos was Russia's oil company, a government owned oil company that was going to go privatize. If you knew some government officials when it was going up for sale and could get the price, and you knew people who worked at a bank, here is what you would do. You find out the day before the sale occurs and go to your friend at the bank and say, "Lend me half a billion dollars," with the promise that you will personally give him 100 million bucks as a bonus after repaying the loan. He smiles and you take the check for 500 million bucks from the bank over to the natural resources department to your buddy who is overseeing the sale of the oil company. You hand him the check for 500 million; he stamps it "sold" and hands you the deed to the oil company. You and your conspirators walk out together the next day as owners of this oil company, and sell it for 2 or 3 billion dollars, its true value, and then you go back and settle up your loan for 500 million with your banking friend and you split the rest. Congratulations! You have joined the ranks of the oligarchs! This was a really seedy big business tactic in the wild west of capitalism where there are no rules. These guys became ultra-millionaires and billionaires overnight.

What is the big deal with this? What's the problem? The Russian people got screwed, that's what. You have state owned resources whose sale should have contributed to the economic growth of an entire country, which instead turned into a billion dollars in a Swiss bank account under one man's name overnight. It was a legal transaction, apparently, but also a massive loss to the government.

In his rise to power, Vlad "the Man" Putin brought some of these crooks to justice—including Yukos head Mikhail Khodorkovsky, formerly the second richest man in Russia—although his motivation for doing so is questionable. It was likely a move designed to consolidate power more than a move to achieve justice.

Another platform in Vladimir Putin's campaign was, "I'm going after those Oligarchs! Going after crime and those chumps that robbed the Russian people!" That is one of the reasons he's such a popular leader. He is seen as a strong anticrime, pro-government force, and he's backed it up because the owner of Yukos is now sitting in jail. As a result, Putin helped Russia stabilize itself a little more. But maybe I am "Put-in" the cart before the horse in this story. . . .

TOP FIVE RUSSIAN OLIGARCHS

Who?	What did he steal from the people?	How much? (USD)	Where is he now?
Roman Abramovich	Oil (Sibneft)	$14.7 Billion	In England, owns big-time soccer team Chelsea FC. Also owns four giant yachts and is building a fifth.
Vladimir Lisin	Aluminum and Steel	$7 Billion	Most likely sleeping in a pile of hundred dollar bills.
Viktor Vekselberg	Oil (Tyumen)	$6.1 Billion	Buying Faberge Eggs. Seriously, he digs them.
Oleg Deripaska	Aluminum (RusAl)	$5.8 Billion	Hanging out with good friend Vladimir Putin.
Mikhail Fridman	Conglomeration of Valuable Things (Alfa Group)	$5.8 Billion	Lost much of his wealth in Russian tort suit—for stealing from the people, probably not enough evidence for criminal case.

PURGIN POPULATION

Another issue that is extremely problematic for Russia is the population itself. What's happening to the people in Russia? Russia's population is one of the classic examples, if not the classic example, of a population in decline: negative fertility rates and dropping birth rates. The population total is declining, with only about 150 million people in Russia; that number keeps getting smaller every year, which doesn't do much for Russia's economy. You need working people to keep things moving. A lot of things are going on that can account for Russia's demographic decline. It is hard to pinpoint a single factor, but nevertheless, the population is shrinking. Shrinkage is just the start of problems.

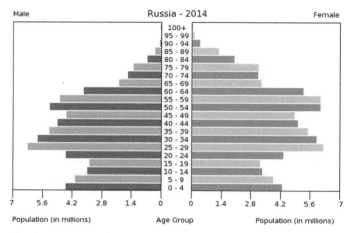

Numbers are going down.

If we looked at some of the social indicators for what's happening in Russia, you would swear that you were looking at an African nation. The life expectancy for males is around 47 years, a little more than half that of the United States. You might guess this life expectancy for sub-Saharan Africa or Central Asia, but you typically would not think of Russia. Several things may be contributing to Russia's decline. One is *lack of health care*. Because of the collapse of the economy and the corruption associated with it, health care has become a huge issue. Access to good health care is almost nonexistent, but again, this may be changing. People aren't as healthy as they should be and don't have access to certain operations or preventive care. Drinking an average of 30 liters of vodka a year probably isn't helping matters either. Other unhealthy attributes include . . .

ENVIRONMENT ENDANGERED

In its rapid industrialization, forced collectivization, and striving to catch up with the West to be a global power phase, the Soviet system completely ignored the environment. There was no limit to the amount of toxic spilling and dumping, or atomic weapons testing. Anything the Soviets could do to get ahead was done with the thought that once they won, they'd go back and clean everything up. Oops, they lost, and it's all still sitting there. Parts of the Kara Sea glow green sometimes from all of the accumulated toxic waste dumped into the river systems. Also, be sure to check out the Aral Sea as soon as you can, because it is disappearing due to misuse and pollution during the Soviet era. There are no less than fifty sites in Russia with pollution of catastrophic proportions. Don't forget to include all of the old nuclear weapons silos that are sitting around, deteriorating. There is impending environmental disaster on a large scale here, just based on past pollutions and degrading weapons. Russia has dealt with this huge problem during the last twenty years, with some international help from the United States and several others.

STOLEN SPHERE OF INFLUENCE: THE REALIGNMENT

The last thing you should be aware of in Russia right now is that a lot of its ex-territories and ex-states in its sphere of influence have now disappeared under radical realignment—much to the chagrin of Russia, because it still sees itself as a world player. What am I talking about? Eastern Europe is now desperately trying to become Western Europe by realigning with the West. Russia is losing its sphere of influence all over the place, though it's desperately trying to hold on. It is still actively courting Central Asia politically because there is a lot of oil and natural gas there. They are also courting China and want to keep good strategic ties there. It is really losing influence over Eastern Europe, and what's happening now only rubs it in their face.

That NATO expansion we talked about in previous chapters is still troubling the Russians. Something quite radical, that would have been unthinkable a decade ago, has happened: places like Poland, Hungary, and the Czech Republic joined NATO very quickly after they declared independence in 1991, mostly to ensure no further Russian influence. The next wave of folks was 2004 when Estonia, Latvia, Lithuania, Romania, and Bulgaria, places that were once firmly in Soviet control, joined NATO. Russia now finds that its former territories, which it sometimes harshly subjugated, are sovereign political entities that want nothing to do with it. They are NATO members and could theoretically have NATO weaponry pointed at Russia. This is a big loss of status and influence for Russia. Think about this in an abstract context. The Baltic States go from being owned by Russia, to ten years later being sovereign states with weapons pointed at Russia. That's a fairly big change. But the scales are starting to tip back to the Russian side. . . .

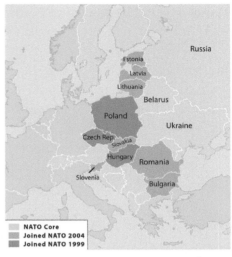

NATO creeps ever closer to the Ruskies.

RESURGENT RUSSIA: THE BAD BOYS ARE BACK!

"I can snap your neck with my bare hand. Enjoy your dinner."

Watch out! Things continue to change fairly rapidly for Russia, and it is the understatement of the century to say that things are looking up. Yes, I've talked about how crappy its economy has been. Yes, I've told you all about its people problems, its environmental problems, its crime problems, its international status problems. But what a difference one man can make, and the Russian man of the century is, with no doubt, Vladimir Putin. Putin, along with petroleum, has brought the Russians from the brink of the abyss back to world power status in less than a decade. There are a few other factors at play here as well that are serving to bring the Russian region back to a starring role on the world stage. Let's wrap up this chapter looking at this incredible reversal of Russian fortune.

PREZ TO PRIME BACK TO PREZ: PUTIN POWER!

This guy is unstoppable! So rarely in life can you truly credit a single individual with changing the course of history but, for better or for worse, that is an apt description for our main man Vlad. It's no wonder that *Time* magazine voted him Man of the Year for 2007—he really has made that big of a difference in his country, and in the world. Vladimir Putin is also easily the most bodacious beast of a leader on the planet: as a former KGB agent and judo blackbelt, he could handily whip the butt of any elected official on this side of the Milky Way. And the people love him! Putin held 60 to 80% approval ratings his entire eight years in office, and stepped down as President with closer to 90% love from the Russians. Stalin would have killed for that kind of popular support—oh, wait a minute, he did kill to get that kind of support. But I digress. Why all the love?

Putin is only the second president Russia has had. He was the hand-picked successor to Boris Yeltsin, a fairly popular leader in his own right. In his eight year run from 2000–2008, Putin oversaw the stabilization of the economic transition process that had thrown Russia into chaos for its first decade. He was tough on the oligarchs, and tough on crime, which helped stimulate international investment. He also took a fairly pro-active government role in helping Russian businesses, especially those businesses dealing with petroleum and natural gas. More on that in a minute. GDP grew six-fold, and propelled Russia from the 22nd largest economy up to the number ten slot. The economy grew 6–8% on average every year

President? Prime Minister? It's all Putin to me!

he was in office. Investment grew, industry grew, agricultural output grew, construction grew, salaries grew; pretty much every economic indicator you could look at has gone up during his tenure.

Unlike most world leaders, Putin was very savvy with this cash flow too. He paid off all Russian debts. Imagine that: Russian National Debt = 0. Dang. That's legit. Putin also stashed billions, if not trillions, in a Russian "rainy day fund" that is set aside for any future hard times. How refreshing! Most leaders usually stash that extra cash in their personal Swiss bank accounts. But not Putin. No wonder Russians love him!

However, the love is not just based on economics. Vladimir Putin re-instilled a great sense of nationalism in his citizens. After the demise of the USSR, Russia was a second rate power with even less prestige, at the mercy of international business and banking and stronger world powers like the US, the EU, and NATO. Russia pretty much just had to go with the flow, even when events were inherently against their own national interest, like the growth of NATO, for example. Putin brought them back politically to a position of strength. He played hardball in international affairs, and re-invigorated a Russian national pride that has long laid dormant.

Back off, NATO! Putin is protectin' his 'hood!

Of course, this has come at a cost: Putin achieved a lot of this by consolidating power around his position, controlling a lot of state industries, cracking down on freedom of the press and free speech, and manipulating power structures of the government. In fact, he stepped down from the presidency in May 2008, and stepped immediately into the Prime Minister position the same day, maintaining a lot of command and control of the system simultaneously. That crafty Russian fox! He is increasingly reviled and even hated by Team West because people see him as leaning back towards totalitarianism, but also because he increasingly clashes with western foreign policy on the international stage. **2014 Update:** After a little constitutional "correction" which changed the term length of office from 4 to 6 years, Vladimir Putin was once again elected President in 2012 and assumed the top office. He is in the Russian hizzle until 2018, and possibly to 2024 if re-elected again. Another Putin decade looms large. . . .

But hold the phone! I keep talking about Putin, and have completely dissed the other guy who held the top slot in the "in-between Putin" years. Let's remedy that right now! Dmitry Medvedev was the hand-picked successor to Putin, and was President from 2008 to 2012 when he basically stepped down by refusing to even run for a second term, allowing Putin to walk back into the position. He pretty much followed and even strengthened virtually all of Putin's policies, and was therefore considered by many in the west as a mere Putin puppet. There could be some truth to that. However, he did have the popular support of the Russian population, and why on earth would he stray too far off of the successful path that Putin had paved? Answer: he didn't. What he did do in office was to start a massive military makeover which seeks to revamp and remodel the Russian armed services into a state-of-the-art ultra-modern military that will certainly be causing consternation for the US and Team West in the future.

My main man Medvedev.

Dig this: there is already speculation that the Russian constitution will be amended to allow Putin to run for president again, in essence, scrapping their term limits. Given the rabid popularity of both men, I think their policies are here to stay for some time. Or perhaps forever. **2012 Update:** Ummm . . . yep. That happened. And Medvedev is now the Prime Minister. A perfect position swap occurred.

In conclusion, Putin power is not to be underestimated. Perhaps 12 more years as President, and if not, he can hold the Prime Minister position indefinitely either way, so he will be around for some time to come. The Russians love him for making them strong, and making them rich and making them proud again. Let's look at some specifics of how he pulled this off and the implications for the future.

Medvedev and Putin: the Russian Dynamic Duo

PETRO POWER!

Putin is awesome. On that point we are clear. However, we do have to at least partially, if not fully, credit his great success in turning the Bear around on this one single commodity: petroleum. Dudes! Russia has made total bank on oil and natural gas in the last decade. To understand the importance of oil for Russia, one need only consider this: when the USSR crashed in 1991, the price of a barrel of oil was about $10. In 2008, the price of a barrel of oil approached $150. The price may have dropped a bit since, but oil is still the preferred energy source of choice for the entire planet. My friends, you should know this: Russia has a ton of oil! Look at the map! You do the math!

Increasingly, Russia has also reeled in all control of the petro industries to the state. They **privatized** a lot of those industries back in the 1990s, but they have been busy **nationalizing** a lot of them back since 2000. Remember those terms from Chapter 4? Russia sure does. It now has controlling interest in virtually all the oil and natural gas businesses in the country, which even more of the profits swing back directly to the state.

To reiterate a few points from this chapter and the last: oil = power for Russia. Economic and political and international power, that is. Not only have they paid off their foreign debt with oil money, they have re-invested that oil money back into their economy, and also set up their rainy day fund with oil dollars. Since the Russians supply one third of European energy demand, this gives them all sorts of economic and political leverage over their Eurasian neighbors. Mess with Putin, and you might not have heating oil next winter. But the Europeans aren't the only ones who need oil. Look eastward and you see Japan and China both vying for petro resources. Russia is really sitting pretty right now. Energy master of the continent!

Russian Oil and Natural Gas at a Glance

Sometimes having a lot of gas is a very good thing: Russia may become the center of a natural gas OPEC-like cartel!

GLOBAL WARMING? NO SWEAT!

I can't get out of this section without referencing the global warming situation. While the rest of the planet may be wringing their hands in a collective tizzy about rising temperatures, melting polar ice caps, and rising sea levels, let me assure you that Russia is not sweating the situation at all. Ha! Not sweating the warming—am I good or what?

Seriously though, think of the strategic benefits that are being bestowed upon Russia as the thermometer continues to creep upward every year. Russia has forever been a land empire due to its lack of accessible and navigable coastlines. Additionally, almost all of its river systems empty into the ice-locked Arctic Ocean, making them essentially useless for transportation and exporting commodities. But just wait! As the permanent ice cap covering the North Pole disappears, Russia will become more open to the world like never before in recorded human history!

"Comrades, the North Pole is ours now. Bye-bye, Santa!"

You may not be paying attention to any of this, but the Russians sure are. In preparations for its ice-free northern coastlines of the future, Russia is staking a claim to the entire Arctic sea east of the North Pole. Seriously, I'm not making this up. The race for the Arctic has begun! Russia stands to be the biggest beneficiary as well. Not only will it open up their territory for greater sea and land access, and greater export power, but it is widely believed that there are massive oil reserves in the Arctic basin, reserves which will become accessible once the ice is gone. Russia is all about the global warming! Let's get this party started! No, actually let's end it with. . . .

GEOPOLITICAL JUMP START

You got your Putin, and you got your petroleum, and you add to that a huge dose of nationalistic pride. Bake for twenty minutes, and you end up with an extremely resurgent and resilient Russian cake on the world political table. These guys are playa's once again. Their huge economy, their huge oil reserves, and their huge potential for the future is fast becoming the envy of the world. They can afford to play the fence when suitors come calling. What do I mean by that?

Russia has been invited to become a strategic partner state of the EU— not a full-fledged member, mind you, but a partner. Europe realizes that the Russian economy and Russian energy is such an integral part of their own lives, that to not have Russia at the table for major decisions would be folly. Russia has a real impact on some EU policy. The same can be said of NATO. While Russia is not a member, the group realizes that almost all major decision-making should involve the Russians since they play such an increasingly important role in Eurasian affairs. Russia is typically invited to major NATO summits now. Why doesn't Russia just join the EU and NATO fully?

Because the Russians are forging economic and strategic political ties to their east as well, mostly with China. They are at a historical east/west pivot point right now; they are straddling the fence about which relationships to make or break. I think they're being extremely savvy and are going to avoid taking sides in anything. Could Russia join the EU? It's already a strategic partner of the EU. I bet it will be invited to join as a full-fledged member, but I also bet it will decline.

Why? It is also strategic partners with China, the other huge economy on the other side of the continent. Russia will not be stymied economically by joining a club that can possibly limit their economic options. The Bear is also still a major player in Central Asia, an area with tons of natural gas and oil that has to be moved out of Central Asia to the rest of the world that wants it. That oil and natural gas usually moves out through Russian territory, supplying most of the fuel for Eastern and Western Europe, in addition to increasingly supplying fuel for Japan, China and India.

Beyond the geopolitical ramifications of energy and economics, Russia is back in full action militarily speaking! Big time! You do know that, in August 2008, Russia invaded the Georgian regions of South Ossetia and Abkhazia, right? It currently still occupies these areas, and has gone as far as to recognize their declared independence, which is infuriating the US and Team West as a whole.

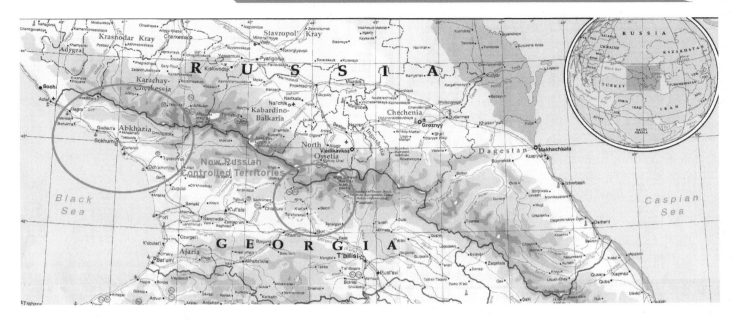

Know this for now: Russia is back! The Russian–Georgian War did not change the balance of power in this region, it simply was an announcement that the shift in the balance of power had already occurred!

Russia is now openly and powerfully reasserting its influence in its own backyard, and that is a big deal. Russia now feels that it has regained a position of strength to not only stand up to what it sees as threatening Team West expansion to its borders, but to stop that advance with force if necessary. With guns! This open, bold, and even proud use of force by the Russians has changed the whole attitude of the US, NATO, and even the EU when it comes to political maneuvers near the Russian border. The bear has lashed out!

At the risk of redundant repetition, what supranationalist organizations are the Russians all about if they don't really want to play with the EU and NATO? Well, remember the BRIC? They dig that one, and it should be noted that BRIC is specifically not a "Western" institution. Of even greater significance is the SCO. Flip back to chapter 6 or ahead to the Central Asia chapter to get the full run-down on this group, but just let me tease you for now by telling you this much: the SCO is going to be a bizarre cross between an Asian version of NATO and a Russian version of OPEC, and Russia has every intention of being the leader of this grand experiment. Stalin would be proud.

DON'T LIKE PUTIN? CRIMEA RIVER!

Holy wowsers huge 2014 update on resurgent Russia! Unless you have been hiding under Vladimir Putin's pectoral muscles, you have heard about Russia invading, occupying, and now completely absorbing the Crimean Peninsula, in what was formerly Ukraine! This is a huge deal. A re-defining sovereignty in the modern world deal. A re-defining of Russia's place in the world deal. And a possible start to an expansionist era of Russia that sees them take back territories that contain ethnically Russian peoples. These peeps are partying like it's 1799, because no one has seen such old-school style empire building for almost a century.

Crimean crisis just the start of the show?

And in that light, let me introduce you to a new organization to be used to expand the Ruskies' reach: The Eurasian Union. While originally an idea posed by Kazak President Nursultan Nazarbayev, the Eurasian Union, is without a doubt the wet dream of Russian President Vladimir Putin, who now openly seeks to reassert a resurgent Russian influence and outright power into areas it lost during the collapse of the Soviet Union in 1991. It is being proposed as a economic/political trade block alternative to the EU . . . and one that has Russia at the center. Belarus and Kazakhstan are already in; Armenia wants in; other states are being pressured hard to join, or possibly face the wrath of Russia.

Yeah. That should do it. Problem solved.

But know this for now: the Crimea craziness is a gargantuan gambit by Putin to expand the power, influence, and control of Russia to other states . . . and so far, so good, for the victorious Vladimir and his legion of fans . . . which includes the vast majority of Russian peeps who support him totally. As in totalitarian. Which is where it looks like this is going. Yikes!

GHOST OF STALIN RETURNS

WRESTLING FOR WEALTH.

Team West taught the Russians too well about that capitalism stuff!

Which is a good jumping off point. I guess any mention of Joseph Stalin is a good point to get the heck out of Dodge. I want to leave you with this: the memory of Smokin' Joe Stalin is actually being softened in today's Russia, and that is a very telling sign of what's happening in the society. Some folks in Russia are looking back with pride at what most of us consider one of the worst dictators in history. Why? This vibrant and growing economy, this resurgent world role, and this overall restored sense of Russian greatness has not been felt since the Stalin era, and quite frankly, the Russians are quite proud to be Russian again.

Add in trillions of petro dollars, increasing empire due to global warming, and a renewed sense of global political power, and you got yourself a region that will be shaping world events of the future. Growl! The Bear is back!

RUSSIAN RUNDOWN & RESOURCES

 ## BIG PLUSES

→ Biggest state by land area, one that is extremely rich in virtually all natural resources

→ Specifically, a huge oil and natural gas producer: energy resources critical to world economy

→ Virtually no national debt

→ Regained sense of national pride and political world leadership

→ Has veto power at the UN Permanent Security Council

→ Has as many nuclear bombs as the US, if not more, as well as a serious military, second only to the US.

→ Is a founding member of some of the most promising future international coalitions like BRIC and SCO

→ Is still considered quite important to the EU and NATO too, enough to get invited to the meetings.

→ Global warming? No problem for Russia! Will increase usable land area and resource base

 ## BIG PROBLEMS

→ Gi-normous size problematic for infrastructure, defense, and cultural cohesiveness.

→ Economy and social structures, like health care, still not fully recovered from Soviet era crash

→ Male longevity rates are lower than most African countries; alcoholism rates among the highest in the world

→ Population overall is shrinking.

→ Organized crime, rampant corruption, and tax evasion have seriously stymied international investment and hinder internal development

→ Economy still heavily reliant on exporting natural resources, mostly energy.

→ Has some of the worst environmental disaster areas on the planet

→ Anti-immigrant, anti-foreigner, and flat out racism are significant issues in Russia

DEAD DUDES OF NOTE:

Tsar Nicholas II: Last of the royal Romanov line, and last monarch of Russia, who was assassinated along with his whole family in 1918. His piss poor rule transformed Russia into an economic and military disaster which helped fuel the revolution which left the commies in charge.

Rasputin: The un-kill-able "Mad Monk" who recommended that Tsar Nicholas get Russia involved in World War I.

Vladimir Lenin: "Father of the USSR" Passionate organizer, orator, and propagandist who led the Russian Revolution of 1917, in which he implemented the first ever communist experiment at the state scale. Then he died.

Joseph Stalin: The semi-psychotic who took over control of the USSR after Lenin's death. Formed the KGB, purged out all political adversaries, and may have been responsible for the deaths of 30 million Russians. But, he also oversaw the industrialization and expansion of the USSR into a world power, and helped kill the Nazis in WWII. So it's a mixed bag.

Nikita Khrushchev: Leader of the USSR at what was perhaps the height of the Cold War. Was the bonehead that decided to put Soviet missiles in Cuba pointed at Florida, which brought the world to the brink of nuclear war aka the Cuban Missile Crisis of 1962.

LIVE LEADERS YOU SHOULD KNOW:

Mikhail Gorbachev: Oversaw the collapse of the USSR, implemented the glasnost and perestroika policies, and helped bring the Cold War to a peaceful end. Is a hero to the West and won the Nobel Peace Prize, but looked upon almost as a traitor in Russia.

Vladimir Putin: Current President of Russia, having previously served as such from 2000–2008, got Russia's mo-jo back, got them rich, and got them powerful again on the world stage. Look for him to be in Russian politics indefinitely.

Dmitry Medvedev: Current Prime Minister of Russia, the Putin right-hand man. Many think he is merely a Putin puppet, but has so far shown himself to be an adept and strong leader in his own right who is continuing the successful Putin policy legacy.

Sergey Lavrov: Current Russian Foreign Minister, which is the equivalent of the US Secretary of State, the head diplomat of the country. This dude is wickedly smart, savvy, and has been behind the scenes making sure Russia regains its power status on the world stage.

KEY TERMS TO KNOW

Demographers

Ecumene

Permafrost

Population Pyramids

Topography

REPUBLICS OF THE RUSSIAN FEDERATION

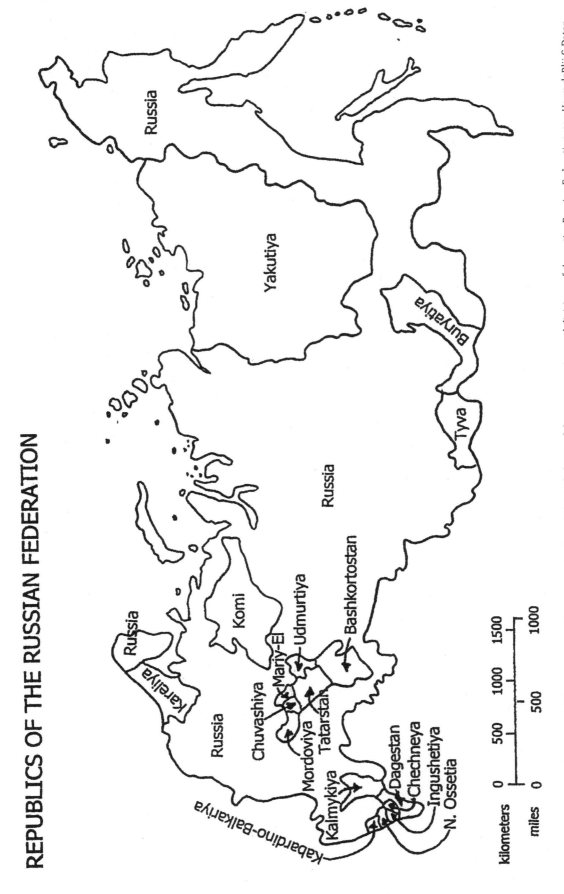

The vast Russian Republic is divided into many administrative units. For a full detailed map of the numerous internal divisions of the entire Russian Federation, see Harm deBlij & Peter Muller, *Geography: Realms and Regions*, 12th edition (New York: John Wiley & Sons, Pub., 2006), pp. 122–123.

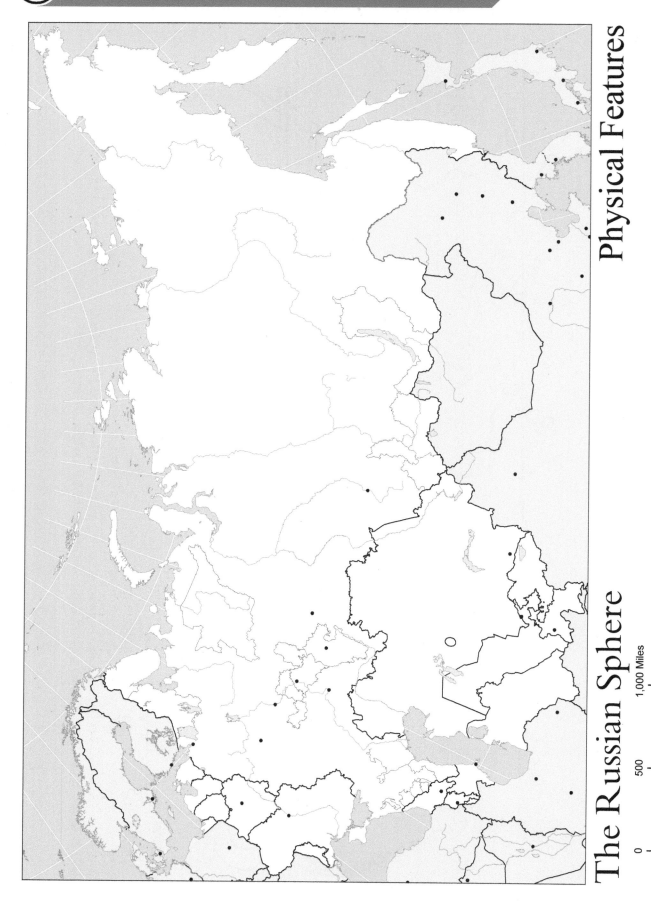

The Russian Sphere

Physical Features

0   500   1,000 Miles

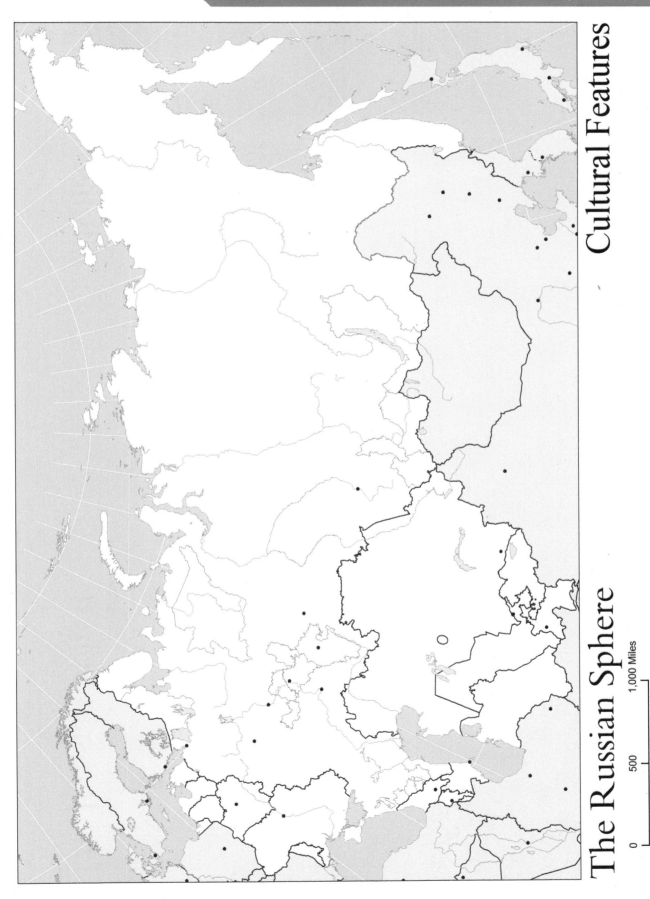

Cultural Features

The Russian Sphere

0   500   1,000 Miles

©2015 BYU Geography
Think Spatial group

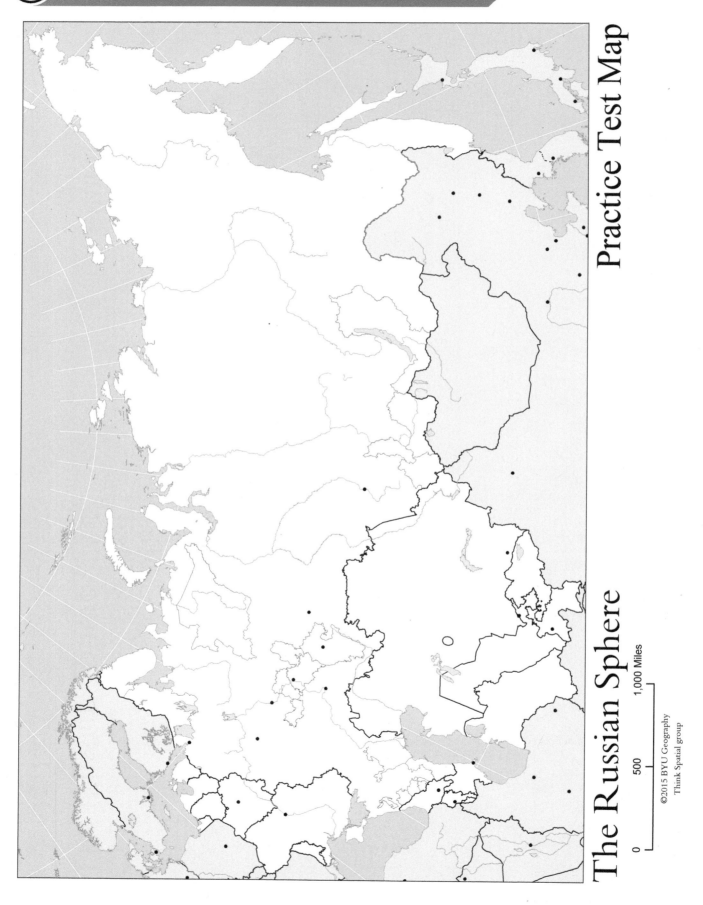

Practice Test Map

The Russian Sphere

©2015 BYU Geography
Think Spatial group

0   500   1,000 Miles

East Asia from Peripheral to Core

9

N

As for the rich in this present age, charge them not to be haughty, nor to set their hopes on the uncertainty of riches, but on God, who richly provides us with everything to enjoy. They are to do good, to be rich in good works, to be generous and ready to share, thus storing up treasure for themselves as a good foundation for the future, so that they may take hold of that which is truly life. (1 Timothy 6:17–19)

CHAPTER OUTLINE:

CHINA has emerged as an industrial giant like the United States. Similarly to the boom in the U.S., times have created certain "growing pains" that provide a ripe opportunity for us as servants of the Lord. One important aspect of this is providing equal rights including the right to life and justice. Remember Psalmist taught in Psalm 72:2 when he said, "He will judge your people with righteousness, and your poor with justice." (NKJV)

IN A NUTSHELL

One day, a veteran of Asian service described his experiences in East Asia to the author as "a totally unique world with a culture so different from ours." When asked to give specific examples the answer was, "everything: the plants, the animals—even the food!" Indeed East Asia is a region of fascinating landscapes and historical extremes of both beauty and darkness. Cut off from much of the western world for so long, this enormous part of the Asian continent developed separately from the Western world while creating various unique cultures whose resources and productivity continuously beckoned the

Landlocked Countries in Asia

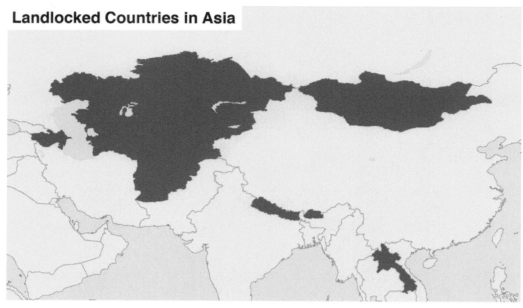

Map courtesy of Katie Pritchard

West with trade and commerce opportunities. In the past as well as today, it seems the West's cravings for the products of the Orient pulled its money to the east. It is believed that a Christian revival as well as unprecedented economic growth is occurring in China today. No doubt this exotic area will continue to make the headlines, and who knows? You or your future family members may live or even marry someone from there.

The area under observation consists of the islands of Japan, the peninsula of Korea, and the large island nation of Taiwan. To the far north are the windswept plains of Mongolia. By good fortune, this generation is alive at an exciting time as we see East Asia emerging scene during one of its periodic forays out into the world. Considered by many to be the

Central Asia.

Map courtesy of Katie Pritchard and NOAA

richest area of the 21st century world, the opportunities for service abound in East Asia, but first one must understand the culture. And to understand the culture's development and dispersion, we must study the topography and other various aspects of physical geography.

To the north of China are the cold and dusty Gobi and Takla Makan Deserts. Standing like a cold sentry to the north are the Pamir Mountains, and to the northwest are the fabled Tien Shan, which translates to "mountains of the clouds." Farther to the south are the Hindu Kush going east and Karakorum Range branching to the west. To the south are the mighty Himalayas and the furrowed brow of numerous mountain ranges with running north to south valleys channeling fast-flowing rivers pushing south taking cultures and people with them.

Dept. of Defense photo by SSgt. Cherie A. Thurlby, U.S. Air Force.

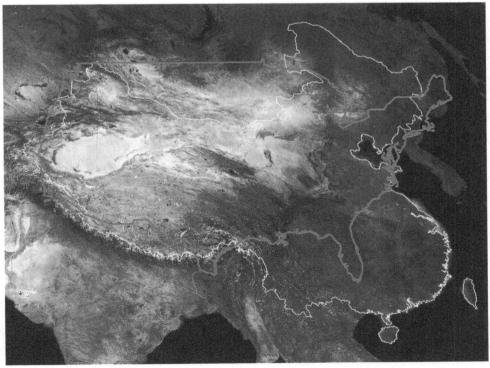

From *Blue Marble: Next Generation* image produced by Reto Stockli,
NASA Earth Observatory (NASA Goddard Space Flight Center)

China: Ethnolinguistic Groups

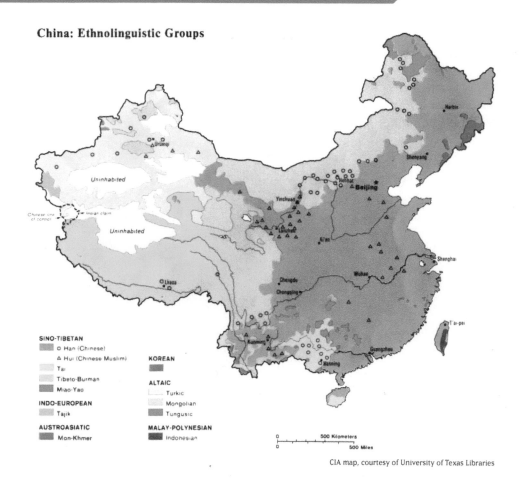

CIA map, courtesy of University of Texas Libraries

These high mountain chains stand as bulwarks against the seasonal tropical rains of the monsoon season and leave China protected like a turtle behind a shell of physical geography. Occasionally this culture has come out of its shell and has ventured abroad only to withdraw itself behind its isolating geography when it felt endangered. This pattern of alternating expansion out of and contraction behind these natural borders has not only protected and sustained China during its long history but also has tended to both protect and limit China's prospects for growing at sea.

A term frequently heard in recent years is **globalization**. Simply stated, globalization is the increasing interconnectedness of the world making it seem smaller. *Smaller* is a relative term but refers to the speed with which one can communicate or travel to distant parts of the world. Globalization has caused many environmental problems; for example, numerous species of East Asian animals and plants have come to the United States with varying degrees of impact on local populations. If you have seen bamboo or kudzu in the American South growing on the sides of the road, you have seen an impact of globalization. The ability to communicate rapidly with people in the different corners of the earth is another reason to study languages and obey the Great Commission of Jesus Christ. As we shall discuss later on, much progress has been made in this regard in recent decades.

SEZ WHO?

We know efforts were made to obtain Chinese silk and other luxury products as far back as the Roman Empire. Quite possibly, attempts to dredge the Euphrates and Tigris rivers and various disastrous military endeavors were initiated by the desire to find a water route through the Indian Ocean in order to bypass the land routes called the **Silk Road**. Later, the Portuguese would exploit the growing weaknesses of the Persian and Turkish empires by what essentially was a maneuvering around,

Image © John Lock, 2013. Used under license from Shutterstock, Inc.

Image © Johny Keny, 2013. Used under license from Shutterstock, Inc.

or flanking, of those empires to gain access to trade. The terminus of this trade was China and some historical geographers even believe the westward expansion of the United States in many ways was a continuation of the search for the elusive "Northwest Passage" into Asia. The capitalist nations of Western Europe and the Americas continued to see their monies drained by buying Asian products. Reciprocity, however, did not occur, as western products seemed to be not so sought after. Recent economic trends have tended to reinforce this general inequality.

In the 1970s, for example, the Chinese began implement a system of reforms, which in many ways reversed the intensely centralized economic planning characteristic of the powerful leader Mao Tse-Tung. Upon Mao's death, a series of five-year plans occurred that had previously been called **The Great Leap Forward**. These plans utilized the collectivization impulse typical of Asian nationalism to break people into groups of **communes** for the production of agriculture and iron. Although this was in many ways disastrous, nevertheless the Chinese nation survived a mercurial dictator and the Korean War (in its infancy) to become an economic powerhouse. The reforms instituted seemed to anticipate the British release of Hong Kong and began a re-characterization of the coastal areas into **Special Economic Zones** or **SEZs**. These areas offering tax incentives for foreign investments are one reason the skyline of Chinese cities are remarkably fresh and new. As China's economy has taken off so has its tendency to directly influence the wider world for the purpose of seeking resources and influence.

HEMMED IN

The famous geographer and traveler Robert D. Kaplan refers to what strategists today call a "first island chain" and a "second island chain" when discussing the various islands that ring around China. The United States has essentially dominated the first island chain closest to China consisting of Taiwan, Luzon (the northern island in the Philippines), and the Ryuku Island chains as well as the Japanese home islands. These islands tend to funnel Chinese attempts at exploration into the Pacific. A secondary island chain consists of Guam and the Marianas Islands.[1] These islands constitute a barrier similar to

---

[1] Kaplan, Robert D., "Monsoon" (New York: Random House), 286.

China: Population Density

Agricultural Regions

China: Industry

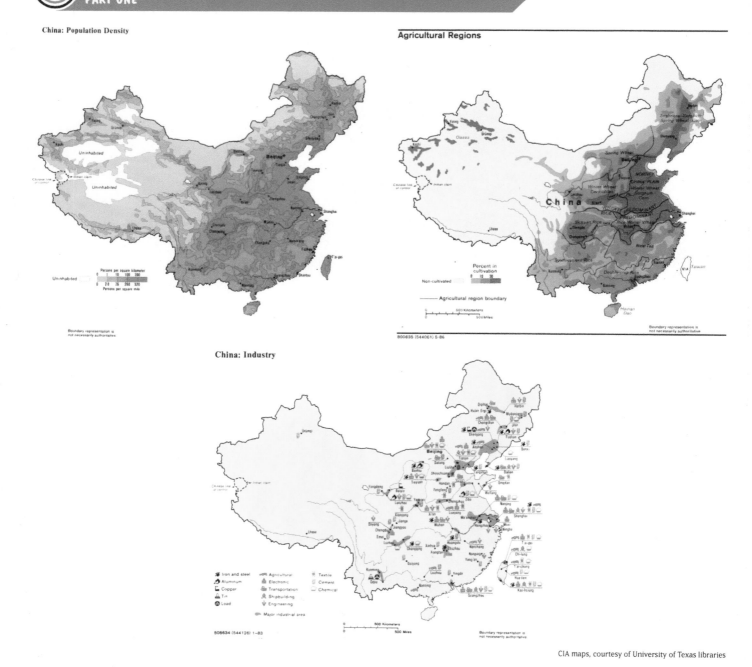

CIA maps, courtesy of University of Texas libraries

those on land, which protect China and yet remain a challenge to the nation as it seeks to emerge from its shell in the future. It is Kaplan's contention that China today is a reflection in part of the Great Wall and its role in the interactions between China and the nomads of Inner Asia.[2] China has periodically grown in varying impulses or dynasties throughout its history and now seems to be re-emerging because it feels secure. China and much of East Asia has also been sought after by various western traders who have attempted to reach it across mutually forbidding geographies of land and the distance by water. China's unique culture despite its geographic isolation has continued to diffuse in unique ways. Religion is a terrific example of a cultural expression that has pierced China's isolating physical geography.

[2] Kaplan, Robert D., "The Revenge of Geography" (New York: Random House), 195.

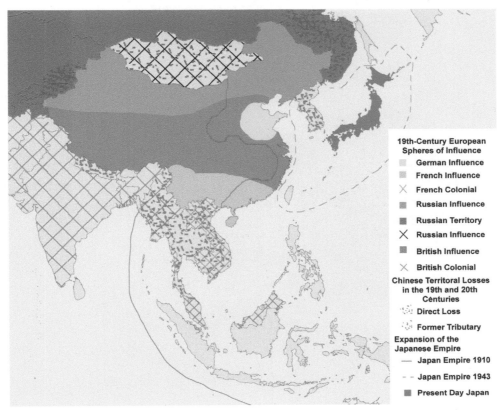

19th-Century European
Spheres of Influence

◻ German Influence

▩ French Influence

✕ French Colonial

▨ Russian Influence

◼ Russian Territory

✕ Russian Influence

▩ British Influence

✕ British Colonial

Chinese Territoral Losses
in the 19th and 20th
Centuries

▨ Direct Loss

▨ Former Tributary

Expansion of the
Japanese Empire

— Japan Empire 1910

- - Japan Empire 1943

◼ Present Day Japan

Map source: *The World Factbook*

LIGHTS PLEASE!

Historically, **Confucianism** has provided an ethical system promoting societal order and concern for establishing proper relationships. The importance of preventing conflict through tact within a traditionally high-population area is obvious. The religion of **Buddhism**, which spread from India, has also historically been a powerful source of hope and remains an important aspect of East Asian cultures today. The type of Buddhism known as **Mahayana** claims that there is a savior that can be the Buddha or any of the **enlightened beings** or **Bodhisattva** who have decided not to go into **Nirvana**, or what we as Christians would call heaven, because they want to help other individuals achieve their exalted state. As you might expect, the geography of China and the direction of the mountains and rivers have helped diffuse Buddhism into Southeast Asia but in a slightly different form, which we will discuss in the next chapter. Another interesting aspect of culture as it has related to China is agriculture.

Today's Christianity has tended to take hold only in a few places as a consequence of the colonial experience when Christians followed the capitalists of Europe to these far-off shores. To many of the colonized, the faithful displayed little difference from the greedy outsiders and were of the same ilk. One such area of Christian expansion is perhaps the most Christian nation in East Asia today, Korea.

THE HARVEST IS GREAT

Today the Korean peninsula stands guard over the northeastern region of China close to the mineral riches of Manchuria and the capital city of Beijing. Divided at the end of World War II almost on a whim by military staff officers tasked with organizing occupation forces from the conquering American army, the line reflected the first major latitude above Seoul and the Han River Valley where the victorious American Army headquarters collocated with the former Japanese

Korean Peninsula

The Korean Peninsula.
CIA map, courtesy of University of Texas Libraries

Photo by U.S. Information Agency Press and Publications Service Visual Services Branch, from National Archives and Records Administration

headquarters. The Soviets honored this line, similar to lines drawn in Germany and Austria and similar to what would again occur in Southeast Asia soon thereafter. Both the Soviets and the Americans thus divided Korea with little consideration of long-term consequences to the Korean Peninsula.[3] The Korean Peninsula now consists of a totalitarian North, somewhat of a relic of the Cold War, and a proudly confident and energetic South, which is much closer to being a democratic republic. With its high GDP, South Korea joins Hong Kong, the Republic of China, Taiwan, and the tiny city-state of Singapore as some of the newest producers of manufactured goods. Despite nearby bellicosity, the South continues to prosper and is a very important factor in spreading the Good News of Jesus Christ to the rest of East Asia. It has been suggested in the past that South Korea could overtake the United States in terms of the numbers of missionaries sent to the field. Though such ideas have proven a bit overly optimistic, the truth remains—the Korean Peninsula has been faithful![4]

---

[3] The War for Korea: 1945–1940 "A House Burning" by Allan R. Millett (Lawrence, Kansas: The University Press of Kansas, 2005), 45.
[4] http://www.christianitytoday.com/gleanings/2013/july/missionaries-countries-sent-received-csgc-gordon-conwell.html

An amazing contrast at night on the Korean Peninsula.
NASA/Goddard Space Flight Center Scientific Visualization Studio

To the south of Korea is the island of Taiwan. Taiwan, though small in size, represents the exiled former government of the mainland of China and is on the island of Formosa. Though only slightly smaller than Maryland and Delaware combined, this nation state of 25 million is a powerful force. At one time, the tallest skyscraper in the world jutted defiantly towards the mainland. The skyscraper Taipei 101 indeed had 101 floors, and, though recently surpassed as the tallest building in the world, it remains a testimony to the energetic peoples who dwell there. Taiwan appears to represent a major thorn in the side of the People's Republic of China. Just as we in America often feel burdened by our past, the Chinese are very sensitive to the presence of a national government claiming authority over the mainland. The emergence of a communist government in China over half a century ago represented a nationalistic impulse that still is extremely emotional today. This emotion can be seen over areas of contention that both appeal to and exacerbate racial and national tensions in various frontier areas.

Map from *The Scottish Geographical Magazine*
(Volume XII: 1896), courtesy of University of Texas

AND IN THIS CORNER...

A legitimate interest in procuring safe routes by sea for vital resources is increasingly causing a shift in relations between China and various nations of the world, particularly in the East and South China Seas and the Indian Ocean. Frontier areas off of the coast of China include the Senkaku Islands (Japanese name) or Daiyou Islands (Chinese name). These islands represent what may be only the beginning of what are becoming different areas of contention as Chinese sea power increases and a simultaneous decline of an American commitment in the South China Sea continues. China's growth has produced another area of dispute in recent years as well, namely, the oil-rich Spratley Islands off of the coast of the Philippines and Vietnam. Such maritime areas of conflict demonstrate a subtle transition occurring as China begins to flex its international muscles. The Chinese have

Image © vadimmmus, 2013. Used under license from Shutterstock, Inc.

Map courtesy of Katie Pritchard and NOAA

implemented what is known as a "String of Pearls" strategy in order to circumvent the Silk Road. This strategy consists of the construction of a series of naval and commercial sea bases in Gwadar, Pakistan and Sri Lanka in order to attempt to secure their access to Middle Eastern oil and the vital **Strait of Malacca** among other things.

Finally to the north of the economic colossus of China is Mongolia. Mongolia was originally the home of fierce nomadic peoples who were walled out of China. These nomads travelled on horseback and dwelt upon the semi-arid plains and grasslands to the north of China. Living in round tents, or yurts, modern Mongolia is a study in contrasts today as electric lights cast shadows on traditional tent cities. Mongolia is in a delicate situation today. A few generations ago, Mongolia was firmly in the grip of a powerful Soviet neighbor. In fact, the capital city of Ulaanbaatar means "Red Hero" after the early communist leader there. In recent years, however, China has been moving ever closer to controlling Mongolia, and as China finds less of a threat from its north, it is increasingly emboldened to take to the sea and explore. Seeking oil and minerals from around the world, the security of a modern China will continue to draw the world to it as it has since ancient times. This increasingly globalized planet will offer us great opportunities to live out our faith and to be living testimonies of hope and faith in a busy world that is preoccupied with riches and security.

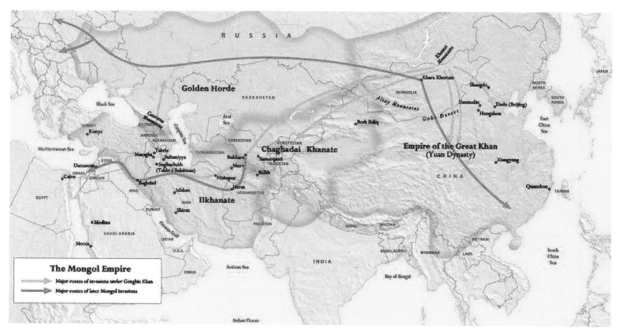

Map source: *The World Factbook*

DEVELOPMENT AND UNDERDEVELOPMENT IN THE WORLD-ECONOMY*

One of the most useful aspects of world-systems analysis with respect to understanding the human geography of the world and its division into "developed" and "underdeveloped" realms is its delimitation of the spatial, political and economic structure of the current capitalist world-economy. According to Wallerstein the capitalist world-economy that has developed since the mid-15th century has three primary structural features:

1. A single world market which operates within the context of a capitalist ideology and logic; this logic and ideology effects economic decisions throughout the entire world-system.

2. A "multiple-state" system in which no single state ("country") is able to dominate totally and within which political and economic competition among various states is structured and defined.

3. A three-tiered economic and spatial structure of stratification in terms of the degree of economic "development" and "underdevelopment" of the world's various states. In this textbook, this basic three-tiered structure is employed as the primary tool, as a model and a theory, for understanding basic differences in the human geography of the world today and in the past. In place of terms like "developed world" and "developing world," or "first world" and "third world," we will use the terms *core, semi-periphery* and *periphery* that are employed in world-systems analysis to differentiate between regions of varying economic "development" in the world. This text argues that an understanding of the differences between these regions is fundamental to a meaningful appreciation of the differences in such things as the nature of cultures, the structure of economies, political organization and population issues and problems around the world.

It should be noted that while the terms "core" and "periphery" are employed to indicate geographical location in common usage, in world-systems analysis they do not *necessarily* refer to a state's geographical location on the globe. Rather, they are used to refer to a state or region's "location" in the world-economy, that is on the "inside" (core), on the "outside" (periphery), or somewhere in the middle (semi-periphery) of the world-economy in terms of relative control. Those states in the core control and direct the world-economy, while those in the periphery are most often controlled and directed by the core states. Further, history shows us that it is clear that "core" and "periphery" are not geographically or temporally static—states move up and down within this hierarchy over time at a non-constant rate. A recent excellent example of this rule is the former Soviet Union. Before its breakup in late 1991, one could argue (using socio-economic data) that the Soviet Union was a core state. Since the breakup, however, economic conditions have deteriorated to the point that Russia is now clearly a semi-peripheral state. Further examples would include several countries in Southeast Asia such as Indonesia, Thailand, Malaysia and Singapore. Peripheral states until the 1980s, they all are now part of the semi-periphery due to surging manufacturing economies and greater political and economic stability.

CORE STATES

The **core states** of the world-economy today include most states in northern, western and southern Europe, the United States and Canada, Japan, and Australia and New Zealand (see Map 1 on the CD that accompanies this text). It is from here that the world-economy is directed and where its primary command and control centers, such as New York, London, Paris, Frankfurt, and Tokyo, are located. Western Europe emerged as the first core region of the capitalist world-economy in the mid-15th century and remained the primary core until the 20th century, when the United States, Canada, Japan and Australia and New Zealand joined its ranks. The wealth of the core stemmed at first from highly efficient agricultural production and

* Pages 291–292 from *Introduction to Human Geography: A World-Systems Approach, 4th Edition* by Timothy G. Anderson. Copyright © 2012 by Timothy G. Anderson. Reprinted by permission of Kendall Hunt Publishing Company.

control of merchant capitalism through long-distance sea trade in valuable tropical agricultural products (such as tea, coffee, sugar, tobacco, cotton and spices) during the era of colonialism. Portugal, Spain, the Netherlands and the United Kingdom were the successive hegemonic powers (dominance in economic, political, military and cultural affairs of the world) of this early world-economy from the mid-15th century until the early 20th century.

The United States emerged as the primary hegemonic power after World War II. From this period until the early 1970s, the wealth of the core was based primarily on dominance in industrial capitalism focused on heavy industry (cars, ships, chemicals, manufacturing and the like). The United States, Canada, the Soviet Union, Germany and Japan all emerged as the major players in world industrial production during this era. With the exception of the Soviet Union, all of these states remain in the core today. Since the early 1970s, post-industrial restructuring in the core economies has taken place, with a switch away from a reliance on heavy industry to a focus on service industries and information technologies. This era is generally referred to as the era of "globalization" that is characterized by a pronounced international division of labor (this will be discussed further in Chapter 15).

In comparison with semi-peripheral and peripheral locations, the core states all have highly developed economies that are oriented toward service industries and post-industrial technologies such as information technology and computer software production. These advanced economies enjoy a very high standard of living, high gross national products and per capita incomes, and are generally the "richest" countries in the world. Other characteristics of the core states include stable, democratic governments with large militaries, low infant mortality rates and high life expectancies (measures of relative health and availability of adequate health care), low birth rates, fertility rates and rates of natural increase (all signs of very low population growth rates), and large urban (as opposed to rural) populations.

PERIPHERAL STATES

The peripheral states of the world-economy today are located primarily in Sub-Saharan Africa, South Asia (Pakistan, India and Bangladesh) and parts of Southeast Asia (e.g. Papua New Guinea, Cambodia, Laos) and some parts of Middle and South America (e.g. Guatemala, Bolivia, Haiti). The peripheral states all have several things in common. Historically, they are all former colonies of core states and, accordingly, are generally located in either subtropical or tropical areas. During the era of merchant capitalism and colonialism, these areas were assigned a specific role in the world-economy by the dominant core powers: producers of valuable tropical agricultural goods such as sugar, tea, coffee, cotton and spices. As colonies, these areas lost political sovereignty to colonial governance by the core powers and traditional economies and societies were replaced by a colonial economy focusing on plantation agriculture linked heavily with the core. Slavery and peonage were employed by the core powers as the primary form of labor control in the periphery. Most of these former colonies regained political sovereignty after the end of the colonial era beginning in the late 19th century but have struggled since then to rebuild economies and societies that were heavily disrupted by colonialism.

Today, the economies of the periphery remain largely agricultural. The majority of the populations in these countries rely on subsistence farming, low-wage labor on agricultural plantations or low-skilled, low-wage service positions in urban areas. Compared with the core and the semi-periphery, living standards are relatively low, as are per-capita incomes—these areas are among the "poorest" countries in the world. Peripheral regions are plagued by a myriad of problems today: weak, inefficient and often corrupt governments, political instability, ethnic conflict, weakly-developed economies and a lack of public services taken for granted in core regions such as regular electric service, availability of clean water and public sewage services. High infant mortality rates and relatively low life expectancies belie weakly-developed health care systems and a susceptibility to epidemics and chronic problems with diseases such as malaria and AIDS. While core states have very low rates of natural increase (or even negative population growth in some countries), the peripheral populations are the fastest growing populations in the world. Compared with the core and semi-periphery, the peripheral states have very high birth and fertility rates and "young" populations—a significant proportion of these populations is younger than fifteen years of age. If core populations can be described as "old" with stable growth rates, then peripheral populations can be characterized as "young" and growing rapidly.

JAPAN*

Konnichiwa! Welcome to the islands! And a most unique set of islands they are—so unique, in fact, that we will classify Japan not just as an independent state, but as an entirely unique region of the world as well. One of the classic mistakes that all textbooks make, and that a lot of people in the West make, is overgeneralizing Asia. They look at this place and say, "Oh well, you know, it's just Asian people there and six of one, half dozen of the other. Japanese, Chinese, Koreans—whatever." This couldn't be further from the truth when we are talking about Japan and mainland Asia. Japan has an extremely distinct culture from all the countries around it, as do the Chinese, the Koreans, and the Vietnamese, but unlike the others, Japan has followed a more western trajectory in the last century that has set them quite apart from even their closest neighbors. They are the most western of the eastern cultures. Huh? Hang on. We'll get to that in a bit.

Just know this for now, we are coming up on the Clash of the Titans of the 21st Century: The Clash of the Asian Titans. The titans are, of course, Japan and China, two states with huge economies, two states shaping global events, and two states with a long history of animosity between them.

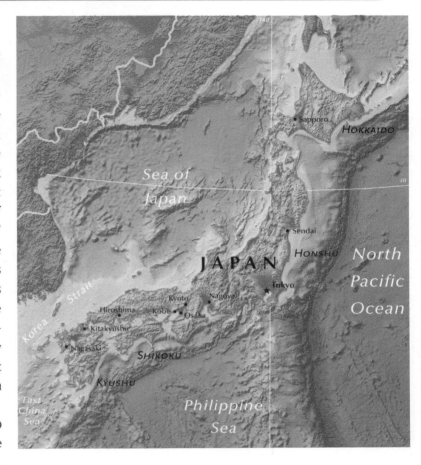

One of the most critical relationships that will be affecting 21st century events is how these two Asian giants will deal with each other as China rises to full superpower status. But I am getting ahead of the story, as usual. Let's get some background on Japan itself, and then look at its evolution into the modern era, and finally finish with that all important relationship with China. Ready? Then Godzilla game on!

Ahhh! Most honorable Mt. Fuji!

PHYSICAL

Japan is a group of four main islands: Hokkaido, Honshu, Shikoku, and Kyushu, from north to south. Japan was created by three or four different tectonic plates coming together in this area. Because of this, Japan is volcanic in origin. Another feature that goes along with this is its mountainous terrain. There are very few flatlands, and this equates to very little agricultural land. Due to plate tectonics, Japan is also earthquake prone—always has been, always will be. It is a cyclical thing and it will not go away anytime soon. It is not a question of if Japan is going to be hit by another major earthquake; it's only a matter of when. 2011 Update: ummm . . . yep. It just happened again.

Why should you care about this? Well, Tokyo's stock exchange, and Japan as a whole, holds tremendous amounts of foreign investment, foreign currency, and bank reserves. When the day comes that a major earthquake shuts them down, depending on the scale of the disaster itself, it could plunge the entire world into recession almost instantaneously. If it is a tremendously bad earthquake, it could plunge the world into a depression in the long term. It's another instance of globalization at its worst, in that a big hit here from the physical world could equate to disaster for the entire planet. However the Plaid Avenger doesn't want to scare anybody; it will happen, they will rebuild, life goes on.

"Honey, I think this section of the interstate is closed."

The other physical thing to consider with Japan is climate. Climatologically speaking, just superimpose Japan, keeping latitude constant, on to the eastern seaboard of the United States (see image below). As you can see, latitude and size being held constant in this image, the climate of Japan is identical

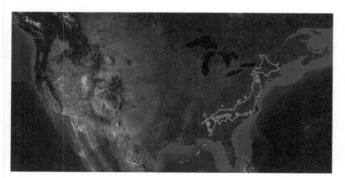

Japan/US: size and latitude comparison.

to the eastern seaboard of the United States. Hokkaido is around the upstate New York area and they get a lot of snow up there. Southern Honshu, as well as Shikoku and Kyushu, reach down into the southern United States and indeed, there are actually smaller islands that go down to what would be tropical Florida. Every place else is similar to some place else in between. For instance, Tokyo is roughly the latitude of Washington, D.C. These two regions, though on opposite sides of the planet, have very similar climates.

One disclaimer here is that Japan is an island nation. As an island, it is surrounded by water, and water is a major modifier of climate. As such, it has a similar climate to the east coast of the United States, though somewhat moderated due to the influence of the surrounding water.

Lastly, physically speaking, Japan is lacking in natural resources. I just suggested that Japan is mostly mountainous, which means it does not have a lot of agricultural resources. The Japanese have found ways throughout the centuries of getting around that by creating terraced rice paddies and fishing like champs. There is still no way it can supply itself with its own food needs at this point due to its lack of agricultural resources, and lack of arable land. More importantly, it lacks just about everything else. Say what?

Japan is a real conundrum in that it is the third largest economy on the planet that, at the same time, lacks virtually everything that most countries with powerhouse economies have. Japan has to import most of its metals, ores, and virtually all of its energy resources. Japan is entirely dependent on the rest of the planet for all of these things. That has something to do with its history.

SAMURAI TO JACK

What about natural resources, Plaid Avenger? Not having any has something to do with Japan in the modern world. I'm going to go through the early history of Japan in fairly short order to point out one thing in particular: In terms of unique Asian cultures and unique ethnic groups, Japan is really a laggard. It's the latecomer. It's the last one to show up on the world scene.

Chinese civilization has been around for 5,000 years. The Koreans have been around for 3,000–4,000 years. Lots of people from Central Asia and Southeast Asia, like the Thais, the Vietnamese, and other ethnic groups with distinct cultures, have been around for a while too. If you search hard, you will find that Japan's culture cannot really be called a distinct culture until about 1400–1500 years ago, and even at that point, it wasn't anything closing in on what we consider a nation-state.

Japan started as a small group of people that migrated from mainland China and through the Korean Peninsula over the course of the previous thousand years. They did not become a unique entity themselves until comparatively late in the Common Era.

Virtually all of Japanese society is borrowed from other societies. All of the writing systems, religious systems, and philosophical systems all have roots over in mainland Asia. However, if you read Japanese literature or government

Bring on the bamboo.

propaganda over the last 100 years, it's all reversed. The Japanese are fiercely proud of their independence and their distinct culture, so given that they have been a powerhouse over the past fifty years, they have rewritten history to make it look like as if they were there first and everybody else borrowed stuff from them. This couldn't be further from the truth.

Move over, UK! Japan has inbred snobs as well!

What we have is this group of folks that started up in 500–700 CE, and it was only at this point that they formed into something vaguely resembling a state. They started to have a head of state, a prince. They started a royal/imperial line, which actually stays in place to the current day. For those of you saying, "What? They still have a royal family there?" Yes, there is still an Emperor and there is still an Empress. They have the longest standing monarchy/royal family in world history, and it continues. You can trace it back to 600–700 CE, same family, same dudes. The faces change, but the DNA remains the same. It is very much like Great Britain and its constitutional monarchy—the idea of royals hanging out. They don't have any real power now, but they are still there.

But let's get back to history. We can fast forward very quickly because Japan turned into, for the next thousand years, something very

similar to a feudal state in medieval Europe. That is, there was the Emperor on top, there were Dukes underneath him, there were Lords underneath the Dukes, and eventually, we get to the peasants that were essentially tied to the lands controlled by the Lords. There are different names and faces in Japan. They are Asians, not Europeans. They have sweet Samurai outfits and swords, instead of knights, and long shafts and jousts, but was all essentially the same system playing out on opposite sides of Eurasia.

The peasants lived on the lands and were essentially owned by the Lords, who paid tribute to the Dukes, who had one master, the Emperor. The easy reference is to what we call **Shoguns**, essentially militaristic governors which were the Dukes and the Lords. Just like in Europe, they competed for lands and ownership of things to gain political influence. They still recognized the Emperor as the main dude, but they all fought for places right below him on the totem pole.

Why was this so great? Nothing was really great about it, unless you're the Emperor. He allowed this feudal system to continue because, as all of his subordinates are fighting each other, no one was ever actually threatening him. That's the point. The kings of Europe did the same exact thing. The point of feudalism was to keep people

Shoguns? Sho' nuff!

beating the crap out of each other so they can't mess with you. This became standard operating procedure in Japan fairly early on, and it continued for most of Japanese history, which we have suggested, began pretty late in the Asian game.

We can tie some dates to this age of the Shogun, this really entrenched take on the feudal system. From around 1000 to 1580 CE, they had multiple competing shogunates all vying for power and influence. After 1580, one group rose to the top. They were called the **Tokugawa Shogunate**. The Tokugawa Shogunate unified the whole country under their single command from about 1600 to 1800. A single Duke or Lord from this line of shoguns became so powerful that they actually stopped all the infighting. Japan had a period of stability and peace where everybody was under the same blanket because of those Tokugawa guys. Here's the big thing: they unified the whole country, united all the smaller armies battling each other into a singular army, but most importantly, they plunged the whole country into global isolation.

These guys were tough. You've seen the movies! They were so awesome that they repelled a Mongol invasion in 1274—no one else pulled that off! They had their own little world going on and they were not interested at all in the outside world. They cut ties with everybody—not that they ever had a lot of contact. Historically speaking, they had always been a subservient state, or **tribute state** to the Chinese, though the Chinese never took them over physically. Japan knew that China was the big power, and therefore, kept itself on the fringe. The Japanese have escaped takeover by all other entities historically, mostly because they were a fringe state of no great strategic value, and had no great economic resources that other powers would want.

What was happening in the rest of the world from 1600 to 1800? Europe was colonizing the world, was taking over other big parts of the planet and seeing the beginnings of its industrialization period. You had a foreign influence making its way over to the Far East, even into Japan. With industrialization occurring, these European guys had better weapons, better ships, and better navigation. By the 1800s, Europe virtually controlled three-quarters of the planet and finally arrived in Japan. Europeans started making in-roads to establish trade with Japan.

During the Tokugawa Shogunate in 1600–1800, missionaries started showing up at Japan's doorstep. The Shogunate said, "What are you peddling? What are you trying to convert us to? Really? Christianity? That sounds great. And now we'll chop your head off! Get out of our country! We are not interested in your beliefs. We are isolating ourselves! We are the Shoguns, we are the Samurais! We don't need your stuff, so go away! If you don't go away when we tell you to, then we'll

Commodore Perry, official Japan-opener.

kill you!" That was the way it went for a good long time, but it was only going to be perpetuated for so long.

This European colonization story is told again and again when we go through the rest of the regions on the planet. As Europe's technology and military hardware became more advanced, neither Asia nor Japan could effectively counter the foreigners forever. Through its self-isolation, Japan got behind the times technologically. It came to the realization that it couldn't compete with the rest of the outside world. The Japanese were still fighting with swords, fighting dudes with guns. Their isolated world came to a quick halt. The Plaid Avenger is not big on dates, but this is a good one to note: 1853, which is not long ago. In 1853, a dude named Commodore Matthew Perry showed up on the scene. Commodore Perry wanted to open up trade so he pulled his ironclad battleship into Tokyo Harbor and came ashore. He said, "Let's meet with the local head-honchos, these Samurai guys with their swords." Perry said, "Hey, we want to establish trade."

For the previous 250 years, the Shogunate answered that with, "Get back on your ship and get out of here or we'll chop your head off!" Perry said, "Hold on, fellows, hold on. Take a look out in the harbor. Do you see that? That's an ironclad vessel. See those things poking out of it? Those are cannons. We can blow up you and all of your people right now and you are completely defenseless." Remember, these Samurai fought mostly with swords; now they were stacked up against modern military hardware. The Samurai figured out pretty quickly that this was going to be a one-sided affair. In one fell swoop, in 1853, Commodore Matthew Perry forcibly opened up trade with Japan. They kind of said, "Yeah, those guns probably do better than our swords. We probably are not going to be able to survive this."

It is important to note which country the Commodore hailed from: the United States of America. It's actually the United States that forcibly opened up trade in Japan, and luckily not a colonizing power. It is historically significant that it was not the land-grabbing, colonizing Europeans who busted Japan open to the world, but the Americans who went over there to establish a trade relationship. This was the beginning of one of the most amazing occurrences in modern history, as far as the Plaid Avenger is concerned.

When a country is confronted with an outside world in which they are severely behind, they can take one of two roads. All other regions faced this exact same choice as the onslaught of European colonization of the world occurred. Road One: "We can fight it out! Let's try to kick them out! Let's have protests and shoot guns at them and do whatever we can do to kick out the European powers!" Most countries went that way. Road Two, which only the Japanese seem to have found: "Let's embrace the West! Let's do it their way! And let's do it better!" This precipitates a fascinating time in world history, not just Japanese history. This is what makes the difference between Japan and China in today's world. This is the crossroads. The Plaid Avenger is telling you, it's rare that you can say, "This is a definitive turning point. Everything is different after this!" The Plaid Avenger is telling you: *this is it.*

MEIJI MANIA!

In 1868, the Japanese decided to undergo something now referred to as **The Meiji Restoration**, named for the Emperor at the time, the Head Honcho Meiji as he came to power. He was known as being a fairly savvy guy, but it was probably the collaborative effort of his counselors and the people around him that, once they were forced to embrace the outside world, understood how far ahead the Europeans and the Americans were ahead of them. They made this radical decision, saying, "Wipe the slate clean. Let's redo everything, EVERYTHING, in light of what we know about the outside world." They embarked on this fantastic re-invention of their society, a "restoration" if you will. . . .

They sent what would be the equivalent of college students, learned men, as well as government workers, all over the planet. They sent thousands of them to Europe, to America, to the Middle East, everywhere there were centers of learning, centers of technology, centers of industry. Their mission was to find out every single thing they could. They not only learned stuff at universities and through businesses, but they also brought stuff back. They brought back military hardware, guns, tanks, whole steam engines, railroad equipment. They shipped it back to Japan, and teams of smart dudes ripped the stuff apart to figure out each individual component. They started the individual industries to make the

Meiji Makeover: Rickshaw to Railways.

pieces, the wheels, steam engines, and everything else on their own, from the ground up. This was a massive innovation, rejuvenation, a catching-up period, for technology in particular, but also for government and military.

The Japanese did this fantastically well. Think about what Japan does today. It's still like this, isn't it? They don't really make anything unique—they make everything better! They started this tradition way back during the Meiji Restoration. They are reverse engineers and efficiency experts. They did this in very short order, in around ten years. In two or three decades, they caught up from a lifetime of technological stagnation. It's insane! It's crazy! It's the Meiji Restoration!

At the same time, they modified a lot of things internally. You can see great movies about this, such as *The Last Samurai* with Tom Cruise. Tommy's Scientology aside, it's a pretty decent movie. The film shows the extremely turbulent times in Japan when everything changed. Not just the technology stuff we talked about, that's easy enough to understand. The social restructuring was the big thing. They actually got rid of the Samurai. They reinvented their society in the likeness of Western Europe and America. They banned the Samurai, which is what The Last Samurai is all about. They wanted to rid themselves of the old ways. They said, "No, we're going to rejuvenate! We are doing everything to catch our military up, which means we're getting rid of the old Shogun ways. We're going to consolidate into a bureaucratic government, just like all these other places. We're going to restructure our banking system. We're going to redo EVERYTHING!" And they did it, in the image of the West.

This is why Japan is a very Western looking nation, as opposed to its neighbors. The Meiji Restoration was the critical turning point in history. Japan became Western-like. Every place else in Asia did not. Japan currently has the number three economy in the world. Every place else in Asia isn't even close, with the exception of China (number two), who has stepped up in the last few decades. But watch out! South Korea will soon break into the top ten as well. This region has three very important 21st century power-player economies. That is a fact worth remembering.

TERRIBLE LIZARD

Back to the story. The Meiji Restoration was the complete, inside-out do-over of Japanese society on all levels, and it was extremely successful. Like I already suggested, they caught up within a few decades. Japan went from the 1850s, where the Samurais were in charge and fighting dudes with swords, to 1895 when, having been a subservient vassal state of China for their entire history, they attacked China. Why? Because once the beast awakened, it was hungry! The beast must be fed! What does it want? Resources! The Japanese adopted another very Western concept in order to get these resources: imperialism!

GODZILLA IS BORN

In 1895, during the Sino-Japanese War, the Japanese went into parts of modern-day Korea and Manchuria and started conquering. They attacked China, its colossal powerhouse neighbor, *and they won.* In 1905, ten years later, they attacked Russia, the largest state on the planet, *and they won.*

Minute 1: battle begun. Minute 5: no Russians alive.

What were they doing this for? They were taking parts of Russian territory on the east coast of Asia. The Russians sent their entire naval fleet around to sink the Japanese fleet because they said, "Hey, we're Russia, we've been around for like 500 years. We're the biggest state in the world! We're European; we know what's going on." They sent their entire fleet around the entire continent, and it was entirely sunk by the Japanese in five minutes. Boom. To put this in perspective: within 40 years, Japan went from having NO navy to having a navy that wipes out Russia without batting an eye—the biggest one-sided naval defeat in all of history! The **Battle of Tsushima** was an embarrassment for the Russians, and established the arrival of Japan in the modern world. Japan goes on to advance further into Korea, and in 1910 it took over all of modern-day Korea. In 1917, it annexed more of Russia, while the Russians were busy doing their Communist Revolution thingy.

We can stop for a second and ask, "What were they doing all this for? Why were they taking all this land, inciting fights with their neighbors?" This gets back to what I was talking about in the physical section: this place had no resources. To become a world power, you need resources. To build railroad cars and tanks and airplanes and industries, you need resources. To build an army, you need resources. What Japan did, as it started to come into power, was acquire resources. Take them from other places. Take over lands. Japan becomes really the only Asian colonizing power in modern history. They were the only Asian imperial power in the Western European sense of the word (i.e., by forcibly taking over other parts of the world). Once the beast is awakened, he must feed. An insatiable hunger develops which inevitably leads to . . .

GODZILLA ATTACKS!

With Japan's acquisition of all this territory and all these resources in mainland Asia, you might say, "That's enough. You've got enough now, Japan." But once Godzilla is awake, he's not going back to sleep very easily! The Japanese continued to meet with success after success. In 1914, before World War I was over, Japan strategically joined the Allies against Germany, though they played no real part in the war. As a result of their maneuver, they gained control of German territories in China and German-controlled islands in the Pacific. Without doing anything, they gained more territory at the end of World War I. We already said they started to pull out more pieces of Russia during Russia's civil war. After Russia got decimated in World War I, the Japanese took over some more of their territory in the interim period.

In 1931, the Japanese took over all of what is now Manchuria, the entire Northeastern part of China, and during the rest of the 1930s, they established more footholds in China and virtually every other Southeast Asian nation. In each new place, they set up shops and military camps.

In this period between World War I and World War II, all the Europeans were worried about themselves. No one was paying much attention to Japan, except the United States. The United States was very worried about Japanese movement and the growth of their Empire. Even before the WWII was officially launched, there was animosity between these two Pacific powerhouses, Japan and the United States.

The United States, being an anti-imperial and anti-colonial force, was looking over the Pacific and saying, "That doesn't look good. How could this end well? This isn't good for us or anybody!" Indeed, they were right. A lot of folks said, "Whoa, we're really surprised that Japan attacked Pearl Harbor." Nothing could be further from the truth. These countries may not have declared war at the time, but they weren't on friendly terms either. In fact, the United States already had an oil embargo against Japan. That's one of the reasons Japan said it attacked the United States. Let's get to this attack.

In 1941, the Empire continued to expand, taking over mainland China and places in Southeast Asia such as Vietnam, Cambodia, Laos, Thailand, Burma, Indonesia. Japan continued to expand its Pacific Island holdings as well. In 1941, it became so big that it decided to attack the United States over in Hawaii. Why didn't it attack mainland United States? That's anybody's guess. In hindsight, it may have been a better idea, but who knows why any of this occurred? What if the Japanese had not attacked the United States? Maybe the United States wouldn't have done anything to them and there would still be a huge Japanese Empire in what is China today. It's hard to tell.

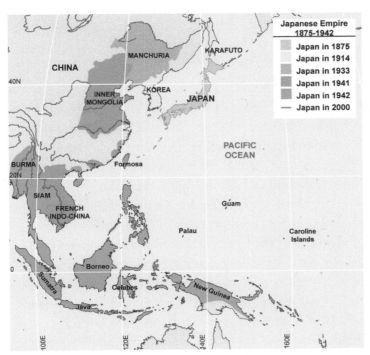

The expansion, and then deflation, of an empire.

Pearl Harbor's significance is it was a warning signal. Japan was setting up a fence around its territory. Japan wasn't trying to take Hawaii over; it was establishing where the fence was. The Japanese said, "Okay, America, you're a power-house; but we're also a powerhouse now. We can bomb you in Hawaii, so don't come on our side of the fence. This stuff over here, this is ours. This is our imperial holding over here, so just stay on your side." Japan apparently didn't reckon that would be a failure. We know the ultimate end to this. Maybe we've talked about World War II too much already in this book. But in the official war years, 1940–1945, Japan invaded every other nation in Southeast Asia and the Pacific: a full-on Asian Empire led by Japan was the plan.

Here's a little known fact that the Plaid Avenger wants you to understand and know: Japan also had designs to invade India and Australia. When I said they had designs, I don't mean some sketches on a cocktail napkin. We're talking about war plans. They had their little Risk board marked with the movements of soldiers, and their troops were in place for this to become reality. It was only a matter of time. They already had small landing parties in Australia, who were camping and scoping things out on reconnaissance.

The other concrete item that still exists in the world today as evidence of Japan's dream to take over all of Asia is the Burma Railway, nicknamed "Death Railway" because 16,000 Allied POWs (mostly British, Australian, Dutch, and American) and over 100,000 Asian laborers died making it. It is the subject of a movie, *The Bridge on the River Kwai*. The laborers were forced to build this railroad system as a supply line for the planned Japanese invasion of India. It ran from Thailand, through Burma, right up to the Indian border.

ANIMOSITY IN ASIA

In these ambitious Japanese plans were a few more details we should flush out in order to understand today's world a little better. How was it that such a small state with a small army ended up roll-ing over all of Asia in the first place?

The Asian Liberators?

Before all the war crimes and atrocities, Japanese forces were often welcomed to intervene in internal affairs of many of these places. The Japanese Army in the 1930s didn't have to fight tooth and nail to take over many of these countries. Why? By and large, the Japanese came in under something called **The Greater East Asia Co-Prosperity Sphere**. The Japanese used this concept to say, "Hey, we're not here to take over; we're here to free you from your Western European colonizers! These guys are jerks! They're white men! Look at us! We look like you, man; we're Asians! We are here to liberate you!" Japan was often welcomed, particularly in French Indochina, where Vietnam, Laos and Cambodia are today.

Japan went into these places and threw off the yoke of European colonialism. However, they simply replaced the colo-nizers, without any real liberation effort. It was all a big sham. This made a lot of folks within Asia pretty miffed because the Japanese ended up being, during the war years, very brutal. The Plaid Avenger can tell you, war is brutal, and war sucks for everybody involved. However, the Japanese in WWII had a particularly nasty style of warfare and the war crimes and atrocities that were committed against a whole lot of Asian people by the Japanese has not been, and will not be, forgotten. Like ever.

A horrific example is the **Rape of Nanking** in China from 1937 to 1938. The Japanese took over a city with millions of people, and since they didn't really feel like having a prison camp, they just killed everybody. They took their bayonets and chopped people's heads off. They lined people up in a row so one bullet would go through more than one of them, so as to save on ammunition. For several weeks, the Japanese slaughtered tens of thousands, if not hundreds of thousands, of civilians. This is something the Chinese will never forget, and actually they haven't: the date of invasion is marked every year in China as a day of remembrance. And that was not the only atrocity that Asians still remember . . . see the Japanese War Atrocities (inset box).

What's the Deal with Japanese War Atrocities?

Bataan Death March

Approximately 75,000 Filipino and US soldiers, commanded by Major General Edward P. King, Jr., formally surrendered to the Japanese under General Masaharu Homma, on April 9, 1942, which forced Japan to accept emaciated captives outnumbering its army. Captives were forced to make a weeklong journey, beginning the next day, about 160 kilometers to the north, to Camp O'Donnell, a prison camp. Most of the distance was done marching, a smaller distance was a ride packed into railroad cars. Prisoners of war were beaten randomly, and then were denied food and water for several days. The Japanese tortured them to death. Those who fell behind were executed through various means: shot, beheaded, or bayoneted. Over 10,000 of the 75,000 POWs died. Check out the old-school films *Bataan* (1943) and *Back to Bataan* (1945) with John Wayne.

Marching to nowhere . . .

Bridge on the River Kwai

Great freaking movie, you have to check it out. Based loosely on these real life events: In 1943, the Japanese used POWs to build a railroad across Burma (this is against the Geneva Convention on treatment of prisoners). This was part of a project to link existing Thai and Burmese railway lines to create a route from Bangkok, Thailand to Rangoon, Burma (now Myanmar) to support the Japanese occupation of Burma. The railway was going to be a critical connection and supply line for the planned invasion of India. About 100,000 conscripted Asian laborers and 16,000 prisoners of war died on the whole project. And the guy who played Obi Wan Kenobi in *Star Wars* is in the movie. That helps.

The Rape of Nanking

The Nanking Massacre, commonly known as **The Rape of Nanking**, refers to the most infamous of the war crimes committed by the Japanese military during World War II–acts carried out by Japanese troops in and around Nanjing (then known in English as Nanking), China, after it fell to the Imperial Japanese Army on December 13, 1937. The duration of the massacre is not clearly defined, although the period of carnage lasted well into the next six weeks, until early February 1938.

During the occupation of Nanjing, the Japanese army committed numerous atrocities, such as rape, looting, arson, and the execution of prisoners of war and civilians. Although the executions began under the pretext of eliminating Chinese soldiers disguised as civilians, a large number of innocent men were wrongfully identified as enemy combatants and killed. A large number of women and children were also killed as rape and murder became more widespread. Final toll: 10,000 to 80,000 raped; 15,000 to 120,000 slaughtered.

Want to know why there is animosity between the Chinese and Japanese to this day? Look no further.

Let's not forget about our friends in the middle, the Koreans. Many still hate the Japanese with a passion as well. Japan took over and extracted all of the resources out of Korean mines. How did they do that? They enslaved the Koreans! The whole country essentially became a work camp for Japan. Everyone in Korea was also forced to change their names to Japanese names. Every single person in the country! This was an incredibly difficult time for the Koreans. Trust me, they have not forgotten about it.

The occupation was also pretty nasty for the Chinese and the Koreans, in that some of the worst documented use of biological warfare took place. The Japanese said, "Let's poison a water supply in a town of five million and see how long it takes to kill everybody!" Here's one of the Plaid Avenger's personal anti-favorites: "Let's get a bunch of plague-infested rats, put them in a bomber, and drop them over the countryside to see how the disease progresses through the country!" They also did some other "little" things like mustard-gassing whole towns. This was pretty nasty business.

Now you know why many Asian countries view Japan with extreme suspicion, apprehension, and/or outright hatred. The Chinese government and some other governments still want more **war retributions**: monetary payments from the Japanese for the atrocities they committed during World War II.

The Plaid Avenger had to elaborate on this animosity that exists to this day between Japan and many other Asian countries, particularly China and Korea, because it continues to play a part of politics in today's world. Every year when the Japanese educational systems releases their history textbooks for the K-12 schools, Korean and Chinese people protest. Why would Koreans care about what school kids in Tokyo read? Just this: the Japanese textbooks continue to gloss completely over the war years, thus marginalizing all the bad stuff that was perpetrated on the other Asians groups. War crimes, biological warfare, and Japanese sex slave operations in Korea are all conveniently missing from the history books in Japan, and that really enrages some Asians.

This animosity also bubbles to the surface whenever the Japanese prime minister visits a war shrine located in Tokyo: **The Yasukuni Shrine**. Why do you need to know this? It will continue to be a point of friction between these two countries who will be the two big power players of the 21st century. It's good to know why they hate each other and this is a great example. The shrine commemorates all of the Japanese soldiers who died fighting for, or in service of, the Emperor, no matter what the circumstances.

This is an actual US goverment wartime propaganda poster. I don't think America was happy about Pearl Harbor. Do you?

You have to know about these bad things in order to get the attitudes that these countries and their peoples have about each other. From the Japanese standpoint, you need to think about how big this goal of dominating Asia proper really was. From the Asian standpoint, think about how far the Japanese demonstrated that they were willing to go to achieve it. This isn't a trivial thing. This gets into why Japan is viewed with suspicion, animosity and sometimes hatred by parts of Asia to this day. The Plaid Avenger started off this chapter by saying China vs. Japan, Japan vs. China for a reason. These are places that do not like each other at all. And it could get worse.

I'm sure you think that all this stuff is ancient history and that it really doesn't affect today's world at all. There hasn't been an official heads of state meeting between China and Japan for almost a decade because of the Yasukuni Shrine business. Trade deals have been blocked between the countries when tensions rise. Lawsuits between the countries on war reparations are still active. Now they may go to war over a group of rocks in the Pacific named the **Senkaku** or **Diaoyutai Islands**. Is that real enough for you? Let's go ahead and get past the war years and bring it on home into the modern era.

What's the Deal with the Yasukuni Shrine?

Yasukuni Shrine (literally "peaceful nation shrine") is a controversial Shinto shrine, located in Tokyo, that is dedicated to the spirits of soldiers who died fighting on behalf of the Japanese emperor.

So what's the problem? Every country honors its war dead, don't they? The Yasukuni Shrine also honors a total of 1,068 convicted Japanese war criminals, including 14 executed Class A war criminals, a fact that has engendered protests in a number of neighboring countries who believe their presence indicates a failure on the part of Japan to fully atone for its military past. 2014 Update: new Prime Minister Shinzo Abe pulled the trigger (no pun intended) and recently visited the Shrine. Thus, his administration is making their attitude crytal clear.

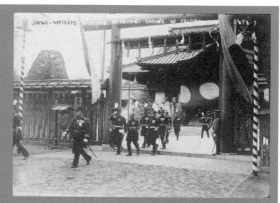

Yasukuni: honoring the dead or enraging the Asians?

It's the equivalent of the Germans visiting a war shrine to pay respects to Hitler and the other Nazi war dead. How do you think Europeans would feel about that? Sound extreme? Yep. I call 'em like I see 'em.

GODZILLA IS DEAD

After the Germans were defeated in Europe, Uncle Sam then turned all his attention to finishing the job in the Pacific theater of war. And it was brutal. Island by island, enclave by enclave, the Japanese were beaten back by the Americans, but at a horrific cost of human lives on both sides. The US was pressing Japan for total, unconditional surrender, something that was not going to come at an easy price.

Nagasaki go boom.

The aftermath.

To finish up, we know in August 1945—Hiroshima and Nagasaki go boom. The other part of World War II history, which is mostly unknown yet crucial, is the fact that the US had been firebombing the entire country for months leading up to the actual dropping of the atomic bomb on Hiroshima from the Enola Gay. After the second bomb was dropped on Nagasaki, the Emperor gave the unconditional surrender, and for the first and only time in its entire history, Japan was not only defeated, but occupied.

Japan, much like Europe after World War II, was totaled. The two cities that the atomic bombs dropped on were gone, but the rest of Japan was leveled as well. There was nothing; it was utter destruction. This is important to note before we go on. Before we get to the rebirth, the rejuvenation of Godzilla!

AFTER THE FIRE

Post-World War 2 was another fascinating time because—just like the Meiji Restoration—the Japanese remade themselves again. Inconceivable! Their country was smashed, destroyed, and occupied by a foreign power. The United States, led by MacArthur, was in effective control of the entire government and they set about redoing it again, in a more democratic fashion. MacArthur and his team helped them write a new constitution, fashioned them a new society, and set about helping them rebuild so that Japan could become a strong, stable US ally in the area.

MacArthur: Japanese breaker, taker, and re-maker.

The Japanese were again faced with the same two paths. "What should we do? Should we take Option 1: resist the foreign invaders or bide our time and wait for them to leave, should we have some underground resistance movement? Or Option 2: should we embrace them, do everything they say and do it better?" Of course, they went with Option 2 again. There was another major restoration, but it had no catchy name like "Meiji." It was just a do-over of everything this time, completely under United States guidance. They went from a monarchy to a constitutional monarchy, which is basically a fair equivalent of a democracy in today's world. In fact, it was exactly like American democracy except they paid tribute to the royal line by keeping their Emperor, just like the British kept their Queen. There are quite a few other British-Japanese similarities of which you should be aware . . . (see box on page 304).

One side note of note: The Japanese themselves added a clause to their new constitution which banned them from establishing a military. Apparently they felt so bad about their wartime activities that they wanted to completely eliminate any possibility that they would repeat their mistake in the future. McArthur and crew were adamantly against this clause, but Japan insisted. Why would McArthur not want this? Well, the old General was pretty savvy: he was already looking over at China as a potential Cold War adversary, and he wanted Japan to be the counter-balance to that growing commie threat. However, instead of getting a militaristic Japan under US direction, he got a pacificist Japan that needed US protection. This sets up the situation which still exists today: the US is the primary protector of Japan should any outside force attack it.

The Japanese embraced capitalism, democracy, re-industrialization and rebuilding of their society with a passion under the tutelage of the US. Uncle Sam gave them billions of Yankee dollars to achieve this goal, much like an Asian version

United Kingdom and Japan: Siamese Twins?

Why are the two countries similar? **Number One:** Island nation. **Number Two:** Off the coast of Eurasia, flanking opposite sides of the continent. **Number Three:** Stand-offish from their neighborhood international organizations: Great Britain is distant from the EU, Japan is distant from ASEAN and other Asian groups. **Number Four:** Lapdogs for US foreign policy; they do pretty much whatever the US tells them to do. The UK does it because they love the US; Japan because they owe the US and are still defended by the US. No one will attack them because the United States is their military. **Number Five:** Both countries are constitutional monarchies which retain their regal, royal figureheads despite their utter uselessness, while both having prime ministers running the real show.

Queen Elizabeth II of England

USA's loyal pets—oops, ! mean allies.

Emperor Hirohito

UK Prime Minister Cameron

Japanese Prime Minister Shinzo Abe

of the **Marshall Plan**. However, one big difference was that the US stuck around to oversee the rebuilding and restructuring of Japan . . . particularly because they had assumed a "big brother" role since Japan had officially banned itself from possessing a military of its own. This is one of the main reasons that these two countries are still tight when it comes to foreign policy issues.

One final thing about the military issue: think of the big bonus and boon to Japanese society due to not having to invest in a military. Most of the countries in the world have to provide for self-defense. They have to spend at least some money on it. Not so in Japan. Although they do, investing billions into their Self-Defense Force. . . .

ECONOMY NOW

This is one of the reasons, among many others, why Japan is so rich today. They've had "big brother" United States there. They had the luxury of investing every ounce of capital straight into the infrastructure to make themselves better and better and better. Let's make a list of why Japan has the number three economy in the world today:

→ **Number One:** They have not *had* to invest in any military expenditures for sixty years now.

→ **Number Two:** They have enjoyed US influence and economic ties with the number one economy in the world for the last sixty years.

→ **Number Three:** Just like BASF, the Japanese don't make anything—they make everything better. They are a value-adding society. While they never invented the automobile, the computer, the video game, television, or cell phones, they make all those things better. They continue to add value in the high spectrum of things.

→ **Number Four:** They are almost completely technology and service sector oriented, and they are very good at it. They're so good at it that they run **positive trade balances** with every other country in the world with whom they trade. They make more money on the sale of the stuff they make than they spend on the stuff they buy. They are always in the plus. I will come back to the list.

What else is going on with Japan? The role of Japan in Asia is very similar to the United States' role in our part of the world. While there is some animosity between Japan and some other Asian countries because of the past, it is still looked to as an economic leader. The US is the number one economic GDP powerhouse, Japan is number three, and both these countries are looked to as models to be mimicked. Just like when the US economy takes a turn for the worse, places like South America and Caribbean follow suit, when Japan's economy takes a dive, all the surrounding economies in Korea, Taiwan, and Indonesia do too, because there is so much trade from one country to the next. In this respect, Japan is to Asia what America is to the Western Hemisphere. Although at this point, China (with the number 2 economy on the planet) is quickly eclipsing Japan in economic importance in this 'hood: a trend that will likely continue the rest of our lives.

We can draw a few more parallels to the United States. Japan is an extremely rich society, so it is industrialized and fully urbanized, which means labor is really expensive, just like in the United States. You cannot get cheap labor in Japan. Like the USA, Japan has minimum wage laws and generally high labor costs. So what? What is the parallel here, Plaid Avenger? This high cost of labor has caused jobs and industries to move out of Japan, just as they have in the United States.

Just like all the car manufacturing companies that were once in the United States are all in Mexico or other nearby places with cheaper labor, Japan's service sector jobs have been moving to places like China, Korea, and Indonesia. Corporations make more money if they manufacture things where labor is cheap. While Japanese corporations might still be the biggest moneymakers on Japanese cars, most of them aren't making them in Japan anymore; they are made in all points south: China, Indonesia, Vietnam, and even India.

POPULATION NOW

Whats up with the Japanese peeps? For starters, the total population of Japan is around 126 million. That is actually quite a few people on a handful of islands, that is mostly covered with mountains and forests. Japan is overwhelmingly urbanized, post-industrialized, and sanitized. They have some of the highest living standards, levels of technology, and GDP per capitas on the planet. Sounds good so far! However, the population of Japan is shrinking. This is a double whammy. People are increasingly coming up short on the 2.1 child **replacement level**, which you remember from Chapter 2. Every year, the population of Japan, while still stable, gets slightly smaller. What does this have to do with anything?

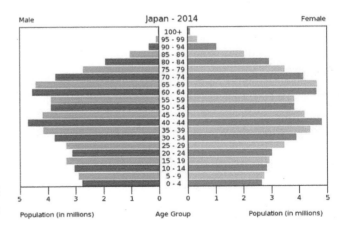

Like the Russians, it's going down.

You need workers in the factories to make stuff, to make money, so that the economy will benefit. If you don't have workers, employers have to go somewhere else. And boy, have they ever! They can't even *maintain* the technology and service sectors there at this point in time because the population is getting smaller. But wait, it gets compounded even further!

As I suggested earlier, the Japanese are extremely, fiercely, culturally independent and ethnically distinct. They're *Japanese,* not Chinese and not Korean. Everybody else in the neighborhood would agree with that statement. By the way, this is not a racist slant by The Plaid Avenger. Both Japanese and Korean societies are almost 100% purely Japanese or Korean. This is even more true in Korea, the most ethnically pure place on the Earth. It's easy to take the census in Korea because there is only one box to check: the Korean one. In Japan, 99 percent of the population is Japanese. They actively discourage immigration there. When you visit there, you will have a great time, but if you move there permanently don't expect many

"Yes Plaid Avenger, I need your help to change Japanese demographic trends."

people to be very friendly, particularly if you are a few shades away from Asian. This may sound blunt, but racist or not, the Plaid Avenger is telling you how it is around the world.

Kabuki theater actor, or psychotic killer?

This is not a place embracing a lot of people from around the planet. That might be all fine and dandy with them, but as I suggested, their work force is going down, birth rates are falling, thus they have a shrinking population. In the United States, maybe people are not having huge families either, but immigrants come from other parts of the world to fill those work spots and keep the economy growing. Japan is coming to a critical juncture at which they need to figure out what to do. These negative demographic trends cannot go on indefinitely. Although, due to strong work ethic and demand, many Japanese are continuing to work into their 70s, 80s, and beyond! Hey Granny Fu, out of the rocking chair, and back to work!

Not only is the declining population affecting the workforce, but it also has impacts on the consumption side of the economy. Many car manufacturers and other retail businesses are re-focusing their sales to accommodate consumers in other parts of Asia. What's the point of having a Honda dealership in Tokyo, where everybody already has a car and the population, and therefore, consumer base—is shrinking? Better to invest in India or China or Vietnam where consumers are plentiful, getting richer, and the potential consumer base is growing. This is affecting development of new products line as well; Honda and Toyota are introducing new lines in India that cost less than $2500 per vehicle! Sweet!

The Japanese government is really struggling with this issue right now. They need workers, they need a tax base to draw from, and they need their economy to stay vibrant so they can maintain their great standards of living. Unfortunately, people don't want to have big families, and they also don't really want to change their extremely restrictive immigration policy. Something's gotta give. Don't be surprised to hear about the government either promoting people to get busy in the bedroom, or "re-educating" their population to be more accepting of foreigners.

MILITARY NOW

Another thing to consider in modern Japan is the fate of the **Japanese Self-Defense Forces**. What's this all about? As I suggested to you, after World War II, under MacArthur, the Japanese signed away their desire and ability to have an active military. Like the Germans, the Japanese conceded that they screwed up bad and that they were going to from that point forward be peace-loving people, and to ensure that

If it looks like a military, and smells like a military. . . .

they could never go to war again, they banned themselves from having a pro-active military. So they do not have a military that has the option of pre-emptively striking an adversary first, but they do have a bunch of dudes that can protect the country from an attack. Their current national Self-Defense Forces are exactly that: the equivalent of the US National Guard that serves as an emergency 911 for the whole country in times of national disaster and national emergency—or, theoretically, to repel any attack on the nation from foreign powers.

Anchors away! But don't leave the harbor . . .

It is important to note that it is implicit in their charter that they do not possess any powers to pro-actively attack anybody. They have absolutely no offensive role, no offensive tactics, no offense, period. No troops abroad, no missions abroad, no action abroad. Only self-defense of the state. It's the "home team" in every sense of the phrase. You do have people wearing helmets and camouflage outfits, sometimes with guns, on ships and in tanks and airplanes and helicopters. It sure looks like a military, but their only function is helping out when natural disasters like earthquakes strike.

The reason that the role of the Self-Defense Forces was so specifically defined was because right after World War II, everyone in the neighborhood was scared that the Japanese might re-militarize and start its imperial rampage all over again. Apparently, Japan was scared of this as well, and that's why they neutered themselves. But everyone was willing to accept the Self-Defense Forces. That's no problem. End of story, right? Well . . .

This force has been evolving rapidly into something that looks more and more like a military lately. Important to note is that this only came to the attention of China and the world a decade ago. How'd they find out? During the 2003 US invasion of Iraq, Japan volunteered its "troops" to go and support the US in Iraq. This was done primarily as a show of support for their US buddies; to be a part of the "Coalition of the Willing." In reality: all they were doing was refueling and resupplying US naval vessels out in the Indian Ocean, but it's the thought that counts. The United States may say, "Oh, that's a good thing; they're our ally and they owe it to us." But this is taken with great apprehension by other Asian nations, particularly China, who has reinforced the notion that if Japan says it doesn't have an army, it shouldn't. They say, "How are the Self-Defense Forces defending Japan in Iraq?" The rest of Asia has also concluded that change is afoot. The fact that Japan has sent its Self-Defense Forces abroad challenges the definition of self-defense. You send armies abroad, not self-defense forces.

With no uncertainty, the Plaid Avenger wants you to know this for sure: While many Asian nations may be apprehensive about the Japanese Self-Defense Forces, there is a country on this planet that wants Japan to throw the anti-war clause out of its constitution altogether and get busy re-arming. I wonder who that could be? The United States is practically *demanding* that they remove it. Why would the United States want this? The United States would prefer Japan be a countering force to the rise of China in Asia. It's a checkmate. In this respect, Japan's role is very much like Great Britain.

Sounds strange, but that's what it is. That's why the Japanese Self-Defense Force status are a big issue. The Plaid Avenger knows that anytime anyone even hints at this issue, China gets enraged that Japan is morphing its self-defense forces into an army, almost as enraged as China gets when any Japanese prime minister visits that war shrine we talked about earlier.

One more point on the Japanese changing their constitution for their military: There are several states in the world that the UN is considering allowing on the UN Permanent Security Council. Japan is one of the candidates to be on this council and the US is its biggest sponsor. The US is all in favor of Japan being on the UN Permanent Security Council. I wonder why? Maybe because Japan will do everything the US says. Meanwhile China is flipping out because it is vehemently opposed to this.

Here's the funny thing about it: The US has told Japan, more or less, that it will not support its candidacy until Japan gets a pro-active military. If it gets one, it would be no surprise if it developed nuclear capacity within days. Japan already knows how to produce nuclear power. It is feasible that if it changes to have an offensive capability military, it would be the next declared nuclear bomb power in the world. That is exactly what the US wants: to have a nuclear station off the coast of Eurasia to be on our team. This is a check and checkmate situation for the US. It is using Japan to keep checks and balances against the rise of Chinese power.

Along with the Chinese checkmate, the US would very much like Japan to reassert itself militarily for much more practical reasons too:

1. The US military has become over-committed, over-stretched and over-used in other parts of the world, and they simply can't keep tens of thousands of troops hanging out in Japan and South Korea indefinitely. Redeployment of troops is already underway, mostly to places in the southern Pacific where countering Chinese influence is becoming a main mission.

2. If Japan gets its army back, then they could actually help out the US in all of its current active conflicts. I'm sure the United States would welcome the change from being their protector to being a joint partner in military activities. The US would warmly welcome any permanent addition to their "Coalition of the Willing" in the War on Terror.

3. If Japan re-arms, especially in the nuclear capacity, then the US could let them deal with the North Korean freaks more effectively. Japan was always the most probable target of a North Korean attack anyway. And a nuclear armed Japan would def be a nice card to play against increasing Chinese power in this 'hood.

Back in 1945, Japan decided that they weren't going to have a military anymore, so they couldn't get into any more trouble. Unfortunately for Japan, trouble has found them. The status and activities of their Self-Defense Forces has them in a pinch between their pro-military, US ally and their anti-military Asian neighbors. Anchors away, my samurai friends! Troubled waters ahead!

FOREIGN POLICY NOW

All that Self-Defense Forces talk is intimately related to Japanese diplomacy in the modern era. As I have suggested earlier and often in this chapter, relations between Japan and its neighbors have been tense, to say the least, since the war years. At times, downright hostile might be a better description. Every year when the school books are released and every time a high-ranking official hits up the Yasukuni Shrine, the Chinese and Koreans get severely steamed and relations get further strained. Japanese troops sent abroad to support the US is just more fuel for the fire.

Koizumi to Abe: Make Japan strong!

As of late, a few of the previous Prime Ministers were pushing for a more positive Japanese relationship within their Asian 'hood . . . and had taken a much more conciliatory stance towards their Chinese cohorts in particular. Their thinking has been that Japan cannot afford to anger the economic giant that China is becoming: both exports and imports have been increasing dramatically between these two super-states, and no one wants to jeopardize that. In fact, China has now superseded the USA as Japan's biggest trade partner . . . which of course makes perfect sense given the proximity of the countries, as well as the sheer ginormous size of the Chinese market. (These statements apply to South Korea as well, since it will soon be the tenth largest economy on the planet.) With Japan still carrying around that bad rap due to their nasty imperial period, China increasingly has the upper hand as the future economic and political leader of the wider Asian region. To stay in the game, Japan simply cannot risk entirely isolating itself from its neighbors in this crucial wider Asian-Pacific region . . .

But. . . .

That was the last few chaotic years in the top slots of Japanese power. As of this recent writing in 2014, Japan has a new sherif in town, and this guy appears to be ready to rumble! The new main man of Japan is one Shinzo Abe, and a more proud and unapologetic and hawkish figure has not held the post since the beloved Junichiro Koizumi (Prime Minister 2001–2006) was in charge. Koizumi was a wildly popular, center-right conservative, a regular Yasukuni shrine visitor, and the dude that sent Japanese defense force ships to help out the US in Iraq. Yeah, that guy. And it looks like Shinzo is shaping up to go even further down that "rah-rah Japan is awesome" pro-nationalist path than even his prestigious predecessor. How so?

Well, for starters, the awesome Abe is a political legacy: his father, grandfather, and even father-in-law were all prominent politicians . . . and all of them fairly adamant about making Japan strong again, even in some cases (his grandpa) wanting to scrap the constitutional clause on having no offensive military, decades ago. And we all know how family influences our attitudes and outlook on life, thus . . .

Shinzo can best be described as a right-wing nationalist, and one who has decided that Japan must take a much more aggressive, pro-active role in the region in order to stay on top. I'm paraphrasing a bit here, but Shinzo is essentially of the opinion that "Japan's era of apologizing is over!" Meaning: he is not going to keep feeling guilty about the WW2 atrocities, is not issuing any more apologies, is not going to talk about it in textbooks, is not going suck up to China, and is likely to work towards changing the constitution to allow for increased military options! Boo-yah! Back up Asia, Japan may be coming back!

And that is going to cause huge tensions. As it is already creating huge tensions. Already, there has been a blow-up over the typical textbook release. Already there has been a fracas over Shinzo supposedly modifying an official state apology to wartime sex slaves in Asia. Already he has visited the Yasukuni Shrine. After two years in office, Abe still has not even met with any leaders of China nor Korea. Anti-Japanese sentiment and protest abound in China and Korea . . . mostly over the biggest issue between these states: a handful of rocks in the ocean!

Believe it or not, these old rivalries have become quite active again as Japan is now embroiled in two differential territorial disputes—one with Korea, and a much bigger one with China—over ownership of some dinky little islands. The rocks in question with Korea are named Takeshima Isles; the MUCH bigger possible flash point to war with China is over the rocks named Senkaku or Diaoyu Islands . . . this issue is important enough to describe in greater details in the "Global Hot Spots" chapter in part 3 of this text. Fast forward to it now if you want to know more asap!

Summary: As as much as I like the guy personally, Shinzo Abe's ardent pro-nationalist policies, at this particular time of China rising to great power, along with these dinky territorial spats, is a ripe recipe for actual Asian conflict. It seems impossible, but my friends, I am telling you, it is getting increasing probable that a major fracas may erupt in the area. . . in which the US is now also involved, since it will likely have to come to the aid of the Japanese if shots are fired, thus making it into a regional Pacific war, if not global one. Troubled waters ahead . . . pun intended. Watch this one closely.

Luckily for the pro-military faction within Japan, the totally whacked-out North Koreans under the leadership of the ever-tubby Kim Jong-Un is rapidly changing the equation in 2014. The increased nuclear activity and missile launching and saber-rattling by the North Koreans is starting to severely swing public support to change that constitutional clause before it's too late! Those asinine antics of the kooky Koreans may be what totally tips the scales and causes Japan to officially re-arm with a pro-active military. The Japanese government has been all for the change; the US government supports this as well. And now, maybe the Japanese public too. Look for this to be hotly debated this year . . . or until the North Koreans attack.

SAYONARA!

With no doubts we can say that Japan will have considerable impact on world affairs for some time to come. As an economic leader in the world and in its neighborhood, it has set the example that other Asian economies are rapidly living up to. That role as an economic powerhouse will hold for a while longer, and it has company coming to dinner! Japan is the number three economy. China is the number two economy and eventually will be number one, South Korea is number eleven soon to be number ten. Wow. Three top ten economies all within spitting distance of each other. This 'hood is hot!

Japan is also staunchly in the Team West camp as well, and one need look no further than the spill-over in commercial and popular culture to see how intimately tied they are to the west. Ninjas, karate, **Atari**, judo, manga, sushi bars, sake shots, Panasonic, Nintendo, Sony, Honda, Toyota, Kawasaki, Speed Racer, Samurai Jack, Tamagotchi, Totoro, Akira, the Matrix, Kill Bill, 4chan, Pocky,

J-Pop, and the godforsaken Pokemon—there was a blasphemous Pikachu blimp floating in the Macy's Thanksgiving Day parade right beside Snoopy, for pete's sake. Japan not only adopted western institutions and technologies, but has the unique ability to morph the Eastern and Western traditions together to come out with interesting products that appeal to consumers in both cultures. They do it like no other.

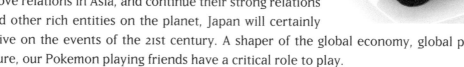

As they try to improve relations in Asia, and continue their strong relations with the US, the EU, and other rich entities on the planet, Japan will certainly have a unique perspective on the events of the 21st century. A shaper of the global economy, global politics, and most importantly, global culture, our Pokemon playing friends have a critical role to play.

Now, only if we can get them to un-invent karaoke. . . .

2014 SERIOUS TSUNAMI UPDATE!

Earthquake! Tsunami! Nuclear Disaster! Unfortunately for Japan, how about all three . . . at once! The titanic 2011 earthquake and subsequent tidal surge severely shocked our Japanese friends and their economy almost into submission. But like so many other times in their history, they are bouncing back fast, and assuredly will ride from the death and destruction of this horrific event to be a major playa again.

Their economy took a huge hit from the disaster, and was stagnant for 2 decades before that, but Shinzo Abe has unleashed an economic stimulus program that has them back in the positive growth column (nicknamed "Abenomics"). In 2012, Japan voted to phase out all nuclear power, but Shinzo and crew are already trying to undo that move as well, since they understand the repercussions of Japan having no home-grown energy in an era that may see oil prices continue to go higher and higher.

So Japan is on the upswing from the disaster, and spirits are high under the Shinzo self-pride program. But, in the end, I want to go ahead and offer this Plaid Avenger assessment: Japan will rebound, but it will not in our lifetime regain its predominant role as the biggest economy and/or most important Asian political power with world impact. China is now richer, and helps Japan out in a crisis . . . that is a significant reversal of fortune to consider. And there are many other Asian titans rising fast too, many of which will soon catch or surpass Japan in international importance and material wealth. This is no diss of Japan (I love the country! And they will continue to be awesome!) but a pragmatic assessment of their future position of one of many Asian powers, not as a leader of Asian powers. Dig?

21st century: Is the last stand of the Japanese economy upon us? Samurai have some experience with this. . . .

JAPAN RUNDOWN & RESOURCES

 **BIG PLUSES**

→ Ninjas

→ Samurai

→ 3rd largest economy in the world

→ Center of technological innovation in electronics, auto industry, robotics, etc.

→ Some of the highest standards of living on the planet; highest longevity rates

→ One of the most solid infrastructures in the world

→ Environmentally awesome; 70% of the country is essentially a forested national park

→ Politically stable, solid rule of law

→ Huge US ally, and would certainly have the entire US military as back-up if they ever get into trouble

 **BIG PROBLEMS**

→ Earthquakes, Tsunamis, & sproadically, Godzilla

→ Population shrinking, and immigration not really accepted

→ Very high cost of living and high wages causing industry/job movement to Asia

→ Heavily dependent on natural resources, fossil fuels and energy in general imported from other regions.

→ Has been economically/financially stagnant for decades; little to no economic growth

→ Has been politically gridlocked and governmentally ineffective for several years

→ Friction still exists between Japan and Asia; particularly China and Korea. This has stymied closer economic/political integration with its nearest neighbors. (Both of which are huge economies)

→ Would likely be the #1 target of a missile if/when North Korea goes completely nuts

DEAD DUDES OF NOTE:

Commodore Matthew Perry: Commodore of the United States Navy that forcibly opened Japanese ports/trade to the West with the Treaty of Kanagawa in 1954, ending centuries of self-imposed isolation under the Shogun.

Emperor Meiji: The main man in charge of Japan during its revolutionary overhaul to western ways, which then set the stage for its imperial rise to world power. Thus, his name is attributed to the entire time period of his rule, and the 'Restoration' itself.

Hideki Tōjō: General in the Japanese Imperial Army and Prime Minister of Japan during most of WWII, Tojo is held responsible for the bombing of Pearl Harbor which pulled the US into the war. Tojo was later tried for war crimes and executed.

General Douglas MacArthur: General of the US Army, played a mjor role in the Pacific Thater of WWII, officially accepted Japan's full surrender at the end of the war, and oversaw the US occupation of Japan from 1941 to '45. As the effective ruler of Japan, he implemented radical economic, political, and social changes to the country. That's why they are kind of like the US!

LIVE LEADERS YOU SHOULD KNOW:

Junichiro Koizumi: Extremely popular former Prime Minister of Japan from 2001–06. Was "the Ronald Reagan of Japan": conservative, pro-business, pro-US, and had a penchant for wild hair and Elvis. Even visited Graceland. Sent Japanese Self-Defense Forces abroad for the first time in its history to help the US in Iraq.

Shinzo Abe: Current Prime Minister of Japan, and an ardent nationalist who is likely to increase tensions with neighboring Asia . . . especially China. Wants Japan to be get past the 'era of apologizing,' has re-juiced their economy, and enjoys widespread support. For now.

KEY TERMS TO KNOW

Bodhisattva

Buddhism

Communes

Confucianism

Enlightened Beings

Globalization

Mahayana

Nirvana

Special Economic Zones (SEZ's)

Strait of Malacca

The Great Leap Forward

Physical
Features
East Asia

0   250   500 Miles

©2015 BYU Geography
Think Spatial group

Cultural
Features
East Asia

©2015 BYU Geography
Think Spatial group

0   250   500 Miles

Practice
Test Map
East Asia

©2015 BYU Geography

Think Spatial group

0   250   500 Miles

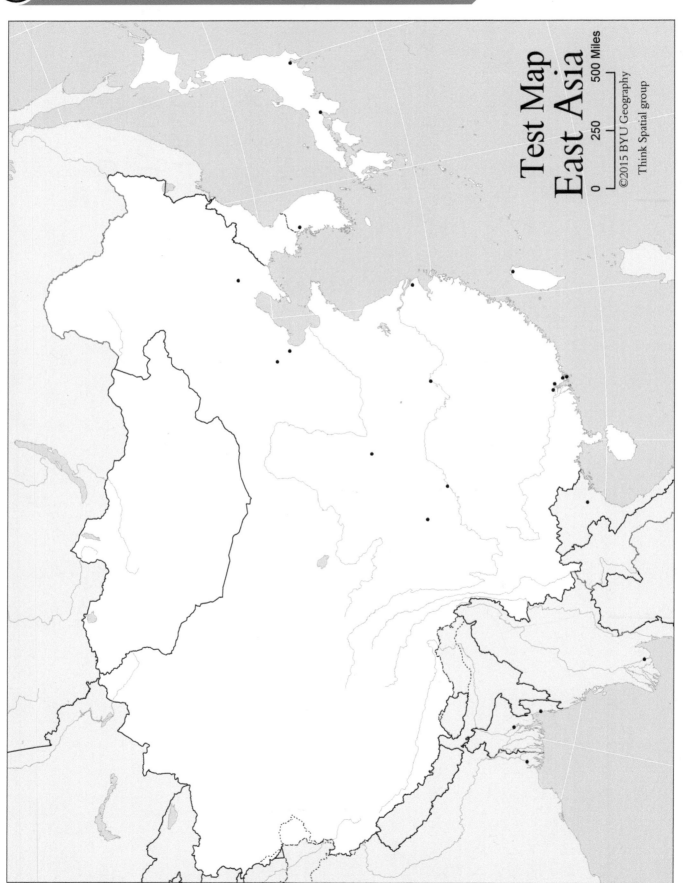

Test Map
East Asia

©2015 BYU Geography
Think Spatial group

0   250   500 Miles

South Asia and the Geography of Language and Religion

(10)

For by him all things were created, in heaven and on earth, visible and invisible, whether thrones or dominions or rulers or authorities—all things were created through him and for him. And he is before all things, and in him all things hold together. And he is the head of the body, the church. (Colossians 1:16–17)

CHAPTER OUTLINE:

IN PROVERBS (23:10), we are exhorted to respect ancient landmarks and to respect the poor. South Asia is a vibrant and wonderful culture and civilization with a terrific future potential. You have many opportunities as a geography student to recognize how these borders, many recently drawn, have provided opportunities for those who would be peacemakers in one of the most populous areas on the planet, and one rife with needs on both sides of India's borders!

Roughly one-third the size of the United States, India is the world's largest democracy. Soon to be the world's most populous nation, it is a reflection of the location and past of its people. Christianity is growing remarkably fast there, and the future of India is exciting and bright!

India is a large, triangular peninsula jutting into the Indian Ocean separating the roughly equal in size Arabian Sea and Bay of Bengal. This relative location has been important over the ages as nations seeking to trade with China have used it as a way station along the route as well as used it as a very important source of natural resources. Located on the southern rim of the gigantic Asian continent, the subcontinent of greater India is greatly affected by the Indian Ocean, it's watery frontier on its southern border. India has seen various land empires, such as Persia and the Mongols, from the north and east move down with the elevation (like a highway off-ramp) into it occupying as far as New Delhi in the Ganges plain to the Deccan Plateau to the south. Simultaneously, the monsoons have created a type of buffer region for trade.

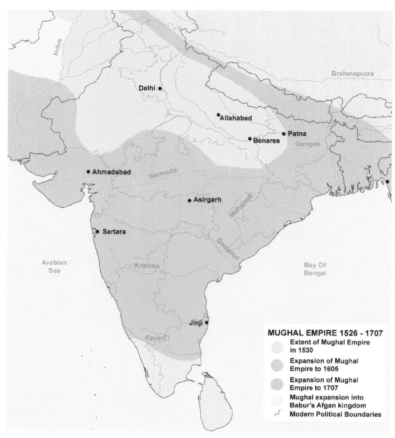

Map courtesy of Katie Pritchard

HERE COMES THE RAIN AGAIN

Because of temperature differences during the year, the Indian Ocean has consistently been a predictable route of travel for the **monsoon** winds. The monsoon winds are drawn off of the Indian Ocean and toward land in the summer because of atmospheric pressure disparities. These winds bring with them extensive seasonal rains. In the winter, the winds reverse direction blowing from the north resulting in a drier season. The tilt of the earth and the disparity of temperatures and pressure systems over continents and oceans ultimately create these special conditions and when high relief such as the Himalayan Mountains become involved, these can be catastrophic. This is often the case in Bangladesh which experiences flooding and consequent disease epidemics due to its tropical nature. In fact, according to Robert D. Kaplan, "The population of Bangladesh is roughly half that of the United States and greater than that of Russia and packed into an area about the size of Iowa."[1] During

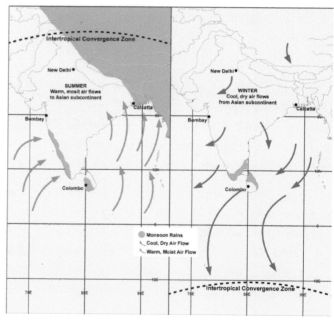

Map courtesy of Katie Pritchard

---

[1] Robert D. Kaplan, "Monsoon: The Indian Ocean and the Future of American Power" (New York: Random House, 2010), 140

© JupiterImages Corporation

From *Blue Marble: Next Generation* image produced by Reto Stockli, NASA Earth Observatory (NASA Goddard Space Flight Center)

the Age of Sail, this seasonally consistent wind pattern from the north and south enabled explorers and traders to either sail into the wind or travel with the wind behind them as they engaged in trans-regional trade between the Middle East and Southeast Asia, and ultimately globally between Europe and East Asia.

As globalized trade continued to develop through history—from Alexander the Great's Ancient Greeks, to the Romans, to the Arabs and Persians—all have mingled along the coasts of the subcontinent that can be seen to this day. India will

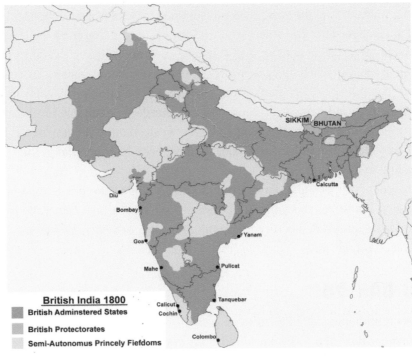

Map courtesy of Katie Pritchard

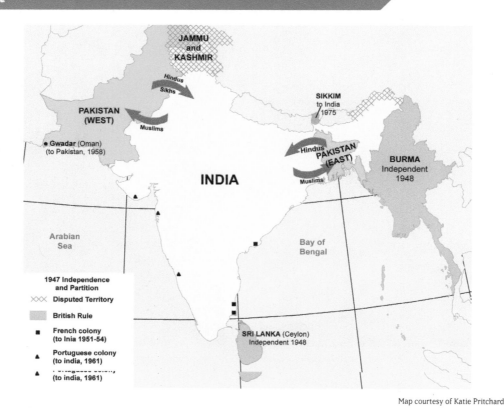

Map courtesy of Katie Pritchard

surely continue to be an extremely important state in the future, for it is situated between the valuable oil supplies of the Persian Gulf and the vital Strait of Malacca to the east, which is China's lifeline to the Gulf. India possesses the Andaman Island chain, which stands guard before Indonesia to the east. So what we have described as Greater India is actually one of many cultures spread over the rim of the continent on the Indian Ocean. Between the volatile Arab and Persian cultures to the west and the exploding economy of China to the east, India will continue to be of extreme importance to the United States.

India has defined natural borders. The term **Greater India** is used to describe these natural borders, which, according to some scholars, have spread from modern day Afghanistan in the west across the inclined elevations of Pakistan to the mountains of Burma in the east. High natural borders in the form of mountains ring Greater India from the Karakorum Range to the Himalayas along the northern flank. Only a few passes functioning as gates from the higher elevations from the west of Central Asia have served as invasion routes for centuries—most notably for the Aryans. The city of Delhi, which is ringed by the Thar Desert to the northeast, seems to be the western limit of many of the historical empires that spread into India. The coastal areas of southern Iran and Pakistan have been instrumental in the spread of culture laterally from west to east corresponding to the rivers.

The Ganges River, great river of India, flows from the mountains of the north and eventually turns east. The other key river, the Brahmaputra, continues this general pattern of East-West trade. Generally isolated but still surrounded by vibrant neighbors, the world's largest democracy today still remains greatly divided between north and south and has seen a renewal of post-colonial nationalism.

DO WHAT YOU CAN DO!

A renewed sense of being Indian is called **Hindutva**. Temples on the cultural landscape today mark religion as a key aspect of India. A relatively new film industry called Bollywood shows the confidence of Indian culture in an uncertain world. Still the world's largest democracy faces challenges from abroad. During the British withdrawal from the

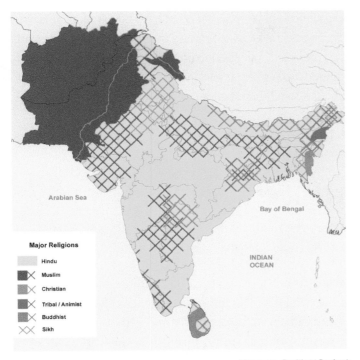

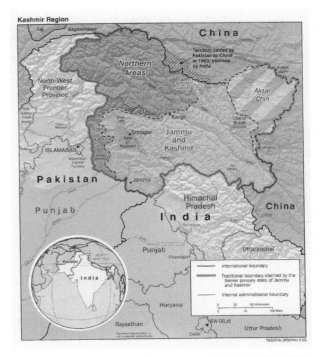

Map source: *The World Factbook*

CIA map, courtesy of University of Texas Libraries

subcontinent, the Great Partition of 1947 divided Greater India by creating East and West Pakistan from the land. Since then, India has had several armed confrontations with its neighbor Pakistan. Pakistan was originally intended to be a Muslim land that was set aside by the British as they withdrew from India. Originally successful, Pakistan has since become what some would call a failed state. The development of nuclear weapons has only increased the tensions between the former residents of greater India. The primary area of tension surrounds the regions of **Jammu and Kashmir** which are located at the head of the Indus River. The conflict surrounding this river, which serves as the key artery of Pakistan, is only increasing as a result of Pakistan's relation with the emerging power of China.

China has also caused some strategic headaches for India with recent Maoist insurgency efforts in Nepal and smaller Bhutan to the northeast. Those living in India are beginning to feel increasingly vulnerable to Islamic extremists who are largely being supported by China. This security concern has become more pronounced since the massive terrorist attack in India's largest city of Mumbai in 2008. Challenged from abroad, India continues to develop technologically. Many have identified the peninsula of Gujarat—the home of the Patel or business families of the nineteenth century—as the seat of Hindu nationalism today.

Many jobs in the technology sector have been relocated in recent years to India. Interestingly, as Christianity has spread in the southern areas of India so have technological opportunities for the Indian

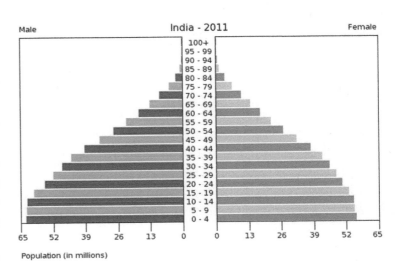

A perfect pyramid indicating the challenges of population growth and the challenges of life in India.

From *The Plaid Avenger's World #6 Nuclear Insecurity Edition*. Copyright © 2008, 2009, 2010, 2011, 2012 by John Boyer. Copyright © 2006 by Kendall Hunt Publishing Company. Reprinted by permission.

people. Many of the people of southern India are of Dravidian background—these people tend to have darker skin. This pattern of skin color has fit into the traditional ethnic religion of India and is known as a defining factor of the **caste system**. Though technically illegal, this system still perpetuates itself and has provided many opportunities for evangelical outreach, particularly among those people considered to be untouchables under the system. It is exciting to see India continuing to grow and becoming stable in uncertain times. Still, as anyone who has served there can tell you, India has a long way to go to become a secure and gigantic democracy!

The population of India is on the verge of surpassing China in terms of numbers. It is indeed a terrific time and to learn more about the rich and ancient cultures of the "subcontinent" to make a real and eternal difference.

THE GEOGRAPHY OF LANGUAGE AND RELIGION*

> *For I say to you, that unless your righteousness exceeds the righteousness of the scribes and Pharisees, you will by no means enter the kingdom of heaven. (Matthew 5:20)*

The development of languages is important to the geographer because it is often integral to both ethnic and nationalistic identifications. Just as languages have diffused, so have the world's religions. Religions like language reflect world view and identify deeply held perceptions of ultimate existence. Indeed, a backward reconstruction of both language and religion demonstrate the truth of Acts 17:26a: "From one man he created all the nations throughout the whole earth."

Students in an introductory human geography course might wonder why the analysis of the distribution of languages and religions is such a prominent component of most such courses given that language is the central theme of linguistics and religion is a major theme of philosophy and history. The answer is really very simple: language and religion are the defining cornerstones of human culture (at least as it is traditionally defined) and identity (ethnic and national or even individual identity). They are two of the most important characteristics that distinguish human beings from all other animals given that no other species possesses the capacity for speech or the ability to perceive of one's own mortality in a spiritual sense (at least as far as most biologists know). Many other kinds of animals have the ability to communicate with each other, sometimes in very complex ways, but none possess the capacity for language. Likewise, it is clear that many other species exhibit emotions and "feelings" but so far as we know no others can think and philosophize about what is going to happen to them when they die. So, if a main goal of human geography is to delineate and understand how human cultures vary over space, then it behooves us to know how the two major facets of culture vary over space. That is, a map of religious and linguistic regions is in many ways a map of culture regions.

THE CLASSIFICATION AND DISTRIBUTION OF LANGUAGES
CLASSIFICATION OF LANGUAGES

Language is defined as an organized system of spoken and/or written words, words themselves consisting of symbols or a group of symbols put together to represent either a thing or an idea, depending on the kind of writing system in use. In **syllabic languages**, the symbols that are used (e.g. "letters") represent sounds. German, English, Arabic and Hindi are examples of syllabic languages. **Ideographic languages** employ ideographs as symbols to represent an idea or thing. Examples of ideographic languages include Chinese, Japanese, and Korean. All human beings are biologically "hard-wired" for language. This means not that we are born knowing a language, but rather that we are born with the *ability* to learn a language or languages because it is imbedded in our genetic makeup. The linguist Noam Chomsky calls this "deep structure."

With respect to other forms of animal communication, human language is unique in two primary ways. First, human languages are *recombinant*. This means that words and symbols can be taken out of order in a sentence and recombined to form a different sentence and thus communicate a completely different, or even subtly different, idea. Dogs, for example, are not capable of arranging barks in different orders to communicate extremely intricate ideas. Second, word formation in all human languages is almost completely *arbitrary*. This means that there is really no rhyme or reason as to why a certain word, symbol or ideograph is used to stand for something. There are, of course, exceptions to this rule. One is *sound symbolism*, in which the pronunciation of a word or the shape of an ideograph suggests an image or meaning. For example, many words in English that begin with "gl-" have something to do with sight (glimmer, glow, and glisten). Another exception to this rule is *onomatopoeia*, an instance in which a word sounds like something in nature that it represents (e.g. cuckoo, swish, cock-a-doodle-doo). But such exceptions are very rare. The vast majority of words in all languages have simply been made up and then passed down over generations although, to be sure, words and languages change over time.

There are roughly 6,700 different languages in use around the world today. But the vast majority of these languages are spoken by a relatively small number of people. This means that a relatively small number of languages have thousands or even millions of speakers. Consider the following list of the top ten languages by number of native (mother-tongue) speakers in 2009 according to SIL International, one of the leading organizations that collect and publish linguistic data (www.ethnologue.com):

Language	Primary Locations	Number of Speakers
Chinese	China	1.213 billion
Spanish	Middle America, Latin America, Spain	329 million
English	North America, British Isles, Australia, New Zealand	328 million
Arabic	Southwest Asia, North Africa, South-Central Asia	221 million
Hindi	India	182 million
Bengali	Bangladesh	181 million
Portuguese	Brazil, Portugal	178 million
Russian	Russia	144 million
Japanese	Japan	122 million
German	Germany, Austria, Switzerland	90 million

Of the nearly 6,700 languages in use today, 389 (about 6%) account for 94% of the world's population. The remaining 94% of languages in the world are spoken by only 6% of the world's population.

Linguists have devised classification schemes that describe and account for similarities and differences between and within different languages. At the broadest level, a **proto-language** describes an ancestral language from which several language families (described below) or languages are descended. No proto-languages are spoken today, but they are theorized to have been in use thousands of years ago. For example, Proto Indo-European was the language theorized to have been spoken in eastern Anatolia (present-day Turkey) and the Caucasus Mountain region 5,000 years ago. These people were the "original" Indo-Europeans (probably some of the first "Caucasians") and the language they developed became the basis for all of the languages that are classified by linguists as Indo-European. If we use the analogy of a tree to represent a group of languages that are linguistically related, then a proto-language is the roots and trunk of the tree.

In this analogy, each branch of the tree represents a **language family**. A language family is a group of languages descended from a single earlier language whose similarity and "relatedness" cannot be the result of circumstance. How do we know that certain languages are "related" to each other? Linguists employ two main methods to determine linguistic relatedness. One is genetic classification, in which it is assumed that languages have diverged from common ancestor languages (proto-languages) and therefore languages that diverged from the same proto-language will have inherit similarities. Compare, for example, the following words for "mother" in selected Indo-European languages:

English	*mother*
Dutch	*Moeder*
German	*Mütter*
Irish Gaelic	*mathair*
Hindi	*mathair*
Russian	*mat*
Czech	*matka*
Latin	*mater*
Spanish	*madre*
French	*mére*

It is obvious that all of these words sound very similar to each another. Given the rule that word formation is arbitrary, it is impossible that such strong similarities are the result of mere coincidence, especially given the fact that some of these populations are separated by thousands and thousands of miles. When we add to this list hundreds or thousands of other words that display such similarities it is clear that these languages have a common ancestor, common linguistic roots. When languages are shown to have a common ancestor, such as those above, they are said to be **cognate languages**.

Reviewing the list above once again it is also clear that some of the languages have even more commonalities to others in the list. Compare, for example, the even more clearly defined similarity between Latin and Spanish, English and Dutch, and Russian and Czech. These groupings of languages whose commonality is very definite are called **language subfamilies**. Think of these groupings as twigs of larger branches on the language tree, while individual languages are the leaves of the tree. Proceeding even further with respect to similarity, linguists recognize **dialects**. A dialect is defined as a recognizable speech variation *within the same language* that distinguishes one group from another, both of which speak the same language. Sometimes these differences are based on pronunciation alone (the different varieties of English spoken around the world or a "southern" or "New England" accent) and sometimes they are based on slightly different words for the same thing (British English "lorry," American English "truck"). Similarly, **pidgins** are languages which develop from one or more "mother" languages that have highly simplified sentence and grammatical structures compared to the mother languages. When such a language becomes the native language of succeeding generations, it is referred to as a **creole language**. Most creoles are either English-, French-, or Spanish-based and are spoken in the periphery of the world-economy, in former colonial areas, where two or more groups of people speaking mutually unintelligible languages were forced to communicate with each other during the colonial era.

Political and ethnic fragmentation is characteristic of many former colonial regions in the periphery of the world-economy. In Nigeria, for example, at least 500 different tribal languages are spoken, many of them not even in the same language family. In such places, governments and businesses often make use of a *lingua franca* to carry out official business. A lingua franca refers to a language that is used habitually among people living in close contact with each other whose native tongues are mutually unintelligible. English and French are common lingua francas in much of sub-Saharan Africa, Arabic is the lingua franca of much of North Africa and Southwest Asia, and English can be thought of as the lingua franca of the internet and of air traffic control and airline pilots around the world.

LINGUISTIC DIFFUSION AND CHANGE

What spatial processes have led to the present-day distribution of languages? How do languages change over time and space? In general, cultural diffusion (both hierarchical and relocation diffusion) and geographical isolation over time and space have resulted in the linguistic patterns we observe today. Relocation diffusion on a massive scale since the advent of the capitalist world-economy in the 15th century, together with the displacement and subjugation of native populations, has resulted in very large linguistic regions, especially in the Americas.

These two processes (relocation diffusion and displacement) explain the fact that the vast majority of populations in North, South and Middle America speak one of three languages (all of them Indo-European languages): English, Spanish, and Portuguese. Before Columbian contact, there were probably as many as 30 million people living in the Americas, speaking literally hundreds of distinctive languages in at least a dozen different language families. In other words, the linguistic map was highly complicated and extremely diverse. Massive relocation from Europe and the decimation of native populations over a 300-year period resulted in a vastly "simplified" and less complex linguistic map in the Americas. This is not to say that no Native American languages survive. Many do, but with few exceptions the number of people who speak these languages is very small compared to the number of English, Spanish and Portuguese speakers. The colonial era also ushered in a period in which many European languages, such as English, Spanish, Portuguese and French, acquired many more new speakers in their overseas colonies than they ever had at home. In part this was a result of the outright extermination of African and Native American languages in the Americas through either severe population decline or cultural subjugation through slavery. In the process, European languages were of course deemed to be "superior" to others, and Europeans forced Africans and Native Americans to learn the language of the colonizers. This was the case not only in the Americas, but in colonial sub-Saharan Africa as well.

The present-day world linguistic map also is the product of centuries of linguistic change over time and space. In the pre-industrial era, for example, migration and spatial isolation and segregation gave rise to separate, mutually unintelligible languages. As populations diverged over time and space, populations became isolated from each other. And as these migrating populations encountered new natural environments and human societies they were forced to invent new words to describe new circumstances, places, and things. It has also been shown that languages change naturally in place over time, even in the absence of outside cultural forces such as immigration or hostile invasion. Take, for example, the case of English and how significantly the language changed between the 9th and 17th centuries. The Old English of 9th-century Britain (*Beowulf* is the most famous literary example) would hardly be recognizable to most English speakers today. But 500 years later, due to influences from Latin, French, and Danish, as well as to natural linguistic evolution over time, the language had evolved into the Middle English of Chaucer (*The Canterbury Tales*), a language that most English speakers today can understand. By the 17th century the language had evolved into the Modern English of Shakespeare. As one of the world's most widely-spoken languages and as a world-wide lingua franca, English is changing more rapidly today than ever before as it incorporates words from a variety of different cultural sources around the world.

THE DISTRIBUTION OF THE WORLD'S MAJOR LANGUAGE FAMILIES

This section lists the world's major language families and important subfamilies and maps the distribution of their speakers.

1. **Indo-European Family (386 languages) (about 2.5 billion speakers)**
   - → Albanian Subfamily—Albania, parts of Yugoslavia and Greece
   - → Armenian Subfamily—Armenia
   - → Baltic Subfamily—Latvia, Lithuania

- → Celtic Subfamily—parts of western Ireland, Scotland, Wales, Brittany
- → Germanic Subfamily—northern and western Europe, Canada, USA, Australia, New Zealand, parts of the Caribbean and Africa
- → Greek Subfamily—Greece, Cyprus, parts of Turkey
- → Indo-Iranian Subfamily—India, Pakistan, Bangladesh, Afghanistan, Iran, Nepal, parts of Sri Lanka, Kurdistan (Iran, Iraq, Turkey)
- → Italic (Romance) Subfamily—France, Spain, Portugal, Italy, Romania, Brazil, parts of western and central Africa, parts of the Caribbean, parts of Switzerland
- → Slavic Subfamily—eastern Europe, southeastern Europe, parts of south-central Asia

2. **Sino-Tibetan Family (272 languages) (about 1.1 billion speakers)**
   - → Chinese Subfamily—China, Taiwan, Chinese communities around the world
   - → Tibeto-Burman Subfamily—Tibet, Myanmar (Burmese), parts of Nepal and India

3. **Austronesian Family (1,212 languages) (269 million speakers)**
   - → Formosan Subfamily—parts of Taiwan
   - → Malayo-Polynesian Subfamily—Madagascar, Malaysia, Philippines, Indonesia, New Zealand (Maori), Pacific Islands (e.g. Hawaii, Fiji, Samoa, Tonga, Tahiti)

4. **Afro-Asiatic Family (338 languages) (250 million speakers)**
   - → Semitic Subfamily—North Africa (Arabic), Israel (Hebrew), Ethiopia (Amharic), Middle East
   - → Cushitic Subfamily—Ethiopia, Kenya, Eritrea, Somalia, Sudan, Tanzania
   - → Chadic Subfamily—Chad, parts of Nigeria, Cameroon
   - → Omotic Subfamily—Ethiopia
   - → Berber Subfamily—parts of Morocco, Algeria, Tunisia

5. **Niger-Congo Family (1,354 languages) (206 million speakers)**
   - → Benue-Congo Subfamily—central and southern Africa
   - → Kwa Subfamily—bulge of west Africa
   - → Adamaw-Ubangi Subfamily—northern part of central Africa
   - → Gur Subfamily—between Mali and Nigeria
   - → Atlantic Subfamily—extreme western part of the bulge of west Africa
   - → Mande Subfamily—western part of the bulge of west Africa

6. **Dravidian Family (70 languages) (165 million speakers)**
   - → Four Subfamilies—southern India, parts of Sri Lanka, parts of Pakistan

7. **Japanese Family (12 languages) (126 million speakers)**

8. **Altaic Family (60 languages) (115 million speakers)**
   - → Turkic Subfamily—Turkey, Uzbekistan, Turkmenistan, Kazakhstan, Azerbaijan, eastern Russia (Siberia)
   - → Mongolian Subfamily—Mongolia, parts of adjoining areas of Russia and China
   - → Tungusic Subfamily—Siberia, parts of adjoining areas of China

9. **Austro-Asiatic Family (173 languages) (75 million speakers)**

   → Mon-Khmer Subfamily—Vietnam, Cambodia, parts of Thailand and Laos

   → Munda Subfamily—parts of northeast India

10. **Tai Family (61 languages) (ca. 75 million speakers)**

    → Tai Subfamily—Thailand, Laos, parts of China and Vietnam

11. **Korean Family (1 language) (60 million speakers)**

12. **Nilo-Saharan Family (186 languages) (28 million speakers)**

    → Nine Subfamilies—southern Chad, parts of Sudan, Uganda, Kenya

13. **Uralic Family (33 languages) (24 million speakers)**

    → Finno-Ugric Subfamily—Estonia, Finland, Hungary, parts of Russia

    → Samoyedic Subfamily—parts of northern Russia (Siberia)

14. **Amerindian Languages (985 languages) (ca. 20 million speakers)**

    → As many as 50 different language families, hundreds of subfamilies

    → North America = ca. 500,000 speakers; 150 languages (top languages = Navajo and Aleut)

    → Central America = ca. 7 million speakers; (top language = Nahuatl)

    → South America = ca. 11 million speakers; (top language = Quechua)

15. **Caucasian Family (38 languages) (7.8 million speakers)**

    → Four Subfamilies—Georgia, surrounding region on western shore of the Caspian Sea

16. **Miao-Yao Family (15 languages) (5.6 million speakers)**

    → Southern China, northern Laos (Hmong), northeast Myanmar

17. **Indo-Pacific Family (734 languages) (3.5 million speakers)**

    → The most linguistically complex place on earth—Papua New Guinea and surrounding islands

18. **Khoisan Family (37 languages) (300,000 speakers)**

    → Three Subfamilies—parts of Namibia, Botswana, Republic of South Africa

19. **Australian Aborigine (262 languages) (ca. 30,000 speakers)**

    → Only five languages have over 1,000 speakers

    → At time of European contact, 28 language families, 500 languages spoken by over 300,000 people

20. **Language Isolates (296 languages) (ca. 2 million speakers)**

    → Languages which have not been conclusively shown to be related to any other language; some examples are:

    → Basque (Euskara)—southern France, northern Spain

    → Nahali—5,000 speakers in southwest Madhya Pradesh in India

    → Ainu—island of Hokkaido, Japan; now probably extinct

    → Kutenai—less than 200 speakers in British Columbia and Alberta

THE CLASSIFICATION AND DISTRIBUTION OF RELIGIONS

CLASSIFICATION OF RELIGIONS

Together with language, religion is a human characteristic that distinguishes us from every other animal species on the planet. As with language, we are also most likely biologically "hard-wired" with the capacity for abstract thought about spiritual matters, our own mortality, and the nature of the universe and our place in it. We are born with these capacities, but not with a certain set of beliefs concerning these things—these beliefs are learned. By definition, a **religion** is a system of either formal (written down and codified in practice) or informal (oral traditions passed from generation to generation) beliefs and practices relating to the sacred and the divine. Who am I? Why am I here? What is my purpose in life? What is my place in the universe? What will happen to me when I die?

These are questions that every human being ponders because we have the biological capacity to think about and philosophize about such things. Human religions attempt to answer such questions through systems of beliefs, practices and worship. As the answers to these questions vary so do religious belief systems; as the answers to these questions vary from place to place, to a large extent so does human culture. As is the case with human languages, there are literally thousands of religions practiced around the world today. In most of the peripheral regions of the world-economy there are as many religions as there are ethnic groups and can vary substantially over relatively short distances. For human geographers, religion is an extremely important aspect of the cultural landscape because religious beliefs and customs have a significant physical manifestation in the form of religious structures. Although religious practices, beliefs and traditions vary substantially around the world, most religions share the following characteristics in common:

→ Belief in one or many supernatural authorities

→ A shared set of religious symbols (iconography)

→ Recognition of a transcendental order—offers a divine reason for existence and an explanation of the inexplicable

→ Sacraments (prayer, fasting, baptism, initiation, etc.)

→ Enlightened or charismatic leaders (priests, shaman, prophets)

→ Religious taboos

→ Sacred structures (temples, shrines, cathedrals, mosques, etc.)

→ Sacred places (pilgrimage sites, holy cities, etc.)

→ Sacred texts

Human religions can be divided into two broad categories and two narrower sub-categories. At the broadest level, we can distinguish between monotheistic and polytheistic religions. **Polytheistic religions** involve a belief in many supernatural (that is, not of this world) beings that control or influence some aspect of the natural or human world. The vast majority of human religions, in terms of number, are polytheistic in nature, numbering in the thousands. Such religions usually, but not always, have a very small geographic distribution, often coinciding with tribal or ethnic boundaries. For most of human history, polytheistic belief systems have been by far the most common. **Monotheistic religions** appeared on the human stage quite late, probably not until around 1,500 B.C. Monotheism involves the belief in one omnipotent, omniscient, supernatural being who created the universe and everything in it and thus controls and influences all aspects of the natural and human world. There are in effect only three monotheistic religions today: Judaism, Christianity, and Islam. In terms of number of adherents and believers, however, two of these (Christianity and Islam) have over one billion adherents each. That is, out of a total world population of around 6.5 billion, fully one-third are either Christians or Muslims. The geographic boundaries of Christian and Islamic beliefs, then, do not coincide with political boundaries but rather supercede and overlap them. We can account for such distributions in much the same way that we can account for the very large distribution of

Indo-European speakers around the world: relocation diffusion on a massive scale during the colonial era, and the acquisition of new members either by force or through missionary activities.

We can also identify two sub-categories of religions. First, **Ethnic religions** are those in which membership is either by birth (you are "born into" the religion) or by adopting a certain complex ethnic lifestyle, which includes a certain religious belief system. That is, an ethnic religion is the religious belief system of a specific ethnic or tribal group and is unique to that group. Most (but not all) of these kinds of religions are polytheistic in nature and have very small geographical distributions, sometimes no larger than a village or group of villages. These religions, therefore, have very strong territorial or ethnic group identity. In most such religions, there is no distinction between made between one's ethnic identity (i.e. one's culture) and one's religion: one's religion is one's culture. Examples of ethnic religions include Judaism, Hinduism and various tribal belief systems that are ubiquitous throughout the periphery of the world-economy. Second, **universalizing religions** are those in which membership is open to anyone who chooses to make a solemn commitment to that religion, regardless of class or ethnicity. Membership in these religions *is* usually relatively "easy", and usually involves some sort of public declaration of one's allegiance to the belief system (baptism, for example). Universalizing religions are also distinguished by the fact that they are often characterized by strong evangelic overtones in which members are admonished to spread the faith to non-believers. For these reasons universalizing religions have very large geographic distributions that cover vast regions of the world, the boundaries of which overlap the political boundaries of individual states. There are only three universalizing religions: Christianity, Islam, and Buddhism, although Buddhism rarely carries with it evangelic activities and therefore has a much smaller distribution than Christianity and Islam.

Finally, it should be noted that the influence of popular culture and post-modern thought and philosophies in core regions of the world-economy, have significantly influenced the growth of secularism. **Secularism** refers to an indifference to, or outright rejection of, a certain belief system or religious belief in general. In its extreme such "beliefs" may become like a religion itself. It is an increasingly characteristic of many post-industrial societies, and thus influences core societies more than any other societies around the world. At least one-fifth of the world's population, by this definition, is secular, and this figure is even higher in parts of northern and western Europe, where the figure approaches 70 percent in some instances.

ATTRIBUTES AND DISTRIBUTIONS OF THE WORLD'S MAJOR RELIGIONS

This section lists the world's major religions, identifies their major characteristics, and maps the distribution of their adherents.

1. **Hinduism (ca. 740 million adherents concentrated in India, Nepal, and Sri Lanka)**
   - → One of the world's oldest extant religions
   - → The ethnic religion of the Hindustanis
   - → Hearth in the Indus Valley ca. 1500 B.C., then spread to India, Nepal, Sri Lanka and parts of SE Asia
   - → Beliefs and practices:
     - ✦ A common doctrine of *karma*, one's spiritual ranking, and *samsara*, the transfer of souls between humans and/or animals
     - ✦ A common doctrine of *dharma*, the ultimate "reality" and power that governs and orders the universe
     - ✦ The soul repeatedly dies and is reborn, embodied in a new being
     - ✦ One's position in this life is determined by one's past deeds and conduct
     - ✦ The goal of existence is to move up in spiritual rank through correct thoughts, deeds and behavior, in order to break the endless cycle and achieve *moksha*, eternal peace

* Life in all forms is an aspect of the divine—hundreds of gods, each controlling an aspect of the natural world or human behavior

* One need not "worship" a god or gods

→ **The Caste System**—a social consequence of the Hindu belief system

* The social and economic class into which one is born is an indication of one's personal status

* In order to move up in caste one must conform to the rules of behavior for one's caste in this life

* This thus highly limits social mobility

→ Sacred texts

* The *Rig Vedas*, hymns composed by the Indo-Aryans after the invasion of the Punjab; the oldest surviving religious literature in the world, written in Sanskrit

* *Brahmanas*, theological commentary, defined different castes

* *Upanishads*, defines karma and nirvana, etc.

→ Cultural landscapes

* Shrines, village temples, holy places and rivers (the Ganges), pilgrimage sites and routes

2. **Buddhism (ca. 300 million concentrated in East and Southeast Asia)**

→ Founded by Gautama Siddharta in the 6th century B.C. in northeast India

→ Diffusion was mainly to China and Southeast Asia by monks and missionaries

→ The primary religion in Tibet, Mongolia, Myanmar, Vietnam, Korea, Thailand, Cambodia, Laos; mixed with native faiths in China and Japan

→ A universalizing religion

→ Beliefs and practices:

* Retains the Hindu concept of *karma*, but rejects the caste system

* More of a moral philosophy than a formal religion

* The ultimate objective is to reach nirvana by achieving perfect enlightenment

* The road to enlightenment, Buddha taught, lies in the understanding of the four "noble truths": 1. to exist is to suffer 2. we desire because we suffer 3. suffering ceases when desire is destroyed 4. the destruction of desire comes through knowledge or correct behavior and correct thoughts (the "eight-fold path")

→ Sects:

* Theravada (Sri Lanka, Myanmar, Thailand, Laos, Cambodia)

* Mahayana (Vietnam, Korea, Japan, China, Mongolia)

* Zen (Japan)

* Lamaism (Tibet)

→ Cultural landscapes:

* Shrines and temples

* Holy locations where the Buddha taught

3. **Chinese Faiths (ca. 300 million adherents in China)**

→ Two main forms: Confucianism and Taoism, both date from the 6th C. B.C.

→ The goal of both is moral harmony within each individual, which leads to political and social harmony

→ Chinese religion combines elements of Buddhism, Animism, Confucianism, and folk beliefs into one "great religion"; each element services a different component of the self

→ The Taoist approach to life is embodied in the Yin/Yang symbol; stresses the oneness of humanity and nature; people are but one part of a larger universal order

→ Confucianism is really a political and social philosophy which became a blueprint for early Chinese civilization; it teaches the moral obligation of people to help each other, that the real meaning of life lies in the here and now, not in a future abstract existence; Kong Fu Chang taught that the secret to social harmony is empathy between people

4. **Judaism (ca. 18 million adherents mainly in N. America and Israel)**

→ The ethnic religion of the Hebrews

→ The oldest religion west of the Indus (ca. 1,500 B.C.)

→ Founder regarded as Abraham (the patriarch)

→ Sacred text = the Torah (the five books of Moses)

→ Beliefs and practices:

   ◆ God is the creator of the universe, is omnipotent, but yet merciful to those who "believe" in Him

   ◆ God established a special relationship with the Jews, and by following his law they would be special witnesses to His mercy

   ◆ Emphasis is on ethical behavior and careful, ritual obedience

   ◆ Among the traditional, almost all aspects of life are governed by strict religious discipline

   ◆ The Sabbath and other holidays are marked by special observances and public worship

   ◆ The basic institution is the Synagogue, led by a rabbi chosen by the congregation

→ Cultural landscapes:

   ◆ Synagogues

   ◆ Holy sites (e.g. the wailing wall, Jerusalem, sites of miracles, etc.)

5. **Christianity (ca. 1.6 billion adherents worldwide, but especially in Europe, N. America, Middle and South America)**

→ A universalizing religion

→ A revision of Judaic belief systems

→ Founder regarded as Jesus, a Jewish preacher believed to be the savior of a sinful humanity promised by God; his main message was that salvation was attainable by all who believed in God (died ca. 30 A.D.)

→ Sacred text = the Bible; Old Testament is based on the Hebrew Torah and is the story of the Jews; New Testament is based on the life of Jesus and his teachings

→ Mission: conversion by evangelism through the offering of the message of eternal life and hope

→ Reform movements:

   ◆ Split in the 5th century between the western church at Rome (Catholicism) and the eastern church at Constantinople (Orthodoxy)

   ◆ Protestant Reformation in the 15th and 16th centuries, led mainly by northern Europeans over moral and political issues

   ◆ Protestantism took hold in northern Europe and spread to North America, Australia and New Zealand

→ Cultural landscapes:

  ◆ Churches, cathedrals, graveyards, iconography

6. **Islam (ca. 1 billion adherents worldwide, but especially in N. Africa, SW Asia, South-Central Asia, Indonesia, Malaysia)**

  → Founder: Muhammad ("Prophet"), born 571 A.D.; believed to have received the last word of God (Allah) in Mecca in 613 A.D.

  → Diffusion: rapidly throughout Arabia, SW Asia, North Africa, then to South and Southeast Asia

  → Organization: theoretically, the state and the religious community are one in the same, administered by a caliph; in practice, it is a loose confederation of congregations united by tradition and belief

  → A universalizing religion

  → Sacred Text: the Koran—the sayings of Muhammad, believed to be the word of God

  → Divisions: two major sects—Sunni (Orthodox) and Shi'ah (Fundamentalist); Shiites mainly in Iran and parts of Iraq and Afghanistan; Sunni are the majority worldwide

  → Beliefs: mainly a revision of both Judaic and Christian beliefs; those who repent and submit ("Islam") to God's rules can return to sinlessness and have everlasting life; religious law as revealed in the Koran is civil law; smoking, gambling and alcohol are forbidden

  → The faithful are admonished to practice the five "**pillars of Islam**":

  ◆ Public profession of faith

  ◆ Daily ritualistic prayer five times per day

  ◆ Almsgiving

  ◆ Fasting during daylight hours during Ramadan

  ◆ A pilgrimage to Mecca at least once in one's lifetime if physically and economically possible

  → Cultural landscapes:

  ◆ Mosques, minarets, religious schools, iconography

The study of both languages and religion reveals key ingredients to human culture. We should study the religions of the world in an attempt to find common ground, whereby we can present the liberating good news of the Gospel of Jesus Christ. While we may not understand the religious actions of others, we can certainly respect other's convictions, since we ourselves have experienced at some time the rejection of truth in our increasingly secular society.

KEY TERMS TO KNOW

Cognate Languages	Language	Proto-Language
Creole Language	Language Family	Religion
Dialect	Language Sub-Family	Secularism
Ethnic Religions	Lingua Franca	Subcontinent
Greater India	Monotheistic Religion	Syllabic Languages
Hindutva	Monsoon	The "Pillars of Islam"
Ideographic Languages	Pidgin Language	The Caste System
Jammu and Kashmir	Polytheistic Religion	Universalizing Religions

FURTHER READING

Melvyn Bragg, *The Adventure of English: The Biography of a Language* (London: Hodder & Stoughton, 2003).

Noam Chomsky, *On Language* (New York: New Press, 1998).

David Crystal, *The Cambridge Encyclopedia of Language*, 2nd ed. (Cambridge: Cambridge University Press, 1997).

Mircea Eliade, *The Sacred and the Profane: The Nature of Religion* (San Diego: Harvest Books, 1957).

Susan Tyler Hitchcock, *Geography of Religion: Where God Lives, Where Pilgrims Walk* (Washington: National Geographic Society, 2004).

William James, *The Varieties of Religious Experience* (New York: Penguin, 1958).

M. Paul Lewis, ed., *Ethnologue: Languages of the World*, 16th ed. (Dallas: SIL International, 2009).

Chris Park, *Sacred Worlds: An Introduction to Geography and Religion* (London: Routledge, 1994).

Steven Pinker, *The Language Instinct: How the Mind Creates Language* (New York: Harper Collins, 1995).

Ninian Smart, ed., *Atlas of the World's Religions* (Oxford: Oxford University Press, 2009).

Huston Smith, *The World's Religions*, 50th Anniversary Edition (New York: HarperOne, 2009).

Roger W. Stump, *Boundaries of Faith: Geographical Perspectives on Religious Fundamentalism* (Lanham: Rowman & Littlefield, 2000).

Roger W. Stump, *The Geography of Religion: Faith, Place, and Space* (Lanham: Rowman & Littlefield, 2008).

Nicholas Wade, *The Faith Instinct: How Religion Evolved and Why it Endures* (New York: Penguin, 2009).

WEB SITES

Ethnologue (www.ethnologue.org).

Sacred Sites: Places of Peace and Power (www.sacredsites.com).

Sacred Destinations (www.sacred-destinations.com).

SIL International (www.sil.org).

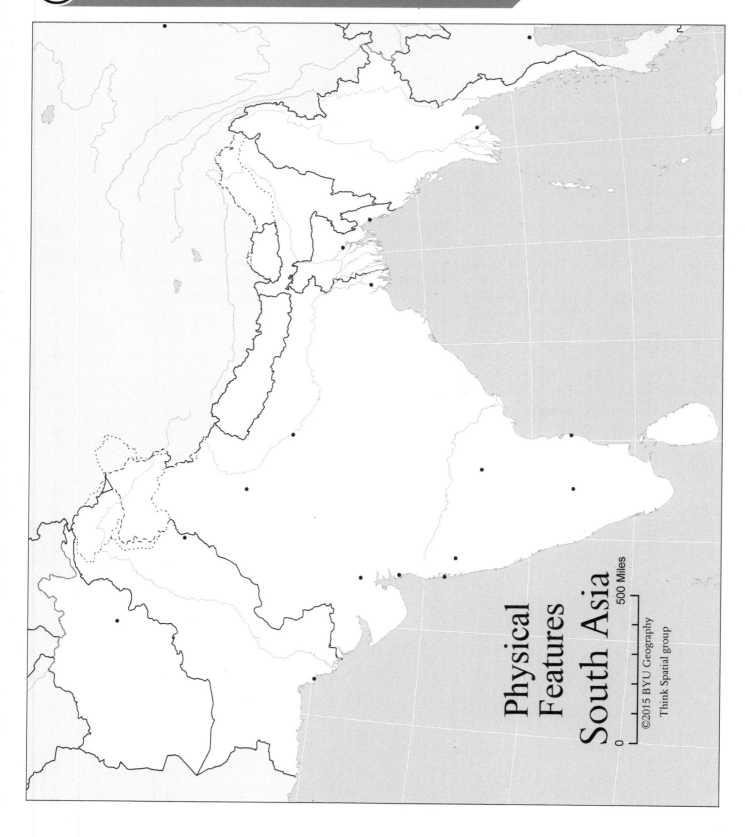

Physical
Features
South Asia

©2015 BYU Geography
Think Spatial group

0 _____ 500 Miles

Cultural
Features
South Asia

©2015 BYU Geography
Think Spatial group

0   500 Miles

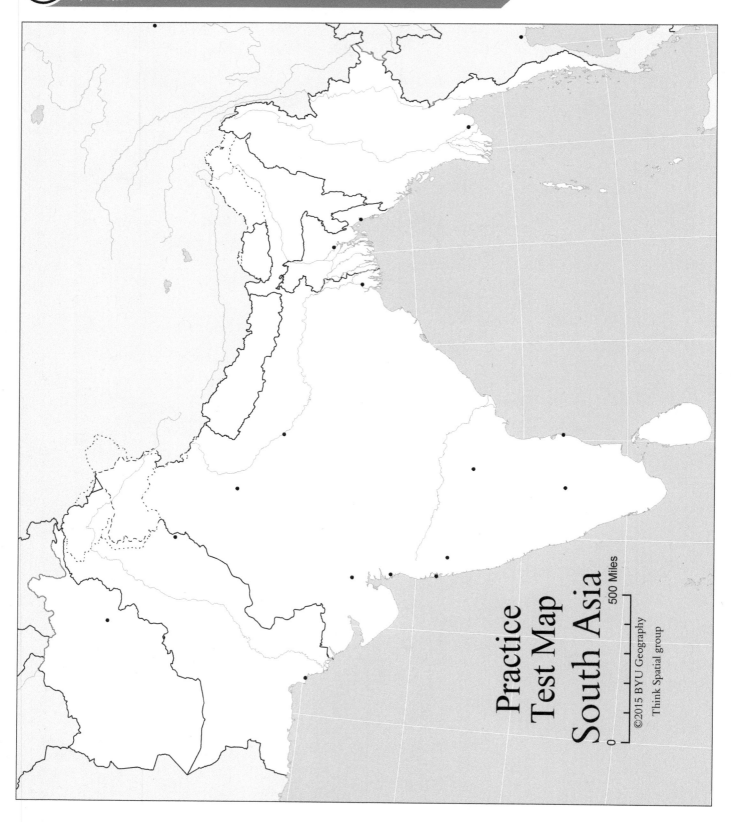

Practice
Test Map
South Asia

©2015 BYU Geography
Think Spatial group

0   500 Miles

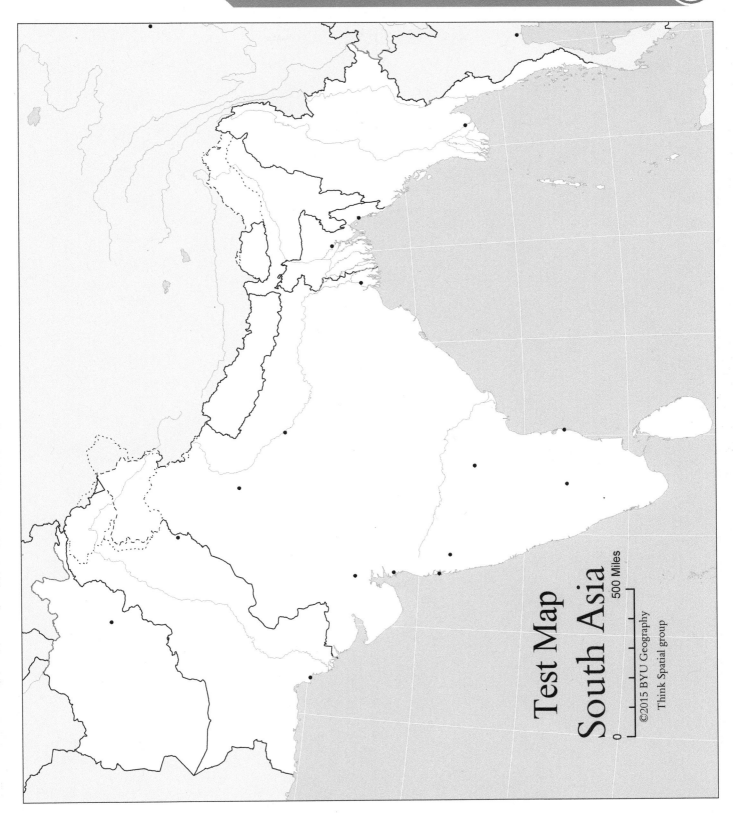

Test Map
South Asia

0   500 Miles

©2015 BYU Geography
Think Spatial group

Southeast Asia

11

And it is my prayer that your love may abound more and more, with knowledge and all discernment, so that you may approve what is excellent, and so be pure and blameless for the day of Christ, filled with the fruit of righteousness that comes through Jesus Christ, to the glory and praise of God. (Philippians 1:9–11)

CHAPTER OUTLINE:

→ Call Your Mommy!
→ Be Good
→ Cultural Mosaic

→ Pearl Power
→ Key Terms to Know
→ Study Questions

IF WE want to reach others with our hope we must share the faith we have been given. Even though the world offers numerous opportunities to share this hope, we must be on our guard to notice and take these opportunities to be the salt we see mentioned in Matthew 5:13. Paying close attention to living a pure life, one without hypocrisy, is instrumental in demonstrating genuine hope to the people of Southeast Asia, many of whom are dedicated Buddhists and Muslims. One thing is for sure; everyone knows when his or her food is salted. So will they notice you when you live a righteous life!

The key to understanding Southeast Asia is to see it as a bridge between East and South Asia. Like the rivers which flow rapidly in a southeastern direction from southern China, much of Southeast Asian tradition is a result of the diffusion of ethnicity and culture from East Asia. In both Buddhism and business, Southeast Asia has roots in China. Southeast Asia is of terrific importance to the United States in particular and to the world in general. A true mosaic of cultural influences can be seen in this vital region.

CALL YOUR MOMMY!

The map of Southeast Asia will demonstrate the key physical aspects that have molded a culture of patience and calm acceptance of the difficulties and joys of life. Positioned between the great and ancient cultures of China and India, this region consists of mostly water on a series of tectonic plates that have caused terrific ocean rifts. A recent tsunami in 2004 claimed hundreds of thousands of lives. Fierce tropical storms called typhoons and tsunamis are just some of the challenges facing an area that consists mainly of high terrain and volcanic cones. The tsunami gave the United States an incredible opportunity to demonstrate good will and its traditionally Christian nature because the United States was the only nation

with fleet assets nearby to provide medical care and emergency services. In an era in which China has been working hard to make cultural and economic inroads into this strategically vital region, the United States was well-served by its powerful effort to help these tsunami victims and their families with seaborne medical aid and supplies.

The climate is, of course, tropical, and the region is almost entirely within the tropic lines of latitude. As you might expect of the location, the rainforest (one of the largest in the world) prevails, and the monsoons bring plenty of rains to the area as discussed in the previous chapter.

So what are the aspects of culture that show an acceptance of the somewhat predictable fate associated with the regular seasonal monsoons?

BE GOOD

Cultural diffusion from India and East Asia is manifested in the Buddhism of the area. The Theravada branch of Buddhism, though different from the Buddhism of East Asia, is still grounded in the peaceful teachings of the great Buddha. Unlike the Mahayana Buddhists of the north, the monks of this sub-region with their bright clothing and colorful temples place a great emphasis on good works and are generally considered to be more concerned with achieving instant insights. The use of nuns and monks appears to suggest the idea that salvation or escape from the samsara cycle can be achieved by good deeds and personal growth. Angels of enlightened beings called bodhisattvas can guide the faithful in their attempts to achieve an escape from life. We would perhaps consider this to be similar to the workings of the Holy Spirit in our lives as Christians. It is as if the peacefulness of Buddhism acts a catalyst in transitioning the Islam being practiced in that area into a more relaxed and mellow form of the religion than many of us have been led to believe by listening to the news.

Evidence of Islam is obvious. Indonesia, the most populous Islamic nation in the world along with neighboring Malaya and Singapore, is comprised of a dynamic people whose religion seems to provide a great measure of stability. The sultanate of oil-rich Brunei on the island of Borneo demonstrates the syncretism of the area and the merging of different religions and traditions. So much history can be understood merely by observing the human imprint on the geography of a region,

Buddhist landscape.

Note the Oceanic influence here on the architectural landscape.

Image © Ekkachai, 2013. Used under license from Shutterstock, Inc.

U.S. Geological Survey map

The Sultanate of Brunei—the toponym reflects a Turkish past while the landscape synchronistically reveals Islam and Pacific influences.
Image © Donya Nedoman, 2013. Used under license from Shutterstock, Inc.

or the **cultural landscape**, and the names of places, or **toponyms**. For instance, the word "sultanate" itself is of Turkish or Central Asian influence. However, despite the influence of Islam, other religions have had an influence on the culture in this region.

CULTURAL MOSAIC

The monsoon winds are important to this area as they brought ships from India to Southeast Asia which resulted in the ubiquitousness of Hindu temples and the diffusion of other aspects of Indian culture. Today the Andaman Islands located on your map by the Andaman Sea are an Indian possession and remind the world of the important role India has played in the past as well as the present due both to trade and the diffusion of Islam. The Philippines to the west, a huge archipelago of over 7,000 islands is overwhelmingly Roman Catholic due to Spanish forays during the age of sail.

From *The Plaid Avenger's World #6 Nuclear Insecurity Edition*. Copyright © 2008, 2009, 2010, 2011, 2012 by John Boyer. Copyright © 2006 by Kendall Hunt Publishing Company. Reprinted by permission.

The largest population of Muslims is in Indonesia in southern Southeast Asia. This is fortunate for the United States for strategic and economic reasons. From a business standpoint, it is hard not to see the benefits of exporting of manufacturing jobs to Southeast Asia. Although Japan is probably the undisputed leader in transnational corporations globally, the rise of developing economies in the world seem to be creating a prospective boom in Southeast Asia. Numerous manufacturing firms and transnational corporations (TNCs) from developing countries have continued to move to Asia, and Southeast Asia is no exception. Of the world's fifty largest TNCs from developing countries, thirty-eight are located in Asia. As one might expect, a large number of these are in Southeast Asia including seven in Singapore alone.[1] The urban skyline of Southeast Asia with its famous Patrones towers in Kuala Lampur in Malaysia reflects the growing economic strength of Southeast Asia as well.

PEARL POWER

Perhaps of more immediate importance to the United States in these uncertain times is the strategic value of Southeast Asia militarily. When one considers the strategic importance of the Philippines with its over seven thousand islands and the legacy of American military involvement, one begins to see why Southeast Asia has become the nexus of interest for the United States, China, and India in recent years.

The United States originally became involved in the Philippines during the Spanish-American War. When one considers the strategic importance of the Philippines with its over seven thousand islands and the legacy of American military involvement, one begins to see why Southeast Asia has become the nexus of interest for the United States, China, and India in recent years. American military leadership fought against the Muslims in the southern island of Mindanao, and Special Forces troops are still there today. The importance of a secure Philippines to the United States goes back to earlier in the last century. With the rise of Japan in World War II, the northern Island of Luzon in the Philippines drew American armies like a magnet. Why? Similarly to Taiwan, the Philippines are of enormous importance because they form a launching point for trade and present a means of restraining China as well. General Douglas MacArthur likened the island of Luzon to an "unsinkable aircraft carrier." Today, another geographic location—the Straits of Malacca—is of vital importance to China as it pursues its "String of Pearls" strategy across the Indian Ocean.

The Petronas Towers in Kuala Lumpur, Malaysia.
Image © Shaun Robinson, 2013. Used under license from Shutterstock, Inc.

---

[1] United Nations Report on Trade and Development, "The Universe of the Largest Transnational Corporations." (2007). http://unctad.org/en/Docs/iteiia20072_en.pdf

The "String of Pearls" strategy is essentially a focused attempt by China to establish a series of ports for the purpose of transporting oil from the Persian Gulf where approximately one-fourth of the world's oil supply flows. To get to this oil, an enormous expenditure of effort has been focused on the Iranian Makran Coast and the island of Sri Lanka to the south of India. Perhaps most important of all has been the demonstrated need to keep the Strait of Malacca open and to prevent any foreign power from blocking the free flow of shipping through the strait and the adjacent micro nation-state of Singapore. Any obstruction to this vital area would certainly result in a near catastrophic situation for oil-hungry China. As such, Southeast Asia has seen an attempt by the Chinese to build a canal across the Isthmus of Kra in Burma. This attempt to build a canal has been compared by Robert Kaplan to America's creation of the Panama Canal, an act presaging a great superpower. Without question, from a Chinese standpoint alone, Southeast Asia will be of vital strategic importance. This makes this business intense climate equally important to the United States who is greatly interested in supporting India.

KEY TERMS TO KNOW

Cultural Landscape   Toponyms   Tsunami
Theravada

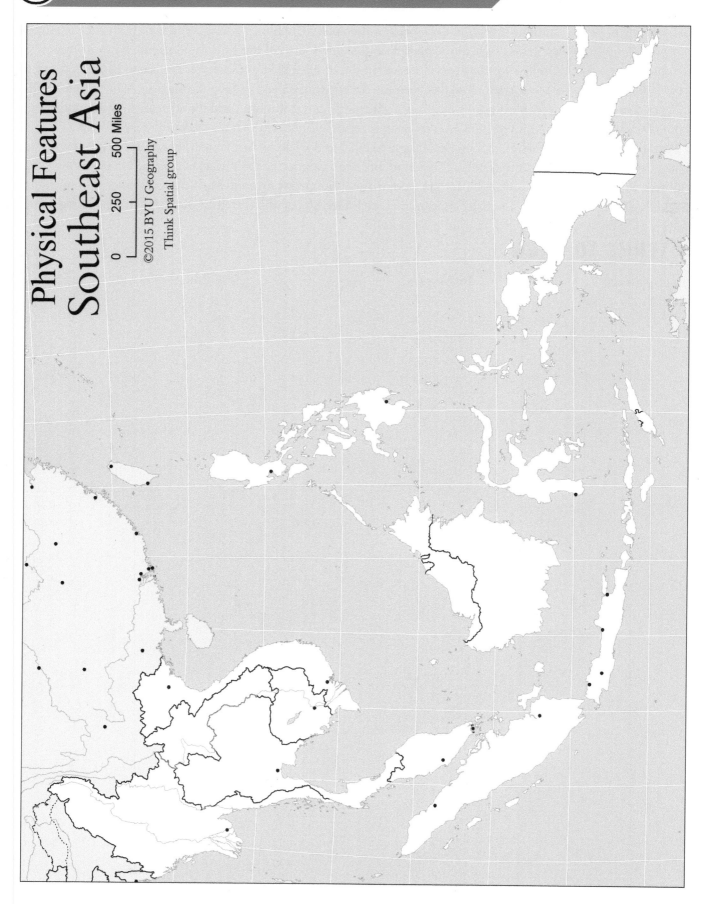

Physical Features
Southeast Asia

0   250   500 Miles

©2015 BYU Geography
Think Spatial group

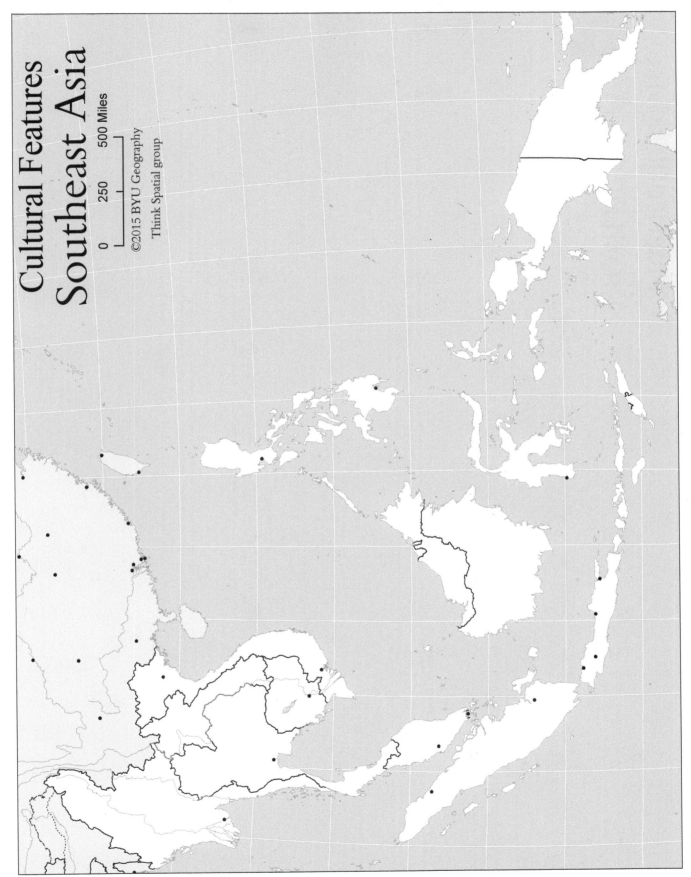

Cultural Features
Southeast Asia

500 Miles

250

0

©2015 BYU Geography
Think Spatial group

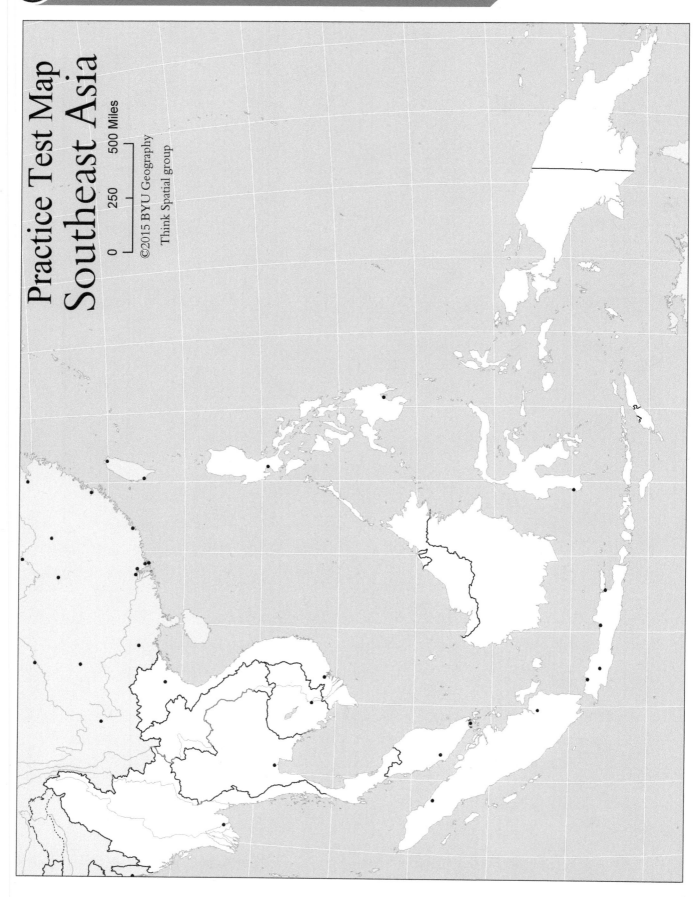

Practice Test Map
Southeast Asia

0   250   500 Miles

©2015 BYU Geography
Think Spatial group

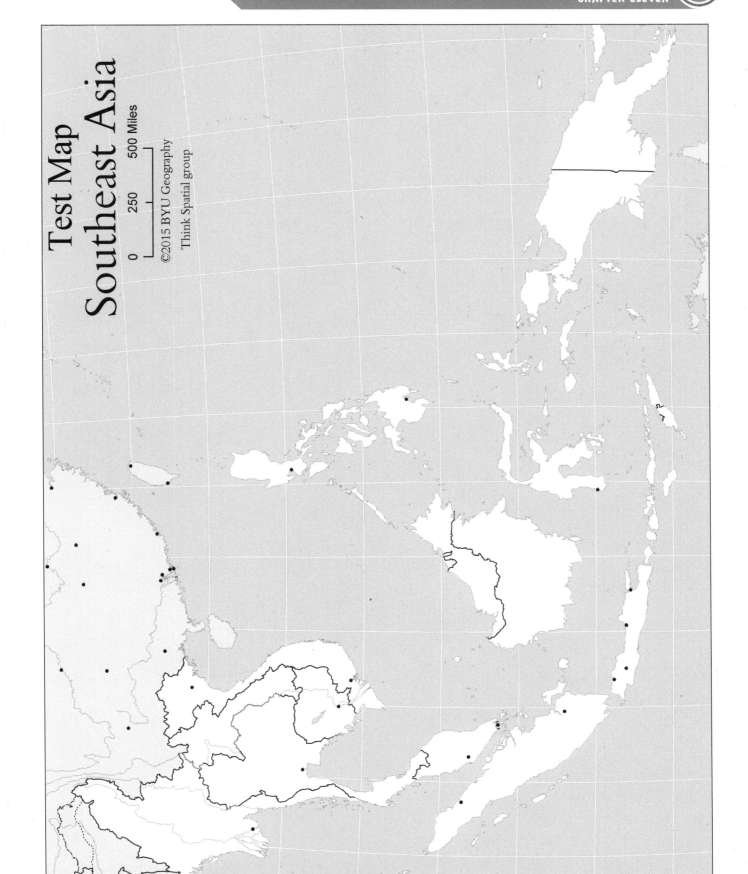

Test Map
Southeast Asia

0   250   500 Miles

©2015 BYU Geography

Think Spatial group

Central Asia

(12)

CHAPTER OUTLINE:

→ Water Please
→ Going Green
→ Floating on By

→ It's Too Oily!
→ Study Questions

EVEN though this area does not receive much rain, the people of Central Asia can still be prosperous by rooting themselves in the Lord. He will provide for those who trust Him. These nations are troubled by many powerful neighbors, but worshipping God will dissolve the violence. When Jesus spoke to those around him, he was able to reach them where they were. We likewise can understand a culture if we honestly attempt to understand it for the purpose of sharing our hope. Central Asia is, indeed, a thirsty land where water is needed and the agricultural aspects of Jesus' word can reach fertile ground!

WATER PLEASE

A common theme you may have noticed throughout this book is how physical geography appears to shape culture. How God created the physical world has without question made a mark on human societies. For this reason the student of geography must be simply amazed at how important the amount of rainfall in the region we call Central Asia has impacted the world. While much of this region only receives about eight inches of rain per year, the impact on livestock has caused dramatic movements throughout history affecting empires from the ancient times to the present. In particular, the Turkish Empire represented in some ways the apogee of a Muslim insurgence into Europe until circumstances enabled the Europeans to emerge as the preeminent power on Earth. Today we see this region again becoming extremely important in terms of oil.

Moving to the east away from the Caspian Sea, you will see the states of Turkmenistan and Uzbekistan and to the far north Kazakhstan. The small mountainous states of Kyrgyzstan and Tajikistan stand between China's ambitions and one of the world's great oil supplies. Rivaling the Persian Gulf oil fields and the oil fields of North Africa are those of the Caspian

Sea basin. As important as these oil fields are, so are the pipelines they are connected to. It is important to understand how and why the Western powers gained advantages over the peoples of Central Asia in terms of business, strategy, and the technology to tap into oil reserves in order to comprehend what the future might hold there.

Another area we will touch upon in this chapter is the region known as the Trans-Caucasus area between the Black Sea and the Caspian Sea. Here the nations of Georgia, Armenia and Azerbaijan stand valiantly against growing encroachments from Russia to the north. As mentioned before, geography is a highly subjective discipline and the definition for a region is a loose one. It is felt that past associations with the former Soviet Union have given these regions a flavor somewhat peripheral to the nation-state of Russia today.

GOING GREEN

Over time, several nomadic peoples dependent upon the thin belt of green grasses in the Central Asian steppes engaged in various movements over time on the basis of their needs for increased fodder for their horses and livestock. The rises and declines of the ancient civilizations was a function of trade and the ability to exchange with the Chinese. The attempts to exploit the manufacturing capabilities and to receive the items created in the Far East led the Egyptians and Mediterranean powers of Greece and Rome to attempt to traverse the Red Sea and the Persian Gulf for the purposes of bypassing the land "Silk Road" used for trade. The Romans, in fact, may well have weakened themselves in their attempts to control Babylonia, or present day Iraq, in various military operations possibly intended to keep the supply routes through the Euphrates and Tigris Rivers open. The great empires of the East and West depended upon mutual trade on the Silk Road of Central Asia. These vital trade routes were vulnerable to the depredations of various Central Asians throughout history. These various Central Asian empires, such as those of the Huns and Mongols, expanded periodically probably as a result of changing patterns of climate and weather. Changing climatic conditions may have forced these nomadic peoples to move towards or away from the threat of the

Image © Oleg Golovnev, 2013. Used under license from Shutterstock, Inc.

great empires of the East and West. Many groups emerged from Central Asia to dramatically crash into the various civilizations throughout history: the Huns, Mongols, and Avars etc. The one group we want to focus on is the Turks. Grouped on the basis of perceived similarities in language, the Turks, at various times, nearly destroyed late-medieval Europe and proved instrumental in taking over the Central Asian area. Why they didn't succeed and how the West gained the advantage in technology can be viewed geographically in terms of land and sea. Just as the water is necessary to sustain life on the Steppes, the need for moisture affects empires in faraway lands as well.

FLOATING ON BY

The water instrumental in marking the apogee of the Turkish Empire is salt water. The maritime Battle of Lepanto between the Holy League and Ottoman, or Turkish, Empire demonstrated the strong desire of Central Asian empires for land. This covetousness brought the merchant class into conflict with the military class known as the Janissaries toward the seventeenth century. The Janissaries grew in strength as the state relied increasingly upon them for loyalty. Initially called the sipahis, or military nobles, they were a landowning class and represented a type of feudal power. When the Turks were unable to acquire more land, however, the sultans were unable to procure additional loyalty by the dispensing of property. In other words, the feudal system was perpetuated.[1] At the Battle of Lepanto, the Turkish commanders demonstrated

---

[1] Getz, Trevor R. and Streets-Salter, Heather. "Modern Imperialism and Colonialism: A Global Perspective" (Boston: Prentice Hall, 2011), 122.

the increasingly rigid, or ossifying, nature of the Turkish civilization when they claimed their bows were superior to the arquebuses of the infidels.[2] This reluctance to explore new ideas also reflected in the lack of support for a middle-class of merchants, or creators of capital. So, while the Turks lost the Battle of Lepanto, more importantly they ceded a waterway from the Mediterranean and across the Indian Ocean preferring instead the conservative leadership of the Janissaries. While the West developed trade with East Asia and the development of capital ensued and with it the opportunity for profits and upward social mobility, the Turks would begin a slow road to relative insignificance in terms of technology.[3] By the end of the seventeenth century, the Industrial Revolution would be underway in Great Britain where it would diffuse to the continent.

IT'S TOO OILY!

Today the West is quite interested in the vital oil fields of the areas around the Caspian Sea. The pipelines for this oil in many ways represent a type of circulatory system for the greater core, or industrialized regions of the world. One very important oil pipeline today crosses Georgia and in part explains Russia's 2008 decision to intervene militarily into the country. It appears the Russian decision to enter northern Georgia translated into an ability to reach one of the few oil pipelines from the Caspian region that did not go through Russian held territory. Now that Georgia is within range of Russian artillery, Russia has been able to monopolize this vital area similarly to how it has isolated oil-producing nations, such as Syria today, from Europe. Georgia is a Christian nation where the Georgian Orthodox Church is the predominant faith.

The Caucasus Mountains, one of Russia's few natural land borders.
Image © My Good Images, 2013. Used under license from Shutterstock, Inc.

To the south is Armenia, a nation still in control of part of Jerusalem, at least administratively. Armenia was at one time a bulwark for the Christian faith against advancing Islam. To the east we come to Muslim Azerbaijan, which has the vital break-bulk point of Baku, which oversees much of the important oil reserves from the aforementioned areas around the Caspian Sea. The nations of Azerbaijan, Georgia, and Armenia have demonstrated throughout history a propensity to avoid the orthodoxy of the adjacent great powers attempting to dominate them. The reason is simply mountains! Towering over Georgia, Armenia, and Azerbaijan are the mighty Caucasus Mountains. This natural border between Russia and the former empires of Turkey and Iran seem to be a psychological border for the Russian peoples who for the most part live in a relatively flat open area.

Due to a tendency to utilize the Aral Sea as a valuable resource for fresh water for irrigation in neighboring Kazakhstan and Uzbekistan, the stage has been set for an environmental disaster of titanic proportions. Any time humans alter the landscape, they run the potential risk of damaging the environment and precautions should always be taken. Good stewardship will remain a challenge for the various emerging economies and nations of Asia.

[2] Rodgers, W. L., "Naval Warfare Under Oars: 4th to 16th Centuries" (Annapolis, Maryland: Naval Institute Press, 1967), 187.
[3] Kaplan, Robert D., "Monsoon" (New York: Random House, 2010), 50.

Australia: The Consequences of Hope from Penal Colony to Great Nation

13

So that we may no longer be children, tossed to and fro by the waves and carried about by every wind of doctrine, by human cunning, by craftiness in deceitful schemes. Rather, speaking the truth in love, we are to grow up in every way into him who is the head, into Christ, from whom the whole body, joined and held together by every joint with which it is equipped, when each part is working properly, makes the body grow so that it builds itself up in love. (Ephesians 4:14–16)

CHAPTER OUTLINE:

→ Outback, Mate
→ From Prison to Production
→ I Will Be a Bilby
→ Drawing the Animal Line
→ Stewardship
→ No Fear Here!
→ Population Theory
→ Australia and New Zealand
→ Let's Go for a Walkabout

→ America, Junior
→ Awesome Ally
→ A Word on Their Weird Way to Wealth
→ Pacific Shift
→ No Worries
→ A Sinister Show and Tell
→ Australia Rundown & Resources
→ Key Terms to Know
→ Study Questions

ALMOST everyone wants to go to Australia! It is such a wonderfully exotic continent and so far away. Despite the terrifically unique land that is Australia and New Zealand, this region faces obvious challenges through its environment and not quite so obvious challenges through immigration and social changes. Like the United States, Australia is learning from its past, and a large part of the wisdom that comes from any lesson learned is to know that God is ultimately going to judge our actions harshly if we are not committed to our faith in His Son's purpose. You can never escape the love and authority of God, and you can't travel farther from the United States to do His work than Australia! Remember Job 28:24 (NKJV). For He looks to the ends of the earth, *And* sees under the whole heavens.

From *Blue Marble: Next Generation* image produced by Reto Stockli, NASA Earth Observatory (NASA Goddard Space Flight Center)

Note the coastal locations of Australian and New Zealand cities.
NASA/Goddard Space Flight Center Scientific Visualization Studio

Australia is three things: an island, a nation, and a continent. This is a good time to review some basic terms. An **island** is a body of land surrounded on all sides by water. A **nation-state** consists of two terms: a **state** and a **nation**. A nation is a location that has a shared history and culture. This is a more human concept, whereas a state has defined borders recognized both within, by the citizenry, and without, by other nations. As you recall, a **citizen** is a member of the state.

OUTBACK, MATE

Regarding physical geography, Australia is a **continent**. It consists of its own tectonic plate, unlike other huge islands like Greenland. Within this continent is a diversity of topography that interacts with the southern currents to create a unique watershed called the Murray Darling watershed in the South Australian state of New South Wales. The vast interior consists of various colors—most noticeably red. The Japanese and Chinese covetously harvest the mineral-rich interior. To the north are the crocodile strewn tropical swamps, to the south is the cooler island of Tasmania, and to the southeast are the major islands of New Zealand. Australia is tangential to the Great Barrier Reef, perhaps the greatest remaining coral reef popularized in a recent children's movie. This brings us to the colorful origins of Australia.

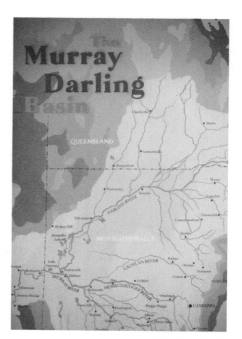

© Ashley Cooper/Corbis

FROM PRISON TO PRODUCTION

Niall Ferguson says in the book *Empire,* "Small wonder the early settlement of Australia required compulsion."[1] When one considers how far Australia is from the rest of the world, one realizes the very remoteness of this wonderful continent. Even with high-speed jet aircraft, it will take about fifteen hours to go from Los Angeles, California to Melbourne, Australia. So why did the early pioneers come to Australia? As Mr. Ferguson points out, many of the early people came to Australia in an attempt to avoid floggings and branding as punishment for crimes. Ferguson again points out in his book the remarkable differences between the origins

---

[1] Niall Ferguson, *Empire.* (Basic Books: New York, 2002), 84.

of America and Australia. America began as an experimental Puritan Utopia and plantation system in North America but became a rebellious republic, whereas Australia, initially colonized at Botany Bay by prisoners that mostly consisted of shoplifters exiled into a penal colony ultimately became a loyal British colony.[2]

I WILL BE A BILBY

Perhaps the things that amazes us most about Australia are the wonderful wildlife and plants. The biomes here are truly unique. Everyone knows about kangaroos, but have you ever seen a wombat or a bilby? The duck-billed platypus epitomizes the strangeness of the creatures found in Australia: this mammal lays eggs and has a beak like a bird, and yet is fur-bearing and possesses reptilian-like poison glands on the rear legs capable of delivering a nasty sting! But the uniqueness doesn't stop here. Even the snakes are different. Most of the snakes are venomous, which is at odds with our experience. There are no boas, just pythons. The turtles are side-necked not like ours in North America, and the lizards are totally different as well. Instead of monkeys or apes, we see the koala bear in the mighty eucalyptus forests. These stately trees like columns holding up the sky are similar in size to our American redwood trees.

The imagination of the Creator's handicraft is evident in this Bilby!

Image © IgorGolovniov, 2013. Used under license from Shutterstock, Inc.

DRAWING THE ANIMAL LINE

Why is the wildlife so different in Australia? This is a good question and one for which there is no easy answer. Wallace's Line is a biogeographic line that reflects the diverging distribution of types of animals, such as placental and marsupial mammals, and the possibly varying sea levels over time. It is also associated with tectonic plate activity. The Line runs generally from southwest to northeast between the islands of Sulawesi and Borneo in Indonesia. Above the line, one finds a plethora of mammalian orders. Below the line, we find marsupials, or mammals that tend to give birth to premature young, which must transfer into a pouch to be carried by their mother. Biogeography aside, Australia is a fascinating place from a human perspective as well.

STEWARDSHIP

Increased travel of people to areas, such as Australia, that were previously ignored due to distance is an unfortunate aspect of globalization that results in the disappearance of unique languages and cultures and damage to the environment. Australia has indeed faced a plethora of natural disasters ranging from the famous fences to contain rabbits or the mice infestations. The incredible damage of invasive species is a frightening aspect of the curse on the planet. Australia's ecosystems seem particularly vulnerable because of the unique niches, or positions, various animals occupy. In addition to the toll increased travel takes on the environment is the abundance of technological devices that quickly alter the landscape such as chainsaws. The lumber industry actually relies on helicopters to lower lumberjacks into forests where they can then cut themselves to nearby roads. Granted, mankind has been given stewardship over the earth, but better stewardship will be required in years ahead by you and other committed Christians. The non-Christian world seems to feel we believe we are to exercise domination over the Earth when, in fact, we have been granted dominion. Just as we are to be good stewards of our bodies, we, likewise, must be considerate of our environment.

---

[2] Ibid., 85.

NO FEAR HERE!

One of the oldest tricks in the world is for a religious class in conjunction with a political leader to use the element of fear as a means of controlling a population. Whether it is the Aztecs with their fear of the sun burning out, or the fear of a celestial object hitting the Earth, we must remember our Lord told us not to worry. We are not given to a spirit of fear, and therefore we can be better stewards of those things we have been given responsibility for because we do not feel the need to hoard or consume more resources than we need. Concerns of global warming, or the proliferation of **chlorofluorocarbons** believed to be responsible for the hole in the **ozone layer**, may or may not be caused by humans beginning in the Industrial Revolution. Regardless, we can study these phenomena and search for solutions not out of a spirit of fear but because we have been granted a spirit of confidence through the peace we have in a loving God who created the earth and sustains it!

The study of climate change often depends on such diverse sources as wine keeper's diaries and the fascinating study of dendrochronology (tree rings) to observe changes in moisture and temperature through history. Many environmental geographers even believe changing climate patterns have been instrumental in the expansion and movement of various nomads such as we mentioned in the earlier chapter of the Middle East. Again, we see how various elements of human culture are, in fact, a function of physical geography.

There are increasingly too many people and not enough resources to go around—not only food resources, but other resources such as fuel and clean water. In the core and parts of the semi-periphery the "problems" are quite different. These societies are in either Stage 3 or Stage 4 of the demographic transition, with low birth and fertility rates and increasingly "older" populations. While the advanced post-industrial economies of the core would be able to cope with larger populations, they are precisely the places where rates of natural increase are the lowest. Here, the most pressing issues related to population involve questions of how to cope with an aging population in which more and more older people who are not working must be supported by fewer and fewer people of working age. This is an especially significant problem in core societies with substantial social welfare systems in which governments are in charge of funding retirement and pension plans.

Different societies are approaching such problems in different ways, with varying results. In India, for example, the population is now over 1 billion and is growing quite rapidly, at just under 2% per annum. In the 1970s India's federal government attempted to take an active role in population reduction by opening family planning clinics, dispersing contraceptives, and appealing to the patriotism of the population through public relations campaigns and advertisements. These efforts, however, have been met with much public resistance because they do not dovetail well with traditional Indian ideas about reproduction and the family. The results of the government's efforts have been mixed at best and India's population continues to grow rapidly. Another example of governmental intervention in population growth is China. In the early 1980s the Chinese government, a very powerful one-party system, took an active and rather forceful role in population reduction. Laws were enacted that gave tax breaks and other incentives to couples who chose to have no children. A one-child-only policy was also enacted and rigidly enforced that limited each couple in the country to only having one child. The government also used public relations campaigns and advertisements to appeal to the patriotism of its citizenry. The results of such policies were quite different than those in India. In 1970 the rate of natural increase was 2.4%, but it had dropped to 1.2% by 1983 and 1.0% in 1997. This success, however, came with some significant social costs. For example, a heavy male gender imbalance now exists in China as a result of an increase in abortions of female fetuses due to traditional Chinese ideals with respect to inheritance.

POPULATION THEORY

The issue of population has attracted the attention of large numbers of writers and social scientists over the past two centuries. Probably the most famous of these was the English writer **Thomas Malthus**, whose *Essay on the Principle of Population* (1798) set in motion a long-running debate regarding population growth that continues to this day. Malthus was writing at a time when England was in Stage 2 of the demographic transition and was experiencing exponential population growth. Malthus argued that while population was growing exponentially, food supplies were only growing arithmetically, and therefore would not be able to keep up with the demand for food. This, he wrote, would at some point result in a crisis punctuated by famine and social collapse.

Obviously, the crisis that Malthus predicted did not come to fruition, for he failed to predict new agricultural technologies and techniques that revolutionized agriculture in the 19th and 20th centuries. These new technologies (such as crop rotation schemes, irrigation technologies, and scientific genetic hybrids) greatly increased the amount of food that could be produced, even in some of the poorest countries in the world. Malthus' failings attracted many critics. Marxist thinkers, for example, argue that the real problem facing the world is not overpopulation, but the fact that the world's resources are not equally shared or distributed and are co-opted by the capitalist class. Another critic, Esther Boserup, has argued that population growth does not *necessarily* produce significant problems; it could in fact stimulate economic growth and better food production technology as it did in Europe in previous centuries. But populations in peripheral countries are increasing at unprecedented rates and many of these countries have more poor people than ever even though food production has in general increased substantially over the past few decades. These alarming trends have caused some experts to reevaluate Malthus' theory, taking into account not just food, but a variety of other natural resources. These so-called Neo-Malthusians argue that Malthus erred in the sense that he wrote only of food and not other natural resources but that his overall idea was correct. They contend that population growth in the developing world is a very real and very serious problem because the billions of very poor people that will be added to the world's population in the coming centuries will result in a ever-increasing desperate search for food and natural resources punctuated by more wars, civil strife, pollution, and environmental degradation.

Children are a blessing not a burden. Despite calamities we are reminded in 2 Timothy 1:7, "For God hath not given us the spirit of fear; but of power, and of love, and of a sound mind." We must also be confident in facing the present opportunities we have to reach the world for the purposes of helping children with our talents and treasure. As His obedient children, we may model His love demonstrated for us on the cross.

AUSTRALIA AND NEW ZEALAND*

G'DAY, MATES!

We're heading south from Japan to our friends in the Australia and New Zealand region. When the Plaid Avenger is traveling around the world and can't make it back home to America, I just stop off in Australia, or as I like to call it, the **Mini-America of the southern hemisphere**. If you ever want to fire up people from Australia, just pass that along. It infuriates them for some reason. They tell the Plaid Avenger that they're distinctly Australian and unique in their own way. Yeah, right! You dudes and dudettes are exactly the same as us: albeit an European and American 21st-century cultural-amalgamation-sensation south of the Equator!

The Plaid Avenger is always straight shooting, so I am not going to lie to you; this region is not one of tremendous significance on the world stage. But like Japan, it does play its part to effect current events and regional activities. Its primary role right now is as a platform of European culture and US foreign policy in the southern hemisphere. Even that is changing quickly of late as the mates from "down under" are shifting their political and economic focus to be more in tune with their Asian 'hood. But I am getting ahead of myself. Let's first focus our attention on the physical traits of the region.

LET'S GO FOR A WALKABOUT

Physically, Australia and New Zealand are two island countries, even though Australia is classified as a continent. If you look at it, it's really just a big island. They are two islands of extremes. In Australia, the vast majority of the interior is desert and steppe. It's very similar to the American Midwest. There's a bunch of cattle ranching going on; it's the outback, after

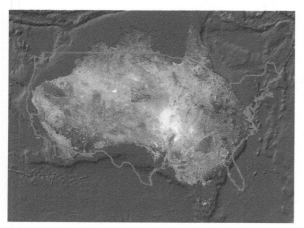

Area comparison.

Lights out in the middle!

Deserted desert interior in Australia.

all, but there aren't too many people there. In fact, a massive percentage of the interior of Australia is so dry, it has promoted a primarily coastal settlement pattern. Everybody's hanging out on the coast and the vast majority of Australia's population is located on these coastal margins, particularly the southeastern coastal margin.

As you can see from the lights at night, virtually everybody is on the east coast of Australia. Before we go any further, let me begin pointing out the parallels between Australia and America. Here is one outlined for you on the map: population primarily in coastal areas, with a high concentration on the eastern seaboard. In New Zealand, we have the same pattern, but for different reasons. New Zealand, much like Japan, is a mountainous country. In fact, Japan and New Zealand are almost identical. They are both island nations, made of several different islands, and both are on the borders of major continental plates; both are volcanic in origin, so the interior of the country is very mountainous. The terrain makes it a no-brainer for most people to live on the coast, especially those little hobbits which seem to be over-running the country here lately. Gandalf, help us!

Climatically speaking, these two countries are opposites. New Zealand is cooler and wetter, with a climate more like the Pacific Northwest section of the US, or even like the UK. Australia has some Mediterranean style climates on the southern coastlines, and even a little tropic savanna-type areas in the north, but the interior is largely steppe and desert; a dry area with limited rainfall, few trees, and scrub and short grass vegetation. What are those climates good for? Cattle and sheep production. There is a lot of grazing action going on in Australia. Big exports of wool, lamb, and beef.

One last physical note: what Australia lacks climatically, it more than makes up for in physical resources. They have tons of coal, iron, copper, opal, zinc, and uranium. Put that together with the agricultural commodities, and you have a serious export base to work with. Think about that for a second; what types of countries mainly export primary products? Hold that thought, we'll be back to it soon enough.

New Zealand: mountainous center spine.

AMERICA, JUNIOR

Why does the Plaid Avenger refer to this region as the mini-America of the south? Where do many people from America who like to travel abroad want to go? Australia! Why? Because it's just like home! People want to see the same stuff, eat the same stuff, and speak the same language as in the US, because they are more comfortable in that setting. It's far enough away that you think you're actually being adventurous! There's actually something to the idea of Australia just being a transplanted southern piece of America or Europe, but just "down under." Down under the equator, that is.

Australia is quite unique in the southern hemisphere of this planet for a couple of reasons. For one, it's the only region in the southern hemisphere that is habitated primarily by white, English-speaking, European descent-type people. It's also the only region south of the equator that is totally rich and developed. It has standards of living and material wealth on par with that of Western Europe, Japan, and the US, and no other region south of the equator can say that, yet. The standard of living thing is another reason why Australia is the Mini-America of the south. But oh, there's so much more.

BACKGROUND

Why is Australia so white? Because it was colonized by the pasty British white peoples, of course! Both the US and Australia were originally British **convict colonies**. In the UK back in the day (in the 1600's and 1700's), one of the subsidiary purposes of Britain's colonies was to get rid of British criminals. Britain was perhaps slightly overpopulated, and laws were very strict. The historical anecdote that explains this fairly well is the establishment of the **baker's dozen**. The baker's dozen is 13 instead of 12. Why? There was actually a law that stated if a baker shorted you a donut, he could be put in jail. So the baker's dozen was born. Throw in an extra one, because who the hell wants to go to jail over a donut?

This was a pretty legally strict society where lots of people got tossed into jail for sometimes menial offenses. Among the options given to convicts were the gallows or a boat ride to a colony. Guess which one people chose? You know there is a 'criminal' element to America's background, and the same goes for Australia. Here's the funny part: the British used to ship their convicts to their American colony, so when did they start shipping them to Australia? Right around the 1780s. Why is that? In 1776, America had a revolution, and we kicked those British tea-sipping, prisoner-exporters out, eliminating the US as their felon-absorbing resource. The British, who were floating around the world, figured out that they had claimed a huge chunk of land down in the southern hemisphere. Look at that, a place to send convicts! Australia was established in 1788 as a convict colony, in part because the British couldn't send them to America anymore.

SOCIETAL STRUCTURE

How else is Australia like America? Let me count the ways. If we look at virtually any facet of life in this part of the world, we can look at America and see that it's pretty much the same. Let's start listing them off, shall we? Australia is a highly **urbanized** society. There are roughly identical urbanization rates between Australia and the US. This point is of significance because everybody has the opposite impression about Australia. Everybody thinks they're all about the outback, the Crocodile Hunter, and Crocodile Dundee. There are all these wild and crazy guys down there! However, if you ask any Australian how much time they spend in the outback, "You've got to be kidding me, mate," is a likely response.

Why is that? Quite simply, the outback sucks. It's crazy that anybody would want to go to the outback. It has all kinds of dangerous creatures in it, spiders that will jump onto your face, the top ten poisonous snakes in the world, and so on. Nobody wants

Peeps are largely coastal urbanites.

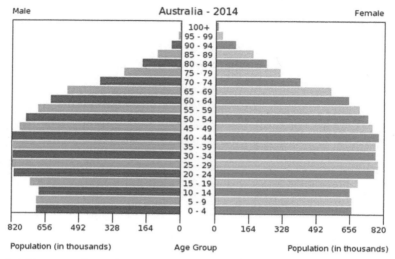

Male · Australia - 2014 · Female

| 100+ |
| 95 - 99 |
| 90 - 94 |
| 85 - 89 |
| 80 - 84 |
| 75 - 79 |
| 70 - 74 |
| 65 - 69 |
| 60 - 64 |
| 55 - 59 |
| 50 - 54 |
| 45 - 49 |
| 40 - 44 |
| 35 - 39 |
| 30 - 34 |
| 25 - 29 |
| 20 - 24 |
| 15 - 19 |
| 10 - 14 |
| 5 - 9 |
| 0 - 4 |

820 656 492 328 164 0   0 164 328 492 656 820

Population (in thousands)   Age Group   Population (in thousands)

A stable yet small population.

Lambs! How cute.

And how delicious!

to go out in "the bush." It is a classic mythology of Australia. It sounds cool, makes them seem all brave and manly, but nobody hangs out in the interior of Australia. Most people live in cities and suburbs, which you are all probably familiar with. They are all content to stay out of the outback, and have the nice little picket fence around their manicured suburban lawns.

And this relates to the standard of living. This is a very rich place. Per capita, it is on par with the US and Western Europe and any other rich place. Standards of living, levels of technology, material wealth: this region is staunchly in the 'fully developed' category. People live there just like they live in the US, and have all of the same creature comforts. This reiterates why nobody is out in the wild, because rich people go and see the wild when they're on vacation, and then they get the heck out of there because the wild sucks.

One of the reasons why they have a high standard of living: there are not a whole lot of people living in Australia. Australia and New Zealand populations are extremely low, only about 28 million folks across this entire region . . . 23 mil in Aussie-land, and less than 5 mil in NZ. New York City, L.A. and Tokyo each have bigger populations than the entire continent of Australia! Look at the population pyramid! As stable as stable gets. Only through immigration will the population of the region increase, much like the US. But there's more!

Australia also has similar dietary habits, which ties back in with standards of living. They are big meat consumers, just like in the US. The only big difference is that the Americans like to slay the cows and eat hamburgers, but the Australians are all about lamb. They love to silence the lambs. And then eat them . . .

They are also big beer and wine drinkers, like the US. In most places that were colonized by Western Europeans, the beer and wine tradition still exists. Australia is one of the booming wine regions now, soon to be the number one wine exporter on the planet.

Talking about standards of living and beer and meat consumption. Let's get back on point. What other parallels are there?

This may just seem like a funny side note, but this may play a part in why the two nations are so similar today. Both countries have a criminal background, a criminal profile; remember, they started as convict colonies. How has that played

American Cowboy.

Australian Cowboy.

See the difference?

into today's society? This is a stretch, but stay with me. What do people like to do in both of these societies? We've already pointed out the beer drinking thing: both are beer drinking societies, and both like to party. What else do they like to do while consuming alcoholic beverages? I don't know—watch brutal sports? Yes! The Americans have their football, Australians have rugby. And the Aussies are even more hardcore than the US: they take the pads off and beat the crap out of each other, while everyone who isn't playing drinks beer and watches.

On a more serious note, each country has a rugged, individualistic archetypal character. The Americans have the Marlboro man, a lone, rugged warrior out on the plains, wrangling cows and looking out for coyotes. That's part of the common symbology, the perception of what it is to be an American. He embodies individualistic pride. They've got the same stuff down in Australia, except they wrangle crocs and look out for dingoes. Do these types of characters represent the majority of either place? No, but these characters are popular symbols in this society. There really aren't these archetypal figures anywhere else in the world. It may be partially due to the fact that both places have large expanses of land with low population density. They have frontiers, and there has to be the rugged individual to explore those frontiers.

Here's another similarity between Australia and the US: the settling convicts exterminated nearly the entirety of their respective native populations. There is a history of repression of native peoples in both places. In the US, it was the Native Americans, the Amerindians. In Australia they had the Aborigines; New Zealand have the Maori. All of those peoples suffered immensely in the onslaught of British colonization in exactly the same ways. And it was particularly bad in Australia.

In Tasmania, an island off of the south coast, there were some indigenous people called the Tasmanians who were openly hunted during a period of the settlement of Australia. The government essentially said, "Hey, settlers, want some free land? Go down to Tasmania! If the natives give you any trouble, don't feel bad about killing them." It became a sport to go out and hunt the indigenous Tasmanians. This was done so efficiently that they killed all of them. Every. Last. One. That's the criminal background manifesting itself in one of the nastier ways in history.

Another historical consideration of why both societies evolved out this frontiersman mythology has to do with how both countries grew. Both started on the eastern seaboard of a large uncharted land mass, and proceeded to expand and conquer the continents in a continuous westward expansion. Interesting, isn't it? Australia even underwent a western "Gold Rush" at almost the exact same time as the famous Californian rush of the same name, but with one slight difference: theirs was more profitable!

Brutal sports, slaughter of indigenous populations, mostly white, cowboy image, and even the English language all play into the cultural baggage of the US and Australia. This is why the Plaid Avenger refers to it as the Mini-America of the south.

Put all these historical, physical, and cultural factors together and you are starting to get a sense of why the Plaid Avenger draws so many parallels between the US and Australians region. But you don't have to take my word for it! There is proof a'plenty that these two states have a tight relationship: let's just take a look at their foreign policies for a minute.

AWESOME ALLY

When it comes to Team West foreign policy, one could not find a better friend than those Aussie mates. They have been there year in and year out for their western allies, for every single international conflict and conflagration. We can be even more specific: first, Australia has always supported every endeavor of the old British motherland, and then later came on board for all things American . This bunch really sticks together.

The 'Roo Brigade will always be ready to serve freedom!

I'm not talking about supporting them on paper, or in a UN vote, or economically. I'm talking about Australians participating side by side with their English-speaking brothers-in-arms in World War I, World War II, the Korean War, the Vietnam War, the Cold War, the War on Terror, which has spawned the two newest Aussie supported wars: the campaign in Afghanistan and the war in Iraq. Every conflict, every time, they are there for the West. Check out one of Mel Gibson's first starring roles in a movie named Gallipoli, about Aussie troops fighting the Turks way back in WWI.

Being a small country with a small population, the numbers of Australian forces that have participated in these conflicts is not a huge number, but it's the thought that counts, especially when the thought is a bunch of dudes with submachine guns. But I digress. Point is, that they have regularly been looked to as a vote of support for western ideology worldwide; most currently, counting themselves as one of the "Coalition of the Willing" in the lead up to the Iraq War. They are the only folks south of the equator you can say that about.

So historically strong are the Australian/American ties that the Plaid Avenger counts them among the three most reliable, solid platforms of US foreign policy in the world. The UK, Japan, and Australia can be counted on to support US endeavors, almost with no questions asked. In the past, this essentially made these countries the American lapdogs in terms of foreign policy, but times do change.

While I stand by the description of Australia as a loyal ally of the US, modern times have forced them to be more pragmatic. Sometimes, being a loyal ally has to be pulled back due to popular opinion. Most Australians were opposed to the US-led invasion of Iraq, and the government at that time did take a popularity hit for its staunch participation. In addition, the Aussies and the Kiwis are having to adapt to a world which sees the rise of Asian powers, as is the rest of the world. But while this region remains entrenched as part of Team West, they have to start being much more sensitive to the feelings of their Asian trading partners . . . which is increasingly putting them between a US rock and a Chinese hard place. More on that later.

But one need look no further than **ANZUS** to prove the tight relationship between these Pacific partners. ANZUS—The Australia, New Zealand, United States Security Treaty—is a binding military alliance between the states signed after WW2 that is extremely similar to NATO Article 5, meaning an attack on any is considered an attack on all. It primarily focuses on the security in the Pacific, although the treaty in reality is related to conflicts worldwide.

ANZUS on the ready, sir!

Unlike NATO, the countries do not have a fully integrated defense structure, nor a dedicated active force waiting on the sidelines ready to go. However, this three-way defense pact is a commitment to defend and coordinate defense policy in the Pacific, and includes joint military exercises, standardizing equipment, and joint special forces training. In addition, the Aussies are host to several joint defense facilities, mainly ground stations for spy satellites and signals intelligence espionage for Southeast and East Asia. Oh China! The Aussies are watching you! Well, maybe its the Americans in Aussie-land that are watching you!

As an interesting side note, New Zealand was "uninvited" from ANZUS from 1984 to 2012, because back in the 80's they declared their country a "nuclear-free zone" and thus refused to let any US navy ships that were nuclear-powered or nuclear-armed come into their territorial waters. Of course that ticked off Uncle Sam mightily, and got the kiwis kicked out of the club. But the world has changed, and the US needs their Pacific allies more than ever, so in 2012 Barack Obama officially ended the ostracization, and the ANZUS crew is stronger than ever . . . and what timing! Why would I say that?

Because also in 2012, the US and Australia announced the FIRST EVER permanent US military base on Aussie soil. It was originally proposed as a small coastal station way up north in Darwin (Northern Territory) that would be home to a few dozen Marines . . . and within 6 months it was expanded to house closer to 2500 US military service personnel. I predict it will likely get even bigger in the future. I also have insider tips from some Marine friends of mine that there are covert bases already in NZ as well. Can you say "countering Chinese influence in the Pacific"? I knew you could! Interesting times my friends!

A WORD ON THEIR WEIRD WAY TO WEALTH

This is where things get a bit trickier, so the Plaid Avenger is going to have to sort a few things out for you. We've made it expressly clear that the US is the powerhouse economy of the planet, but China, at number four, is catching up. Japan is at the number two slot and Germany's at number three. Where's Australia? It's in fifteenth place. In terms of total GDP, how rich is this place? Not very. However, in levels of material wealth per person, they are totally awesome. How could this be? How could they have a standard of living identical to ours if they are not a huge exporter of manufactured goods or technology or other rich stuff? It is a strange circumstance, and perhaps a unique circumstance to Australia, that when we think of features that make it the most unique, it's not an exceptionally wealthy place as a total.

Australian number one export. Talk about a return on investment! Ha! Boomerang? Return? Get it?

What do they produce? What do they really do? What do you buy from there? Do they make computers? Cars? Linens? DVD players? Software? Video games? I don't think so. What have you ever bought that said, "Made in Australia"? Here's the quick answer: didgeridoos and boomerangs. That's not much of an export economy to put them among some of the wealthiest countries in the world. Well, that's not what has made them wealthy.

Australia is perhaps the only country in the fully developed world that has gotten rich from the export of primary products. And it is still a big part of what they do. Remember that primary products are things extracted from the earth like coal, oil, minerals, diamonds, uranium, wool, cattle, and meat. What is all this stuff worth? When it comes right down to it, not worth much. They are a bunch of unprocessed materials. How can the Aussies be rich if they are doing stuff that poor countries do? Well, for starters, there are only 23 million peeps in the entire country, and it's the size of the US. They export so many goods because of the resource-rich land, and the wealth generated from that is spread between so few that they are rich overall. A pretty big GDP based on natural resources, divided by a small number of peeps, makes for a really high GDP per capita number.

But don't let the Plaid Avenger lead you too far astray. If you look at the numbers, over 60% of the workforce is in the service sector, but who are they serving? Pretty much just themselves. Seriously, no innuendo implied. They work in malls and mailrooms and breweries and bakeries and accounting and telemarketing centers. They produce computers and cars, but many of the manufacturing facilities are owned and operated by multinational companies (my favorite example: the Subaru Outback—a car named for Australia, made by a Japanese company), and the products are only built to be consumed in Australia. In other words, they make Apple computers there, because people there want Apple computers. They are not exported out of Australia. There are no cars or refrigerators exported from Australia. That's because they are too busy making serious bank on exports of primary commodities. Their internal economy has enough juice to keep them going and growing, but it is a matter of low population and lots of primary resource wealth that makes for such high GDP per capita. They are the only place on the planet that I know of that has gotten away with that. Bottom line: they may have 65% of the people in the service sector, but they make 65% of their total GDP on exports of primary products.

You'll often see information about sheep and lamb being a huge export to places like the Middle East, since they consume more lamb than they do things like beef and chicken. You'll also see information about the coal and oil that Australia has being exported to places like China and Japan, both of which are hungry for those energy resources. Australia also signed a uranium deal with China on April 4, 2006, to export raw uranium to be processed and used. No other "rich" country on the planet is exporting raw commodities at this scale. Everybody else processes things, because that's where the money is. It's fascinating that their primary source of income is from raw products. They are definitely an economic conundrum of the rich/fully developed world.

Throw him on the barbie, I dare you.

PACIFIC SHIFT

What is in Australia's future? To answer that we might ask ourselves what's changing, because while we've been saying that this is the mini-America of the south, things are definitely on the move.

ECONOMICALLY

As China and other Asian states increases in world power and wealth and are requiring more resources, Australia is reorienting its economic focus to nearby countries in Asia—instead of trading primarily with other white English-speaking countries like the US and Great Britain. In the past, these two countries were the primary trading partners with Australia. Seriously, what a pain it is to ship exports that far. It's a long commute for meetings and it's difficult to maintain trading relationships with their white, English-speaking allies. They're on the other side of the planet! Nothing, other than fine Australian wine can be shipped all the way across the world anymore and still turn a profit.

Also, consider what impact longitude has on trade relationships. If you are at the same longitude, you are in the same time zone. Now, Australia is doing business in real time, during their business day, instead of having to deal with the massive time difference between it and other trading partners like the US and Europe. It's a tremendous complication to be awake when your trading partners are asleep.

They have reoriented their trade relationships to more local areas are neighboring regions like southeast Asia, China and Japan. "Why continue trading with the white people that look like us, when we can make lot more money trading with the Asian people that don't look like us?" Money talks. This is the way it is. Plus, the booming economies and large populations of Asian countries make for a much bigger demand

Aussies really on Tokyo time.

on Aussie resources. Team West economies have been stagnant or slow growing, while China and ASEAN countries have been on fire with economic growth. And they are closer too! So there is much more money to be made right now supplying Asian demand, not propping up old trade ties on the other side of the planter.

In conclusion: Aussie economic future lies in Asia. Over the next few years, look for the portion of Australia exports of raw goods to Asian countries to increase, and dramatically. Raw material exports are not the only thing on the menu, my friends. Many Australian businesses, like so many others across the planet, are moving manufacturing facilities as well as research and development operations to Asian locations to take advantage of both cheap labor and the large local talent pool. I suppose the future of the continent may also be as a giant national park serving tourists from around the world serving shrimp on the barbie after you scuba on the Great Barrier Reef. And there is something to that: tourism is a big money maker, particularly since Australia and New Zealand are beautifully scenic areas, with the added bonus of being fully developed and politically stable . . . which means people love to come visit to live well, be safe, and buy great stuff.

Those same reasons are why many multinational companies (even Asian ones) are located their corporate headquarters in Australia/NZ; headquarters that run their global operations in Asia. Say what? Heck yeah: corporate heads want to do business in Asia to make their companies rich....but they want to live in a awesomely nice and clean and stable place like Australia while they do it! High end sectors like finance, banking, and research are taking hold in Oz as well, since a critical mass of highly educated, internationally-connected peeps has evolved there.

Even more so than America, Australia's future is tied closely with Asia. Asia is a booming region on the planet, a place where a lot of things are going on economically, and Australia is right there to help them along and make some profit at the same time. As we will see in the Mexico chapter, it is good to be next to a gigantic economic engine, and Australia's proximity to China, who will be the largest consumer of raw goods on the planet in the coming century, will enable Australia to turn huge profits. Former Prime Minister John Howard (in office 1996–2007) was quoted in the past about Australia refocusing their efforts and economy to better compliment China's rising power. He has called Australia "an anchor of stability" in this region. He also said, "When we think about the world, we inevitably think of a world where China will play a much larger role."

POLITICALLY

Let's stay with Former Prime Minister John Howard for just a minute more, so I can explain to you how Australia's role in the vicinity is changing politically as well. To do this, I must elaborate on the most hilarious "story of the sheriff" which ruffled many feathers. It goes a little something like this:

A decade ago, former US President George W. Bush *complimented* Former Australian Prime Minister Howard during a press release by referring to Australia's status as the "sheriff of Asia." What the heck was that supposed to mean? Most Asian countries interpreted this as "the white dudes in the area are in charge," and that the long arm of the US law was being stretched through Australia to keep order in the area. This seriously ticked off all of Australia's neighbors. Malaysia, which is a awesome state in its own right, nearly declared war over it. To paraphrase the Malaysian prime minister, "Think that all you want, but if you ever set foot on our soil we will declare war on you instantaneously, no matter what the US says about it."

Even John Howard was a little miffed. He was probably thinking, "Thanks for the compliment, Mr. Bush, but try not to ever say anything about me in public again." This was a problem especially since Australia is reorienting itself toward Asia. These are the guys with whom they are trying to buddy up! The last thing they need is negativity about their foreign policy ties to the US. They still want to be allied with the US, but they don't want to be throwing it in people's faces.

The Plaid Avenger suggested in an earlier chapter that Europeans are slightly apprehensive to support US foreign policy, as they have an increased risk of terrorist threats due to their proximity to the Middle East and Central Asia. Australia is in the same boat. They're closer to other regions that may be slightly hostile to the US foreign policy. Australia is right there beside Southeast Asia, parts of which are known hotspots for extremism and terrorism. There are some seriously extreme fundamentalist Muslims in this area; even my Muslim friends would concur. Terrorist cells are known to exist in lots of places in Southeast Asia (e.g., Indonesia, the Philippines, or Malaysia), and Aussies have been targeted by terrorist attacks in the past (look up 2002 Bali Bombing for an example).

This is why John Howard was quick to offer a modification of Bush's comment, because there is a real threat nearby. The US is far enough away to be safe from imminent threat, but it's not hard for terrorists to make it from Southeast Asia to Australia. Because Australia is seen as the face of American foreign policy in the area, it is very surprising that Australia has not seen any terrorist activity on their soil already. However, there have been several foiled terrorist plots in recent history and this situation is not likely to let up anytime soon. The Aussies remain ever vigilant.

Bush was quoted as saying that he didn't think of Australia as a deputy sheriff, "but as a full sheriff." In fact, Australia does supply soldiers in most UN activities in the countries surrounding them. In many cases, they are actually doing patrols and peacekeeping directly for the UN in places like the Solomon Islands, Indonesia, and other hotspots where the UN needs peacekeepers. The idea of them as sheriff does have some weight, and is going to cause them some problems in the future. Many countries in the 'hood remain perpetually pissed that the white man is on their soil at all, even as peacekeepers.

To end this rant, know this: this is changing fast. Another Former Prime Minister Kevin Rudd (who served after Howard) was a bit more liberal and conciliatory than his conservative predecessor, and he made moves to soften the Australian image in this part of the world. He decided to pull Aussie troops from Iraq, and was much more engaged with local Asian countries' leaders to head-off any friction that might arise due to Australian presence, as UN representatives or otherwise. Oh yeah, and he spoke fluent Chinese, and not just to order from a take-out menu either. Given Australia's changing focus, I assume it may be a common attribute for future leaders to be just as fluent. . . .

THE PEEPS

First off: there hardly ain't none of them! Say what? Yep. There are only about 23 million Aussies, and 5 million Kiwis (plus or minus a few Hobbits.) Altogether this region has less peeps than New York or Tokyo or Rio, but spread out over an area the size of the continental US! And it's not like the equation is going to change much in the near future either, as both countries have full-on stabilized, low pop growth rates. As in the US and Europe, the only way their pop gets any bigger is via immigration from other places, and (just like in US and Europe) this inflow of people is changing their ethnic and demographic outlook here in the 21st century . . . but how?

Because Australia is doing a lot of business with China, Japan, Indonesia, ASEAN, etc, they will become more like those countries culturally, as interaction with and immigration from Asian countries increases. This is changing very rapidly just in the modern era, because quite frankly it was not allowed to happen any earlier. How so, Plaid Avenger?

At the expense of aggravating my Aussie friends, I will have to tell you that a racist streak has run through the society since its inception—a bad habit that they probably inherited from the Brits. Perhaps racist is too harsh, but certainly they had a superiority complex when it came

Aborigines: Not held in high regard.

to other peoples in their neighborhood. The Australian treatment of the **Aboriginals** was bad from the get go, mostly treating them as third rate citizens, at best. There was also an official state policy from 1870 to 1970 which made it perfectly legal for the state to take Aboriginal children from their parents to re-educate and 'civilize' them in white families. These folks are now referred to as the **Stolen Generation**.

I guess that is better than what the Aussies did to the Tasmanians, which was to kill them all in an open hunting season. On top of that, Australia had a white-only immigration policy in place for most of its history. What? Yeah, I'm

DISTRIBUTING SHIP CARGO OF STANDARD BUGGIES COAST OF AUSTRALIA

What is this madness? No wonder Rudd apologized.

afraid it's true. If you were of European or American descent, then you had an open door to the country; but that same door was slammed shut for Asians or Africans of any stripe. Pretty nasty business.

To be fair to our friends "down under," they have made great strides in the last few decades to make amends. Much has been done to alleviate the impoverished plight of many Aboriginal communities, and even more has been done to overcome the negative attitude and stereotypes of the group. In fact, Australia picked an Aboriginal woman to carry the flag into the introductory ceremonies for the 2000 Olympics. That brings me back to what former Prime Minister Kevin Rudd did his first month in office: in February 2008, he made an official government apology to the Aborigines for all past reprehensible deeds of the Australian government, specifically citing the Stolen Generation. This was big news, and perhaps a critical turning point for the society.

Back to the story: Since 1976, when their historically strict immigration laws became more relaxed, the country's Asian population has increased. Surprise, surprise. More people that are close by, from regions adjacent to Australia, are coming there because it is a rich place where there is more opportunity to succeed than there may be in their home countries. Again, it's America, Jr.! All of the poorer people nearby want to get there so they can set up shop. Australia may not be full-on encouraging it, but at least they are allowing it. They are changing direction, changing focus, and becoming more Asian. When you visit Sydney now, you can bump into a very vibrant and growing Chinatown for the first time in its history. These changes are occurring nationwide. As you may remember from the international organizations chapter, ASEAN is now a rallying entity for inter-regional action in this part of the world, and Australia is a part of the ASEAN +3 dialogue. They are also a new member of the EAS: that East Asia Summit that is a open trade/politics/planning talk shop that everyone in the Asian 'hood is on . . . and now so are Australia and New Zealand, despite their lack of Asian-ness.

Who else have we talked about that is becoming more Asian due to increasing immigration from Asian states? That's right: America, Sr.!

NO WORRIES

This is a region that plays a dual role as a US foreign policy anchor and as an economic player with China. They have a lot of natural resources and stable population growth, that is being bolstered by Asian immigration too. There is still the terrorist threat, however; we'll just have to see how they balance their security against their economic interest.

Australia would like to be a power broker between China and the US. This is promising; they have far fewer complications to worry about when dealing with China, as opposed to Japan, which has loads of historical emotional baggage. Japan may be a little more abrasive than Australia when dealing with Asian relations in general, and this gets the Aussies ahead.

Australia is going to have a leg up dealing with China and all the other Asian economic giants, because of proximity as well as the fact that they don't carry a lot of cultural baggage. What do I mean by that? They haven't dissed. They haven't invaded anyone. They never colonized anyone. They are pretty antiseptic all the way around. That is a big plus in today's world, particularly for any state wanting to get in on the action in Asia, which is, of course, everyone. And it sure don't hurt that they have an absolute ton of resources and services that those growing Asian economies will be snapping up indefinitely. Dudes! The koalas and the kiwis are sitting pretty down there right now! Down under is the place to be! Down under Asia, that is.

Uncle Sam's little brother down south is still a staunch ally of the US . . . and the awesome ANZUS pact makes them secure, all while US troop presence is increasing across the region. Luckily, they remain just distant enough to dodge a lot of the negative press associated with such a role. Of course, the Aussies are distancing themselves slightly from US foreign policy of the last decade . . . but dang, who isn't? Australia is a pretty chill place all the way around: politically, economically, socially. That's why we like to go vacation there, because Australia is a laid back sort of place. Throw a shrimp on the barbie, crack open a Foster's, and enjoy the prosperous future, America Junior.

A SINISTER SHOW AND TELL

Here's a great game that all the kids love to play! It's called "Quickest Kill." Known for it's exotic and often lethal wildlife, the Australia/New Zealand region is hot to a whole slew of the world's deadliest animals. Of the following creatures listed below, which one do you think could take out the kid sitting next to you in geography class the fastest? And which one do you think is responsible for the most deaths? Have a round table debate in your class, and you get bonus points for letting one or more of these creatures loose while defending your choice. It's the Aussie version of 'Hunger Games'! Good luck!

Salt Water Croc   Eastern Brown Snake   Stonefish   Red-backed Spider   Death Adder   Great White Shark

  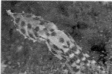 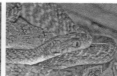  

Funnel-Web Spider   Inland Taipan   Blue-ringed Octopus   Tiger Snake   Box Jellyfish   Gollum

AUSTRALIA RUNDOWN & RESOURCES

 ## BIG PLUSES

→ Koala bears and kangaroos and hobbits. Who doesn't love them? Ok, besides Sauron.

→ Small population that is about evenly balanced with resource/industrial base

→ Location, location, location! In the neighborhood of the hottest, fastest growing Asian economies, to which Australia sells tons of goods

→ 13th largest economy in Australia; very high standards of living across region

→ Resource/agricultural product rich

→ Stable democratic governments

→ No real enemies on the planet; no international conflicts with anyone

 ## BIG PROBLEMS

→ Crocodiles, the Sydney funnel-web spider, 9 of the top 10 most deadliest snakes in the world. Crikey!

→ National legacy of racism: still has friction with its Aboriginal groups, and immigration is becoming a hot button issue much like it is in US

→ Economy not very diversified; heavily dependent on natural resource exports

→ By mere association with Team West, combined with its proximity to SE Asia, Australia is a high-risk target for international terrorism

LIVE LEADERS YOU SHOULD KNOW:

John Howard: Popular former Prime Minister of Australia from 1996–2007. Kind of the "Ronald Reagan of Australia": Center-right conservative, pro-business, pro-US, and presided over a period of strong economic growth and prosperity. Strongly supported US in Iraq and Afghanistan. Mostly lost in 2007 simply because voters were bored.

Kevin Rudd: Prime Minister of Australia from 2007–10, and the polar opposite of Howard. Center-left liberal who was more focused on establishing stronger ties with Asia, and possibly not following US foreign policy so adamantly. Speaks Mandarin Chinese fluently, supports Kyoto Protocol, made official apology to Aboriginals, and pulled troops out of Iraq.

Tony Aboott: Current Prime Minister since 2013, center-right conservative, thus swinging the power pendulum back to the other side. Is more "business friendly" and no fan of Kyoto Protocol nor a carbon tax, is anti-immigration, and re-started a program to appoint Australian knights and dames. Seriously.

KEY TERMS TO KNOW

Biomes
Chlorofluorocarbons
Citizen
Continent
Crude Birth Rate
Crude Death Rate
Crude Population Density
Demographic Transition Model

Dominion
Doubling Rate
Esther Boserup
Island
Nation
Nation-State
Neo-Malthusians
Ozone Layer

Physiologic Population Density
Population Pyramid
Rate of Natural Increase
State
Thomas Malthus
Total Fertility Rate
Zero Population Growth

Physical Features
Australia and the Pacific

0   500   1,000 Miles

©2015 BYU Geography
Think Spatial group

Cultural Features

Australia and the Pacific

©2015 BYU Geography
Think Spatial group

0   500   1,000 Miles

Practice Test Map

Australia and the Pacific

0   500   1,000 Miles

©2015 BYU Geography

Think Spatial group

Test Map
Australia and the Pacific

0   500   1,000 Miles

©2015 BYU Geography
Think Spatial group

Antarctica and Oceania

Jesus said to her, "Everyone who drinks of this water will be thirsty again, but whoever drinks of the water that I will give him will never be thirsty again. The water that I will give him will become in him a spring of water welling up to eternal life." (John 4:13–14)

CHAPTER OUTLINE:

→ Whale Tale
→ Drowning in Thirst
→ Going Nuts

→ Miles of Isles
→ Study Questions

MOST people think water is a cheap and easily procured resource. When our Lord spoke to the multitudes in Matthew 6:25 telling them not to worry about what they would eat or drink, they more than likely understood the reference to water in an intense way that might elude us today. An arid climate can indeed lead to a worrisome existence and make life harsh. A tough existence is the calling of many Christians throughout the globe and we have a duty to encourage and support them in any way we can. No place on Earth is likely to be more extreme than Antarctica or more pleasant than much of Oceania. Even though our lives have peaks and valleys, we can be consistent in our approach towards our loving and eternal Father who loves us without qualifications. Remember, regardless of the extremes we face in this life, the Lord's admonition not to worry is always good to reflect upon. Let us, therefore, start to accomplish the works He considers important now without delay!

Commonly known as the South Pole, Antarctica is really a continent. Famous for its severe conditions and interesting animal life, few people know what potential it really holds. So why is Antarctica a continent?

Antarctica is actually part of a tectonic plate and as such is a continent. Largely unexplored, it is nevertheless one of the most peculiar places on Earth. The severe cold and accompanying wind and dryness make it seemingly uninhabitable. There are, however, a few sea mammals that have proliferated in the area in addition to the birds. Specially adapted to the cold, these generally large and bulky creatures depend on the fish-rich seas for their diets. Whales, seals, and penguins are generally what one thinks of when one thinks of the animal life on Antarctica. Other than simple plant life that appears on the tundra, these are the most noteworthy life forms on the continent.

WHALE TALE

The Southern Sea is the name of the sea that flows around Antarctica. Continually flowing eastward, this ocean is one of the most tumultuous in the world. Its northern boundary empties into the other oceans of the world, and there is a marked difference between the icy continent of Antarctica and the relatively warm (but still cold!) waters surrounding it. This contrasting temperature accounts for the roughness of the seas, and this, in turn, seems to provide nutrients to the abundant whale populations of the world. With little resistance from landmasses, this current flows fast and can be littered with icebergs of varying sizes. So what resources does Antarctica hold?

Although many mineral resources including coal and oil have been found in Antarctica, it remains a continent of untapped potential. Numerous international treaties designed to prevent greedy people from competing with one another to be the first to tap these potentially valuable areas hinder any meaningful exploration. Isn't it amazing how God continually provides us with the resources we need, such as heat, food, clothing, and water? Unfortunately we seem to never have enough and often fail to give thanks. Perhaps if we were more thankful we would need less, and then there wouldn't be so much difficulty finding more resources without being burdened by regulations and restrictions. Perhaps you will be the one to find a resource and you will be able to share it providing a Christian testimony in the process.

DROWNING IN THIRST

Imagine living where water is everywhere, but there is not enough to drink. This is the situation the early travelers faced throughout the Pacific Ocean region we now refer to as Oceania. Where these people came from is still a mystery; theories argue that the peoples of Oceania originated in China or Siberia have argued that the peoples originated in South America. The truth is there is that a terrific amount of mystery here. What are not in question are the remarkable accomplishments of those men and women who migrated across half of the earth's surface to create unique cultures on islands of varying hospitality. Some writers believe this circumnavigation may be the single greatest and one of the most recent accomplishments in the history of mankind.[1] Equipped only with simple catamarans, or boats with outriggers to provide additional support, this movement was a terrific accomplishment no doubt contingent upon understanding currents both atmospheric and at sea.

Image © Kamira, 2013. Used under license from Shutterstock, Inc.

GOING NUTS

A quick look at a selection of flags will reveal the importance the people of this region place on celestial objects such as stars and the sun. These might be insights into the deep-seated awareness of the environment that led these early explorers. A common color for these flags is the blue of the ocean. Without question, the influence of the physical environment on the human landscape can be seen in the vexillology.

An important biological aspect of the biomes of this region is the simple coconut. The coconut, which is able to float on the ocean for great distances before coming ashore in many different places, was an important aspect of soil creation enabling life forms to thrive. And thrive is exactly what some of these creatures have done. The Christmas Island crabs

---

[1] Borthwick, Mark, "Pacific Century" (Boulder, Colorado: Westview Press, 2014), 11.

cover the streets with bright red shells as they run to and fro. Crabs have even developed that are dependent upon the coconut for their source of food and have specially adapted bodies for this purpose. Therefore, the lowly coconut probably served to sustain human life and enable the early travelers to cross this amazing area.

MILES OF ISLES

Some of the important sub-regions of Oceania today include Melanesia, Micronesia, and Polynesia. These places are loosely divided on the basis of varying interpretations of language and ethnicity as well as their relationship to tectonic activities and the resulting landforms and volcanoes. The enormous distances involved make the distinctions among these peoples extremely controversial. The people of Melanesia

Image © Luca_Luppi, 2013. Used under license from Shutterstock, Inc.

appear to be very similar to those on the island of New Guinea. The Micronesians may appear in many ways to be descendants of East Asia, while the Polynesian peoples are best characterized by the Hawaiians. As you probably learned, the Columbian Exchange, or Age of Discovery, unleashed numerous diseases that make it difficult today to draw historical distinctions of origins based solely upon ethnicity since these people groups died out before we could record their history. Additionally, many of the islands became rapidly depopulated so we can not study their cultures today. Since we all originated from Noah's ark, it is probably best to simply reflect on the importance today of the strategic significance of this area, in particular to the world's great powers in the last century.

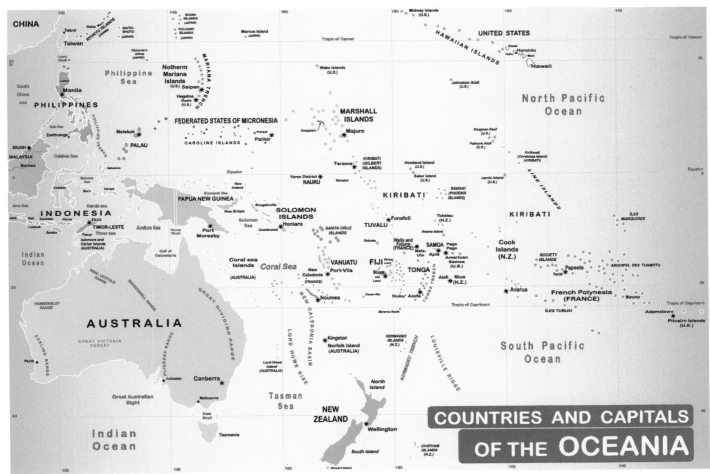

Image © Zolta_11, 2013. Used under license from Shutterstock, Inc.

America in general and the West Coast in particular have been viewed as increasingly interconnected with the Pacific Rim, particularly in terms of business. This area was of extreme importance during an age of steam as ships required coaling and supply bases every thousand miles or so. Various developing nations attempted to stake out "turf" across the Pacific for the purpose of developing secure transport routes to fulfill that ancient yearning—trade with China! The United States became involved in the Philippines at the beginning of the last century during the Spanish American War. This involvement in the Philippines would later bring the U.S. into conflict with the Japanese at first in Polynesia when Pearl Harbor erupted into flames during World War II. The Central Pacific drive by the Naval forces of the United States would bring American interests in a type of continuing "Manifest Destiny" or, westward expansion, to the areas of Melanesia. It would be on the islands that U.S. troops would fight to defend sea routes from the U.S. Micronesia would be taken by the U.S. in an effort to bring force to bear directly upon the Japanese home islands. American involvement in the Pacific has continued through the Vietnam War and into the present, and, in fact, the recent presidential administration has attempted to "pivot" in its efforts to capitalize on American interests in the areas adjacent to East Asia. This will no doubt continue to make the Pacific Ocean a road to the back door of the United States in years to come.

Physical Features

Australia and the Pacific

0   500   1,000 Miles

©2015 BYU Geography
Think Spatial group

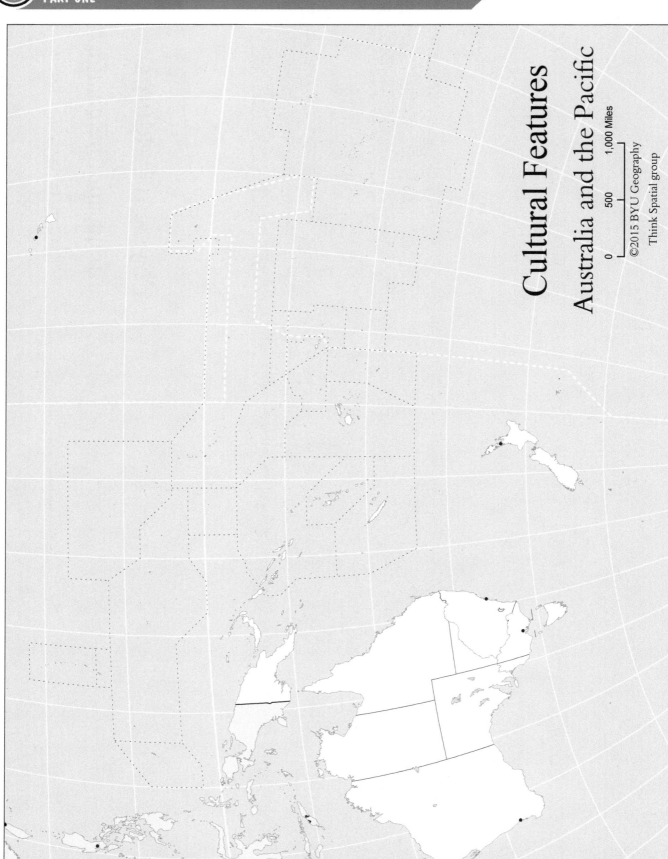

Cultural Features

Australia and the Pacific

0   500   1,000 Miles

©2015 BYU Geography
Think Spatial group

Practice Test Map

Australia and the Pacific

©2015 BYU Geography
Think Spatial group

0   500   1,000 Miles

Test Map

Australia and the Pacific

©2015 BYU Geography
Think Spatial group

0   500   1,000 Miles

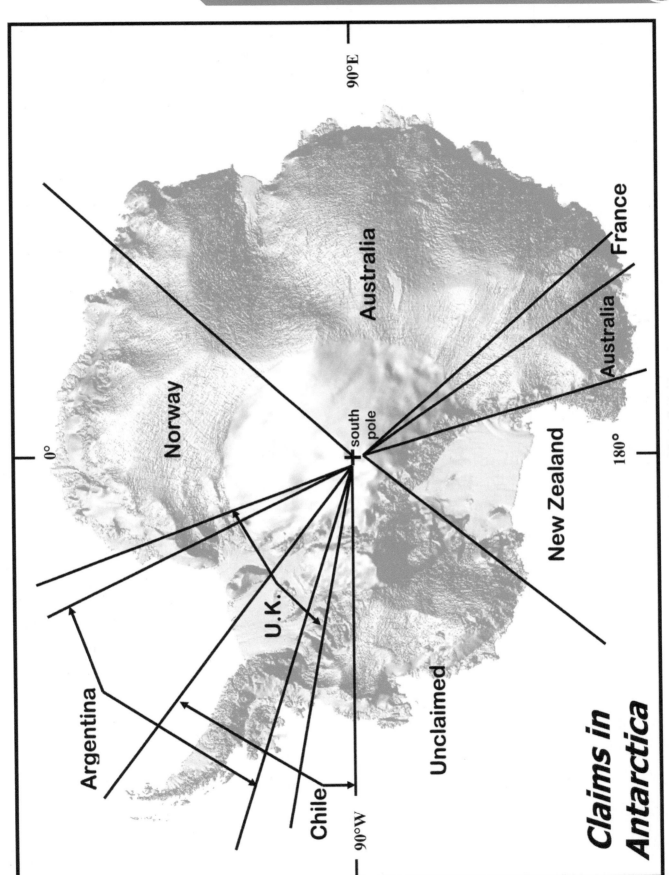

Claims in Antarctica

90°E

France

Australia

Australia

Norway

south pole

0°

180°

New Zealand

U.K.

Argentina

Unclaimed

Chile

90°W

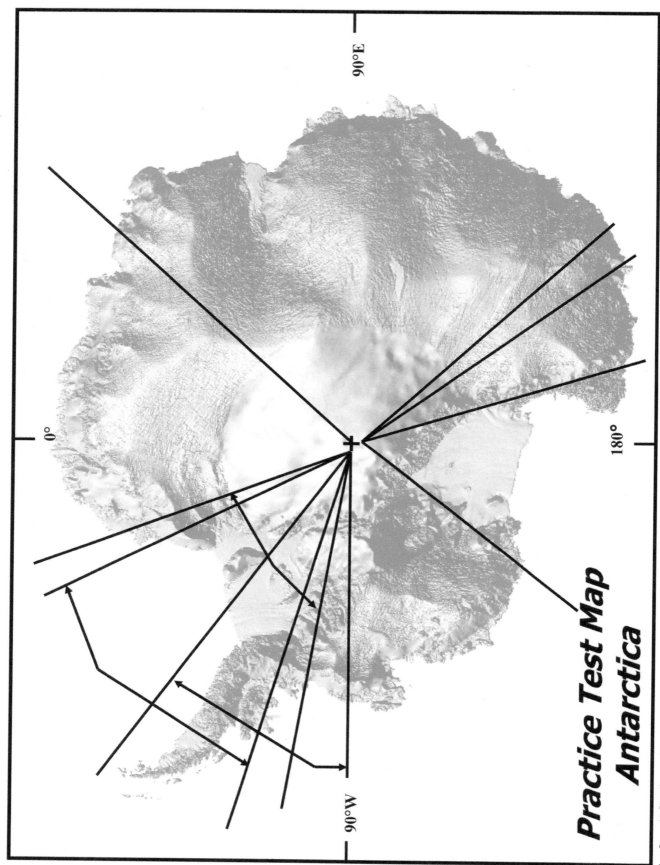

Practice Test Map
Antarctica

90°E

0°

180°

90°W

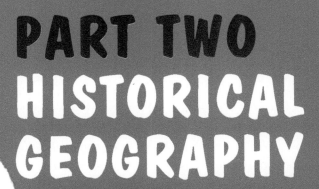

PART TWO
HISTORICAL GEOGRAPHY

PROLOGUE

Psalm 19:6
Its rising is from one end of heaven,
And its circuit to the other end;
And there is nothing hidden from its heat. (NKJV)

THE world exists in time and space. No historical survey is complete without considering geography. This section argues the world has existed through four arbitrarily chosen periods or ages. This delineation is made for the purpose of identifying recurring themes and patterns over time utilizing a spatial approach. All cultures and civilizations are essentially a function of the physical geography, climates, and locations chosen before time even started. The ability to utilize time to its best advantage and in conjunction with available topography and resources has allowed empires to sustain themselves and grow; the failure to adapt to time-saving technologies towards trade and to further increase speed and transport has caused states to rise and to decline. Where an empire stagnates, it becomes weaker in relation to other competing states. **Frontiers** tend to emerge where the expansion and decline of different states come into contact. Frontiers are characterized by fluid zones of control; the extent to which a state or empire funnels resources into these frontiers or bypasses them is often critical to the vibrancy or health of the empire and may determine that state's chances for success or failure in the future.

God created the world through time and space with the ultimate purpose of allowing some on Earth to return to a perfect condition or state of existence in total communion with God. After the curse came upon mankind in the Garden of Eden, the geographic theme of movement represented one of mankind's first acts. Adam and Eve left the garden but instead of continuing to populate and spread over Earth, mankind instead settled in the plains of Shinar and began to develop early economies replete with records, numbers, and coinage for the purpose of trade. Choosing to settle, capitals rose to control outlying areas. Elites controlled these areas with graven images, statues, and cultural landscapes to create the impression of security. We see throughout scripture that God scatters His people in an attempt to force them to be dependent upon Him. We see a human propensity today to travel and seek change or luxury. Opposed to this travel through dependence on God is the inherent attempt by localities to maintain control over others through the consolidation of power. As states inevitably grow, eventually one area maintains control administratively over another greater area. This capital, or **metropole**, serves as the administrative center for the state's leadership and directs the maintenance of the empire.

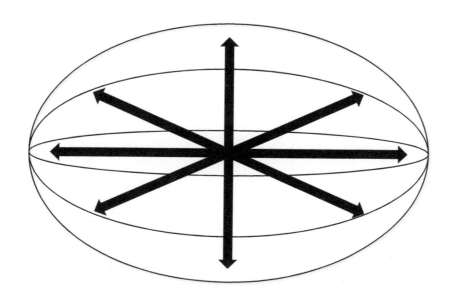

An **empire**, for our purposes, is an area encompassing different nations or peoples. The challenge for the empire is to keep the people unified, joined by trade, religion, or through reverence for the state. One of the first identities of civilization or the differentiation of labor is coinage depicting loyalty to a leader. These graven images both reflect and reinforce the authority and control of the state. The term empire was first used in the late middle ages when Henry VIII battled the pope in a divorce case. Arguing for independence, the term imperial justified an independence from Rome. Revived from the ancient world, empires tend to be unintentional entities covering large areas and with recognizable **capitals** where the governing authority sit. The inhabitants of most empires are willing to let go the burdens of costs laid upon them by the people in charge. If a family has the power, the empire becomes a **hereditary empire**.

Elites in governments in the last millennium in the West have tended to claim a justification for authority on the basis of being the successors of the Roman Empire. Looking to the greatness of the past tends to denote a conservative state while an idea of optimistic progress may reflect an aggressive nation seeking to grow. The invocation of the revived Roman State is an example of a glorious past and perhaps a bright future and has been a technique used to claim sovereignty in more recent European history.

These metropoles on a smaller scale tend to reflect the core and periphery relationship described in an earlier chapter. The metropole controls the surrounding countryside and is a repeating theme regardless of scale or size. On a local scale, patterns of location include a manor house or even a regional state with a national capital. Once a state's boundaries exceeds several regions, it becomes an empire. When the metropole exercises too much authority, the state sees a loss of initiative and tends to stagnate; yet if the state does not maintain a modicum of control, consequent challenges emerge to the authority of the state and the resulting internal divisions can be exploited and absorbed more easily by adjacent competing states. A balance must be maintained between the control of the state and the independence of its peoples. This balance is reflected spatially in the size of the state: when the size exceeds the ability of the state to defend or control its borders, it will inevitably decline. If too much state control of the borders exists, a crushing economic and cultural burden will exist, usually in the form of over-taxation or the stifling of individual creativity and initiative. The ability to maintain control or to tax is a vital source of revenue for the state and strengthens loyalty through the bonds of its members and also creates salaries for its soldiers. The frontier is the perfect gauge of the health of nation-states and can determine if they grow or if they go.

The state's leadership often uses defense for its justification. This brings us to the concept of a **frontier**, an area of uncertain control, often at the boundary of an empire or on the fringes of several empires and often an area of contention by the state. (GSS-2) A frontier is often represented by a non-contiguous transportation or movement terminal. For example, control may be maintained over an ocean or sea because of the control of a local point of entry or harbor. Once one begins to go inland, the **friction of distance**, or cost of travel, may result in a loss of control or an inability to procure tax revenues justifying the efforts involved in travel. [Venn diagram] Where key resources exist, an ability to control and develop these resources often determines the health of the state. It is the stated goal of an empire to utilize a **colony** and settlers to extend this control. As resources are revealed and technology demands different materials, the direction of empires and states will change over time. It is the argument of this book that various empires have risen and fallen due to their ability or inability to maintain control over vital areas by maintaining a healthy relationship between the capital, or metropole, and the surrounding areas. The degree to which this control is in balance can be determined by the use of technology to enable movement to occur with greater speed and precision. The themes of growth described in this book over time can be broken down into the Age of Muscle, the Age of Sail, the Age of Steam, the Age of Air, and the Age of Space.

The **Age of Muscle** represents the achievements of the state to expand and to exploit natural borders and topography maintaining a relatively low cost of security to the state. Mountains, rivers, and seas are utilized to maintain cost-effectiveness by limiting the number of soldiers necessary.

By the **Age of Sail**, we see natural landforms and global physical phenomena finely put to use through the use of currents and sails along with the technology to navigate and deliver firepower. This transregional experience initiated true globalization where empires emerged over various regions, uniting the peoples through trade and defense.

The **Age of Steam** was the beginning of the manipulation of chemical energy and demanded terrific resources to maintain the appetites of the machines while the world continued to move even faster and synchronization occurred.

The **Age of Air** began with the use of the internal combustion engine and great circle routes to circumnavigate the atmosphere, eventually pushing into another frontier—outer space.

The **Age of Space** sees a competition for control of Earth's orbits in outer space. It is the beginning of an invisible empire utilizing the inner space, or air around each one of us, through the increasingly precise use of the electromagnetic spectrum and manipulation of sub-atomic particles. The result is an increased interconnectedness, or globalization, characterized by faster and more accurate communications networks and increased precision in the guidance of weapons systems and speed of commerce.

Throughout these ages it is important to note the various themes described are not new but rather have existed from the beginning only becoming faster and more complex.

KEY TERMS TO KNOW

Age of Air

Age of Muscle

Age of Sail

Age of Space

Age of Steam

Capitals

Colony

Empires

Friction of Distance

Frontiers

Hereditary Empires

Metropoles

A Spatial Survey of the Ages

CHAPTER OUTLINE:

A THEME to watch for in this survey is a distillation of geographers through the ages. Many writers have claimed there is an east-west component to history. The great Professor Samuel Huntington described this in his epic book, *The Clash of Civilizations and the Remaking of World Order*. Dr. Huntington argues that the world consists of various cultures bound to clash in the future.[1] This somewhat unpopular thinking tends to grapple with the fact that the state's relationship tends to vary with that of the peoples. Control of water in particular has led many geographers to refer to oriental despotism when describing the empires of the Near East and the total authority of the states in Asia in comparison to Europe. Another interesting divide of interest to writers on the subject is the methods of warfare. A fluid and fast-flowing cavalry type force tends to be at odds with a western style of holding ground with infantry. This temporal speed versus holding territory has been a contributing factor to moving frontiers throughout history on both land and sea.

Another interesting theme to the historical geographer is the idea of peripheral control. The frontier tends to energize and necessitate the emergence of a high energy and aggressive people and culture, which in turn tends to expand its influence over the state from the edges. A good example of people representing this phenomenon would be Austrian Adolph Hitler taking control of Germany or Joseph Stalin in the Soviet Union; in the United States, Barbadian Alexander Hamilton was a powerful advocate for royalty in the new world. A final example of this theme would be Napoleon Bonaparte, a Corsican leader in France. It is almost as if an ultra-nationalistic identity emerges from the periphery or frontier with a vision seized upon by the state seeking new leadership.

A whale versus lion approach to geography reveals the advantages of sea power over land power. Cost of travel, or friction of distance, is overcome greatly by sea powers, but the technical aspects of sea travel require a high degree of education relative to soldiering. The Delian League of Greece or the various empires of exploration during the Age of Sail represent examples of largely commercial enterprises in contrast to the Spartans or the Turks. In many ways, frontiers between these types of powers continue since neither can finish the other off in a fight.

---

[1] Huntington, Samuel P. *The Clash of Civilizations and the Making of World Order* (New York: Simon & Schuster, 2011).

Another interesting phenomenon as it relates to frontiers is the idea of **nomadic versus civilized** populations. Today, this can be seen in a manner similar to the peripheral control theme described above, where the hard and toughened outdoorsman stands in marked contrast to the soft city-dweller. In many cultures, corruption is identified with softness and a sedentary life, causing gender issues to come into play. It is almost seen as a duty in some cultures for the leadership of the state to prove its toughness and prowess, therefore proving its fitness to lead or its purity by maintaining control. In many areas, an insurgency continues to challenge control. A state of equilibrium exists between the state and the warlord or gang, which is often the case in Middle America and South America. This concept of a challenge to power is still actively at work in the world today, particularly in the Arab and Islamic nations, including those on the expanding African Sahel. The challenge to the state is to create institutions capable of funneling a challenge to power through the political process.

Empires appearing to last tend to have natural borders easily defended and a proximity to trade routes, the life-blood of any enterprise. There is leadership in the state, but not to the degree that total control is dictated from the capital. Efficiency is maintained where total control exists, but new ideas are rejected while other states adapt to changing times. The first example we will use to illustrate some of the above points will be the various empires that emerged during antiquity in the proximity of the **Mediterranean Sea**. We will then move our analysis to various points on Earth over time using the themes but taking into account the changes in technology referred to in the four ages above. As each period of time is described, it is important to recognize that each has built upon an earlier time. We may have nuclear power today, for example, but somewhere electricity is still obtained from burning coal and this resource has traveled on the old sailing routes and over the vertical borders where modern day badge-wearing legions protect its travel. The first major empires saw humans unknowingly yet increasingly seeking each other while using and exploiting natural borders and frontiers in the Age of Muscle.

The Mediterranean represented the middle of the Earth. A central location in terms of latitude and in large measure longitude served to create a natural base for an empire. The tell-tale shape of tectonic activity runs roughly parallel to the **Black Sea** and somewhat perpendicular to the Red Sea and the Persian Gulf. The utilization of these features would both limit and funnel through global trade networks. The Euphrates and Tigris rivers flowing down from the **Anatolian plateau** of Turkey created a natural funnel for the developing empires of Sumer, Babylonia up to Assyria. Throughout the first millennia before Christ, one notices the increasing rise of peripheral states and how the rivers of Mesopotamia seem to act as a funnel for invaders with few natural borders.

The east versus west school of geography would be quick to note the intensity of control of early resources as well as the lack of natural borders and subsequent fast moving fluid cavalry-type attacks leading some determinists to believe violence and total control are characteristic of this region today. Water is the ultimate resource needed for life. The ability to control water supplies throughout history would suggest a cultural tendency toward autocracy. Insecurity and a fluid frontier on almost all sides would be seen as leading people towards a fast maneuvering, cavalry-type of military at odds with cultures regarding holding ground as the key objective. The opposite of this fluidity in the open plains or steppes would be where vertical borders exist, such as in the mountains of Greece.

The Greeks remind us of other mountain peoples who tend to be less dogmatic in their adherence to religion or the state. Note the Islam of the Caucuses and how it is much less intensive than in the flatlands of Iran. Mountain areas tend to lend themselves to conceptions of self-identity that are hard for a state to control due to friction of distance because of elevations. The mountains and climate of ancient Greece created an ideal environment for the open exchange of ideas, and only with great difficulty could the tyrant emerge. The Greeks eventually would find themselves at odds with various states on their frontiers. It is instructive to see how some attempts at expansion succeeded and others failed.

Drawn abroad, the Greeks attempted colonization. Expeditions to Syracuse in Sicily were a disastrous attempt at overstretch. (The term **imperial overstretch** comes from the book, *Rise and Fall of the Great Powers*, by Paul Kennedy and is used throughout this text.[2]) By 491 BC, attempts to reinforce this colony failed due to the Greeks inability to convert

---

[2] Kennedy, Paul. *The Rise and Fall of the Great Power* (Random House: New York, 1987), 438.

their sea power into land power. Armed with the knowledge of the terrain, the native Sicilians defeated the Athenians who could not transport cavalry on ships. The Athenians had tried to control too much space for their ability to project force through time. The rate of travel on the island of Sicily required horses, but the fodder and ability to supply cavalry over such a distance did not yet exist. In the Age of Muscle, it is important to understand horses and mules could carry three times the weight of a man but ate seven times and drank eight times as much weight and required twenty pounds of fodder a day.[3] In an Age of Muscle where galleys at sea depended on men rowing and on land beasts of burden to pull supplies, speed required resources to be transported great distances. The time needed to traverse sea and land in the Age of Muscle did not favor the state's attempt to increase great distances spatially and resulted in an imbalance of expenditures of space against time. Overcoming friction of distance by utilizing sea power and the natural frontiers of the sea represented a better investment.

Greece represents the foundation of a western society. The rise of the Delian league a generation after the ill-fated attempt at Syracuse would see the procurement of the vital resource of grain from the Black Sea wheat-growing regions at lower costs. Historically sea transport is between 1:35 and 1:100 times less expensive than land transport.[4] The Peloponnesian War with the land power of Sparta would see Athens weakened due in part to imperial overstretch unable to use the technology and speed of the day to overcome the friction of distance or space it sought to occupy on land. An attempt to expand on land ran into difficulties as in Syracuse and resulted in the eventual weakening and destruction of Athens by the Spartans; lion beats whale. Eventually, however, some rough-hewn Macedonians from the periphery would be instrumental in launching the first transregional Greek Empire.

Eventually an example of peripheral control would occur in Greece with the rise of Alexander the Great. This Macedonian conquered the independent Greeks who resided in their mountainous peninsula and confronted the Persian Empire whose periphery reached the **Levant**. The classic east-west confrontation only resulted in a peripheral area controlling a larger adjacent empire as the ink spots seemed to spread together. The Persians had maintained control over their subjects, who obeyed them, but note the relative lack of loyalty within the population and the rapid willingness to accept leadership by another despot. When the state exercises too much control over its own people, there seems to be a lack of resiliency to withstand an outside invasion or perhaps a lack of loyalty. Unlike the Greeks' earlier conditioning by living in mountains and valleys where little central control existed, the Persians' culture had accepted absolute obedience to the ruler. Once the ruler was replaced, another was quickly obeyed. Another geographic explanation for Alexander's success is the recognition of his father Phillip and limiting the amount of equipment used. The Macedonians only carried the necessary equipment, such as siege equipment for the baggage trains. The reduction of equipment to transport enabled the army to travel light and fast, utilizing light infantry and cavalry in tandem. Taking advantage of diplomacy by accepting locals stretched Greek tolerance but nevertheless proved instrumental in overcoming distance in proportion to time. The eventual lack of rivers would mark the greatest expansion of Alexander's empire as transport times reduced.

Alexander's conquering took him to the **Oxus River** and the **Indus River Valley**. Both of these are natural frontiers today between Islam and Russia in the case of the Oxus and between Islam and Hinduism at the Indus River. Perhaps the real geographic gift of Greece as its culture continued to drift west was the rise of the Roman Empire.

The geography of Greece had allowed for an autonomously free people to accept new ideas while they learned to utilize the natural borders of the sea. Absorption of the ossified Persian Empire demonstrated many of the previous themes. Though globalization was nothing new (British tin had been traded for Afghan lapis, a semi-precious blue stone, as far back as 4000 BC), the Greeks would mark the beginning of a European power being pulled inexorably towards the east.[5] Taking up dredging the **Red Sea** as the Egyptians had, immediate attempts to deepen the Euphrates River would be an alluring draw to the Romans. The Greeks were blessed with a Black Sea dumping water into the Mediterranean causing a generally

---

[2] Addington, Larry H. *The Patterns of War Through the Eighteenth Century.* Univ of IND Press: Bloomington: IN, 1990, 9.

[4] *Special Series (Part 1): Assessing the Damage of the European Banking Crisis / STRATFOR*

[5] Stearns, Peter N. *Globalization in World History* (NY: Routledge, 2010), 9.

counterclockwise movement (opposite of what one would expect from the Coriolis effect) which harmonized with the natural inclination of human culture to move along latitudes from east to west or vise-versa. The currents brought those early Phoenician global traders so they eventually would follow new ideas and the Christian faith through trade. The Greeks epitomized the adaptability of an early western power germinated within the secure natural boundaries of mountains and through mistakes had discovered the failings and success of following the path to empire along the natural borders of the sea. Lacking rivers in southern Europe the Greeks entered the east where fewer natural borders lay. As Alexander the Great demonstrated, a dynamic people carefully controlled but enjoying a measure of freedom quickly absorbed an intensely autocratic state and exploded in size until it reached its natural limits.

Note: As of this writing Greece is facing possible economic default. Interestingly, the lack of rivers in southern Europe stands in sharp contrast to those of northern Europe. What we shall see may dramatically explain why the southern European nations are indebted to the northern European nations today.

The Romans also illustrated how the equilibrium between the power of the state in proportion to its ability to control space through time resulted in its greatness, but also how a state can overstretch to the point where the healthy vitality of its people become overburdened by state control. The increased state control begins to overregulate and control its borders out of proportion to its ability to project control over space within a time frame fast enough to maintain the frontiers.

The Romans initially followed a policy of strong diplomacy and allowed for half-citizenship to those who would agree to be their colonies. The colonies would receive the advantage of trade and remained responsible for their own defense. Hence, a somewhat self-sufficient republic of free farmers continued to maintain a healthy resiliency that demanded individual sacrifice of its people and expected the same of its allies. Responsible for local defenses, these foreign colonies did not need to look to Rome for defense. Roman legions could be kept in reserve centrally located on the Mediterranean from where reserves could be floated at relatively rapid speeds. Ideally situated approximately midway between north and south on the Italian peninsula in the center of the central sea, Rome's legions could also reinforce political control at home and, if necessary, be sent to the borders rapidly. The brilliance of the Roman Empire lay in its ability to use the land intelligently as the limit of its expansion.

As Edward N. Luttwak points out in his book, *The Grand Strategy of the Roman Empire: From the First Century A.D. to the Third*, the Romans used natural landforms, or **vertical or natural borders**, to ensure their legions were effectively used and in expedient manners allowing the empire to rise. The subsequent failures of the Romans essentially existed when they traded in this system.[6]

Using the **Rhine** and **Danube Rivers**, the Romans effectively used natural borders for centuries only requiring a minimum of manpower thereby saving expenses and not overburdening the state with military control. Even as far north as Britain, the geographers had the sense to draw a line close to the narrowest point of the island to build the artificial border known as Hadrian's Wall. It was more expensive perhaps than a natural border but still one that could be maintained at minimal cost. As we have seen, the key to success for the state is to not overstretch its ability to control relative to the time it takes to reinforce and resupply. Look at the peninsula of Korea. Had a line been drawn more carefully, utilizing the available natural borders, the cost in manpower and treasure to support a division of United States troops for the last sixty years there would have been unnecessary. Rome became subject to imperial overstretch when it began to deploy its forces to the frontiers, utilizing the state and not its allies' and colonies' troops and leaving behind the cost savings of natural borders.

In the **Levant** and along the north coast of Africa, the minimal use of troops resulted from the effective use of the **limes**, or series of block houses with prepositioned food and water, and an effective network of communications allowing

---

[6] Luttwak, Edward N. *The Grand Strategy of the Roman Empire: From the First Century A.D. to the Third* (Baltimore: Johns Hopkins University Press, 1976), 192.

the Romans to use manmade "natural borders" of the walls and ramparts to defend against arriving enemies. Tired and dehydrated troops would be met by reserves of legions stationed intermittently along the desert frontiers. The Romans avoided settling the deserts aware of the cost of transport with friction of distance and realized there simply was no profit in leaving the relatively green coastal areas.

Beginning with the deployment of a legion under Titus in 70 AD to modern day Israel, the Romans subsequently and increasingly over the next four centuries began to deploy their own troops to the borders of the empire. Surrendering to the expediency of sending loyal troops instantly to the frontiers initiated a precedent that would see the finite number of legions deployed with smaller returns. The enemies of Rome began to grow unified in their hatred of the empire. As troops increasingly saw deployment, the cost in manpower represented a disequilibrium and eventually resulted in a proletariat or landless veteran class. Expensive to maintain, the self-governing farms looked to the state and the state's expansion increased tax rates. Hence, a vicious cycle began. A lack of individual initiative and fresh ideas may have resulted in poor decisions regarding the use of the frontiers. Additionally, long battles in **Mesopotamia** bled the Romans white as their legions long-accustomed to using the deserts and a lime system of defensive forts and messengers along the natural borders of the deserts in the Levant and North Africa had to be deployed more and more to protect projects in the east.

Some say luxury killed the Romans, but this nomad versus civilized type of argument would be better replaced with the thought of hemorrhaging funds and trade imbalance. Never having met the Chinese, the Romans quickly learned to enjoy the foodstuffs, silks, and luxuries coming across the Silk Road and Indian Ocean from China. Similar to today, the need to debase coinage resulted in excessive production of the coinage since specie no longer limited its production. This rise in inflation may have led to the government's attempts to stop the flow of specie from the Roman Empire by securing the trade routes to the east. People have always sought luxury and new pleasures, and eventually overseas trade would drive the Roman Empire into a weaker position at a time when its enemies had continued to grow more numerous. Concurrent with increased dependence upon Roman central authority the population began a long but steady decline. Human movement has always been a push-pull proposition whereby poor crops push people into areas looking for work while profits tend to lure. The vacuum of population decline and the subsequent uniform recognition by Rome's enemies of the value of working together against a recognized enemy instead of themselves would enhance the decline. The Roman legions found themselves toward the end of their empire bogged down in numerous wars using their military rather than depending upon the client states. Exhausted and broke, the Romans abandoned the vertical borders and forests of Europe they had held for so long and had slowly bled to death in the Middle East and in debt to the Chinese they had never even met. The control of the Roman state led to disequilibrium, and its people declined resulting in a vicious cycle of more central control eventually weakening the state further.

The Roman Empire actually would continue for almost another thousand years after the fall of Rome. When Diocletian had divided the empire, the eastern part remained. The Byzantines continued to thrive with their capital of **Constantinople** located midway between the Black Sea and the Mediterranean Sea. Like Rome, the city had a terrific location on a north-south axis but now at the center of the eastern end of the Mediterranean. Ideally situated on a peninsula where security could protect the city, the control of the vital Bosporus Strait into the Sea of Marmara and through the **Dardanelles** maintained the empire's life blood. A **Byzantine way of war** represented a terrific example for relatively weaker nations surrounded by enemies. Perhaps the strongest factor preserving the state was its foreign policy of only fighting when it could be fairly confident it would win and the use of lightning fast cavalry raids when military action did occur. Using an intelligence network capable of spreading disinformation, it used the earlier Roman methods of diplomatic alliances to offset its own need to use force. Hitting unexpectedly, hard and fast, carefully picking its fights, taking on only those they could win against and ensuring its allies did their fair share, the Byzantines were able to maintain the equilibrium of control over its peoples. A keen intelligence service that understood and thought like its surrounding enemies used natural borders to minimize costs and carefully not over-expanding its ability to use time to project and sustain force. Utilizing natural borders and state equilibrium through a smart foreign policy and the strengths of its geographic position, the Eastern Roman empire continued to survive, even against the energized armies of Muhammed.

The Chinese had then, and seem to have now, what everyone wants. Conversely, the world has seldom had little that the Chinese want. This abundance of natural resources and luxurious materials combined with severe state control has prevented China from leaving its natural borders for long. Despite the brilliance of the Chinese civilization, arguably the longest continuous in history, the Chinese have tended to have a top-down centralized authority. Confucianism sees obedience to the leader as proper as the leadership has the mandate of heaven. The idea of cosmic order and a unified state under firm control had been a prevalent theme throughout Asia from the Persian Empire to China. A culture subservient to the greater whole never seemed to possess the individual initiative or innate curiosity leading to a desire for exploration such as the Europeans would have. Long supplying the Europeans with goods and the original creators of such technologies as gunpowder and clocks it would be left to the Europeans to develop these to their full potential, China meanwhile continued to beckon exotically to a Europe embarking in sailing ships where individual ship's captains commissioned by states sought the individual profits possible from reaching sources of Chinese resources and on the way the Indies.

So why did the great Chinese civilization seem to lack motivation to explore? From a geographic standpoint, the answer again may be explained by natural borders. The Chinese leadership saw itself able to maintain total control over its population generally crowded into an area where the corral could be easily maintained by imperial troops. A cyclical and fatalistic view of time possibly stood at odds with a European linear perspective with an optimistic and progressive view of history; this European perspective was the result in large measure of the eventual printing and distribution of the Bible with its admonitions to be ready for the master's return. **European landforms** did not enjoy the isolation effect of natural physical borders as the civilizations of East Asia did; instead all Europe is bound by seas and oceans. Europe essentially is a peninsula of peninsulas crossed by important rivers. In terms of biomes, the peninsulas of Europe generally saw heavy forestation. The vertical structures of the forest acted in part like the mountains of Greece thereby allowing for easier defense and a heightened sense of individualism. The rivers tended to be deep and served to connect the peoples through these wet highways. Located in cool temperate latitudes ideally suited for labor, the Christian tradition of separation of church and state or rendering to Caesar what was Caesar's resulted in cultures generally less controlled by the state. With independent cities often frequented by the arrival of the Jews whom the Romans had tossed out of their homeland, the idea of profits and independence in chartered and independent towns became both a refuge from an overpowering state and a source of life blood for the rise of absolutism within the nation state.

Peninsulas prevented a complete monopoly of power by any one European leader over the continent. Often separated by physical and natural borders, a rising collection of nation states engaged in geographic competitions. The competing peninsulas also tended to provide for rising nationalistic identities with homogenous populations and cultures. Foreign discourse depended on ports or harbors connected to gateway cities on deep waters reachable by ships. The rise of capitalism in the noisy towns and the combination of competing separated cultures and leaders in part explains why, for example, Columbus could go from his home in Italy to the Iberian Peninsula of Spain for support. Ships could pull into the free and charted cities on the deep water ways and various products crafts could be floated off for trade. Blessed with harbors beyond comparison with any other continent, the Europeans had deep rivers leading to ports on competing peninsulas where a balance of centripetal and centrifugal forces prevented monopoly and encouraged competition.

Time, too, served its purpose. Awakening to the bells in the morning and summoned to sleep by them, the hard workers of Europe learned how to master time. Europe attracted the best and the brightest immigrants using time, efficiency, competition, and individual creativity in the town combined with a profit motive thanks to burgeoning capitalism and an understanding of the mathematics of lending. The Chinese approached South Africa as the Europeans did but withdrew, somewhat bored with what they had seen. The Europeans drove on seeking profits and setting up various empires of enterprise on the high seas.

THE AGE OF SAIL

Work smarter not harder is an admonition often heard these days. During the Age of Exploration, distance continued to shrink and the world moved faster by effectively utilizing time over distance and by the diffusion of new ideas and technology. Circumnavigating the Earth required the use of an increased knowledge of geography and currents. Newly blazed water trails scrupulously followed records of discovery in ships logs. Financial compensation and individual initiative had found ancient alternatives and ways around the Silk Road and true globalization had indeed occurred. The Portuguese, Spanish, Dutch, French, and British each successively invested in this great adventure while other mostly landlocked nations continued to withdraw and fall behind those of Europe.

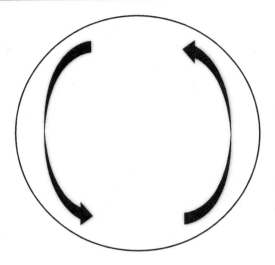

Portugal tamed the ocean currents before any other European nation. Located on the Iberian Peninsula and jutting out into the Atlantic, this tiny attenuated coast naturally attracted a sailing element. The Portuguese set up a trade empire utilizing minimal force in the process and sailed as far as India at **Diu** on the **Gujarat** Peninsula and **Goa** to the north of the **Malabar** coastline to the south. The key contribution of the Portuguese may have been their discovery of the currents dividing at the **Azores** where the westerlies and the northeast winds generally diverged. This vital intersection pushed the North Atlantic current and the Canary currents. Normally the Portuguese would take a turn to ride the currents back home. Knowledge of geography had given the Portuguese a decided advantage and showed how time could change the importance of an empire's relative location on the earth's surface. The real breakthrough for the Portuguese to create an empire came after a decade of trial and error at the Canaries Islands when the Portuguese refused to quit and instead sailed down the coast of West Africa. Using techniques of navigation and understanding the geometry of latitude had brought to them, as David Landes says, "Here Iberia's position as frontier and bridge between civilizations paid off . . . Arab and Jewish astronomers there prepared convenient tables of solar declination for the use of navigators."[7]

Little incentive actually existed for the Portuguese to travel south and even less for the Chinese to go anywhere. The stretch between West Africa and the Cape Verdes Islands had little to offer the Portuguese between **Guinea** and the **Cape of Good Hope**. Distances seemed simply out of proportion to the riches of the east and surely more than simple profits initially drove these bold explorers. What is incredible to note here is the century of perseverance that Portugal had exhibited. Eventually headed east or west to follow their parallel back home, the Portuguese suffered a severe shock when their Spanish neighbors discovered the New World. Contrast this Iberian determination and competition to Asia.

China continued to be directed from the top. While the Chinese by 1500 only granted permission to use canal transport, any attempt to trade outside the nation grew increasingly prohibited. An entrenched Mandarin (a Portuguese term and description from this time) bureaucracy, recently removed to the interior of Beijing, had grown landlocked physically and transformed into uninspired visionaries. The state-directed Chinese increasingly denigrated past adventures as mere wastes of time. The Chinese even faced the death penalty for creating any ship with more than two masts.[8] An official and pragmatic emphasis on short-term investments had tended to stifle initiative and individual creativity under the inertia of state control. With no missionary zeal or opportunity to individually profit or enjoy upward mobility, individual ships' captains followed orders directly from the emperor's throne.

Eventually the Spanish would take the right turn at the Cape Verde Islands and would bring sugar cane with them probably from Sao Tome and Principe off of the coast of West Africa. The capital for these ventures soon ran out for the Iberian

---

[7] Landes, David. *The Wealth and Poverty of Nations* (Norton: New York, 1998), 86.

[8] Ibid, 97.

powers as the movement of capital from the emigration of the conversos and Jews forced the Spanish to seek investments from Genoan banks at fantastic rates.[9] While treasure fleets returned from the New World, the Spanish continued to sink in wealth. Boldly successful, men, such as Cortes armed with Salamanca forged iron weapons, literally tore apart both the Aztec and Incan Empires. Often forgotten in the horrors of the European arrival is the boldness Cortes displayed in burning his ships to prevent sailors from attempted returns home. Only the fiercest individualism described this courage and the upward mobility of such a personality. Conversely, the Native American empires represented extensive but fragile structures governed through fear. Like the ancient Assyrians such fear could easily be turned towards helping any outsider willing to undermine the central authority by enticing ambitious and frustrated individuals. Alexander the Great in the ancient world had exploited the brittleness of the Persians in one of the most rapid transformations of power in history. Likewise, the Native American empires paid the price for their authoritarian control. The Native American empires rapidly collapsed in large measure because of the disloyalties of many of those working within the system.

Eagerly accepting the capital and the capitalists arriving into **Amsterdam** from **Antwerp**, the trend for banks and insurance companies to gather on the banks of European rivers during the Age of Exploration left indelible marks on the cultural landscapes. A relatively free urban class lent money to the Dutch who bypassed the Spanish right turn in the mid-Atlantic and instead followed the older Portuguese routes across the Indian Ocean using the same predictable ocean currents. These monsoon winds had earlier facilitated ancient Arab trade routes ranging from Oman at the edge of the Saudi Peninsula loaded down with their frankincense and confident new religion to Indonesia. Arriving at what was then called the East Indies, the Dutch capitalized on trade from China to India to Europe and protected the vital straits of Malacca. The power of relative location had moved yet again! Controlling the city of Batavia in the vital **Sunda Strait**, the Dutch by the end of the sixteenth century had managed to leapfrog Portuguese ports and eventually capitalized on the spices and treasures of the East long sought by the Europeans.

The Dutch could not compete with the newly emerged French and British competition for two reasons: sugar and cotton. The British Empire essentially, according to Niall Ferguson, "robbed the Spanish, copied the Dutch, beaten the French and plundered the Indians."[10] Though this is perhaps a bit cynical and simplistic, the truth is that the British cashed in on individualism with piracy and enjoyed the most commercially profitable aspect of trade—sugar. Seizing **Gibraltar** in 1715 after the War of the Spanish Succession, the British demonstrated an innate almost Roman sense of the practical in taking essential landforms and controlling vital choke points as carefully as had the Byzantines in Greece. The Seven Year's War of 1756–1763 would deliver India to the British. The British whale would continue to contend against the French lion for almost a generation until the end of the Napoleonic Wars in 1815 defeating a foe with twice as much population and land.[11] Nearly stretched to the breaking point and alienating its North American colonists by its attempt to pay for the wars through taxes, the British nonetheless succeeded in maintaining the necessary equilibrium of control and freedom by their use of time.

At a time when travel across the Atlantic took between four and six weeks, the British had maintained the requisite equilibrium necessary for the empire by extending their reach at profitable opportunities while using natural borders within the effective grasp of their ability to utilize time.[12] As joint-stock companies in Antwerp had accompanied the growth of banking in Amsterdam, the central bank would arrive in Great Britain at a time of relative decline in Central Asia. Remember the old adage, "Time is money." There is a good amount of truth to this according to some monetary theorists today. Utilizing specie (valuable metal) to represent elements of time, one's labor received compensation, and the freedom to invest attracted industriousness, which was often the result of greed. If controlling a boundary of space over time represented a successful empire, then the use of money and the balance of accounts represented a type of time-space balance the British had successfully exploited through transportation and capitalism despite the potential of overstretch as reflected in the Seven Years wars.

---

[9] Chua, Amy. *Day of Empire* (Doubleday: New York, 2007), 135.

[10] Ferguson, Nigel. *Empire: The Rise and Fall of the British World Order and the Lessons for Global Power* (New York: Basic Books, 2002), 43.

[11] Ibid, 30.

[12] Ibid, 140.

As the invisible and numerical profits of capitalism spun their web across Africa, initial excitement over the gold found in **Mali** where the top secret and highly protective "City of No Return," **Timbuktu**, controlled the flow of West African gold into Europe and Asia. The Europeans, on occasion distracted by mythical figures such as Prester John or an El Dorado, expanding neatly through recorded coordinates of latitude and longitude into a trade empire where people who otherwise would never have met traded to the sounds of the ringing bells in the busy European sea ports situated on deep gateway-city rivers and legally independent from totalitarian royal control surrounded by competing peninsulas. Central and South Asia receded from the high water marks of their earlier empires.

The Turks, like all empires, had started strong and genuinely attempted to keep it so. The Turks had sought a means to reinvigorate their peoples for years. As trade grew increasingly difficult on the old Silk Road to China, the customers began to leave, preferring wet feet and cost savings at sea. Again, throughout history, sea transport ranged from 1:35 to 1:100 times less expensive—the more so during the Age of Sail—and perhaps even more importantly, one could at least outsail pirates at sea.[13] The states of Asia—in another east-west comparison—demanded the total allegiance of their people. These nations based on loyalty had not seen the rise of recorded drawn borders reflecting the creation of the nation state at Westphalia in 1648. Bottom line, they simply could not continue to command loyalty. The central control of Asian economies necessitated the purchase of loyalty with land. This feudalistic type of enterprise stood in marked contrast to the wealth being both built up and relented upon European ledgers. Think about it, the Europeans had borrowed culture while the Asian nations tended to remain introverted. Europeans recorded transactions on paper purchased from China, recorded with accurate time from clocks also probably borrowed from the East floated on increasingly new ship designs supplemented with rot-proof wood from the East Indies piloted by nimble ships. Even the guns had been cast by bell manufacturers willing to change over to new production methods of metallurgy. The great Asian powers of the Persian Savafids, the Turks, and the Mughuls of India just could not compete with the capitalism of Europe and, though they continued to engage in global commerce by conquering, they increasingly moved slower and therefore continued to fall behind European enterprise.

The Seljuk Turks initially relied upon the Sultan's "men of the sword" and Persian or Arab lawyers, bureaucrats, and merchants. The military aristocracy, or landowning class, received the right to tax the people in return for service. Nomads had an interest in stability and the recently moving Seljuks established themselves in Baghdad. Following a now-familiar peripheral pattern funneled by the Euphrates and Tigris Rivers, these hard-battened riders of the steppes had burst through into the Middle East and slammed on the front door of the Byzantines during the eleventh and twelfth centuries. The Seljuks took control of the Levant where the Roman limes had protected the eastern Mediterranean 500 years before; a new group of nomads had taken the old Persian Empire once refurbished by Alexander the Great. Reinvigorated during the 14th to 15th centuries by the arrival of a newer Turkish-speaking group of invaders, the Ottomans saw the rise of the **sipahi** (soldier-nobles), a group prized for their loyalty. This "middle class" would eventually decline under government attempts to consistently squeeze them for tax dollars. Eventually, a military class arose to replace the **Ayans**, but land is simply finite. Feudalism did not profit because no one made more land. There could never be enough arable land to buy sufficient loyalty. With a declining merchant class in a culture of central control, frustration seemed to exist in the Middle East as a theology permeated the region demanding dedication to the leader who best personified the religious beliefs a warrior-led, nomadic-type culture. Stifled initiative resulted and disequilibrium over space and time. By not allowing people to move up or fail as their abilities and imagination should dictate, social systems instead ossified. With no give, loyalty remained to be raised along military lines and the result continued to be inconsistent tax revenues, artfully found ways to hide their money from the tax collectors. The Ottomans began to slowly lose their ability their grips on new peoples and places.

European ships, which nimbly plied the seas, often committing terrible atrocities and seemingly always other, grew rich off of the global trade. West African **Luanda** slaves acted as trade items for **Zambezi**

---

[13] *Special Series (Part 1): Assessing the Damage of the European Banking Crisis* / STRATFOR

system eventually spread to the Caribbean sugar plantations, where all profited except the slaves. As the Europeans engaged in the African and Atlantic trade capitalizing off of the insatiable hunger for sugar, the Turks continued to rely upon the loyalty of the conservative janissaries and continued to decline relative to the Europeans as had the Chinese. Eventually the British and French found themselves in India.

The Mughals of India had also once been a vibrant peripheral people riding through the world, becoming Islamized in the process. Like the Turks, these masters of Northern India continued to see growing merchant classes but with international loyalties helping speed up decline. The Mughals emulated the Turks with a Mandsubdari class similar to the Ottoman sipahi. These land-owning and militarized elite could not continue to support Aurangzeb's leadership in a manner similar to the sipahi and the Turks for the same reason: there was no more land to take. The Mughals, like the Chinese, enjoyed the advantages of natural borders but could not remain isolated from the competition for land empires in Asia. Along the Oxus River and the area of modern day eastern Afghanistan, the Persians and the Turks continued to take more and more land on this eternal frontier.[14] At this crucial moment, the British had moved into India.

The British East India Company (EIC) moved into Bengal, Bihor, and Orissa by 1765. The Mughals had lost the allegiance of the Mandsubdari class at a vulnerable time. The Mughals and Sunni Muslims, could not replace them with loyal merchant-nobles before a diffusion of merchant-empires took wealth from the very lands the Mughals had plowed. The Moghals, like the Turks and Chinese, responded to a perceived threat of invasion by Christian merchants with extremely conservative movements similar to the Safavids (Persia), Wahhabis (Arabia), and neo-Confusicanists (China). As these Central and Southern Asian states closed themselves up to the world, Mughal traders found themselves increasingly undercut and bypassed by a different race of people working through an international cabal of business. It appeared loyalty to the state could indeed cost one business.[15] A symptom of the imbalance created by a leader's intense control of the people and attempts to "squeeze them" through taxation revealed a vulnerability to the Europeans.

The tradition of the Indians to hide their wealth by investing in jewels and gold is still strong today. Passing these treasures along to their children hidden from the state and its tax collectors offered about the only opportunity to move up in Indian society. The British once again found ambitious courtiers they could afford and gave those titles. These Nawabs represented a new zamindari class in furthering British interests from Calcutta. The British offered investment opportunities while moving in from their ports. The British had indeed learned a lesson in North America where they had lost the colonies to a people who said, "No taxation without representation." The Turks and many of the Asian powers had for the short-term solidified their cultures and the grips on their peoples by the 1850s, but they had continued to fall behind in relation to finance as western banking continued to create European inroads by buying off the loyalty of indigenous peoples and incorporated them into a global empire. The French and Dutch had lost control of India to the British but profited immensely from the resource of cotton.

The use of cotton would result in worldwide demand. It was comfortable and hygienic because it could be easily washed. ⁓ the tropical climates the fabric found particularly great use in the Caribbean where sugar plantations had diffused from Sao ⁓ Principe. Those sugar workers had originally been funded by the Calvinist Dutch who busily attempted to invest ⁓ French in Haiti and the British in Jamaica both would reap enormous profits in the sugar business. Said to ⁓tive substances on Earth, the European demand for sugar again seemed insatiable. The plantation ⁓d with labor supplied by the horrors of slavery moved a globalized economy into a new ⁓s and seeking more cotton, Europe readied for the Age of Steam.

⁓ the relative decline of their continental rivals, the Turks, Mughuls, and Chinese. ⁓ur, Mongolia, and Siberia by the early 1800s. The Manchu's of China were declining ⁓e early 1800s. Russia, like the powers it would expand into, maintained strict con- ⁓nder the Romanovs, were able to harness the middle-class merchants into their own

⁓. *Modern Imperialism and Colonialism: A Global Perspective* (Boston: Prentice Hall, 2011), 29.

control. Those who did not obey or follow instructions were simply crushed and forbidden to do business.[16] The Russians used the religious elites to help control trade and spy on nobles. They carefully controlled the movement of Russians and Ukrainians into Siberia.[17]

The Russians also used the Eastern Orthodox Church to control their subjects. The western tradition of a separation of church and state rule did not extend to the Byzantines; the Russian Orthodox Church also remained beholden to central authority. The Russian government offered military aide to one group versus others as it sought to divide and increase control of its subjects and foreign peoples. Use of movement of people to confuse loyalties and the destruction of livestock reflected the totalitarian control of the Russian state. Similarly to the Central Asian states, the Russians could only expand their holdings in their attempts to grow. In Poland, for example, noble loyalties were bought by absorption into the system but total replacement served for control in Kazakhstan.[18] To this day, the challenge for Russia is to raise revenue by expansion since it does not have access to sea trade. Likewise, the test for China and Iran is the need to develop blue water navies and shipping capable of negotiating the seas; it appears both China and Iran are working hard toward this end in the South China Sea and the Mediterranean respectively. As of this writing, strategists are pondering Russia's next move for expansion. Future movements must focus on the Baltics or the Mediterranean if the Russian regime is to survive. Without the access to free ocean trade long enjoyed by the western Europeans, the Russians cannot grow financially secure.

As the Russians, the Persians, Mughuls, and Chinese continued to control their subjects through various means, the British continued, forcing their way into China using bread and opium from **Bengal** as weapons of trade. The Chinese now, in an opposite or converse condition to that of the Romans earlier, saw their specie drained and peasants starved. The Dutch also found themselves forced out of the East Indies and the British took charge of Malacca and set the stage for their domination of the water routes to the east.

The Age of Sail saw the development of the first true global empires where families of power controlled non-contiguous lands in distant shores. An equilibrium existed between the space of the empire and the amount of expenditures it took to maintain these enterprises primarily due to the relatively inexpensive use of water-borne commerce, transport, and the geographical understanding of the world's currents and winds. The time needed to traverse these watery roadways consisted of hour glasses at the beginning of the Age of Sail, but, by the end, clocks had reached levels of such accuracy that a pocket watch would have a second hand. Intelligently using the natural borders and phenomenon of the globe in a manner compatible with the time it took to traverse would demonstrate the power of the invisible realm of numbers that provided banking tables and charts, ballistic accuracy, and free enterprise. Capitalism and the promise of upward mobility bought loyalty with money. Indeed, a measurement of time and increased transportation speeds would only be enhanced by science and technology during the Age of Steam.

THE AGE OF STEAM

The Age of Steam would see the world move faster and people moving more than ever. The Europeans would dominate a world growing smaller but, in many ways, more distant. Among the European nation-states, the United Kingdom would dominate the world. During the Age of Sail, the British followed the Roman Republic and the early Roman Empire's tendency to maintain an equilibrium of control of its frontiers, ensuring time and space aligned through the use of a commercial sea empire operating at a minimum of cost and maximum speed. The British, and subsequently the Americans, would likewise similarly master the industrial revolution for numerous reasons, but eventually would themselves become subject to imperial overstretch. Restated for the reader; overstretch is an imbalance between a state's ability to control its borders and space in proportion to time. Competition between the imperial powers of Britain and France with

---

[16] Getz, Trevor, and Heather Streets-Salter. *Modern Imperialism and Colonialism: A Global Perspective* (Boston: Prentice Hall, 2011), 126.

[17] Getz, Trevor, and Heather Streets-Salter. *Modern Imperialism and Colonialism*, 135.

[18] Ibid, 130.

a newly-emerged land power, Germany, would eventually lead to the First World War and set the stage for an American empire of the air and space.

The world of the Industrial Revolution moved faster and faster. The Industrial Revolution diffused through Europe from **west to east** and from **north to south** at an uneven rate. Why the Industrial Revolution spread from the island of Great Britain to the mainland the way it did can be explained through the perspective of physical and cultural geography. Europe is essentially divided in the center between a maritime climate to the north characterized by cool moist weather, and a Mediterranean climate to the south, typically warm and dry. The divergence in rainfall is about **30 inches** and grain production and literacy drops precipitously as one goes from the northwest to the southeast. With the exceptions of **Catalonia** and the **Po Valley** of Italy, the southern and eastern portions of Europe do not reflect the cultural traditions of biblical literacy abounding in Protestant countries. Though it may be extremely unpopular in the present time to point out these differences, the truth remains that the British received the Industrial Revolution first, and geographers seek to determine how relative location factored into this. As David Landes points out, this approach is at odds with the trend towards studying a core and periphery, but there is also a function implicit in the diffusion of the Industrial Revolution from the educated to the illiterate, from representative to despotic institutions, and from those cultures of equality to those with hereditary traditions.[19] Wages differed by as much as eighty percent between western and eastern Europe in the last century. So what exactly made the Industrial Revolution, and why did Britain lead it?

The ability to harness resources, particularly minerals beneath the earth's surface, set the British apart from the other European governments where the crown could claim propriety to what the Creator had laid below. The British tradition of private property ownership had also played a part in the use of fossil fuels going back as far as the 16th century. It was iron that made the real difference. At a time when the forests of Britain had been largely depleted for use in the long Napoleonic wars whose rivaling navies had hungrily devoured trees for masts and beams, the availability of iron and coal for refinement would give the British the "leg up" on casting iron. Once iron was cast, it could be used for transport. Building upon the method of Sir Henry Bessemer, refining steel had begun in 1856.[20] It was steel that enabled man to harness the inherent power of steam, and the world would never be the same because of it.

The railroad and steam engines would give the British the time they needed to secure their empire against rivals in the world race for resources and coaling stations. Between 1870 and 1900, shipping tonnage tripled[21] and travel time decreased dramatically. By 1825 the British could reach Calcutta in only 103 days.[22] By the 1830s, the British could cross the Atlantic in two weeks' time. By the 1880s, this would be reduced to ten days. The ability to transport commodities such as cotton at such increasing rates of speed occurred with the development of the separate steam condenser which had made steam profitable by 1768 and provided the power necessary to move the engine on wheels by 1783. It was high pressure that made ocean trade possible, but ships could not go in reverse until the **steam turbine** emerged replacing the reciprocation system.[23] The installation of a **surface condenser** combined with the steam saver allowed ships to steam indefinitely on water by maintaining a nearly continuous circulation of fresh boiler water. With the technological advantages of resource minerals, the British could harness the potential of companies and banks to minimize risk and make the capital investments necessary to make the pursuit of heavy industry profitable.

Now the world would transport people on a stupendous level and history would see the greatest mass movement of people in human history. Additionally, steam technology would create steam shovels so canals of unprecedented size could be built. Following these steel transport lines at sea would be the railroads and telegraphs on land. Because the cost of sea travel had dropped by as much as two-thirds the cost in the 1820s, the world could afford to travel, and travel they

[19] Landes, David S. *The Wealth and Poverty of Nations* (Norton: New York, 1998), 252.
[20] Boot, Max. *War Made New: Technology, Warfare, and the Course of History* (New York. Gotham Books, 2006), 176.
[21] Stearns, Peter N. *Globalization in World History* (New York: Routledge, 2010), 99.
[22] Johnson, Paul. *The Birth of the Modern* (HarperCollins: New York, 1991), 201.
[23] Landes, David S. *The Wealth and Poverty of Nations* (Norton: New York, 1998), 188.

did. Between 1815 and 1840, approximately one million people immigrated into British America. According to Paul Johnson, someone with only ten pounds could afford a trip to New York and had to produce no paperwork before doing so. Such was the ease of travel that led to this huge human movement.[24] With time on their side, the world system had collided with traditional western values. In the last chapter, we studied the social changes that accompanied the growth of capitalism. What are some of the cultural changes we see during the Age of Steam?

The beginnings of what today we call international non-government organizations (NGO's) occurred during the Age of Steam. By 1878, the Universal Postal Union's creation had allowed any nation's postage stamp to be internationally recognized regardless of destination. It's hard now to realize that the postal system at this time presented a much greater opportunity for interpersonal communications than did the "talking wires" of the telegraph which traced its development to 1832. And as these telegraphs followed the new steam engines, since U-turns on railways are scarce, the trains had to run on time. Standardized in 1884 with the International Meridian Conference, the world set its clocks to the one in Greenwich England. A level of syncretic global behavior cooperating on a scale never before seen had arrived. Various Geneva conventions allowed for the humane treatment of soldiers, and various feminist organizations beginning in the 1880s would eventually attempt to stop foot binding in China and would attempt to eliminate the "White Slave Trade."[25] Some of the reactions by other nations to this explosion of generally western, and particularly British, presence accompanied social confusion that presaged troubles for the future. In Japan, for example, the people seemed to be unable to understand how an adversarial political body such as the United States Congress could operate and still be friends. In Egypt, confusion set in over whether women should wear veils or not. Men's western wear began to diffuse over the world, and by the 1890s, suits could be seen in China.[26] Parisian department stores would affect the American way of shopping as would Van Gogh be affected by Japanese art. The Japanese become infatuated by baseball towards century's end.[27] Cultures coming into contact created a different type of frontier whereby elements of one society diffused into another in some ways and conflicted in others.

With the diffusion of cultures globally and with an understanding of the pattern of European moments of industry, how did America fit into the Age of Steam and why? Since America will dominate the next chapter's discussion of physical geography we will wait, but in terms of culture, it is important during the Age of Steam to see how the machines in Europe came to America and more specifically New England. Less beholden to a cottage industry of individual craftsmen working from home, Americans willingly sent their daughters to work.

Standardization appealed to an egalitarian America not burdened by a long tradition of social classes. The tomato is an example of the differences between Americans and Europeans. Americans would happily eat any food regardless of what others thought. Long memories in Britain, however, recalled the nightshade poison and fearful acts of regicide by these plants. Tomatoes are a fruit related to the nightshade family of plants, and potatoes are likewise a vegetable of the same family. Botanical trivia aside, because of the hierarchical diffusion of acceptance, the better-off folks in Britain simply wouldn't eat tomatoes for quite a while since it remained feared by royalty and the upper classes. In the United States (and Ireland) these foods represented fair game and proved quite tasty! How did this culture so open to the practical manifest itself in the Age of Steam? It proved to be the fastest to accept new ideas and peoples!

After the War of 1812 a spirit of egalitarianism had swept through the nation and maritime traffic on the American Great Lakes would see America become the fastest moving nation in a very fast world.[28] American and Canadian lumber prices represented the lowest in the world. The availability of resources thereby enabled America to develop shipping at much less cost, and finally American lakes did not require the investment in canals such as overseas. Most of Europe had to dig canals for transport except the British, who had generally been able to depend upon river travel. With the Great Lakes,

[24] Johnson, Paul. *The Birth of the Modern*, 202.

[25] Stearns, Peter N. *Globalization in World History*, 109.

[26] Ibid, 114.

[27] Stearns, Peter N. *Globalization in World History*, 114.

[28] Johnson, Paul, 194.

Americans had a transportation advantage due to only minimal need for the development of canal infrastructures. By utilizing resources, transport corridors on the lakes, a propensity towards standardization, and a willingness to work, we see the most biblically literate society in the world particularly adept in the manufacturing of firearms.

New England manufactured firearms, but it was in the South and the expanding western frontier where they remained so eagerly sought after. American firearms manufacturing was led by brilliant men, such as Eli Whitney and Samuel Colt as well as all of the inventors of virtually all machine guns: Isaac Lewis, Benjamin Hotchkiss, John Browning, and Hiram Maxim. In fact, Americans seemed to have an innate intelligence of staying out of fights and selling weapons to the Europeans who seemed equally unable to stay out of brawls with each other. Times have changed. The United States built its industrial potential during the Age of Steam but would appear in a grand way in the future. The scramble for world resources to feed the insatiable machines continued.

The Industrial Revolution created amazing machines and the Age of Steam saw mankind in many ways tasked with feeding these mechanical behemoths. The need for resources required coaling stations that would dot the great expanses of the oceans, providing ships fuel and water every few thousand miles and up to 3.5 thousand for extremely large vessels. With about thirty-three of these holding prepositioned supplies throughout the world, the British were able to minimize expenses.[29] Rivaling room on the ship was precious cargo and the break bulk points required the building of canals and railroads. In particular, the **Suez Canal** would draw the British in the same direction and in the same part of the world as the Romans had been pulled by globalized trade: the Indian Ocean. Originally, Napoleon had been attracted to Egypt, but it would be the British who would build the canal. Like the Romans, however, the British found themselves constantly drawn toward the edges of frontiers.

Winning overwhelming victories, the British used **"butcher and bolt"** techniques very similar to those of the Byzantines by sending in shock troops of infantry armed with superior firepower to "pacify" any cultures seeming to interfere with Britain's attempt to secure the canal. Initially using only 2.5% of their defense expenditures (compared with about 20% of the United States in recent years), the British successfully utilized client states and client peoples like the Romans, but eventually London would succumb to an arms race with catastrophic consequences.

After the Indian Massacre of 1857, the British had been impressed by the loyalties of certain people who stuck with them throughout the ordeal. While the British government pondered reforms and Victoria became the Empress of India, the British from **Waziristan** north through **Khyber** and **Malakand** used lightly armed militia and friendly tribal auxiliaries as their first line of defense, exactly as the Romans had done against the Germanic tribes even as they built forts and improved roads to expand their influence. The British used Indian troops because of the costs in maintaining whites in the tropics due to losses caused by disease.[30] The British Army, using a "butcher and bolt," or rapid-deployment force, kept India securely within the empire at minimal costs. Eventually, the British would blunder into Afghanistan in attempts to prevent the Russians from achieving diplomatic victory with the Persians in reaching the Indian Ocean. This "Great Game" began to reveal that the British had begun to fortify the frontiers of their empire and by the end of the 19th century, the situation would be right for imperial overstretch with the arrival of a perceived threat from Germany. By the end of the nineteenth century the Germans not only had mastered organic chemistry, but metallurgy as well. Subsequent inventors added alloys of nickel, chromium, and manganese and utilized spray methods, thereby allowing the German Krupp manufacturers more space for guns on battleships since the armor could be made thinner. A rivalry at sea had been created where the German lion would attempt to swim after the British whale.[31]

Steam ships eventually needed to be replenished with fuel on land. Transport of resources required railroads and enormous areas of land as intermodal transport nodes arose. Africa offered an opportunity to build a railroad from the

---

[29] Crick, Timothy. *Ramparts of Empire: The Fortifications of Sir William Jervois, Royal Engineer 1821–1897* (University of Exeter Press: London, 2012), 126.

[30] Kaplan, Robert D. *Imperial Grunts: The American Military on the Ground* (New York: Random House, 2005), 253.

[31] Boot, Max. *War Made New: Technology, Warfare, and the Course of History* (New York: Gotham Books, 2006), 176.

Cape of Good Hope to the south in South Africa to the Suez Canal in the north. Involvement in the "dark continent" led the British into various wars as they developed a new frontier in the Age of Steam. The deployment of troops in "butcher and bolt" operations transitioned into more permanent employments of troops and native auxiliaries whereby a type of over-stretch occurred and an arms race with Germany started.

The arrival of German battleships off of **Zanzibar** in East Africa contributed to the failure of the British attempt to build a transcontinental railroad from the Cape to Cairo. Britain's perceived threat to its colonial empire from Germany forced an attempt to match the Germans who now had experienced the Age of Steam. Embracing industrialization in a big way, the Germans had the second largest economy in the world and the largest in Europe. Leading the way by 1913 in steel production and chemistry, the German population of 65 million represented an unprecedented explosion in the growth of a nation-state with unification only occurring by 1871.[32] It seemed by the dawning of the 20th century, the British, on the hunting fields of the elite and the cricket courts of the people, had forgotten the lessons learned from the Romans that had served them so well. To quote the great Robert D. Kaplan, the "legacy of Rome-like Britain . . . [represented] a 'doctrine of inevitable reversibility.'" As Rome could not stand everywhere, it had likewise emphasized rapid strategic reaction rather than continual presence in too many areas.[33] With the rise of Germany, it would appear the British greatly feared the time had arrived to step down from global leadership in the Age of Steam.

A unified Germany represented a fantastic phenomenon militarily. The year before Germany emerged, 350,000 German troops had been sent on railways into France at the rate of 2,580 per day.[34] These troops represented reservists, not an expensive standing army. As the world's population exploded by over **70%** between 1871 and 1914, in large measure due to the better diet resulting from increased global trade and the ability to transport food during the Age of Steam, the Germans also continued to plan how to transport troops more effectively. The Germans studied the problems of pre-staged supplies and scrupulously learned these lessons until it could **mobilize** and move to the battlefield 1.5 million men by 1914 within ten days to the front at a rate of 11,530 per day.[35] Against this incredible efficiency it is not hard to see why the British used the pretext of the violation of Belgian neutrality to justify its entering into the war against Germany before this industrial juggernaut could possibly grow even larger. It is certain no one in Britain wanted to fight another Napoleonic land power ever again.

Psalm 14:34 tells us that "righteousness exalts a nation and sin is a reproach to any people" (NKJV). It would seem the fate of empires is to grow as long as their resources and use of time are employed in balance with the size of an area they seek to govern. Once a state of disequilibrium occurs we have a situation ripe for "imperial overstretch." The British people had hitherto successfully utilized their God-given resources and through thrift, honesty, and diligence had used these wisely through the Ages of Sail and Steam. Eventually, the fears of an island people dependent on foreign goods brought the United Kingdom into war with Germany, beginning the decline of the British Empire and the emergence of an American empire focused on air and space.

THE AGE OF AIR

This text has argued throughout that a state is like a circle with a circumference representing frontiers or limits of expansion. To maintain a healthy empire throughout history, space and time must be proportional; if either gets out of balance the risk is a type of "imperial overstretch," or stagnation as we have seen. Throughout the Age of Muscle, natural borders helped to maintain this equilibrium by saving resources, the representation of time-money. The use of rapid deployment-type forces characterized by small military footprints had sustained the British as it had the Byzantines. The natural frontier of the sea, and the cost savings and speed inherent in the medium enabled these powers to maintain their empires for centuries.

[32] Crocker III, H.W. *The Yanks are Coming* (Wash D.C.: Regnery, 2014), 17.

[33] Kaplan, Robert D. *Imperial Grunts: The American Military on the Ground* (New York: Random House, 2005), 13.

[34] Van Crevald, Martin. *Supplying War: Logistics from Wallenstein to Patton* (Cambridge: London, 1980), 112.

[35] Ibid, 113.

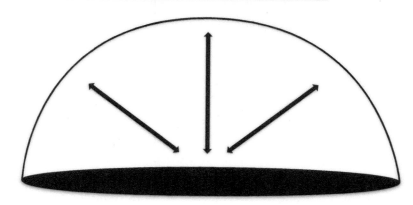

With the Age of Sail we saw the successful empires again engaged in commercial endeavors exploiting the medium of the atmosphere and utilizing the power of capitalism and expanding at minimal cost while other states atrophied due to their inability to adapt to the new reality. The power of geography worked with capitalism to reveal winds and currents offering energy for transport at minimal costs. The British and the United States represented the ideal spirit of individual initiative unbridled by excessive state or empire bureaucracy and control through regulations. The Age of Air took the model of the circle we have used in earlier chapters, but it can also be represented by a hemisphere. Like a circle, the model of a hemisphere represents the ability to project control of the state at a rapid speed to protect frontiers, but, in this case, it is more of a "three dimensional effort" with the aircraft moving through an air medium surrounding the spherical earth. By the time World War I occurred in 1914, the United States had already embarked upon a course to become an empire of the air.

The United States, due in large part to its geography, pioneered aviation and used the air platform to become the premier leader throughout the Age of Air from 1914 until the present. As air technology improved during World War I, the United States utilized the air medium to its fullest potential and would eventually use it to win essentially a resource war against the Axis powers in World War II. By the end of World War II the seeds had been planted for an imperial overstretch of a Pax Americana. The ability to transcend both inner and outer space simultaneous to the development of Air explains how the United States continues to manage a slow decline. Before we get ahead of ourselves we must answer the question, why did America lead the way into the air?

The American Civil War still represents the seminal event in American history. The remarkably peaceful end to this terribly bloody war allowed the nation to quickly heal. The sectionalism of the United States represented a typical difference of northern and southern latitudinal cultures such as we have earlier seen in China and Europe. The difference in North America would be the **Mississippi River**. Flowing from north to south, the control of this vital interior waterway, the result of a natural funneling of the winds and rains from the **Gulf of Mexico** between the Rocky and Appalachian Mountains, ensured the domination of northern culture over southern when northern industrial culture exploited its perpendicularity to the westward-expansion of southern cultures. The Union actually emerged richer than the South by war's end and the factory system discussed earlier ensured that an intelligent and moral work force dedicated to egalitarian principles would emerge. A millionaire class with available capital and freedom of the laborer's imagination would merge in Ohio.

Cincinnati had long been visited by the Germans in the late 1800s where American tools of the highest quality had enticed customers from all over the world to line up to purchase items "made in America" on the banks of the **Ohio River**. In **Columbus, Ohio**, the Wright brothers developed the genius idea for an aircraft. The genius did not lay in the design of the aircraft so much; it was in the propeller. This spinning type of wing could create the lift necessary for a lighter than aircraft using an internal combustion engine utilizing aluminum. **Aluminum** quickly became a resource of great value that was seldom found and only available in a few places in the world. The Wright brothers' engine would represent an even greater need to procure resources. The frontier of the Age of Air similar to the Age of Steam would essentially be marked by the ability to procure and deny resources to the enemy of the state. Fortunately for the United States, World War I started without it. The United States had time to grow in a relatively free and uncontrolled environment. With numerous harbors, a terrific coastal water way for transport, and the aforementioned Mississippi river ideally suited to transport farm products and industry to the world, America grew into a two ocean superpower at a time when sea travel predominated and in time to harness the power of steam while staying largely out of foreign wars.

World War I would see the eventual development of the potential of the aircraft to create an empire of the air. The inherent strength of air is its ability to be used to travel great distances. Able to detect German armies moving on Paris, aircraft delivered messages, appropriately organized into the U.S. Signal Corps. A key power of the air is still the ability to see, transmit, and receive communications through the atmosphere, or inner space. The fixed wing aircraft would emerge as the premier platform while the resource of oil would simultaneously grow in importance to a world increasingly dependent on both motorized transport and the internal combustion engine. The means to winning the First World War would sow the seeds for an even more terrible war to follow and one where aircraft would continue to play a key role. The United States, slow to enter World War I, would emerge richer as the European powers bled each other white.

On September 2, 1916, the first fixed-wing aircraft shot down a lighter-than-air aircraft or dirigible over England to delighted crowds below. Many theologians today see World War I as the beginning of the birth pangs of the end times discussed in scripture, and one can imagine the fear the British must have had of these giant silent aircraft lurking behind shadows on moonless nights waiting to drop the infamous 500 pound "blockbuster" bombs upon the hapless civilians below. The aluminum girders of the zeppelin could only come from the Caribbean island of Jamaica in the Greater Antilles. Other than Australia and Ghana in West Africa, only three percent of the world's aluminum could be found elsewhere.[36]

The Zimmerman Telegram aside, the U.S. fear of German strength in the Caribbean can be easily overstated, but the key is to see how the fear of the development of the German navy discussed in the last chapter could easily be extended to the New World and the air above. Other issues between the contending powers in World War I likewise saw continued challenges in the air.

While the Germans pursued aviation with fixed-wing aircraft as well as lighter-than-air aircraft, they did so in a different manner. The autodyne would have an advantage over speed, but the autostat would be able to remain on station virtually permanently and required a minimum of energy. The rise of the autostat in the current times can be seen with the unmanned aerial vehicle (UAV). Beginning in World War I, the Germans would tend to view aircraft in a tactical close-air-support role. The most dangerous of tasks was flying above enemy trenches dropping bombs and strafing with machine guns in a canvas covered, wooden craft. Lining their cockpits with lead, the Germans early on saw the aircraft potential as "flying artillery." This is not to suggest the Germans neglected strategic bombing, however.

Following up their zeppelin attacks with heavier-than-air and fixed-wing bombers, the Germans led in strategic bombing as late as 1918.[37] The Germans, however saw the aircraft more as an extension of the army and projected its strength within the context of the ground scheme of maneuver. Never moving beyond a two-engine bomber, it would be left to the U.S. to develop the first truly strategic bomber, the B-17, capable of traversing the Pacific and able to reach U.S. territorial acquisitions previously gained in those heady days of expansion after the Civil War. The speed with which aircraft could move also dictated the changing frontiers of this time, going from 120 miles per hour in World War I to 469 miles per hour by 1939.[38] While the nations fought in the air, things also speeded up on land.

The automobile in the United States doubled in numbers between 1916 and 1918. The need for oil drove the Dutch Shell company into Persia seeking to increase the amount available, and quickly the combatants realized the importance of oil. The American proclivity towards egalitarianism produced a market for standardized automobiles made on the assembly line by Ford's five dollars a day; workers received twice the wages of other rival manufacturers and were now able them-selves to buy the car of their dreams. The real saving grace for the Allied powers in World War I was America's oil supply,

36 Collins, John M. *Military Geography: For Professionals and the Public* (Washington, D.C.: National Defense University Press, 1998), 156.
37 Yergin, Daniel. *The Prize* (New York: Simon and Schuster, 1991), 172.
38 Boot, Max. *War Made New: Technology, Warfare, and the Course of History* (New York: Gotham Books, 2006), 208.

however. Possessing two-thirds of the world's supplies until 1941, the United States had much more to offer than just workers and new methods of manufacturing; it had resources.

The British had learned the importance of strategic bombing and how it could affect communications. During the **Second Battle of Armageddon** in September of 1918, General Allenby's forces defeated the Germans and their Turkish allies by initially knocking out the telephone switchboards. Here we see an important concept: the air has an invisible element. Just like the wind filling the sails of the explorers was supplied by the interest rates on European accountants' ledger sheets, so too did an invisible element accompany the ability of the state to control the air: space. Space both inner and outer will be discussed in detail in the next chapter, but it is important to understand how the theoretical scientific and mathematical approach to technology had reached a point where science could, for the first, time begin to anticipate the manufacturer's next move. This **"invisible realm"** of numbers would sustain America as it transitioned from the Age of Air and into the Age of Space, but this puts us ahead of the story.

After the British Grand Fleet (whale) knocked out the German High Seas Fleet at the Falklands in 1914, the Germans could only fight back with submarines in what had essentially become a modern war of resources. No game is ever won on the defense, and with the British fleet able to ply the seas to overseas markets, it received resources needed for the ultimate victory on the ground. America's rise to greatness resulted. The end of the war would bring crushing debts and high taxes would increase the dependence of victorious populations on the government. With this control of the population the disequilibrium of imagination that allowed empires to slow down relative to others would begin. As the Germans turned to National Socialism, the Americans likewise would see their fund raising and control measures from World War I under the War Industries Board (WIB) essentially converted into the New Deal during the 1930s. America's advantage remained an ability to harness the imaginations of the individuals and institutions since the government had much less control. America readied itself to fly ahead of the competition between the wars.

During the period between the wars, the balance between government control and a free people resulted in great gains in the ability to pioneer the potential for air power. While America explored efficiency and new flight routes, other nations began preparing for the next war of resources to come. American pilots at the end of World War I found themselves able to go into a private airline industry. Barnstorming and delivering mail throughout the country, an unregulated air industry emerged. Obtaining massive flight hours of experience and discovering **great circle** flight patterns (see the book introduction for an explanation of the great circle concept) these entrepreneurs worked for the customer, not the government. Because the government ran the airlines in Germany for example, inefficiency tended to restrict flight hours and opportunities to devise new and better means of air travel.

In the United States, a culture of the air began to grow and air travel was facilitated by the outstanding DC-3 aircraft, a dependable model used highly in World War II. The world's first truly strategic bombers developed in Russia with its enormous land empire and in Britain which had, after all, launched a strategic airstrike on the zeppelin pens of Germany at war's end. It would be in the United States, with its new continental and Pacific empire, that a truly strategic platform emerged. It would be the American B-17 produced in massive numbers that would represent the ability to project force and control over space in the most remarkable time of up to 287 mph; it could attain a flight ceiling of 35,600 feet and could go 2,000 miles, and the airlines industry had maintained a supply of pilots ready to fly it.[39]

World War II started from a spatial context because of the need for **resources**. The Second World War would see rivaling states attempting to overcome space by denying resources to their competitors. The Japanese attacked the United States at Pearl Harbor because of resources, and Germany epitomized the initial advantage of intense control but would eventually lose due to its inability to maintain control of the air. The predominate effort to control the frontiers of Western Europe and the littoral of East Asia would begin at **Pearl Harbor, Hawaii**.

---

[39] Boot, Max. *War Made New: Technology, Warfare, and the Course of History* (New York: Gotham Books, 2006), 208.

Japan consists of four major home islands and had a growing population almost half that of the U.S. population today but in an area the size of California. The amount of arable land, however, is less than that of the state of Iowa.[40] Japan's need for resources had taken them east across the **Sea of Japan** and the **Tsushima Strait** to **Korea** and into **Manchuria**. The United States had limited exports of scrap metal into Japan and established a precedent of economic coercion. It was not, however, until the Vichy government of Vietnam in July 1941 allowed access to the Japanese army that the U.S. intervened by hitting the Japanese with embargos on oil. The British and Dutch followed the U.S. and forced a war in the process. Strategic stockpiles of up to five years' worth of required industrial supplies represented what nation's sought in the twentieth century. Japanese stockpiles of required resources for its growing population had declined to a one year supply (what most nations have today). The Japanese Navy General Staff ordered Commander Minoru Genda, an air war expert, to attack the American base in the central Pacific at Pearl Harbor with the hope that it would take a year for the Americans to be able to counter attack after a knock-out blow. Time would allow Japan to gather and utilize the resources it needed to create an unassailable Pacific that would require the Americans to sue for peace. The U.S. had expected and geared up to fight the Germans but had failed to see just how important resources had become to the Japanese. The Japanese also underestimated the power of resources.

The naval base at **Pearl Harbor** had oil lines crisscrossing the island of Oahu, Hawaii. Most naval bases, unlike airbases, have refined storage tanks for petroleum connected to the piers. Even more vulnerable to explosion than coaling stations, these vulnerable lines appeared invisible to Japanese pilots who in a traditionally symmetric fashion attempted to hit their enemy counterparts—the aircraft carrier. The Mitsubishi "Betty" bomber was an amazing weapon that could fly 3,700 miles by the late 1930s, and the new line of Japanese Aircraft carriers could plow the ocean for 11,000 miles without refueling.[41] This ability to get to Hawaii and back made the attack on Pearl Harbor possible. The myopia of focusing on like systems so common to the military with a desire for symmetry had led Japan to miss the potential of hitting the American oil supplies. Had they done so, their war aims may have been accomplished.[42] The American carriers happened to be gone, and the Japanese settled on hitting the battleships and ignited what would essentially be a war of extermination in the Pacific.

Submarines would eventually win the war in the Pacific. Japan is an **archipelago** of islands limited by resources. The Japanese mysteriously failed to see the need to defend an ocean empire from the very technology that could starve it. An air campaign of the most ferocious nature imaginable would burn virtually every Japanese city while American ground forces would simultaneously advance under cover of air. Using both close-air support and strategic bombing, the war for resources would be won by America. Likewise, the Germans would be defeated through the war of attrition and the eventual destruction of resources in its two front war with Russia. The most noteworthy aspect of the air campaign against Germany for the purposes of this study was the pushing into higher elevations. The frontier in the Age of Air consisted not only of reach but of altitude as well.

Prevailing winds favored the island of Great Britain over Germany. The wind blowing in the face of the Luftwaffe (German Air Force) was as serious an impediment to empire during World War II as wind currents had been in an earlier age. By using radar, the British greatly enjoyed the advantage of gaining altitude before the Germans could hit. With the winds at their back, the British could dive upon the desperate German pilots during the Battle for Britain. For reasons similar to the Japanese failure to hit American oil supplies at Pearl Harbor, the incredibly vulnerable radar towers stood relatively unscathed by German aircraft and could have spelled the difference between victory and defeat. As the contending air forces desperately sought to more concisely control distance through time, the means of destruction of enemy resources shifted from attempting to hit strategically important resources such as **Ploesti, Romania**'s oil to Stuttgart Germany's

[40] Collins, John M. *Military Geography: For Professionals and the Public* (Washington, D.C.: National Defense University Press, 1998), 159.

[41] Boot, Max. *War Made New: Technology, Warfare, and the Course of History* (New York: Gotham Books, 2006), 255

[42] Perret, Geoffrey. *A Country Made by War* (Random House: New York, 1989), 364.

ball-bearing plants in the industrialized **Ruhr Valley**. Eventually turning to terror such as employed against Japan, a two-front war would destroy Germany. The earlier mentioned ability of the Germans to master chemistry came in handy as synthetics proliferated to meet the needs for petroleum. The most successful use of air against Germany came against the rolling stock from rail yards, the shape of which virtually guaranteed success for an aerial platform along a single axis. By destroying vehicles, German army tanks and vehicles became immobilized, and, as the United States struggled across Europe facing its own energy shortages due to an inability to capture adequate ports, the German army eventually became overwhelmed in a two-front war. The dying Nazi monster still produced aviation technology that pushed the air frontiers.

The use of a guided missile off of **Salerno, Italy** in 1943 represented the increased control of the air in years to come. Missiles would be a German specialty as they regularly hit **Amsterdam**, **Antwerp**, and **London** with the V-1, and later the V-2 rockets, reaching a maximum height of about 110 miles while traveling over 3,500 miles per hour at a distance of 200 miles. This pioneer of the Age of Space was really nothing new. The desperation to preserve their state and its access to resources had led the relatively impoverished Japanese to employ a bushido-warrior culture to put an emphasis on the offensive. Neglecting oil at Pearl Harbor and ASW (anti-submarine warfare) in the Pacific, the aircraft tended to have little armor for defending the pilot. The pilot seemed to be viewed as a sort of expendable resource by the state and the result was a high loss of trained pilots. The need to maintain the offensive attack continued with a man-operated "cruise missile" in the form of the kamikaze. While **Doppler effect** directed V-2 rockets fell by the thousands over northwest Europe, the Japanese launched kamikaze suicide pilots by the thousands. This disparity in technology but the same essential mission is as lost upon our society today as was the opportunity to hit Pearl Harbor's oil missed by the strategic planners of Japan. Societies often seem doomed to view the world through their own eyes. Regardless, the societies at war in World War II quickly saw airpower as their salvation.

The atomic bombs dropped on Hiroshima and Nagasaki (the one place Christianity had traditionally been allowed to exist in Japan) appeared to represent the end of an age and the complete dominance of air power. To the western mind the bomb seemed to represent the old Manichean dualistic worldview of absolute good triumphing over evil once again. Depending upon one's viewpoint, the aerial platform saved a million U.S. casualties and lives in what would have been an inevitable Allied attack on the home islands of Japan, or it represented an apocalyptic scenario where science would end the world. This did not hold up under scrutiny.

The Strategic Bombing Survey conducted after the war determined air power did not actually win the war. According to Geoffrey Perret, "Strategic bombing had not been decisive. For every one million dollars' worth of damage done to the enemy the United States had been forced to spend one million dollars. . . It was just another form of attrition warfare, but in the air instead of trenches."[43] Indeed, it had been a war of attrition but one of resources. Lost in the action had been the real breakthrough of the use of radar, Doppler Effect, and variable-timed fuses. The key to winning the Second World War had not been in dropping bombs or destroying cities, it had been the ability of the state to rapidly control its people and resources faster than its opponents. The true means of control lay in the invisible realm of space and it would be here where the danger of disequilibrium in the state's attempt to control its people would go too far.

THE AGE OF SPACE

Many are offended by the use of the description of an American empire. The start of the United States of America coincided with a liberation from an empire and it is difficult, therefore, for many Americans to reconcile this seemingly conflicted view. Precisely what defines an empire discounts intentionality. Empires are seldom planned, and often the transregional nature of the empire escaped the understanding of its citizens at first. Once actual hegemony exceeded the previous frontiers of the state then it often seemed the empire's enemies better understood the extent to which a foreign system, not of their own origins, had controlled them. In many ways, this is the case with America today. The expression, "History does not

---

[43] Perret, Geoffrey. *A Country Made by War*, 437.

repeat itself, but it does rhyme," is often attributed to Mark Twain, and it is this rhyme which describes the development of the American empire of air and space and the means necessary to sustain it.

The United States of America is unique in that it represents many of the historical and geographic themes discussed previously and new ones as well. A dynamic attenuated colonial coast constantly expanded with a long record of experience with a western frontier. American growth had required an imaginative use of tactical forces over large areas of the earth's surface, and the American military is essentially continuing to attempt to bring into the globalized fold many frontiers in a similar manner to this day. America has increasingly relied upon a military force consisting of a volunteer class combining ideological crusader zeal of the North with a conservative hunting military tradition of the South. The United States was well blessed with natural borders and developed at an important time during both the Ages of Sail and Steam. The oceans served to protect the fledgling republic from enemies as had the Mediterranean protected the Roman Republic during the Age of Muscle and likewise served to pull America into the expanses of global trade. The mastery of a New England sailing tradition further established the U.S. Navy as it led to a national culture bent on both commercial exploration and expansion.

The digging of the Panama Canal to procure strategic waterways and supplies necessary during the Age of Steam was one of America's first major acts upon entering the world scene after the Spanish-American War. The Age of Air began with the Wright brothers in the United States, and the ability to master the air finished the Second World War. The Age of Space would likewise see the Americans utilize aspects of geography to maintain the necessary speed over the globe to maintain a healthy equilibrium and an empire in proportion to size and speed. After the Age of Space began (for the purposes of this book) in the 1950s, Americans would see themselves responsible for the free world and would find themselves burdened by a permanent military and in a top down system with seemingly unlimited government spending directed towards increasingly complex and expensive defense platforms. Eventually the cost of governance would result in the rise of myriad challenges from peripheral areas to an American global empire of air and space in the now familiar pattern.

If vertical limits and natural borders marked a western tradition versus a fluid eastern one, then the American empire represented a rise of a peripheral British colony to essentially take over a seaborne British commercial empire but expanded to include the air and space of the globe. The model of a sphere is appropriate here. As you have seen, we have gone from a model represented by a circle, to a hemisphere to a total sphere. The earth's shape is actually an oblate spheroid, and it is indeed three dimensional. With a curvature of sixteen feet for every five miles traveled, the shape of the earth and its surrounding atmosphere and various orbits have provided a new frontier that the American empire has seized and controlled effectively for over half a century. What is most often overlooked is the invisible space between physical objects on the earth's surface where the often invisible light of the electromagnetic spectrum washes up and even penetrates the entire earth. The ability to take the "high

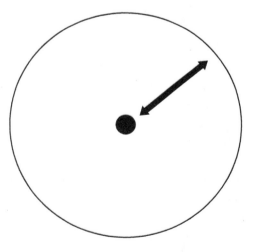

ground" of space and the ability to manipulate the electromagnetic spectrum has resulted in both unprecedented speed and the ability to accurately project power by the state. Conversely, the autonomy and individual initiative of the people may have been sacrificed to the numbers, rubrics, and spreadsheets so typical of the technology dictated by numbers that created this empire. Rubrics and spreadsheets are rational, but humans are not. Humans unfortunately often want to attack systems they feel threatened by or alienated from, and the result is they irrationally and willingly attack people they don't know and haven't met.

By the end of World War II, the United States had obtained German scientists and computers. The impact in the area of German cultural diffusion launched the United States into a military doctrine that would see its forces designed to utilize mechanization and technology to slowly retrograde back across Western Europe. Guided by electronic airborne systems capable of categorizing and prioritizing fires, the West planned to use its technically superior army in the defense. The

American military and its NATO allies planned to battle numerically superior Russian tanks expected to come riding in like some eastern Parthian force on horseback, eastern hordes once again threatening to put out the light of civilization.

The world had, indeed, become bipolar at the end of World War II and the United States and the Union of Soviet Socialist Republics looked at each other apprehensively over the **Arctic Ocean** only **fifteen minutes** apart in terms of the range and speed of the intercontinental ballistic missiles (ICBM's). Lurking under polar ice caps for up to a year without surfacing, rival submarine forces engaged in a deadly game of hide and seek. The USSR represented a primary challenge to the United States, but just as a unified Germany in the 1870s had thrown off the balance of power in Europe, so had China thrown off the world's power structure.

The Chinese had finished their revolution in the late 1940s. After being allies to the United States during World War II, a fear of holocaust led the United States in its **Manichean** worldview of good versus evil understandably assumed China had closely allied itself with the USSR by virtue of its claim to communism. What the U.S. in Asian cultures, however, is the significance of Confucian systems demanding obedience in the **Mandate of Heaven**, a concept basically at odds with democracy. With loyalty to the state seen as a virtue, communism probably represented an initial step in Asian nationalism where a two-party system such as in America just didn't exist. The time between the late 1940s and the realization of the actual differences between China and Russia by the late 1960s cost the United States dearly. For example, by reacting to Chinese actions off of their coast, the United States humiliated China by interposing the 7th Fleet between the mainland and Taiwan at **Quemoy** and **Matsu Islands** during the first Taiwan Strait Crisis of 1954. By causing the Chinese to suffer loss of face, the fear of an apparent loss of the Mandate of Heaven forced Mao to lead a buoyant China fresh from a tie on the Korean battlefield against the United States to decide to go it alone without expected help from the Soviet Union. To this day, the People's Republic of China (PRC) fears the Republic of China (ROC) located offshore in Taiwan. Many Americans do not understand this, but the PRC is extremely insecure in its loss of control of the newly burgeoned capitalism. Ever mindful in the minds of Chinese leadership is the knowledge that coastal prosperities have resulted from borrowed Western culture due to the proximity to **Hong Kong**, **Macau**, and **Taiwan**. Nations consist of people and are irrational. One must attempt to see the world through the eyes of others, and China, like many post-colonial states today, is very sensitive to the painful memories of its past. China's methodical approach to foreign policy must be understood before we can trace past American conflicts in the region.

Chinese foreign policy since the mid-twentieth century has been extremely pragmatic. Avoiding direct conflicts in non-contiguous areas, the Chinese have been the model of patient practicality, sometimes buying shares in both sides of a fight where it suits them to ally with a winner. This is important to understand. China sees very little moral purpose behind its growth, just the natural inclinations of a great ancient power to move into blue water (deep sea) in a manner similar to the Americans, and to make money. The Chinese have their own frontiers, and of course the United States sits powerfully astride one of them, hemmed in by the Gobi desert to the north. The **Xinjiang Province** to the far west is a frontier where **Turkic Uighurs** despise the pretentious Han Chinese. The Mongolians have even reached out to America in recent years in their attempts to stem the Chinese tide. Still, with the rise of a Chinese-Russian relationship, the Chinese are much freer to explore into the Pacific Ocean. Hemmed in by the first island chain of the Japanese **Ryukus** and **Taiwan**, the Chinese, in order to be a truly great world power, must use the Indian Ocean trade routes sought by all empires of the past. This means getting past the first island chain, and this requires bypassing the Philippine archipelago as well as Kra Isthmus of Malaysia. Strange insurgencies have accompanied the attempts by the Chinese to build a canal across the **Isthmus of Kra**. The canal is an insurance policy in the event the Chinese cannot secure the vital **Straits of Malacca** adjacent to the rich city-state of Singapore. Practical, indeed! The problem for the Chinese has been the unpredictable Americans; the new empire on the block, from a Chinese perspective, seems so irrational and spontaneous—almost childlike!

A lack of predictability can be an asset; unfortunately, however, spontaneity can have its costs too. When Harry S. Truman decided to intervene in Korea in 1950, he abandoned an offshore defensive strategy favored by the Joint Chiefs of Staff (JCS), one promising security at low cost potentially forever, and instead ordered the 7th fleet into the **Taiwan Strait** where the Chinese had been on the verge of launching a junk navy to take Taiwan with hundreds of thousands of troops.[44] By sending U.S. troops into Korea without a declaration of war, Truman essentially grabbed a short-term ticket into a war but one that would begin the United States on a long road of wars with China that would bleed it in similar measure as had the Greeks dredging the Euphrates and the Egyptians in the Red Sea. Now China would be an oozing sore for the American empire of the air and space in a way the Parthians were to the inflation-racked Romans. A horrible precedent had been established. No longer did a check on the hemorrhaging of funds and indiscriminate use of force by the state exist. Now endless wars could be waged under the guise of multi-national coalitions that would ensure future victories would be virtually unobtainable.[45] Holding South Korea and committing itself to placing one of the ten precious active divisions in the U.S. Army on a new land frontier paralleled the later Roman strategy in the Age of Muscle and began the hourglass of money flowing. China had learned that the United States could not be defeated on the battlefield in a conventional struggle and instead would concentrate on continuing to draw out American power in Vietnam.

By 1954 the U.S. had also committed itself to supporting the southern frontier of China in **Vietnam**. Fresh off of its "tie" with the United States, the Chinese felt buoyant at their return upon the world scene against these upstart Americans. Once again drawn into a limited conflict like Korea, further American legions found themselves making a stand in the long war of communist containment. Using weapons and a doctrine designed to fight in Europe, instead low intensity asymmetric warfare obliged American dollars and self-confidence in a triple canopy jungle of little strategic merit. America's facade of invincibility and weaknesses had been revealed: great speed had to be maintained to support the new American empire.

If money is time, America it seems had almost unlimited time and money. Beginning with the international standards for gold values in 1944 the way was set for the American dollar to become the international medium of exchange. The ability to gain the world's trust by basing the dollar on a recognized value for gold created an international market essentially under the control of the U.S. government. The creation of the International Monetary Fund (IMF) in 1947 represented an attempt to prevent a return to high rivaling tariff levels leading to international competition such has seemed to have occurred during the 1930s. The **IMF** would be an instrument to lend money and spread American financial influence, particularly after the decision in 1971 by the U.S. president to create a **"fiat currency"** backed no longer by gold, but only by the faith in the U.S. government. With the end of the Vietnam War and the unsuccessful air war based on gradual escalation, it must have seemed the American empire would be short lived, but out of the failings of American technology would come great promise for the future.

The air war in Vietnam represented an attempt to use fighter-bombers, such as the F-100 series and heavy bombers as in World War II, in a desperate attempt to rationally demonstrate American resolve with air power. Operation Rolling Thunder represented more than a losing strategy in many ways guaranteed to fail; it also showed the American propensity towards bigger, heavier, faster, and higher aircraft better suited for the doctrine of fighting the Soviets in Europe than it did a jungle war in Southeast Asia. Thwarted by geography and costs, the control of the state seemed to stifle initiative but eventually out of the high flying bay doors of rockets and shuttles emerged an empire of space.

When President Eisenhower left office in 1960, he warned of a danger from a military-industrial complex. Ike knew that the American strength of character and a sense of purpose emerged from free peoples. The perfect foundation for developing American ingenuity had been laid in a society that was energized by controlled immigration favoring bright people and freedom from barriers that had prevented both upward and downward mobility. Ike recognized the bungling nature of a state controlled populace as epitomized by the Air Force who had pioneered refueling in the 1920s but had not developed it for jet aircraft until near the end of the Korean War thirty years later. Through the use of DUST, the Dual Use Science and Technology, an attempt to strike a balance between commercial innovation and state control pushed the frontier of America into the invisible envelope of space and into the increasingly small universe of the electron. In most cases

44 Perret, Geoffrey. *Commander In Chief* (Farar, Straus and Giroux: New York, 2007), 137.
45 Ibid, 148.

Americans had no idea to what extent their real national power had been developed behind the scenes between the 1950s and the 1990s as they went about the business of life.

ENIAC, with its vacuum tube technology, filled rooms with sweltering heat and represented the first step of the American nation choked in Korea and Vietnam, flirting dangerously with the crushing centralization of the state. The Western love of the rational, the orderly, the mathematically perfect had been met by the arrival of th computer, with its ability to perform up to 5,000 calculations per second using binary on/off signals.[46] The first computers had been used in the Second World War to recreate the parabolas of artillery, increasing accuracy and helping counter-battery fires. By calculating bomb paths, German scientists quickly learned how to manipulate the electron for the purposes of increasing the reach of their new missiles. The army soon became beholden to the brilliance of the systems analysts and the new, smarter breed of man, the computer operator. By the time of the late 1940s, a team of Bell Labs engineers had developed the transistor, or semi-conductor, and by the late 1950s the integrated circuit had appeared. The result would be increasingly tiny urban-like landscapes of solid-state electronic chips holding various devices enabling computers to become smaller. Microprocessors arrived in the 1970s and would alter American culture with the arrival of games and desktop computers as well as calculators and wristwatches.[47] Thanks to the transistor, **Silicon Valley, California** had been born. This high-technology corridor would contribute to the American empire by allowing it to go into space and outside of its historical borders eventually achieving its limit of expansion into the Middle East.

A race for the soul of the American empire had begun. The decision to maintain a large standing and permanent American army burdened the state whose growth appeared to threaten to stifle creative imagination. The computer with its amazing speed and potential for organization allowed the American empire to maintain an equilibrium of control over space through time as money turned from coinage to digitization and transportation speeds competed with communications. As early as July 1950 James Reston of the New York Times said, "the United States was going to have to live with permanently higher defense budgets, less spending on nondefense programs, and a large peacetime army . . . Whether we like it or not, we have inherited the role played by the British . . . this role must be organized, not on a temporary, but on a permanent basis."[48] With the death of American isolationism, restraints on the use of the expenditure of American military force seemed to continue to grow smaller, but a faster, smarter empire controlling the air and space had also arrived. The challenge for the United States would be to maintain the equilibrium of Dual Use Science and Technology (**DUST**), and it would be this idea of dualism, or balance, that would bring dividends in outer space during the period of the 1960s through the 1980s to the United States.

The key to the frontier of space is to understand it is both around us and it extends to outside our atmosphere. The limits of turbojet and ramjet technology is about **thirty statute miles** above Earth's surface. From this elevation upwards outer space becomes colder, silent, and increasingly dangerous. Varying degrees of radiation threats accompany the individual seeking to achieve the geosynchronous orbit altitude of about 22,300 miles above the earth's surface. It is between this orbit and the orbit of the moon around the earth and about 240,000 miles from Earth we are most interested in. It is here that the majority of craftborne fuel must be expanded. It is control over this area and its limited numbers of orbits and increased amounts of space debris that the ability to harness American energy into an American empire is maintained.

Despite the greatest public relations disaster of all times, the first satellite in space being Russian in 1957, the United States better harnessed the free energies of its peoples much more effectively and the end result by the 1980s saw an American ability to use the space shuttles to place satellites into orbit and to maintain them. The geography of space has been particularly kind to the United States since the relative location of Florida allows launches east concurrent with the earth's spin. The speed of the earth's spin is about 1,040 miles per hour at the equator and only equals about half of this at 60 degrees latitude. Space programs must, therefore, be located near the equator to take advantage of the speed necessary

---

[46] Boot, Max. *War Made New: Technology, Warfare, and the Course of History* (New York: Gotham Books, 2006), 309.

[47] Ibid, 310.

[48] Jager, Sheila. *Brothers at War* (WW. Norton: New York, 2013), 309.

to break free from Earth's gravity and to allow entry into outer space.[49] The European Space Agency utilizes launch platforms in **French Guiana**, and China is forced to use Jiuquan space station in **Gansu Province** in inner-Mongolia.

With the development of global positioning satellites technology, mostly behind the scenes and with the continuing build-up of the American military, the United States won one of the fastest and least costly wars in human history and set itself on the course of attempting to establish a new world order. Though it is trendy to say now that the space program is dead, it is far from it. Just because we are not using manned platforms in outer space does not mean the American empire is not sustained by the technology and use of space. The United States would, in fact, with its new found military confidence and war machine find itself pulled into undeclared wars in over forty places during the 1990s and increasingly dependent upon precision guided-missiles (**PGMs**) in places like Serbia. PGMs and precision guided-munitions, or smart bombs, had created a super state, one rescued from military overreach in Asia by the computer and the use of the space shuttle's trips into space to set up the requisite satellite system of control of the Global Positioning System (**GPS**).

With only one-third of the munitions guided, the Serbian government of Slobodan Milosevic surrendered after only seventy-eight days of air action. The questionable legality of a NATO (North Atlantic Treaty Organization) air war notwithstanding, the multinational coalition format, now so popular with American interventionists, had arrived in time to ensure the unprecedented victory over Serbia would continue to give the Americans excuses and opportunities to intervene throughout the globe. The accidental destruction of the Chinese embassy and the resulting loss of face to the Chinese government would go far to turn the clock back in favorable relations with China enjoyed since 1972 when Nixon triumphantly had visited China at the end of the Vietnam War. Additionally, the Serbians had always been viewed very paternalistically by the Russians; this view began during World War I when Austria-Hungary threatened to invade causing the Russians to come to Serbia's defense. It would also be at this time the Russian autocracy would begin to grow alienated to the American Empire as now would China after the Soviet Union's fall at the end of 1991. As of this writing, it would appear an attempt to create a unipolar new world order might have represented an imperial overstretch for the United States as it sought to use its precision guided-missile (PGM) technology without a declaration of war. Perhaps future historians will identify overconfidence after the First Iraq War of 1990–1991 as an American imperial overstretch. The key here is to see the power of the satellite in an American empire of the air and space.

With an army using satellites to an amazing degree of efficiency, a simultaneous threat revealed itself. The speed of the American forces had reached roughly seventeen miles a day for the invading force. Compare this speed to the six miles per day for the highly controlled North Korean Army to move towards the Pusan Perimeter during the summer of 1950.[50] This dramatic increase in speed is still not proportional when one looks at the Roman legions. Including time to dig in for fortifications, it took sixty-seven days to march from Rome to Cologne (Northern Germany) at a rate of thirteen miles per day-through the Alps.[51] Speed is important but so is tactical control. Whereas the legions adapted their defenses to meet terrain and risk factors, the modern U.S. military quickly realized it had a shortage of satellite bandwidth for its network centric strategies. Most bandwidth must come from commercial satellites in geosynchronous orbit around the earth. Two simultaneous factors had emerged for American forces, vulnerability and dependence.

When the United States on 9/11 faced the most costly attack on its own soil in history, it understandably sought to strike back at the perpetrators through the use of its technology. Strike back it did with amazing speed and using a whole new family of weapons and systems dependent upon GPS and satellites.

During blinding sandstorms, Blue Force Tracking enabled American commanders to monitor and deploy forces to maximum potential. With new technology there appears to be a balance that must be maintained very similar to the balance of control necessary to enable an empire to survive. Too much control and atrophy of imagination and initiative occurs

[49] Collins, John M. *Military Geography: For Professionals and the Public* (Washington, D.C.: National Defense University Press, 1998), 143.
[50] Perret, Geoffrey. *A Country at War* (Random House: New York. 1989), 452.
[51] Luttwak, Edward N. *The Grand Strategy of the Roman Empire: From the First Century A.D. to the Third* (Baltimore: Johns Hopkins University Press, 1976), 82.

such as did with the slow moving North Koreans mentioned above. Again, if the civilian or commercial sector does not balance with the military DUST there is either a stifling of new ideas, or there is a lack of funds to develop. The government is instrumental in the distribution of funding for new technologies such as has occurred with electronics and computers, but only the commercial or business side truly succeeds in terms of innovation such as has been seen with the commercial airlines industry between World War I and II in the United States. The other concern with use of technology to overcome the friction of distance with speed is the inherent dependency upon this system.

During the Iraq War the American military used forty-two times the amount of bandwidth it had in the First Iraq War in 1991. Commanders still had problems obtaining access. Space reconnaissance assets proved unable to provide the necessary intelligence on the fluid battlefield. The use of Blue Force Tracking enabled the execution of precision airpower of incredible accuracy in the urban/freeway environment of Iraq, but there remained insufficient band width for all units involved in any given endeavor. The lesson learned by potential adversaries is the American empire has an Achilles' heel of vulnerability. The modern military cannot operate without satellite or space assets and for an opponent to defeat the U.S., one need only interfere with those assets.[52]

Since most satellite bandwidth is commercial and for sale, the current dependence on foreign companies to sell bandwidth in time of war will continue to be an issue. Every GPS, cell phone, and smart bomb dropped on the battlefield today is gobbling up bandwidth. A recent exercise with Taiwan even saw the U.S. military unable to rehearse its potential mission of supporting the island against a Chinese invasion because the Chinese government had bought up all available bandwidth making the U.S. military unable to participate. The need to secure the geography of outer space is just as important today as a secure Mediterranean was to the Romans.[53] To maintain secure space transmissions will be a paramount need for the Pax Americana in the years ahead. Cybersecurity and the need to continue innovating space and air systems will be the primary challenge in the future.

The drone incorporates many aspects of the Age of Air and Space. It can float in some cases seemingly perpetually, and it can provide instant information. Utilizing satellites for navigation, these robots of the air can maintain the frontiers of the American empire at minimal cost compared to aerial platforms, but will demand a new type of soldier to maintain them. In Iraq, this writer witnessed drones reputedly being operated from around the world. These drones had been launched in response to an IED detonation probably initiated by someone on a cell phone himself far from the scene. Here sat an expensive to maintain human being complete with healthcare and retirement benefits between two platforms directed through outer space between belligerents potentially thousands of miles apart, at least. To maintain this empire of the air and space two major shifts will need to occur: the nature of security needs will need to be identified, and a force capable of providing those needs must be developed at an affordable cost.

The United States is broke and vulnerable in many areas at once. A primary threat to American control of the air and seas could come from China. The answer lay in the new description of the world seen through the eyes of the American joint force commanders. The world is drawn up into six borders. The largest of these is PACOM, and, if history is any guide, this is where the likely action of the future will be. To secure needed sea lanes, the Chinese have embarked on a "String of Pearls" strategy from Hainan Island off of the coast of China to the port of Gwadar in Baluchistan on the Makran Coast of Pakistan. Along the way, Sri Lanka, Bengal, and the vital Strait of Malacca are being developed in a sort of reverse pattern from the coaling stations during the Age of Steam. It is the Pacific Command (PACOM) theater where drones are increasingly playing roles in the Spratley and Paracel Island chains. Runways for UAV's (unmanned aerial vehicles) and drones are being built at alarming rates. A return to the offshore defensive strategy mentioned in the last chapter appears to be in the works, but the role of such drones as the RQ-170, with its real-time video and audio uplinks and ability to employ deadly force, will become paramount.

---

[52] Johnson-Freese, Joan. *Space as a Strategic Asset* (New York: Columbia University Press, 2007), 42.

[53] Johnson-Freese, Joan. *Space as a Strategic Asset*, 95.

Security will require the United States to understand the vulnerabilities of its drone fleet. It is suspected that on the night of September 7, 2007, Syrian air defenses became compromised by an Israeli drone. The Israeli's might have fed false images into a Syrian computer system tracking one of their stealth drones on September 7, 2007. Utilizing the return beam from the aircraft, entering the ground system may have offered the opportunity for the distribution of a false image showing an empty night sky, belying the fact that aircraft had breached the system and would begin bombing a suspected nuclear facility. Like the Battle of Britain in 1940 in reverse, the control of the sky by radar won the battle. This scenario presents another possibility to accompany the vulnerability of space systems.

Various Chinese leaders have openly stated their interest in defeating American air and space networks in their attempts to break out into the world with a blue water navy. At the end of the last millennium, the Chinese wrote openly of rectifying qualitative inferiorities through control of America's battlespace of satellites and cyber networks.[54] The reason the U.S. system is so vulnerable is because of the way the internet has been developed. Designed to be an open exchange medium for packets of information shared between computers and networks of academics and theoreticians, the system had been designed with a hippie-type attitude towards security. A rampant egalitarian view suggested utopia around the corner—truly a system without any controls. Society quickly became hooked on an insecure system and so have America's adversaries. Teams of computer programmers and various government agencies including U.S. Cyber Command scan lines of code constantly looking for prospective breakthroughs. Attempting to patch holes in this open network in turn can attract more villainous activity, and the ultimate result of a worm or virus planted can create a botnet of computers, usually without the computer operator's knowledge, to overload and cripple through denial of DDOS, or distributed denial of service.[55] U.S. Cyber Command is expected to have 6,000 active-duty "cyber warriors" by 2016 and should guarantee cyberspace superiority.[56] Security needs in an age of air and space range from satellite to drones to the computer but can even be of a social nature.

Just like social networking has been used in dramatic fashion, so have recent cyber-attacks that have presaged military action. From the coordination of cyber-attacks in tandem with Russian protests against the Estonians, likewise Georgia saw attacks before the Russians invaded in 2008.[57] The Israeli's probable launch of the mysterious Stuxnet virus into the Iranian nuclear program and the use of bot networks to do anything from moving money out of banks to causing energy supplies to cease or threaten transportation networks are vulnerable.[58] The danger to networks from without is obvious, but the danger of a purely government or military approach to security is demonstrated by the Iraq invasion where to receive space intelligence assets, one commander reported having to get permission from between twenty to twenty-five different levels of command.[59]

To protect American air and space assets a potential in cost savings must be acknowledged as well as a pending demographic shift. To overcome crushing expenses and stifling of initiative from the government, use of civilian contractors should be increased. America has a long tradition of contractors, from the discovery of America by the contracted Christopher Columbus to the survival of the first English-speaking colony at Jamestown under the leadership of the mercenary Captain John Smith at Jamestown.[60] The use of contractors can take advantage of the coming demographic shift throughout the world with increased numbers of young people accustomed to staring at screens and living in a digital universe. To fail to employ these armies of potential cyber warriors would be a grievous mistake. Modern day barbarians seeking to bring down the empire can employ these legions of aggressive youth skilled at technology from around the

---

[54] Clarke, Richard A. *Cyber War* (HarperCollins: New York, 2010), 50

[55] Ibid, 284.

[56] http://thediplomat.com/2015/02/the-us-military-wants-to-train-more-cyber-warriors/

[57] Bowden, Mark. *Worm* (Atlantic Monthly Press: New York, 2011), 48.

[58] Clarke, Richard A. *Cyber War* (HarperCollins: New York, 2010), 70.

[59] Singer, P.W. *Wired for War: The Robotics Revolution and Conflict in the 21st Century* (Penguin: New York, 2009), 202.

[60] Prince, Erik. *Civilian Warriors: The Inside Story of Blackwater and the Unsung Heroes of the War on Terror* (Penguin: New York, 2013), 60.

world.[61] The United States must recognize that a new type of warrior is at work and should invest in this asset through private contractor providers, allowing the market to pay for their recruitment. What better source for future computer fighters than the open market?—And at costs both affordable and in line with the latest and most secure practices.

While recognizing the nature and vulnerabilities of the invisible multifaceted empire of air and space, the United States must remember the ages that have preceded it. An increased emphasis on geography will continue to reveal an ability to determine sub-state political entities not content with both grievances and needs capable of being reached by joint special operations teams. Without Indian guides, Geronimo would never have been captured. Likewise, small elite teams of joint special operators must also be employed with the necessary language and cultural knowledge to places on the map reflective of the various ages discussed.

Key strategic natural borders and terrain must be carefully chosen on the basis of suitability to doctrine and the structure of military forces utilizing the inherent strengths of an air/sea empire while avoiding those areas such as Asia that play to our weaknesses. The need for resources such as in the Age of Steam will require maintenance of the **Bab-al Mandeb** waterway between the Arabian Peninsula and the African Coast as well as the other straits so sought after during the Ages of Sail and Steam: Malacca, Gibraltar, **Hormuz**, and Suez will require "butcher and bolt" or rapid deployment forces of sea/airborne infantry to secure strategic choke points as situations dictate.

To maintain an empire of the air and space, the United States must see beyond drawn borders. So many borders are vestiges of a colonial legacy (such as Kuwait); they are not even recognized by the tribes of people who pass them without even being checked. Many so-called nation states are, in fact, failed states and are truly borders and must be seen as such. One of these is Pakistan. An understanding of the sub-states is key to increasing geographic knowledge. Mired down in unceasing fighting in Afghanistan, the Obama administration rightly concluded the real enemy occupied "Pashtunistan." After understanding that the enemy could walk across the border with impunity and kill American troops, it only made sense to use PGM's and drones to find, fix, and eliminate the intruders. The risk to acting without lawful precedent is great but so is war. Understanding culture, customs, and ethnicities is more important to securing a frontier than holding to an international agreement. America must engage in unceasing diplomacy with disaffected groups such as the Baluchis (**Baluchistan**) in southern Pakistan if it will compete with the Chinese.

Another example of smart diplomacy would be Thailand. From Thailand any nation can be flanked and turned in Southeast Asia, and any attempt to cross the Kra Isthmus could be threatened if the Chinese acted provocatively. Again using the lessons of the ancients, seizing terrain and holding it is key. The lesson from the ages is to adapt technology and institutions to this end. Ceasing to hold bases, the United States can return to a policy of maintaining a central reserve within the United States and use its elite forces and downsized Marine and Army combined arms teams for short missions, relying more on drones and robots to maintain the frontiers of air and space. Launched from Alaska and using the shape of the earth and the great circle concept, or launching stealth bombers and aircraft from the Marianas in the Pacific, using the natural borders of ocean such as in the Age of Sail, the United States can sustain its defense at minimum cost.

Use of contractor teams would additionally save money enabling security funds to go further and would help ensure the maintenance of the equilibrium between state control and the inherent creative freedom of the individual. State-sponsored teams of cyber-warriors can likewise be recruited in cooperation with private industry utilizing global youth. At low cost, strategically placed teams could continue to maintain an empire over time in a manner similar to the empire the Byzantines maintained for a millennium against the zealous forces of Islam led by Muhammed himself. By recognizing geographic opportunities and limitations, intelligent decisions regarding the control of space over time can harness an energetic and hopeful people seeking to use their God-given talents towards the end of maintaining a still shining city upon a hill for future generations to come.

---

[61] Singer, P.W. *Wired for War: The Robotics Revolution and Conflict in the 21st Century*, 284.

KEY TERMS AND CONCEPTS TO KNOW

30 inches
30 statute miles
70%
Amsterdam
Amur
Anatolian Plateau
Antwerp
Archipelago
Arctic Ocean
Australia
Automobile
Ayans
Azores
Bab-el Mandeb
Baluchistan
Bengal
Black Sea
Butcher and Bolt
Byzantine Ways of War
Calcutta (Kolkata)
Cape of Good Hope
Caribbean
Catalonia
Constantinople
Danube River
Dardanelles
Diu
Doppler Effect
DUST
East versus West
Estonia
European Landforms versus China
Fiat Currency
Fifteen minutes
Fixed-wing
French Guiana
Gansu Province
Georgia
Ghana
Gibraltar
Goa
GPS
Great Circle

Great Lakes
Greater Antilles
Guinea
Gujarat
Gwadar
Hainan Island
Haiti
IMF
Imperial Overstretch
Indus River Valley
International Meridian Conference
Invisible Realm
Jamaica
Khyber
Korea
Kra Isthmus
Levant
Lighter-than-air
Limes
London
Luanda
Macau
Makran Coast
Malabar
Malakand
Mali
Manchuria
Mandate of Heaven
Mandsubdari
Manichean Worldview
Matsu Island
Mediterranean Sea
Mesopotamia
Mobilize
Mongolia
Natural Borders
Nile River
Nomad versus Civilized
North to South
Oil
Oxus River
Pakistan
Paracel Islands

Pearl Harbor, Oahu
Peripheral Control
PGM
Ploesti, Romania
Po Valley
Prevailing Winds
Quemoy Island
Red Sea
Resources
Rhine River
RQ-170 "Predator Drone"
Ruhr Valley
Ryukus
Salerno, Italy
Sea of Japan
Second Battle of Armageddon
Siberia
Silicon Valley, California
Sipahi
Spratley Islands
Sri Lanka
Steam Turbine
Strait of Hormuz
Straits of Malacca
Strategic Bombing
"String of Pearls" Strategy
Suez Canal
Sunda Strait
Surface Condenser
Taiwan
Taiwan Strait
Timbuktu
Tsushima Strait
Uighurs
Vertical Borders
Vietnam
Waziristan
West to East
Whale versus Lion
Xinjiang Province
Zambezi
Zanzibar

BIBLIOGRAPHY

Abulafia, David. *The Great Sea: A Human History of the Mediterranean*. Oxford: Oxford University Press, 2011.

Anderson, Malcolm. *Frontiers: Territory and State Formation in the Modern World*. Cambridge, MA: Polity Press, 1996.

Barraclough, Geoffrey, ed. *The Times Atlas of World History*. Maplewood, NJ: Hammond, 1988.

Boot, Max. *War Made New: Technology, Warfare, and the Course of History*. New York: Gotham Books, 2006.

Bowden, Mark. *Worm*. Atlantic Monthly Press: New York, 2011.

Chua, Amy. *Day of Empire*. Doubleday: New York, 2007.

Clarke, Richard A. *Cyber War*. HarperCollins: New York, 2010.

Collins, John M. *Military Geography: For Professionals and the Public*. Washington, D.C.: National Defense University Press, 1998.

Crocker III, H.W. *The Yanks Are Coming!* Washington, D.C.: Regnery, 2014.

Diamond, Jared. *Guns, Germs, and Steel: The Fates of Human Societies*. New York: W. W. Norton, 1999.

Docherty, Paddy. *The Khyber Pass: A History of Empire and Invasion*. New York: Union Square Press, 2008.

Getz, Trevor, and Heather Streets-Salter. *Modern Imperialism and Colonialism: A Global Perspective*. Boston: Prentice Hall, 2011.

Ferguson, Nigel. *Empire: The Rise and Fall of the British World Order and the Lessons for Global Power*. New York: Basic Books, 2002.

Hourani, Albert. *A History of the Arab Peoples*. Cambridge, MA: Belknap Press of Harvard University Press, 1991.

Hutchinson, Robert W. "The Weight of History: Wehermacht Officers, the U.S. Army Historical Division, and U.S. Military Doctrine, 1945–1956. *Journal of Military History* 78 (Oct 2014): 1321–1348.

Jager, Sheila. *Brothers at War*. WW. Norton: New York, 2013.

Johnson-Freese; Joan. *Space as a Strategic Asset*. New York: Columbia University Press, 2007.

Johnson, Paul. *The Birth of the Modern*. HarperCollins: New York, 1991.

Kaplan, Robert. *Hog Pilots, Blue Water Grunts: The American Military in the Air, at Sea, and on the Ground*. New York: Random House, 2007.

———. *Asia's Cauldron*. New York: Random House, 2014.

———. *Imperial Grunts: The American Military on the Ground*. New York: Random House, 2005.

———. *Monsoon*. New York: Random House, 2010.

———. *The Revenge of Geography*. New York: Random House, 2012.

———. *Warrior Politics: Why Leadership Demands a Pagan Ethos*. New York: Vintage Books, 2002.

Kearney, Milo. *The Indian Ocean in World History*. New York: Routledge, 2004.

Keegan, John. *A History of Warfare*. New York: Alfred A. Knopf, 1994.

Landes, David S. *The Wealth and Poverty of Nations*. New York: Norton, 1998.

Kleveman, Lutz. *The New Great Game: Blood and Oil in Central Asia*. New York: Atlantic Monthly Press, 2003.

Luttwak, Edward N. *The Grand Strategy of the Roman Empire: From the First Century A.D. to the Third*. Baltimore: Johns Hopkins University Press, 1976.

Overy, Richard. *The Times Complete History of the World* 8th ed., London: Harper-Collins, 2011.

Parsons, Timothy. *The Rule of Empires: Those Who Build Them, Those Who Endured Them, and Why They Always Fall*. Oxford: Oxford University Press, 2010.

Perret, Geoffrey. *A Country at War*. New York: Random House, 1989.

Perret, Geoffrey. *Commander In Chief*. New York: Farar, Straus and Giroux, 2007.

Prince, Erik. *Civilian Warriors: The Inside Story of Blackwater and the Unsung Heroes of the War on Terror*. New York: Penguin, 2013.

Riedel, Bruce. *Deadly Embrace: Pakistan, America, and the Future of Global Jihad*. Washington, D.C.: Brookings Institute, 2011.

Singer, P.W. *Wired for War: The Robotics Revolution and Conflict in the 21st Century*. New York: Penguin, 2009.

Stearns, Peter N. *Globalization in World History*. NY: Routledge, 2010.

Tanner, Stephen. *Afghanistan: A Military History from Alexander the Great to the War Against the Taliban*. Philadelphia: Da Capo Press, 2002.

Van Crevald, Martin. *Supplying War: Logistics from Wallenstein to Patton*. Cambridge: London, 1980.

Woodward, Bob. *Obama's Wars*. New York: Simon and Schuster, 2010.

Yergin, Daniel. *The Prize*. New York: Simon and Schuster, 1991.

WEB SITES

http://thediplomat.com/2015/02/the-us-military-wants-to-train-more-cyber-warriors/

PART THREE
THE GLOBAL WRAP UP

The Cover Story: Ukraine Unraveling, Resurgent Russia

CHAPTER OUTLINE:

A PRO-EUROPEAN Union (EU) west faces off against a pro-Russia east! A President once more thrown from power! A bitterly divided country! The last battle of the Cold War . . . or the first battle in the new Cold War? A redefining of sovereignty in the modern age that will reverberate for decades to come! It's all about Ukraine!

Yes, my friends; all of that and even a President Putin judo-chopping his way into a peninsula. Say what? Where is this scene of such serious shenanigans? That would be the unraveling state of Ukraine in 2014, which is displayed on the cover of this edition, and is thus our cover story of the year. 2014 is the year of Ukraine on the brain! Ukraine in revolutionary pain! Ukraine gone insane! But with so many other conflicts and chaos afoot around the world, why would the Plaid Avenger choose this story for the feature? Because the repercussions of the conflict within Ukraine will have far-reaching consequences for not just the state, but for all the regions around it, as well as the EU, Russia, NATO, and the very nature of sovereignty itself on the world stage. In other words, this ain't no local issue, it's a global one. So what the heck caused this state to splinter in 2014? Who are the major players? And what are the repercussions for the future?

Let's delve into it in a bit more detail. . . .

THE SETTING . . .

First off, the country is called Ukraine—not "the Ukraine"—and it's a big ol' chunk of Eastern Europe that has been a pivotal space, long tussled over by bigger empires. And by "big ol' chunk" I mean that it is sizable in terms of area and peeps: roughly twice the size of Texas, with over 45 million citizens . . . making it one of (if not the) largest state of Europe in size. And its importance goes well beyond its borders, as Ukraine has forever been a "breadbasket" of Europe due to its extensive, fertile farmlands and huge grain production. It also has been a source of industrial/manufacturing power, as well as a corridor for billions of metric tons of oil and natural gas that flow from Russia and Central Asia to Europe. It has always been in the middle ground between the European "East" and "West" . . . which is part of the reason for today's crisis. A real battle for the soul of this state has been brewing for centuries because. . . .

The political setting is just as complex. Culturally, linguistically, and historically, Ukraine has been a key center of East Slavic culture . . . thus making it intimately related to the evolution of the language and culture of fellow Slavs in Belarus, Poland, Serbia, Slovakians . . . and most importantly, Russians. Ah! And not just the evolution of language, but the evolution of empires, as Ukrainians have been subjugated by foreign powers for most of their history . . . specifically, they have been incorporated into an entity named "Kievan Rus" hundreds of years before we even started calling it Russia. Meaning that for most of the last 700 years, Ukrainians have mostly been a part of an evolving empire, then a state, named Russia since its inception! Heck, back in the 11th century, the current capital of Ukraine, Kiev, was the most important city of the fledgling Russian Empire and it laid the foundation for the national identity of both Ukrainians and Russians! Are you starting to see how intertwined these societies are?

Eastern Europe in 1560 . . . where is Ukraine?

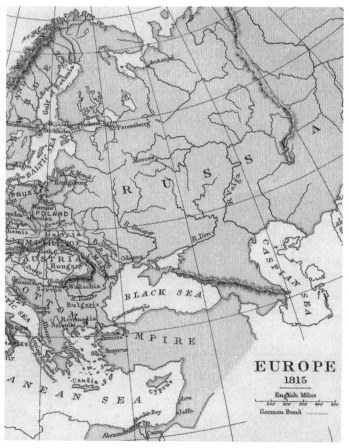

Eastern Europe in 1815 . . . where is Ukraine?

But that is not to suggest that Ukraine has always been a part of Russia politically. On the contrary, parts of Ukraine have at times been held by the Romanian Empire, the Lithuanian Empire, the Polish Empire, the Crimean Khanate (Mongol Empire), and the Ottoman Turkish Empire. The state that we know today has seen a thousand years of intermittent bloodshed as competing empires battled for control of this vital piece of European territory, with its awesome agricultural lands and Black Sea ports on the Crimean Peninsula. But as Russian power grew in the last couple of hundred years, little by little, all the areas where Ukrainians lived came to be controlled by Mother Russia. . . .

But enough of the ancient history! I just wanted you to understand that this area called "Ukraine" and the peoples called Ukrainians and Russians have an extremely intimate relationship and historical evolution . . . do not think of this place in the 21st century as some distinct nation–state that is entirely unrelated to Russia. That would be nonsense! Most Russian leaders would have always considered the place part of their territory, and certainly part of their cultural 'hood . . . if not outright part of the same family. Really, in the last 500 years, the only time that Ukraine has been an entirely independent sovereign "state" in the modern sense was for a few brief years just after World War I (WWI), when Russia surrendered the territory to the Germans, and then the Germans ended up losing the war anyway. Remember all that stuff from back in the Eastern Europe chapter? Russia was in the throes of its 1917 Commie Revolution, and Lenin just wanted out of the war, so he gave up territory held by the Germans at the time. Once the Germans fell, a power vacuum existed in which places like Estonia, Latvia, Lithuania . . . and yes, parts of what is now western Ukraine . . . suddenly found themselves in a position to declare independence from everybody! Huzzah! Huzzah! Ukraine is free! Ummm . . . hold on to that thought for a minute. . . .

And that, my good friends, is when the trouble you are seeing really started. What? Post-WWI? Yep. Because that taste of freedom for Ukrainians was not to last, as you well know from reading this book already. I said that *parts* of Ukraine declared independence back in the day . . . the parts that mostly contained ethnic Ukrainians. In the west. But there were always lots of ethnic Russians around as well! And there were some Ukrainians who actually wanted to stay as a part of Russia. In fact, during WWI, some Ukrainians fought on behalf of the Austro–Hungarian Empire against Russia, and some Ukrainians fought on behalf of Russia against the Austro–Hungarians! What a mess! This even spilled over into a civil war between these very Ukrainians, as several groups fought for control of their short-lived "independent" Ukraine from 1917 to 1921. Over 1.5 million Ukrainians died in that chaos.

So divisions within this state between a pro-Western west and a pro-Russian east have very deep roots . . . and are roughly based on the concentrations of ethnic groups, from even back in the day. About 1/3 of the place speaks Russian as their first language, the rest Ukrainian as their first language . . . but probably most speak at least a little or a lot of the other language. And the place is roughly 70% ethnic Ukrainians, about 20% ethnic Russians, with the greatest concentrations of those Russians on the Crimean Peninsula and in the east . . . basically, the closer to Russia you are, the greater the concentrations of ethnic Russians you will find. Surprise, surprise. So therefore it should come as no surprise when understanding the rift we are witnessing today, because it has always been there. . . .

Significant concentration of ethnic Russians

Scattered presence of ethnic Russians

SOVIETS ASSUME (OR RUSSIANS RE-ASSUME) CONTROL

Let's play catch up now: Once the Russians got their act together and became the Soviet Union, they had every intention of getting those lost territories back, and western Ukraine was the first to be pulled back in. The "independent" Ukraine was busted up among its neighbors, with the Soviets getting back the lion's share, which became the Ukrainian SSR in 1922. After originally siding with Nazi Germany in WWII for their Pact of Non-Aggression, then siding with the victorious allies, the USSR gained back all those other territories like Estonia and Latvia by the end of the war in 1945. On a separate note, the Crimean Peninsula, which you have heard so very much about this year, has always kind of been held by Russia since they grabbed it from the Ottoman Empire back in 1783, so originally it became its own SSR as well. However, in a bizarre twist of fate that no one can really explain, Soviet leader Nikita Khrushchev transferred the peninsula to the Ukrainian SSR in 1954 . . . a move that seemed insignificant at the time, since both republics were a part of the Soviet Union . . . crazy crap, huh? But I am getting ahead of myself. . . .

With me so far? That is how Ukrainian peoples, and a territory named Ukraine, came to be part of the USSR. From the start, it had a distinct Russian population in it, and the Crimean Peninsula specifically became the "heart of Russian Romanticism" and a destination for Russian vacationers, thinkers, poets, the rich and famous, and eventually even the Russian military moved in to make it a seriously strategically important naval port—like it is one of their most awesome military ports. Thus, that little peninsula is a much bigger deal in the Russian mind than you can imagine as an outsider.

And now for some insight into the Ukrainian mind: Life under Soviet rule was initially pretty crappy. Ok, like horrifically crappy. Ok, like millions died or were deported crappy. Especially under insane Soviet ruler Joseph Stalin, who early on wanted to keep those uppity Ukrainians in their places, and likely wanted to teach them a lesson for having declared independence in the first place. Stalin enacted a forced agricultural collectivization policy in Ukraine, in which the locals could not even eat the food they grew until export quotas had been met. Resistance met with visits from the notorious KGB Soviet Secret Police. Stalin intentionally starved out millions of Ukrainians (google Holodomor or "Great Famine" of 1931–32); hundreds of thousands (including the Tatars of Crimea) were deported; and mass killings took place during the "Great Terror" . . . all told, per-

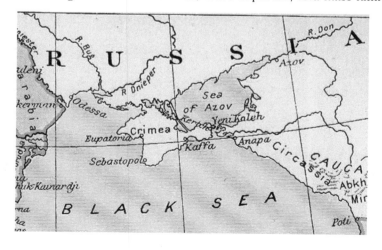

haps 8–12 million Ukrainians died in the first two decades of Soviet rule. Let me repeat that, lest you think it was a typo: 8 to 12 MILLION Ukrainians died, in a period of time that had no active war. Don't think for a second that many ethnic Ukrainians have forgotten that period of their history. Or ever will. Many are now calling this era a Soviet-constructed genocide, and it is easy to make the case.

Did the Soviet era do anything positive for the Ukrainian SSR? Sure it did. The SSR was massively industrialized, especially in the east. Agricultural productivity eventually rebounded and grew, especially in the east; nuclear power plants and energy production grew, especially in the east. Crimea was developed as a great naval base and tourist destination. And as a result of all of this, ethnic Russians started migrating in, especially in the east and Crimea. Migration is perhaps too strong a term, since all these peeps were at the time inside a single sovereign state called the USSR. The Soviets were keen on keeping ethnic identity alive, though, since the very naming of most SSRs was based on incorporated an ethnic group into this grand multi-ethnic commie experiment . . . thus the Estonian SSR was mostly Estonians, the Kazakh SSR was mostly Kazakhs, and the Ukrainian SSR remained mostly Ukrainians.

Comparative Ethnic Groups in the Former Soviet Union, 1989

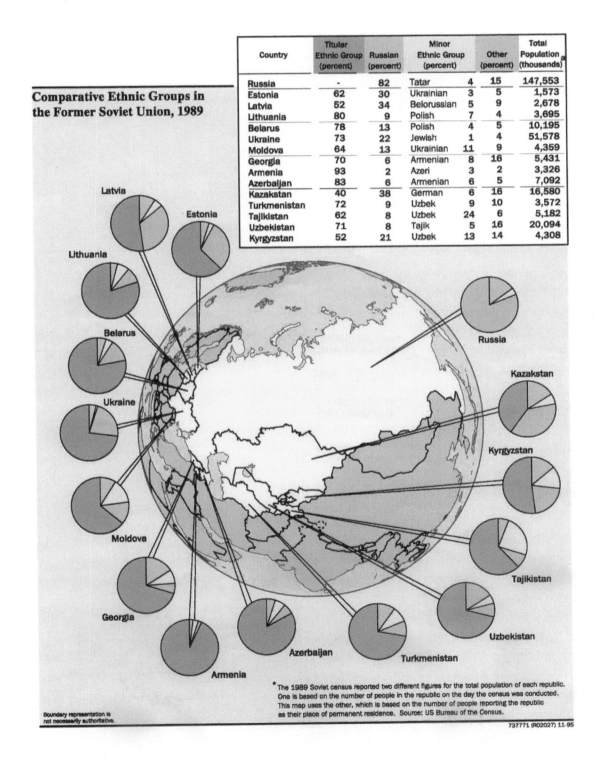

Country	Titular Ethnic Group (percent)	Russian (percent)	Minor Ethnic Group (percent)		Other (percent)	Total Population[a] (thousands)
Russia	-	82	Tatar	4	15	147,553
Estonia	62	30	Ukrainian	3	5	1,573
Latvia	52	34	Belorussian	5	9	2,678
Lithuania	80	9	Polish	7	4	3,695
Belarus	78	13	Polish	4	5	10,195
Ukraine	73	22	Jewish	1	4	51,578
Moldova	64	13	Ukrainian	11	9	4,359
Georgia	70	6	Armenian	8	16	5,431
Armenia	93	2	Azeri	3	2	3,326
Azerbaijan	83	6	Armenian	6	5	7,092
Kazakstan	40	38	German	6	16	16,580
Turkmenistan	72	9	Uzbek	9	10	3,572
Tajikistan	62	8	Uzbek	24	6	5,182
Uzbekistan	71	8	Tajik	5	16	20,094
Kyrgyzstan	52	21	Uzbek	13	14	4,308

[a] The 1989 Soviet census reported two different figures for the total population of each republic. One is based on the number of people in the republic on the day the census was conducted. This map uses the other, which is based on the number of people reporting the republic as their place of permanent residence. Source: US Bureau of the Census.

Boundary representation is not necessarily authoritative.

737771 (R02027) 11-95

COMMIE CRASH: UKRAINE UNLEASHED

Well enough of the commie sugar-coating, 'cuz you all know how this Soviet story ends. The USSR system sucks: They lose the arms race with the United States; they end up flat broke and riddled with corruption and inefficiency; environ-

mental destruction was everywhere . . . and they can no longer hold on to or control the Soviet satellite states or even most of their ethnically based SSRs that begin demanding independence from the Union. When Mikhail Gorbachev (the last Soviet leader) became President in 1988, he introduced the policies of glasnost ("openness") and perestroika ("restructuring"), which spelled the beginning of the end of the USSR. All those that wanted to leave the Union were allowed to go, before the Union itself was voted out of existence in 1991. Thus, a single sovereign state (the USSR) became 15 new sovereign states . . . including one particular place called Ukraine!

Woo-hoo! Now we are all caught up on history, and Ukraine is free and happy!!! Right? Well . . . not

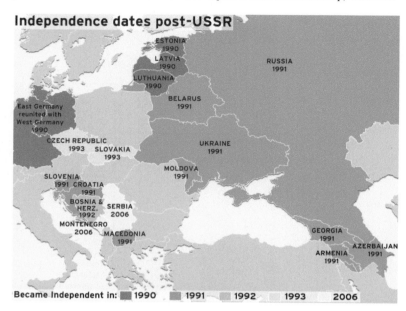

Independence dates post-USSR

Became Independent in: 1990 1991 1992 1993 2006

so much. Because that linguistic/ethnic/historic divide we have discussed in great detail is still there, and there are those that feel more comfortable with the old ways and with Mother Russia, and others who have really hated the Soviet experience so much that they want to move away from Russia as fast as possible. So in 1991, Ukraine starts life as an independent state completely broke and corrupt (like all the other post-Soviet states), and with a population that is slightly unsure about what path to put the state on for its future. Let me make one thing perfectly clear before going any further: in 1990, virtually EVERYONE in the Ukrainian SSR voted to become independent of the USSR, and thus also to become independent of Russia. In that, the entire population was united. And in those initial hectic years after independence, there was a sense of optimism, hope, and agreement that change was going to come by becoming a modern democratic state that would be geared toward market capitalism, and thus engaged with "Team West," specifically the European Union. Which was a sentiment shared by most of Eastern Europe that was striving for a policy of. . . .

RE-ALIGNMENT WHILE RUSSIA IS WEAK

The 1990s into the 2000s was an era of transition for the whole Eastern Europe region. And now it becomes quite obvious what they were transitioning from and to—from that Soviet (aka Russian) era of occupation and control, and mostly realigning and transitioning to become adopted into the capitalist democracies of "Team West." This happened at a brisk pace after 1991 when the Soviet Union officially voted itself out of existence, and the entity broke up from one huge power into fifteen sovereign states, as already pointed out.

We also already pointed out that many of the states, Czech Republic, Hungary, and Poland, were quick to embrace the West; Estonia, Latvia, and Lithuania were right on their heels, by the way. Because they used to be independent sovereign states in between WWI and the end of WWII, but then were reabsorbed by the Soviets. Those guys were always chafing under Soviet rule, so Estonia, Latvia, and Lithuania were the second round of states that jumped ship into the arms of the West. Many others have since followed. All of this realignment and transitioning of these countries of Eastern Europe was occurring right after the Soviet crash . . . quite frankly, when Russia (and Russia became a new country, as well, at this time) was excessively weak. I mean, they just lost the Cold War, and they mostly lost the Cold

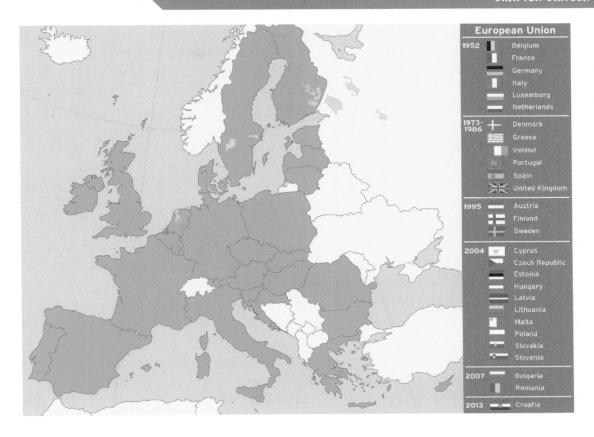

War because they were freaking broke. Their economy was in shambles, their power structure was shattered, and their government was in chaos. It was precisely in this period of Russian weakness that so many of these Eastern European countries just ran to the West.

This transition, of course, happened in a couple of distinct ways by some distinct entities we've talked about many, many, many times already in this great text. Number one was the EU—the European Union. Most, if not all, of these countries immediately applied for EU entry status. Many of them were soon granted it. As you can see on the map, there has been a progressive wave of Eastern European countries entering the EU since the very beginning of their "freedom" from Soviet domination.

That brings us to the other avenue for the Team West love embrace by these newly independent Eastern European countries in the 1990s, and that was NATO. No surprises here. In fact, you can essentially look at the story told by that map of EU entry and see that the NATO entry tells the same tale. As already suggested, Czech Republic, Hungary, and Poland were quick to escape the Soviets and jump under the NATO security blanket. Estonia, Latvia, and Lithuania were right on their heels—just like with the EU. Many of the other countries who joined the EU are also now NATO members.

Hang on, Plaid Avenger! You have spouted for several paragraphs at this point without even mentioning Ukraine! What up with that? Good call, my astute amigo. Like almost all other Eastern European states, there was popular support to jump into the EU as soon as

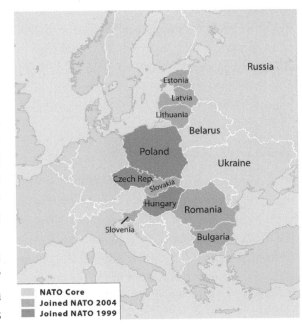

possible. And given Ukraine's huge area, its big population, and its immense economic potential, it would become a huge power player were it ever to join the EU. So why didn't it join the EU already? Well . . . let's just be straight up honest: they have not yet been up to the standards to gain entry into the club. Political corruption, financial corruption, infrastructure issues, poor economy, high unemployment, inflation, electoral fraud . . . the list could go on. Ukraine has been in talks with the EU since the 1990s about cleaning up its act enough for entry, and it has only been in the last few years that the talks have gotten more serious, even though Ukraine still has a ways to go in order to be up to snuff. These more serious talks about entry into the EU is what sparked the recent fury and started the unraveling of the state . . . but I will get back to that in a minute.

When it came to Ukraine joining NATO, we find an even stickier situation. Many EU countries were against taking on the economic burden that a poor-performing Ukrainian society would burden them with . . . but when it came to NATO countries, hardly any of them wanted to take on the military risks and possible reignition of the Cold War that would happen by admitting Ukraine into their club, which they all understood would infuriate Russia. Hardly any of them, that is. So who did want Ukraine into NATO immediately? That would be the USA!!! I distinctly remember both U.S. President George W. Bush and Vice President Dick Cheney visiting Ukraine many times in the early 2000s, mostly talking about NATO, and the United States led the charge within NATO ranks to get both Ukraine and Georgia into the group. You can love or hate the Bush/Cheney team, but they understood perfectly well the advantage of getting all these countries into the Team West orbit while Russia was weak. . . .

But the general consensus was that Ukraine was not militarily or economically advanced enough to meet NATO standards. More importantly, most NATO states thought that the era of expansion had perhaps gone far enough, and that admitting Ukraine or Georgia would destroy all relations with Russia, which everyone was keen to develop further in a positive way. Keep in mind, Russia had long been invited to NATO summits, and some analysts were hopeful that Russia would itself join Team West, and therefore could become a NATO member itself someday! But even if that were never to happen, Russia was pretty much pissed all throughout the 1990s and beyond because it interpreted the expansion of NATO as a direct threat to their own national security. Russia was none too happy about any of its former territories joining NATO, and especially not the ones it shares a direct border with like Estonia and Latvia . . . the idea of Ukraine joining, with a HUGE shared border with Russia and the capital Moscow just hundreds of miles from that border, was from its inception extremely agitating to the Russians. Add to that the concept that Ukraine has historically been a part of Russia proper. Add to that the foreign policy concept that "Russia without Ukraine is a country; Russia with Ukraine is an empire."

And add to that the fact that the Russian's most important naval base was located on the Crimean Peninsula . . . which of course had become part of Ukraine at independence! Back to that pesky peninsula! It's a huge deal! Here is Russia, a world power, whose main military naval base is located in another country . . . and that country is contemplating joining NATO, Russia's arch enemy from the Cold War! Unthinkable! Insane! Infuriating! Insulting!

Russia could not do much to prevent EU or NATO expansion from 1991 to 2004, but all that Russian frustration and impotence was all about to change because . . .

RUSSIA REBOUNDS

Well, this eastward expansion of Western institutions, like EU and NATO, is now being slowed, stalled, or stopped, depending upon your point of view. And the current Ukrainian crisis is case in point. The battle for influence between East and West is still alive and well, but it appears that the Ruskies are winning this round. What am I talking about? Hey, I'm talking about Russia, man! These guys are back! Mostly because of the Russian man of the century: Vladimir Putin. This guy is unstoppable! So rarely in life can you truly credit a single individual with changing the course of history but, for better or for worse, that is an apt description for our main man Vlad. It's no wonder that *Time* magazine voted him Man of the Year for 2007—he really has made that big of a difference in his country, and in the world.

Putin was only the second president Russia has had. He was President (2000–2008), then stepped down to become Prime Minister (2008–2012), and now back by popular demand to be President again, for possibly a long time to come (2012–???). In his first run as President of Russia, from 2000–2008, Putin oversaw the stabilization of the economic transition process that had thrown Russia into chaos for its first decade. He was tough on the oligarchs, and tough on crime, which helped stimulate international invest- ment. He also took a fairly proactive government role in helping Russian businesses, especially those businesses dealing with petro- leum and natural gas . . . thus making Russia into an international power player for energy, and one that Europe relies on heavily. Oh

My name? Putin. Vladimir Putin.

yeah, so does Ukraine! GDP grew sixfold, and propelled Russia from the 22nd largest economy up to the number ten slot. The economy grew 6–8% on average every year he was in office. Investment grew, industry grew, and Russia started to get its mo-jo back. And the mo-jo is actually the most important part for our story. . . .

See, Vladimir Putin re-instilled a great sense of nationalism in his citizens. After the demise of the USSR, Russia was a second-rate power with even less prestige, at the mercy of international business and banking and stronger world powers like the United States, the EU, and NATO. Russia pretty much just had to go with the flow, even when events were inher- ently against their own national interest, like the growth of NATO, for example. Putin brought them back politically to a position of strength. He played hardball in international affairs, and reinvigorated a Russian national pride that had long laid dormant. You got your Putin, and you got your petroleum, and you add to that a huge dose of nationalistic pride. Bake for twenty minutes, and you end up with an extremely resurgent and resilient Russia on the world political table. And Putin wants to get his country's place back at the head of that table as a serious player to be respected, or feared if necessary. One of the most frequently repeated quotes from Putin's leadership in Russia, when he said, "The breakup of the Soviet Union was the greatest geopolitical tragedy of the 20th century." Make no bones about it: Putin and crew want to re-establish Russia's position of dominance in its historic zone of influence, which is so say its immediate neighborhood . . . and the way to do that is perhaps by using force. . . .

RUSSIA–GEORGIA WAR

Beyond the geopolitical ramifications of energy and economics, Russia is back in full action militarily speaking! Big time! And it didn't even start with the Ukrainian/Crimean mess! It started next door in that other country that was think- ing about going NATO: Georgia. To see when Russia really turned the corner on its new-found assertiveness, we have to go back a couple of Olympics ago (yes, in a bizarre twist of fate, everything seems to turn on the Olympic games.) Dig this: On the first day of the 2008 Olympics in Beijing, Georgia launched a large-scale military attack against the self-proclaimed Republic of South Ossetia, which was within their sovereign state. The Russians, who totally hated the Georgian government at the time, were totally prepared for this action and already had thousands of troops massed on the border . . . and invaded Ossetia, thus launching the 2008 Russia–Georgia War between Georgia on one side, and Russia together with most ethnic Ossetian and ethnic Abkhazian on the other. The Russians proceeded to crush all opposition in 5 days, and "liberate" these two enclaves in Georgia. Note: Abkhazia is on the Black Sea, just like Ukraine . . . hmmmm . . . coincidence?

The Russian–Georgian War did not change the balance of power in this region; it simply was an announcement that the shift in the balance of power had already occurred! Russia is now openly and powerfully reasserting its influence in its own backyard, and that is a big deal. Russia now feels that it has regained a position of strength to not only stand up to what

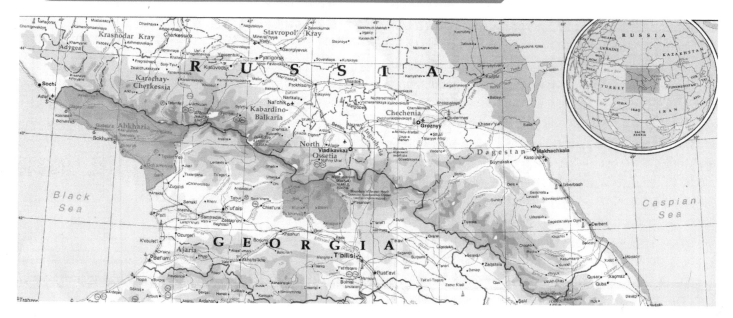

it sees as threatening Team West expansion to its borders, but to stop that advance with force if necessary. With guns! And they are doing it in areas where they have historic ties. And that brings us back to Ukraine! What the burnt borscht does all of this history have to do with Ukraine???? Now that you understand the re-rise of Russian power in this 'hood, let's catch back up with our Ukrainian friends. What was happening there in the last decade, just as Russia was getting its mo-jo back?

UKRAINE YEARNING: THE ORANGE REVOLUTION

In a word: floundering. Unlike many of those Eastern European states that got their acts together and joined Western institutions, Ukraine has pretty much just bombed. Ukraine's economy has mostly tanked for two decades, and only seen sporadic grow spurts that ultimately fail. Corruption in the financial sector, business sector, and political life is so rampant that the average citizen has given up on the entire rotten apple. Ultimately, the deeper issues here is unresolved national identity . . . who are they as a single group of citizens? Are they pro-Russia, or pro-West? What direction do they want to go in? Whose camp should they join?

This divide has been a challenge for Ukraine since it became "free" in 1991. Almost every election has been evenly split between the two halves, pulling the country in opposite directions. One could make the argument, as most succinctly put by Ukrainian political scientist Leonid Peisakhin, that Ukraine "has never been and is not yet a coherent national unit with a common narrative or a set of more or less commonly shared political aspirations." That is a problem. I would argue the core of the problem. And that problem has allowed outside powers to exploit the situation to try and get Ukraine on their teams. The first time Ukraine publicly became a poster child for this Team West vs. Russia battle was the Orange Revolution. . . .

Due to political mismanagement and the aforementioned corruption and general economic sucky-ness, Ukrainians were fired up about changing their fate during their 2004 presidential election. The two main candidates were openly expressing two different paths: one to the EU, the other to Russia. The scenario included two dudes. The first was Viktor Yushchenko: the pro-democracy, pro-EU, pro-NATO candidate. A real poster child for the West. The EU supported him, and the United States thought he was awesome. They loved this dude! His opponent was Viktor Yanukovych: an old-schooler, very conservative, more Russian influenced, and in fact, he's the candidate that Vladimir Putin from Russia came over and campaigned for. To try to help lock up the election before it even was held, former elements of the KGB even unsuccessfully

Bush

Yushchenko

Yanukovych

Putin

Bush supported Victor Y. while Victor Y. was backed by Putin. Y ask Y?

poisoned Yushchenko! Talk about a street fight! How hilarious is this? I can almost hear the boxing announcer: "In the red, white, and blue trunks, fighting for Team West, hailing from western Ukraine, it's Viktor Yush! And his opponent, in the red trunks, representing the red Russians, from eastern Ukraine, it's Viktor Yanuk!" Hahahaha dudes, could you make this stuff up? Viktor vs. Viktor—I wonder who will be the victor?

When this election went down, the pro-Russian candidate won, and everyone in the world said it was a fraudulent election. There were massive street protests, and this ultimately turned into the 2004/2005 Orange Revolution. So much heat got put on the Ukrainian government, they threw out the election results, re-ran it, and then the pro-Western Yushchenko won. We can see this as part of an ongoing battle for influence in Ukraine, which is representative of the broader battle for influence in the region as a whole.

The Russian candidate was down because the pro-Western candidate won—down, but not out. In the post-Orange Revolution Ukraine, there were still a lot of pro-Russian people. There was still a pro-Russian political party, and in March of 2006, the same party gained enough seats to win back control of the Congress. The Plaid Avenger knows that those in the United States think democracy's great, that the "good guys" won, and that's that. Not so! This is an ongoing battle for control. It's not over! Most other countries have gone the way of the West, but it's definitely not over yet for the Ukrainians.

The East/West divide we have referenced all chapter long translates exactly to political life as well: it's a western pro-West and an eastern pro-

Official Breakdown of Ukraine's Pro-Western vs. Pro-Russian Districts
(as per the 2006 presidential election)

won by Victor Yushchenko
won by Victor Yanukovych

Russia, as you can clearly see from the results of the 2006 election cycle in the map to the right. Could the divisions in this country be any geographically clearer? Because of the pro-Russian political party gaining the majority of seats in their Congress, they got to choose the Prime Minister position . . . which promptly went to Yanukovych! (The President is voted in by direct election; the Prime Minister is appointed by majority of Congress.) So after the 2006 election, you had this bizarre situation where President Yushchenko was pro-West, and Prime Minister Yanukovych was pro-East! Heck, Yanukovych spoke Russian as a first language, and barely spoke any Ukrainian at all! Talk about a country with a split personality! I mean, really, it can't get any crazier than that, right?

YANUKOVYCH, REVOLUTION, TAKE 2

Hahahahaha; this is Ukraine my friends! So, yes it can get crazier! I hope you are not Yanuk-ed out yet, because that dude is crucial to the rest of the story! This internal division EXACTLY symbolized the fight for influence in Eastern Europe between Team West and Russia. From 2005 to 2010, President Yushchenko continued to push for EU and NATO entry, while Prime Minister Yanukovych actually announced that no way in Hades would his country join NATO. The United States sent aid to Yuschchenko and his allies; the Russians sent aid to Yanukovych and his party. The Russians and pro-west Ukrainians even had several economic battles over Russian supplies of oil and natural gas to the country. And then things get crazier. . . .

In this period, the economy got even worse, corruption became even more obvious, and pretty much everyone blamed the pro-Western leaders for botching the job, and dashing the hopes of the Orange Revolution with total political mismanagement and in-fighting. The Ukrainians went back to the presidential polls in 2010 . . . and take a wild stab who won. Victor Yanukovych, of course! Seriously? Yep, seriously, and apparently legitimately that time. The first order of business for Yanukovych was to solidly state there would be no NATO in Ukraine's future, which coincided with Russia giving a huge price break to Ukraine on oil/gas contracts and promising them a bunch of foreign aid as well. Isn't it funny how coincidences like that happen? Yanukovych also immediately extended leases on critical Russian naval bases located on the Crimean Peninsula, a move that infuriated pro-Western politicians who retaliated by throwing eggs at the Speaker of the House when the bill was passed. No. Really. Google it.

Whoa . . . I just realized we are now getting dangerously close to current events! So what else happened from 2010 to 2014 under the second round of Yanukovych leadership that sparked a second round of revolution? Well, it actually gets pretty easy to describe events from here on out, because it is a repeat of the past. The Yanukovych government cozies up with Russia more in order to get cheaper energy, allows the Russians to have their Crimean military base for another century, jailed a former pro-West Prime Minister (Yulia Tymoshenko) on trumped up charges, and even forwards a law to recognize Russian an official language of their state. Ummmmmm . . . starting to get the picture of what direction Ukraine is being led to? In making all these plays, Yanukovych starts to move away from any aspirations of actually joining the EU. NATO is completely off the table by this time as a viable possibility. Let's lower the bar even further, though: In this time frame, corruption got even worse; unemployment went up; inflation went up; the economy stagnates; and laws are passed giving more power to the Presidency. Authoritarianism is on the rise. Societal tensions were once again mounting in Ukraine, and could blow at any moment. . . .

But Yanuk played his cards very wisely during all this: while all his concrete moves cemented Ukraine's relationship with Russia, he simultaneously gave a lot of lip service to the idea of continuing talks to get into the EU. And the EU by this time was wanting to work a lot harder to actually help Ukraine get in (likely on the request of the United States). FYI: What the EU and Ukraine were working on all this time was not actually actual entry into the EU; they were working on an EU ascension agreement for Ukraine . . . in other words, an agreement about what Ukraine needed to do to START the process of getting into the EU, and to be associated with the EU. That make sense? Since Ukraine sucked so bad at the time, that process of preparing to get into the EU entailed a lot of painful financial reform, a lot of painful budgetary reform, and a lot of painful corruption . . . meaning that all the current corrupt politicians and businesspeople would take a hit, but also that the Ukrainian government would have to tighten its belt and clean up its act big time, which would mean the Ukrainian people would also be taking a hit in terms of decrease in government services, and so on. The idea is that once they got their house in order enough to join the EU, and then finally did join the EU, they would generate a ton more economic interaction and thus finally become stable and rich like the other European countries.

So it wasn't like this was a no-brainer for Ukrainian politicians to agree to, because it would equate to short-term pain in exchange for long-term gain. And let's be honest: that is something that no politicians anywhere on the planet want to do, because short-term pain means they don't get re-elected. Corruption reigns supreme, and no one at the top wanted the system to change. However, the average Ukrainian did want the long-term gain, and did want to get into the EU, as did many powerful businesspeople, so Yankuovych basically pretended to go along with these ascension talks with the EU for years . . . grudgingly going along as the slow, bureaucratic meetings muddled on year after year. But a funny thing happened on the way to nothing happening. . . .

By late 2013, the European Union representatives had reached a pretty significant deal with their Ukrainian counterparts, and the time had finally come for this agreement to be signed and implemented. And that my friends is when Yanukovych had to finally show his cards in this high stakes poker game he had been playing . . . and bluffing at, by the way. On November 13, 2013, he rejected the deal out of hand, thus publicly announcing (using no words) that his intent was for Ukraine to draw closer to Russia, not integrate with the EU/Team West. The fuse had been lit. Public protest broke out instantaneously . . . and as you might have predicted, those protests occurred largely in the western side of the state. In Kiev, the capital, clashes between protesters and security forces have become violent, killing several people . . . the "Euromaidan" was born. . . .

EUROMAIDAN

Euromaidan was a wave of demonstrations and civil unrest in Ukraine, which began on the night of November 21, 2013, with public protests demanding closer European integration. The term "Euromaidan" was initially used as a hashtag on Twitter, and a Twitter account of the same name was created on the first day of the protests. The name is composed of two parts: "Euro" is short for Europe and "maidan" refers to Maidan Nezalezhnosti (Independence Square), the main square of Kiev, where the protests were centered. What started as a peaceful political pro-EU protest of a couple of thousand swelled to 10,000 or more within a week, with even bigger congregations of protestors in other western cities like Lviv. The government violently cracked down on the protestors on November 30th, sparking rioting and further violence from the protestors themselves. It is at this point you may have heard something about it on the nightly news in the United States, but most of you were just waiting for Santa to deliver you presents, so you didn't pay much attention. But the world was now watching. And Yanukovych? He was conveniently out of the country, specifically for a meeting on December 6th with Vladimir Putin in Sochi, Russia, to discuss a whole slew of sweet energy and industry deals that Vlad the man was hooking up Ukraine with. Oh, and Vlad also promised to give Ukraine something like $15 billion in aid, effective immediately. Again, what a coincidence. Why in Sochi, Russia? Russia was preparing for the Winter Olympics in Sochi . . . I told you! The Olympics! Beware the Russians when the Olympics are happening!

This meeting infuriated the opposition, obviously, but mostly the daily protest in the Euromaidan and other western towns had quieted down by the holiday season, and it appeared for a while that things were going to normalize. Appearances

The Maidan Nezalezhnosti, pre-protests.

can be deceptive. The protests may have decreased in the numbers of people actually on the streets on a daily basis, but a wave of dissent was sweeping across pro-EU western Ukraine, and this dissent evolved to include calls for the resignation of President Yanukovych and his government altogether and "a will to change life in Ukraine." Tensions mounted. Everyone waited for the next spark to fly, and they got it on January 16, 2014, when the government accepted Bondarenko-Oliynyk laws, also known as Anti-Protest Laws, which restricted free speech and freedom of assembly . . . basically in an effort to crush all the protesting via legalistic avenues. Now people got really fired up! Even peeps who were not prone to protest started to worry

about this move toward dictatorship and degradation of individual rights. Violence escalated on all sides. Some police and military started to "defect" to the other side and join the protestors. Rage was fueled by the perception of "widespread government corruption," "abuse of power," and "violation of human rights in Ukraine" . . . this is now getting way bigger than just a pro-EU demonstration! This is getting borderline political revolution! And now Russian troops were starting to mass on the Russian/Ukraine border. . . .

2014 REVOLUTION

By this time, tens of thousands start to protest across the country . . . again, mostly in the west, but some even in the east! The Euromaidan protests reached a climax on February 18th, after the parliament did not accede to demands that the Constitution of Ukraine be rolled back to its pre-2004 form, which would decrease presidential power. Police and protesters fired guns, with both live and rubber ammunition, in multiple locations in Kiev. The riot police advanced toward Maidan later in the day and clashed with the protesters but did not fully occupy it. The fights continued through the following days, in which the vast majority of casualties took place. But the writing was really on the wall for Yanukovych after February 20th: that was the day that all hell really broke loose, and Yanukovych sent trained, ski-mask-wearing snipers into this situation to shoot at people. Many of these government-sponsored snipers were caught on film shooting unarmed civilians. When any government starts killing its own unarmed people, it is usually game over for them. And it was. Game. Freakin. Over.

Virtually all of the pro-EU western protestors, the USA, the EU, and most of Team West saw the perpetuation of Yanukovych's power as completely propped up by Russia, and many felt that Yanukovych's decision to crack down on the

Kiev becomes battleground.

protest with overwhelming force likely came as an order straight from Moscow. On the night of February 21, 2014, Maidan protestors vowed to go into armed conflict if Yanukovych did not resign by 10:00 A.M. Subsequently, the riot police retreated and Yanukovych and many other high government officials fled the country. Protesters gained control of the presidential administration and Yanukovych's private estate. The next day, the parliament impeached Yanukovych, replaced the government with a pro-European one, and ordered that Yulia Tymoshenko be released from prison. Putin then canceled that $15 billion he was going to give to help float the Ukrainians . . . what a shocker. Anti-Russian sentiment swelled, and many ethnically Russian people started counter-protesting across the east to protect themselves and their communities, but also to fight against what they perceived as an illegal and undemocratic coup . . . remember, their boy Yanuk had been legitimately elected, so they felt dissed and dumped on when he was ejected from power. In the aftermath, the Crimean crisis began amid pro-Russian unrest.

Keep in mind the Olympic timeline here, too: The eyes of the world were focused on Russia hosting the XXII Olympic Winter Games that ran from February 7 to 23 in Sochi, a mere 900 miles away from the action in Kiev. Why would I point that out? Because Russia was kind of busy putting on a pretty face for the world, in what was supposed to be a show of its technological prowess and national pride and world power position by hosting the Games. It could not simply act at the time to physically stop the turmoil in Ukraine, nor to forcefully stop Yanukovych from being ousted. In addition, Putin and his crew interpreted this timeline in a very different way: they were all convinced that this entire Ukrainian Revolution was being sponsored by the USA, and the CIA, at the exact time that Russia could not act. Thus, it was feeding into Putin's paranoia that the revolution was not only not legitimate, but partially done just to embarrass Russia during the Games.

Perceived Putin Power

But then the Olympic Games ended, and the Crimean games began. . . .

BACK TO THE CRIMEA . . . WHICH GOES BACK TO RUSSIA

Bringin' it back to cold reality now: remember there are a significant number of ethnically Russian peeps, mostly in the east, and a very high concentration of them on the Crimean Peninsula. And Russia has an extremely important naval base there, and used to outright own the whole peninsula back in the day. And I want you to think about this whole mess from the Russian perspective for just a minute: what they were seeing (and still are) is a totally dysfunctional Ukrainian state, in borderline chaos, overthrowing democratically elected leaders, while trying to "side up" with Team West . . . the very group that was the cause for Soviet Union collapse. If you were Russia in this situation, what would you think? In addition, NATO was perceived by most Crimeans and Russians in general as encroaching upon Russia's borders. This weighed heavily upon Moscow's decision to take measures to secure her Black Sea port in Crimea.

As mentioned, these pro-Russian Ukrainian citizens were not only feeling threatened by what they perceive as a hostile take-over of the government that they elected, but also believed that anti-Russian sentiment, then at a fever-pitch, would result in a Russian genocide . . . of them! Make no mistakes about it: Russian propaganda back in the motherland was certainly feeding into that fear, and indeed there were/are small extremist groups in Ukraine that are calling for "death to the Russians," but those groups are tiny and do not reflect the popular will of the people in any legitimate way. Although Putin himself credited the entire revolution to "Russophobes and neo-Nazis," who want the expulsion of all

Russians from Ukraine, this potential genocide of Russians within Ukraine has not materialized in any way at all. But fear makes people act in strange ways . . . so . . .

In the wake of the collapse of the Yanukovych government and the resultant 2014 Ukrainian revolution, a secession crisis began on Ukraine's Crimean Peninsula . . . which is 70% ethnically Russian. Large protests erupted there in February calling for independence from Ukraine, and absorption into Russia proper! This is unheard of in the modern age! Well open your ears and hear it now, 'cuz it just happened. Unmarked, armed Russian soldiers began being moved into Crimea on February 28th. The next day exiled Ukrainian President Viktor Yanukovych (still hiding somewhere in Russia to this day) requested that Russia use military forces "to establish legitimacy, peace, law and order, stability and defending the people of Ukraine." Later that same day, Vladimir Putin requested and received authorization from the Russian Parliament to deploy Russian troops to Ukraine and took control of the Crimean Peninsula by the next day.

On March 6th, the Crimean Parliament voted to "enter into the Russian Federation with the rights of a subject of the Russian Federation" and later held a referendum asking the people of these regions whether they wanted to join Russia as a federal subject, or if they wanted to restore the 1992 Crimean constitution and Crimea's status as a part of Ukraine. Though passed with an overwhelming majority, the vote was not monitored by outside parties and the results are internationally contested; also, it is claimed it was enforced by armed group that intruded

Hmmm . . . no markings on their uniforms. That's odd.

and forced voting according to their demands. Crimea and Sevastopol formally declared independence as the Republic of Crimea and requested that they be admitted as constituents of the Russian Federation. On March 18, 2014, Russia and Crimea signed a treaty of accession of the Republic of Crimea and Sevastopol in the Russian Federation.

Later that day, President Putin gave an emotional speech in the Grand Kremlin Palace announcing that Crimea had been peacefully returned to Russia, reversing a "historic injustice" of the Soviet Union. He also peppered his speech with references to the restoration of Russia after a period of humiliation following the Soviet collapse . . . and blamed most of that nasty era on the domination of the "global superpower and its allies," which of course means the United States, the EU, and NATO. In addition, Putin at this point starting referring to the duty of the Russian government to protect Russian people, everywhere . . . not just Russian citizens. Do you understand the difference? "Russian people" is not the same as "Russian citizens," and implies that Putin's Russia is giving themselves the authority to protect and possibly unite this proud tribe of people, wherever they are. This speech was given to thunderous applause, standing ovations, and tear-filled eyes of people chanting "Russia! Russia!"

Wow.

Okay. That is kind of a big deal. I will come back to that. But what about the rest of Ukraine?

A CATCH-UP ON THE REST OF THE COUNTRY . . .

What did the United States or the EU or the United Nations (UN) or the world in general do about all this chaos in Ukraine, and the absorption of Crimea into Russian hands? Quick answer: not much. Team West refuses to acknowledge that the Crimea now belongs to Russia, and has been encouraging Russia to stand down from the whole affair, and for them to pull back the massive troop buildup on the border, which Putin actually has done as of June 2014. Team West slapped some sanctions on Russian officials in an effort to pressure them into not invading Ukraine, but those amounted to nothing. Team West has also been in a hurry-up offense to push through billions of dollars of aid to the newly elected leadership of the rest of Ukraine to keep them afloat and functioning, since Russia bailed out on that deal after their boy got ejected. At this point, I'm not even sure Russia recognizes the new President of Ukraine . . . oh yeah, I forgot to tell you that on May 25th, Ukraine held an election to re-do their government, and the pro-European Union candidate named Petro Poroshenko won over 50% of the vote. He was inaugurated on June 7, 2014, and announced that his immediate priority would be to stop the civil unrest and protests in eastern Ukraine and mend ties with Russia. What? Civil unrest in the east?

Yep. While all of the Crimea secession and Presidential election was happening from March to May, protests and civil unrest began in the eastern and southern regions of Ukraine. In several cities in the Donetsk and Lugansk regions, armed men, declaring themselves as local militia, seized government buildings, police and special police stations in several cities of the regions. Basically, that stunt worked in Crimea, so pro-Russian forces have been trying it in the east as well! Get a group of your pro-Russian buddies together, don some ski masks, grab some guns, and take over government buildings . . . then declare your town/region independent and ask for Russia to absorb you! Seems to be all the rage lately. Try it in your hometown and see how it works out.

For months, it was feared that Russia would fully invade and absorb even more areas of Ukraine, if not the whole darn country. And those pro-Russian, building-seizing peeps are not playing around, and have become super violent. There are open pitched battles between the Ukrainian military and some of these groups, who are armed to the teeth, and may even include the local police/military forces of the areas that are attempting to secede. Ukrainian military planes and helicopters have been shot down and death tolls mount. So this mess is far from over, although Putin has called for these rebel groups to stand down, and has moved many of his troops away from the border, so it looks to be cooling down a bit here in July 2014 . . . but who knows what tomorrow will bring?

And that chaos in the east/south is not the only place where protestors are still active. Despite the impeachment of Yanukovych, the installation of a new government, and the signature of the political provisions of the Ukraine-EU Association Agreement, the protests have been ongoing to sustain pressure on the new Ukrainian government, to counter pro-Russian protests, and to reject Russian occupation of Ukraine. The general area of the pro-Ukraine and pro-Europe protests has shifted from Kiev and western Ukraine to include the eastern and southern areas of the country as well. Whew. That is it. That is the cover story. Let's finish with some . . .

REPERCUSSIONS

In some ways, the Ukrainian crisis is/was about popular anger against a crappy president who mishandled an already crappy economy and whose attempts to quash protests have pushed the state into an increasingly authoritarianism craphole. But it's also about Ukraine's long-unresolved national identity crisis. From the outsiders' points of view, whether you are Western or Russian, the story is framed as Ukraine being pulled apart by Moscow on one end and Europe/United States on the other . . . and that is certainly part of the plot. But Ukrainians themselves have been the ones doing most of the pulling: it's been a two-decades-long tug-of-war between two halves and two identities of what was supposed to be a single state. Will it go ahead and split into two? Will it be in a perpetual state of conflict and chaos? Perhaps not so much now that it came to a head and the pimple got popped. Now that the world is involved, maybe things will be sorted out, and a better era of life for Ukrainians (no matter what their ethnicity) is about to dawn. Of that, I am unsure. But I can give you a quick run-down of impacts this nasty business has had on the regions around it, and on the world. Let's start with our old friend sovereignty. . . .

SOVEREIGNTY REDEFINED?

Yes, loss of life and societal chaos are terrible. And, yes, I guess it's a big deal whether Crimea is part of Russia or part of Ukraine. And, yes, the future of Ukraine is important. But . . . none of those things has a big a global impact and will affect the future events on our fair planet as much as the part where, Vladimir Putin, wildly popular President of Russia, suggested that he had the duty to "protect Russian people." I mentioned above that this turn of phrase is quite important. Prior to this Putin, and every other world leader, would say that their duty is to protect the *citizens* of their sovereign state . . . not the members of their particular tribe. Because that is what he is saying: that he as the President of a sovereign state of Russia not only has the duty of protecting Russians in his sovereign state, but all Russian, the entire tribe called Russians, no matter what other sovereign state they happen to be in!

Dudes! Dudettes! That is the old school definition of sovereignty! For most of human history there have been kings, or rulers, or leaders of *peoples*, not of geographically defined political spaces. Think back to, let's say, the French. Throughout most of history, the king or leader has been referred to as the King of the "Francs," not the King of "France." But times changed, and we all accepted this concept of sovereignty attached to states, that is, defined geographic spaces . . . but not so much to peoples anymore. The leader of every sovereign state has complete control over all the peoples in their defined geographic area, their state . . . but no one outside that state! Those outsiders are controlled by the leader of whatever state they are in! Well, not for Russians, according to Vladimir Putin. All the Russians, everywhere, are his state's responsibility, and it is his duty to protect them . . . wow, he is so old school that he rides a dinosaur to work.

And why would he do that? For starters, to justify Russian intervention in Ukraine, and justify them reabsorbing Crimea . . . after all, it was the Crimeans who asked for it, and it just so happens they are ethnically Russian, so to not help them would be to turn his back on fellow Russians . . . at least in Putin's interpretation. But it is also about something bigger: legitimizing Russia's dream of expanding its influence back out into the world it lost when the USSR collapsed. A Putin quote from that same Putin speech I referenced above, and he is referencing the end of the USSR: "Millions of Russians went to bed in one country and woke up abroad. Overnight, they were minorities in the former Soviet republics, and the Russian people became one of the biggest—if not the biggest—divided nations in the world." And he wants to protect his Russian brood, wherever they reside. And check out the map below to see where that would be. Think Kazakhstan or Latvia may be a little worried by Putin's declaration and actions in Crimea? I would be. Think they will be going out of their way to kiss the butts of their ethnically Russian citizens? I would be.

In one swoop, Putin has gone old school, and that may seriously upset the delicate balance of sovereignty and statehood worldwide in the future. Think about this, and chat it up with your teachers: Can you find any other places on the planet where an ethnic minority exists that may have ties to a larger tribe in another territory? Yep. They are all over the place in today's world. Will Putin's move trigger a redefining of state borders based on ethnicity? Who knows. Tibetans would be all for it, as would the Basques. And even China might dig it, as there are millions of ethnically Chinese people across the border of northwest China in Russia . . . oops, might China claim protection status over its peoples in that area as well? Putin may not have considered that. . . .

A "TEAM WEST" ALTERNATIVE OFFERED?

According to the current Russia's National Foreign Policy Concept (basically, an annual "white paper" or report on the objectives of Russian foreign policy), a major priority of Moscow's diplomacy for the next decade is the strengthening of regional integration in the post-Soviet territories, and the grand formation of this Eurasian Union among the countries of the region. Is this a Soviet Union, take 2? USSR 2.0? Soviet Union Reboot? Eh. Doubtful. But it is 100% being offered as a Team West alternative that Putin wants to build into something akin to the EU, or perhaps more. And with the show of force by Russia in Ukraine, it looks to be a grouping that many countries will feel compelled to join just to stay on Russia's good side! So what is it again?

While originally an idea posed by Kazak President Nursultan Nazarbayev, the Eurasian Union, is without a doubt the wet dream of Russian President Vladimir Putin, who now openly seeks to reassert a resurgent Russia's influence and out-right power into areas it lost during the collapse of the Soviet Union in 1991. Putin for his part has actually referred to the demise of the USSR as "the greatest geopolitical catastrophe of the 20th century,". . . and it is a catastrophe he now seems intent on fixing. Through economic coercion, intimidation, and possibly force if necessary, Russia looks to be regaining its old sphere of influence here in Central Asia and Eastern Europe. Is this an attempt to rebuild the Soviet Union proper? Is this a new Cold War? Eh. I'm not sure it is all that, at least not yet, but here is the deal as of 2014:

The formations of it are entirely logical and benign, at least on paper: In November 2011, the presidents of Russia, Belarus, and Kazakhstan signed an agreement to create the Eurasian Economic Union: a customs union (like a free trade zone) with partial economic integration among the states, with the intent of becoming something very close to the European Union (you know, the EU) in the future with full-on economic, political, and military cooperation. And supposed to be fully functional by 2015. Just like the EU, it is sup-posed to safeguard regional economic inter-

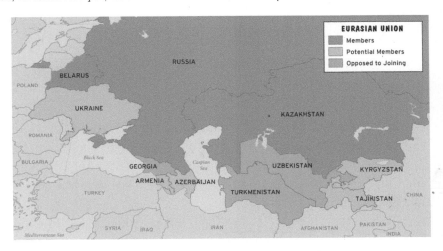

ests and facilitate trade and business within member countries. Since Russia is obviously the powerhouse player in all of this, Putin could not be happier. But he isn't satisfied just yet. . . .

Putin's grand gamble here is that his union (and, yes, it is referred to as his personal grand strategy at this point) will become a successful alternative to the European Union, and that states of Central Asia, Eastern Europe, and even the Caucuses Region will want to sign up for it. I suppose it is being marketed as a great way to boost economic growth, without having to bother with all that pesky democracy and human rights stuff. And it is definitely being pushed as an alternative to joining Team West. And it definitely is Russia trying to regain "influence" in areas it previously held as its own. The Ukraine was to be a jewel in the crown of all of this, thus Putin's interest in keeping that country in the Russian orbit, and the Eurasian Union specifically. That whole taking over Crimean Peninsula stuff? Yeah, that was the kick-off of the game, not the finish of it. And what game would that be? Ah! Back to the Great Game, of course! This is a now fervently fired-up Russian response to the last 2 decades of the EU and NATO creeping closer and closer to its borders. The Bear wants its backyard back! Game on!

Will it work? Who knows yet, but everyone in the 'hood is worried, as Russian pressure to get on board this Eurasian party train is growing. The EU, the United States, and even China are watching these developments warily. States that have lots of ethnic Russian people in them (like Kazakhstan) are starting to fear a Russian take-over of parts of their territory, much like what happened to Ukraine with Crimea. Belarus may be concerned about being reabsorbed altogether. But the other Central Asian states are now also stuck in this battle for control, as Kyrgyzstan and Tajikistan are highly dependent on Russian foreign aid and economic trade . . . they will likely join. The other -stans seem to be hedging their bets, as they like their sovereignty more than they like Russia. Armenia has announced it will join soon, and the fate of Ukraine as a whole is (as of this writing) still unsettled . . . a full Russian takeover of that state will see it bolster the Eurasian Union's ranks into a fully legit power block. But will Putin want to go that far to see his 21st century Ruskie dream realized? We will soon find out. . . .

IS THE COLD WAR BACK ON?

Oh no! Is this East vs. West throw-down alive once more? Is the Russia vs. United States bipolar world battle back on over this Ukrainian unraveling?

Once again, doubtful.

For starters, the new animosity between the United States and Russia over this business can't really be compared to the Cold War for several reasons:

1. It is not an ideological battle for global dominance like the capitalism vs. communism was;

2. Generally speaking, the rest of the world doesn't care, and is not going to "side up" with one of the big boys over the issue; and

3. Business will still be conducted between most Team West states and Russia, despite whatever happens (and I am thinking mostly of the Europeans, which simply must have Russian energy in order to keep the lights on).

So it is a very different world than the one in which the original Cold War was fought, and no matter what happens to Crimea, Ukraine, or the Eurasian Union, a cold war or even a hot war seems highly unlikely.

However . . . that is not to suggest that there are not some Cold War themes present here! And some definite repercussions that will take on a Cold War flavor! And there are some obvious winners and losers that we can point out to finish up this special "Cover Story" report.

Putin puppy love! Happy Days are Back!

For starters: Russia is riding high! This event has restored Russian pride, happiness, nationalism, and even tenacity to a level that has not been seen in decades. They feel as if they are a world power again, for real! Putin now has the popular support and political clout to unite the country like never before, and rebuild their society into some sort of happy-go-lucky, law-and-order, family values, ultra-socially conservative throwback to their 'good ol' days' . . . whatever the heck that means to them. Putin is going to rebuild the military into a significant 21st country fighting force, military parades will abound, and enemies will cower. Well, if Russia can afford it.

And I'm not making that stuff up about preserving tradition and bringing back the good ol' days: Putin's Russia is using good ol' fashioned fear tactics to destroy the press and terrify anyone who speaks out against the government. Anyone inside of Russia that is critical of Russia's current path will be viewed as a traitorous, Western-sponsored spy. It's going to be like the McCarthy Era in Russia for years to come. On top of that, they are doing everything in their power to "ban" gay people; they are banning the use of curse words in literature and communications; and they are now promoting Eastern Orthodox Christianity like never before. Seriously, this is like Rush Limbaugh's wet dream. No offense to Limbaugh fans out there . . . but what part of Putin's agenda do you disagree with? Hit me up and let me know. Long story short, this is a radical new age for Russia, for better or for worse. Personally, I think they are going to go economically and morally bankrupt again in this endeavor, but that's just my call.

Second, this whole episode has served to galvanize and energize Team West as well! Let's not overlook that little fact! Just at a time when most European countries have been questioning the whole concept of NATO, and just at a time when they are wondering if any further expansion of NATO should happen, and exactly at a time when both the United States and Europe are decreasing military spending on NATO because they are broke due to the recession . . . they are presented with the worst-case scenario of why they should keep NATO alive! And alive they shall keep it: the threat of further Russian expansionism has fired up the military alliance, which has already increased spending, increased visibility and troop deployment in Eastern Europe, and increased active military exercises. Take that, Russia! NATO has just been given a huge shot of steroids, and their future and their mission is no longer in doubt.

Speaking of which, I would be remiss in my duties as a global educator if I failed to point out that a huge winner in all of this mess is going to be the defense industries, worldwide. Russia is going to invest billions in its military now. So will Ukraine. NATO will outspend them both. It's a good time to be making high-end, high-technology weapons. Vladimir Putin is "man of the century" for the defense industries, who all likely have a gold-plated portrait of him in their boardrooms.

Third, this episode finally and authoritatively—in bold italic underlined 1,000 point font—signifies the absolute shift of Russia from Team West's orbit.

Defense Industries: Happy Days are Back!

For years I have been saying the Vlad the Man Putin has been slowly reorienting Russia's future toward Asia, and in particular China, at the expense of turning its back on Europe/EU/the West. Well, that day has come my friends: it is over. As long as Vladimir Putin lives and breathes (which could be forever, as he is possibly an immortal or a vampire), Russia will not join Team West, will not be partners with Team West, will not strategically plan with Team West, and won't even be friends with Team West. But once again, do not mistake this for a new Cold War: Russia may not be friends with the West, but that does not necessarily make it a mortal enemy either. Russia wants to first and foremost be considered a significant world power in its own right, a leader within Eurasia, and after that, a strategic partner of China and others in Asia. Not in Europe. Russia is now Asian in outlook.

And that, my inquisitive readers, is what the cover story for 2014 is all about. Centered in Ukraine, but affecting the outcome of multiple regions and issues on our planet. If you know your history, then current events are usually no great surprise, and speculating about future events becomes second nature. I hope that Ukraine can pull it together and survive the coming era, and we will be watching the moves of this resurgent Russia, and its Eurasian Union, quite carefully. . . .

Plaid Avenger's Hotspots of Conflict

CHAPTER OUTLINE:

→ Ideological Showdowns & Territorial Throw-Downs

→ Flailing, Failing States

→ International Intrigues

WHAT CAUSES HUMANS TO KILL EACH OTHER?

Since the end of the Second World War in 1945, there have been over 250 major wars in which over 23 million people have been killed, tens of millions made homeless, and countless millions injured and bereaved. In the history of warfare, the 20th century stands out as the bloodiest and most brutal—three times as many people have been killed in wars in the last 90 years than in the previous 500.

The nature of warfare has also changed. From the set-piece battles of the earlier centuries, the blood and mud of the trenches in the First World War, and the fast-moving mechanized battlefields of World War II, to the high-tech "surgical" computer-guided action in Iraq and Afganistan,

> One from the west: "And there went out another horse that was red: and power was given to him that sat thereon to take peace from the earth, and that they should kill one another: and there was given unto him a great sword."
>
> —*The Revelation of St. John the Divine: 6:4*

> And one from the east: "Now I am become Death, the destroyer of worlds."
>
> —from *The Bhagavad-Gita*

war as seen through our television screens appears to have become a well-ordered, almost bloodless, affair. Nothing could be further from the truth. During the 20th century, the proportion of civilian casualties has risen steadily. In World War II, two-thirds of those killed were civilians; by the beginning of the 1990s, civilian deaths approached a horrifying 90 percent.

This is partly the result of technological developments, but there is another major reason. Many modern armed conflicts are not between states but within them: struggles between soldiers and civilians, or between competing civilian groups. Such conflicts are likely to be fought out in country villages and urban streets. In such wars, the "enemy" camp is everywhere, and the distinctions between combatant and noncombatant melt away into the fear, suspicion and confusion of civilian life under fire.

Also, people kill each other because of ethnic differences, religious differences, cultural differences, and any other difference that distinguishes groups of humans. Wars are also fought over resources, land, and political control of areas—particularly where states or empires have dissolved into multiple entities, such as British India, or Yugoslavia. Now we have new wars with new dimensions the likes of which the world has not seen before: Battles of ideologies that span the entire planet.

Perhaps you are thinking the Cold War was just such a conflict—and you would be right. But it was cold, as its name suggests. Here in the 21st century, we have the War on Terror, the War on Western Imperialism, and the War on Drugs, which are global ideological battles—and they are hot ones! People are actively shooting at each other in these wars, wherever they are sponsored, all over the planet. That's the new part.

Conflict within or between states, and ideological conflicts in which the entire globe is the battlefield: we've got it all in our day and age, but we can't possibly cover all the active conflicts happening. The Plaid Avenger wants to pick out a few handuls situations that have global implications, as well as a few more than may be heating up in the coming year.

Go ahead; make my millennium.

IDEOLOGICAL SHOWDOWNS & TERRITORIAL THROW-DOWNS

Our first round of this global boxing match will cover a handful of places where good old fashioned fighting over control of territory is taking place. These actual or potential spats are basically just sovereign states (or groups of peoples) squabbling over actual chunks of land that each thinks is theirs to control . . . and which in these circumstances have led to actual fighting, or have the promise to lead to bullets flying in the next year. To make it just slightly more complicated in some cases, there are ideological differences between the warring parties which exacerbate the tensions . . . for example, the spat between Russia and Ukraine is a a territorial challenge, which has at its root a fight between Ukraine staying in Russia's orbit versus joining the European Union. FYI: there are all sorts of borders that are disputed by sovereign states across the planet, but in this day and age those disputes are largely settled through legal and diplomatic means. But not the ones below. . . .

SRI LANKA

2009 CONFLICT STATUS DOWNGRADE!!!

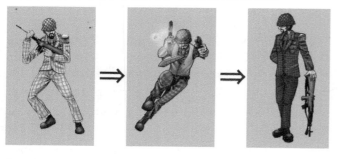

Sri Lanka is a paradox. On one hand, it has the highest per capita income in South Asia, with an ancient culture, and progressive economic and industrial sectors. On the other hand, they have been courting a decades-old civil war. The conflict in Sri Lanka has arisen because of differences between its two main ethnic and religious groups: the Sinhala-Buddhist majority and the Tamil-Hindu minority. For years, Sri Lanka was under British rule; the Sinhalese complained that the British were giving the Tamil preferential treatment. The Tamil claimed that there was no preferential treatment; they just got better jobs because they were better. When British rule ended, the Sinhalese majority decided to get back at the Tamil with the Sinhala Act of 1956, making Sinhalese the only language allowed in Sri Lanka, effectively blocking the Tamils from getting any good jobs.

To fight back, the Tamil Hindu minority formed the Liberation Tigers of Tamil Eelam (LTTE), or Tamil Tigers for short, to gain independence. This terrorist organization has successfully assassinated several Sinhala prime ministers, used claymore mines in guerilla attacks, and continues to be responsible for attacks on the island. The Tamil Eelam pretty much invented and patented the practice of suicide bombing, and even has a group known as the "Sea Tigers" who have clashed with the Sri Lankan navy. To my knowledge, they are the only terrorist group that has their own Navy, air force, and military uniforms with a wicked looking tiger logo emblazoned across the chest.

Since 1983, the two sides have fought on and off in civil war. A 2005 cease-fire between the LTTE and the government held fairly well for a little while, but the situation was once more complicated because both sides accused each other of carrying out covert operations against the other. The government claimed that LTTE rebels were killing opponents and government soldiers, while the rebels accused the government of supporting paramilitary groups against the organization and assassinating Tamil journalists and civilians. The conflict there has claimed the lives of 50,000 and forced millions to flee to India on both sides.

Why did it heat back up again to a boiling point? The 2004 tsunami disaster caused a breakdown of everything—including the peace process. Accusations that disaster relief and aid were withheld from Tamil populations by the government incited rioting and the civil war was back on. With the national government unwilling to secede any territory or grant independence to a Tamil country, and the Tamils steadfast in their fight for independence, violent conflict increased steadily over 2008 to 2009, and then it all hit the fan! The Sri Lankan government reached a breaking point and decided to go balls to the wall and finish the conflict once and for all. It withdrew from all previous cease-fire

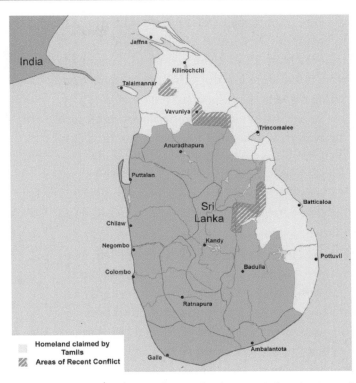

Homeland claimed by Tamils
Areas of Recent Conflict

agreements and started a massive offensive campaign against the Tamils, which ultimately resulted in total destruction of the rebel/terrorist force and the deaths of most of its top leadership by May 2009. War over.

Maybe. All of Sri Lankan territory is now controlled by the Sri Lankan government, and the Tamil Tigers are no more. That is why I have downgraded the conflict from red to blue. BUT, as we have seen with al-Qaeda and the Taliban, you should never count these guys totally out of the game. Look for a small re-grouping of Tamil forces and terrorist-bombing style attacks to increase in Sri Lanka in the coming years. You heard it here first.

THE BALKANS

Once known as the "powder keg of Europe," the Balkans once more may be expected to explode back into war at any time. The roots of the conflict are very deep, yet they can be directly linked to the independence of Slovenia and Croatia in 1991 from the state known as Yugoslavia. The area consists of three main ethnic groups and several loosely connected factions: the Serbs (Orthodox Christians), the Muslims, and the Croats (Roman Catholic) with links to Western Europe.

Once upon a time, there existed the Socialist Federal Republic of Yugoslavia. Within that republic were six constituent states, Bosnia-Herzegovina, Croatia, Republic of Macedonia, Serbia-Montenegro, and Slovenia. The conflict begins when Croatia and Slovenia announced their independence from Yugoslavia. This gave rise to surrounding states in the former republic in ousting their one-party communist states systems and electing officials with nationalist platforms. Here's the main problem: all of those regions are comprised of several different ethnic groups who believe they are misrepresented. Croatia still has issues, and quite frankly it's a miracle that Bosnia has not re-ignited back into bloodshed given its ethnic tensions. However our best example of Balkan bravado that has both local and global repercussions is Kosovo, a region comprised of 88% Albanians with an ethnic minority of Serbs. The Kosovo-Albanians want to leave the Serbian Republic, while the Serbs within Kosovo fear what will happen to them if they stay in Kosovo.

On February 17, 2008, Kosovo formally declared independence from Serbia. Serbia rejected this notion and upheld that it was illegal under U.N charter; Albania, who strongly supported the KLA (Kosovo Liberation Army) who fought guerillas wars against Serbia, was the first to support Kosovo.

Are you ready, 'cause here's where it gets interesting. Technically speaking, Kosovo is a region within Serbia, which is an internationally recognized state by the U.N., which then gives Serbia the right to preserve its territorial in-tegrity. The wrong words and the wrong moves could push this region over the edge and start a war. No biggie, right? Well, except for the fact that this is the same region that sparked WWI.

Dig this: The US and a dozen other EU countries, including the UK and France, recognized the Kosovo independence claim. Unfortunately for Kosovo, both Russia and China, which are both veto-wielding members of the UN Security Council, side with Serbia on this question of sovereignty. I'm starting to see a larger world showdown over this tiny little province. Which ties directly to our next section on a couple of other tiny little provinces causing calamity over similar claims of independence, but these two are backed by the Russians!

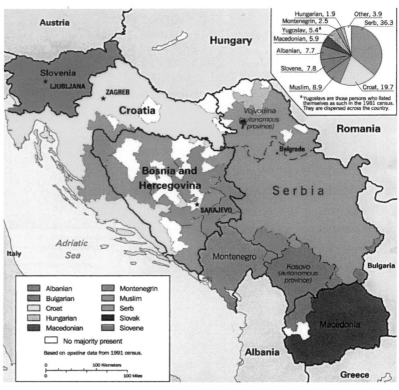

Former Yugoslavia and the ethnic division therein.

RUSSIA-GEORGIA WAR

Let's get to that rascally Ruskie move right now! The 2008 Russia–Georgia War was an armed conflict in August 2008, between Georgia on one side, and Russia together with most ethnic Ossetians and ethnic Abkhazians on the other.

See, this whole area used to be a part of the mighty USSR, but once it crumbled in 1991, Georgia claimed independence and claimed these two ethnic enclaves as part of their territory. To be fair, it historically had been counted as part of Georgia for a long time. However, many ethnic Ossetians (with a smattering of Russians in the area) actually fought a war for their own independence from Georgia in 1991–1992 which left most of South Ossetia under control of a Russian-backed internationally unrecognized regional government. Some Georgian-inhabited parts remained under the control of Georgia. This was pretty much the exact same situation in Abkhazia after the War in Abkhazia (1992–1993).

With the story so far? These two dinky provinces were technically part of Georgia, but were "protected" or "administered" by Russians. By the way, in what is a richly ironic claim, the Russians always said that they were there simply to "protect" these locals from genocidal tendencies of the Georgian government . . . much the same way the US and NATO was protecting the Kosovars from the Serbian government. Ah! Starting to see the connections here?

Let's pile kindling onto this pre-fire: then Georgian President Mikheil Saakashvili was a hugely pro-western, pro-US leader who had been petitioning for EU and NATO membership for years, mostly to shield his country from Russian influence (just like Estonia, Latvia, Lithuania and Poland did back in the late 1990's). Of course, that really rankled the Russians. That Georgia might join NATO was an option utterly unacceptable to Russia, and they were very vocal about this, so tensions rose even further. Then in 2006 US President Bush gave a speech at a NATO conference where he openly pushed for member states to put Georgia on a fast track for NATO membership. Pressure built more.

For reasons which continue to elude the entire planet, on August 8, 2008, Georgia launched a large-scale military attack against the self-proclaimed Republic of South Ossetia. Within seconds, Russia reacted and deployed combat troops

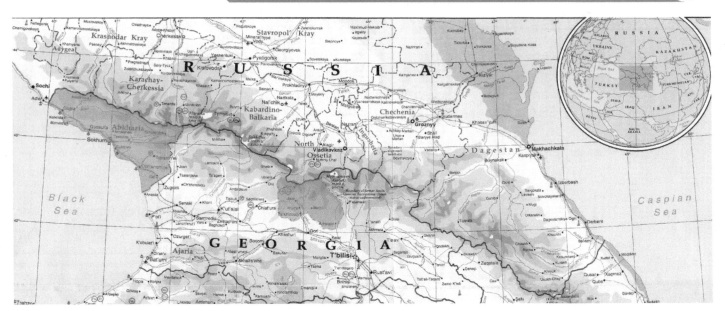

The newly independent states of Abkhazia and South Ossetia???

in South Ossetia and launched bombing raids farther into Georgia. Yeah, if you haven't picked up on this yet, be sure to take note: the Russians were totally prepared for this attack to happen. Russians landed support craft in Abkhazia and opened up a second front there.

Long story short: After five days of heavy fighting, Georgian forces were totally ejected from South Ossetia and Abkhazia. Russian troops entered Georgia, easily kicking the crap out any resistance. Fighting ended and in the subsequent peace agreement the Russians withdrew from Georgia, but remained as "guardians" in the two small enclaves. Oh but here's the fun part: On August 26, 2008, Russia recognized the independence of South Ossetia and Abkhazia. The US and Team West said, "Hey! You guys can't just invade a small ethnic province, help them secede, and then recognize their independence! That's not cool!" To which Russia responded: "Oh really? You mean like what you guys did in Kosovo?" Ouch! That stings!

2014 UPDATE: Since the 2014 Ukrainian meltdown and Russia's subsequent annexation of the Crimean Peninsula from them, the peeps in Georgia are rightly worried that Russia is feeling its oats once more and could ratchet up the pressure on them significantly. And they could be right. Given Russia's current attitude and new found mission to 'protect' Russian peoples in other countries, I think it is safe to assume that Abkhazia and South Ossetia will NEVER revert back to Georgian control. And perhaps the Ruskies will be even more mischievous in the coming year if Georgia makes any moves to join NATO again. . . .

Now this international fracas is not settled yet, and it certainly is calling the whole perceived concept of sovereignty into question, but I want you to know this for now as we watch the future events of this area: Russia is back! This invasion has announced the return of Russia not just as a military power re-establishing control over its neighborhood, but also as a military power that is not afraid to use force to stop what it perceives as threats to its security, and by threats I am going to specifically call one by name: NATO. This Georgian debacle could well have been started by former President Saakashvili because he truly believed that NATO would come save him, and that is why it's an international hotspot, even though based on local factors. Georgia is not the only place that Russia may be facing off with NATO and the West, which brings us to . . .

RUSSIA-UKRAINE: TERRIBLE INTERNAL TENSIONS

Ukraine has now become the center stage for the ideological battle for control between Team West and Russia. The Plaid Avenger believes that to determine which world power has the upper hand in this part of the world, simply keep your eye on the events in Ukraine. (2014 UPDATE: Yep. Called it.)

Like Georgia, Ukraine used to be part of the USSR, with Russia as the supreme overseer. Like Georgia, Ukraine declared independence when the USSR collapsed. Like Georgia, Ukraine had a mini-revolution that put pro-democracy, pro-Western leaders into power. Like Georgia, Ukraine petitioned for EU and NATO membership. And like Georgia, former President Bush lobbied to get NATO to accept the Ukraine.

Unlike Georgia, Ukraine has yet to be invaded by Russia. Yet. (2014 UPDATE: Yep. Called it.)

That is exactly why I want you to know about this potential flashpoint. It could happen, but in much more sneaky ways than the incited invasion of Russian troops in the Georgian example discussed above. (2014 UPDATE: Yep. Called it.) Because unlike Georgia, the country itself is internally divided on which "team" to join. To be brutally honest, the stakes are way higher in Ukraine than they ever were in Georgia. Why?

First, check the map below to see this pattern: there are a lot of ethnic Russians living in Ukraine—approximately 20% of the total population. Look where they are really predominant: in the eastern half of the country, you know, near Russia, and specifically packed into that little area referred to as the Crimean Peninsula. This is where it gets interesting . . .

Official Breakdown of Ukraine's Pro-Western vs. Pro-Russian Districts
(as per the 2006 presidential election)

won by Victor Yushchenko
won by Victor Yanukovych

The Crimean Peninsula has been a major repository and home base for the Russian navy for decades, perhaps centuries depending on your historical perspective. When Ukraine declared independence, Crimea ended up as part of their country, much to the chagrin of many Russians living there, as well as the Russian government itself. There are still dozens of naval facilities on Crimea leased and operated by the Russian navy. Those pro-Western, pro-NATO peoples in Ukraine would like the Russians to pack up and get the hell out. The pro-Russian peoples, and mother Russia herself, have an entirely different view of the matter.

This division between pro-West and pro-Russia can be seen most clearly in the 2006 presidential election map above: the pro-Western west, versus the pro-Eastern east! Ha! Too easy! Which brings me to the second big point about this hotspot: unlike in Georgia, Russia has a big support base inside the country! Especially in that Crimean section, which has tons of ethnic Russians in it! How do I know? When Ukraine invited NATO to conduct a joint military exercise there back in 2006, massive protests broke out when US Marines arrived, mostly to chants of "Occupiers go home!" Yikes!

Is the Plaid Avenger suggesting that Russia might invade? No, not really. They don't have to. They are already working very hard to secure support with ethnic Russians within Ukraine, and to monetarily support pro-Russian political candidates there too. The pro-Russian politician, Viktor Yanukovych, was the guy thrown out of office during the pro-West Orange Revolution of 2004, then became Prime Minister of the country in 2006, then became President again in 2010. How is that for a flip-flopping country that is unsure of its direction? At the same time, there are pro-Western politicians calling for canceling/not-renewing Russian lease contracts for their Crimean naval bases. The stage is set for a show down.

When the pro-Russian Viktor Yanukovych just won the presidency in 2010, he immediately announced that Ukraine would not be joining NATO, and also immediately extended the Russian leases on their Crimean naval bases. That in turn got Russia to reduce the rates on all the oil they sell to Ukraine. Everything is chill for now, BUT times do change, as do politicians, so this one isn't over yet either.

2012 Update: The fight is still alive! Further evidence of this West/Russia divide in Ukraine erupted badly in May, as fist-fights broke out in their Parliament when the ruling party put forward legislation to make Russia one of the country's official languages. Seriously? Congressmen beating the hell out of each other over a language status. Yep. That's the level of passion that exist in this country that is fighting itself over its very own soul. More to come, for sure.

2014 UPDATE: Yep. Called it! I intentionally did not edit this section which was written years ago, so that you can see that having a good sense of global topics makes predicting the future pretty easy! Even for the Plaid Avenger!

ISRAEL VS. PALESTINE

The Israel-Palestine conflict has very deep roots—stretching back thousands of years. In ancient times, the area that is now Israel was inhabited by Jews who had a degree of autonomy and self-governance. In 70 AD, the Roman Empire, who controlled the land, burnt down the Jewish temple and kicked all of the Jews out of Israel. The next 1900 years were known as the Jewish diaspora, when Jews were scattered around the world and often persecuted for being Jewish. After a particularly atrocious bit of persecution in World War II known as the Holocaust, European countries and the United States decided that it was time to help the Jews to return to Israel and have their own homeland. The state of Israel was created in 1948; Jews from all around the world migrated to Israel to be free.

The problem: a new group of people were living in that area who were calling themselves the Palestinians. These Palestinians were mostly Arab and Muslim; and by "new," I mean that they had only been living there for the 1900 years since the Jews were ejected. To make room for the rejuvenated state of Israel, some of these Palestinians had to go move out or adjust to being a monitory in a new Jewish-dominated state. This was based on the proposed UN Partition Plan of 1947, which was to divide up the territory into two new states: one for the Jews named Israel, and one for the Arab Muslims named Palestine. On the eve of this partition plan taking effect, Israel unilaterally declared independence. Say what? What was up with that? Well, all the surronding Arab countries had boycotted the vote at the UN on this whole business, and had massed troops on the border with the intent of wiping Israel out as soon as the partition was to become law . . . so Israel pro-actively declared independence in 1948, and WAR ON! The result? Israel won, actually increasing their territorial gains from the original partition plan, and 700,000 Palestinians left Israel and settled in the countries bordering Israel, including Jordan, Egypt, Syria, and Lebanon. To make matters worse, all of the countries bordering Israel were Arab and Muslim, and their governments wanted to look tough on Israel to make their people happy. For this reason, the Arab countries demanded that Israel take the refugees back. Israel said "no" and these refu-gees quickly faced a Catch-22 situation: Israel would not let them come back to their former homes and the Arab countries they migrated to would not find them places to live either.

Israel and the occupied territories (West Bank + Gaza = Palestine, but also Golan Heights is "occupied" but officially still belongs to Syria).

As you can imagine, some Palestinians turned violent quickly and began forming terrorist organizations to attack Israel. Again, to look tough on Israel, the surrounding Arab countries went to war with Israel multiple times in the last 60 years, ultimately losing each war every time, and thus limiting outside influence ever more, each time. After each of these wars, Israel took possession or control of more of these lands, establishing de facto control of the West Bank from Jordan in 1967 and the Golan Heights from Syria in 1967. Israel still controls these areas and claims that it needs these lands because of the geographical and strategic locations of the sites for their national security.

Currently, the areas referred to as the "Occupied Territories" consist of the West Bank and the Gaza Strip. These areas were not originally part of Israel in 1948, but were taken control of by Israel after a 1967 war. These areas are predominately Palestinian and the Palestinians in these areas want their own state. However, each part wants different leaders. The Palestinian Authority (PA) has been the pseudo-government of this pseudo-state, and has been dominated by the political party Fatah since its inception. However, Fatah and the PA have become unpopular with the Palestinians because of rampant corruption and their inability to make any real progress on the big issues.

For this reason, a more radical political party named Hamas continued to gain popularity in parliament. A major rift occurred in the summer of 2007: the democratically elected Hamas completely took over the government of the Gaza Strip, and kicked out their Fatah opponents. Now this pseudo-state of Palestine is even divided amongst itself: The Fatah dominated PA controls only the West Bank, while the Hamas party controls the Gaza Strip. Let's make it even messier: Hamas has a history of supporting terrorism in Israel, and many western governments (the US, the EU, Israel) consider Hamas a terrorist organization and, as such, will not deal with them.

The situation only looks like it will worsen since Israel closed off the Gaza Strip to all flows of goods and traffic. This is a move intended to punish the more militant Hamas, while Israel, the US and the EU are holding up the PA as the true leaders of the Palestinian people in a desperate bid to try and normalize relations. But even that is not working out so well: the Gaza embargo, which is starving civilians to death, has become a debacle for all parties involved, and is stymying the peace process further. Expect this one to get even messier. Heck, you don't have to expect it: just watch it unfold.

To summarize this soup of names and nonsense: West Bank + Gaza Strip = Palestine. Palestine is not an independent state, but may become one eventually, however it is currently under Israeli control—that is, it is "occupied." To make this worse, Palestine used to be of a singular voice in negotiations with Israel and the world, but now it has splintered; The Fatah political party controls the West Bank, and the Hamas political party/terrorist group controls the Gaza Strip.

Hold on friends, the fun ain't over yet: for decades, the Israeli government has allowed/ encouraged Jewish folks to start "settlements" in the West Bank as well, thereby making the occupied territories even more under Israeli control. I bring this up because in June 2009, US President Obama made a speech to the Muslim world in which

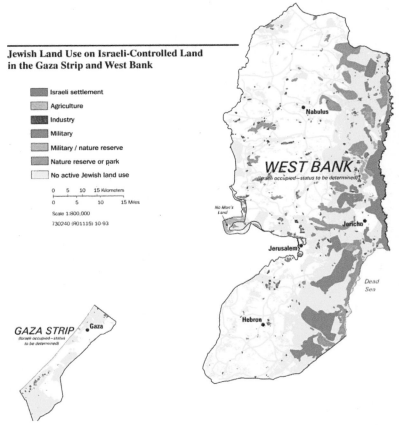

Jewish Land Use on Israeli-Controlled Land in the Gaza Strip and West Bank

- Israeli settlement
- Agriculture
- Industry
- Military
- Military / nature reserve
- Nature reserve or park
- No active Jewish land use

0 5 10 15 Kilometers
0 5 10 15 Miles
Scale 1:800,000
730240 (R01115) 10-93

WEST BANK
(Israeli occupied—status to be determined)

Nabulus

No Man's Land

Jericho

Jerusalem

Dead Sea

Hebron

GAZA STRIP
(Israeli occupied—status to be determined)

Gaza

Jewish settlements in West Bank are specifically what Obama asked Netanyahu to stop.

he specifically asked Israeli Prime Minister Netanyahu to stop these settlements in order to foster goodwill towards a peace settlement. Let's make it hotter: in 2011, the Palestinians are planning to declare full independence at the UN, thus forcing a showdown on the global stage on who will recognize them, and who won't. The issue is now being force to a climactic conclusion. . . .

INDIA–PAKISTAN: HOT SPOT: KASHMIR

In 1947, after India gained independence from Great Britain, it was divided into two separate states because of large differences in their religious populations. India retained most of the subcontinent, with its predominately Hindu population. Pakistan and East Pakistan, now known as Bangladesh, were formed as a home to South Asia's large Muslim population. From the beginning, friction has existed between these two giants of the subcontinent. Bloody clashes broke out the very day of independence, as millions of folks were on the move, shifting to the "proper" country that housed their religious inclinations.

So in today's world, everyone in South Asia is now on the correct side of the fence, right? We're forgetting about that tricky place that makes ridiculously comfortable fabric: Kashmir. While most of the other borders between these states is settled and accepted, open fighting still sporadically erupts over this state of India named Kashmir, a small mountainous region about the size of Kansas, located in the northwestern part of the Indian subcontinent where the two countries meet. The Kashmir region was originally ruled by a monarch, or Maharaja, who gave India control of the region shortly after the nations split. The majority Muslim population of Kashmir did not agree with this move. Afterward, war erupted and continued until the UN arranged a cease-fire in 1949, by which time Pakistan gained one-third of the Kashmir territory. After India formally annexed the rest of Kashmir in 1956, the Muslim population was once more provoked to rioting, and fighting has continued ever since with a repeated pattern of attacks and cease-fires for the last fifty years.

The situation intensified in the 1990s. Both sides began testing nuclear weapons and gained global attention because neither state has signed the nuclear nonproliferation agreement. Although recent meetings have shown signs of a peaceful resolution, this conflict must be considered a political hot spot as India and Pakistan remain the only neighboring nuclear powers with hostilities toward one another. In fact, the historical hostilities between these two states, which often erupt openly on the border, is one of the reasons why Pakistan is currently losing control of its own country. Say what? Am I suggesting that India is

Jammu and Kashmir: Ethnic Mix of a Disputed State

Buddhist
Hindu
Muslim

China

Pakistan

Line of Control

Kashmir

Jammu and Kashmir State

India

Pakistan

Source: Kashmir Study Group: *Kashmir: A Way Forward, February 2000.*

Pakistan

India

Indian Ocean

Area of main map

somehow undermining the Pakistani government in these volatile border regions? Nope, I am not: the Pakistanis are doing it to themselves. See, the Pakistani military has spent the previous six decades preparing itself for confrontation with India; therefore, the country has been ill-prepared to handle internal conflict coming from its western regions: you know, the Afghan side.

Why would I mention this in the Kashmir section? Because in the summer of 2009, India intentionally drew down their military forces in the border region, specifically in Kashmir, in order for the Pakistanis to feel secure enough so that they can deploy their military FROM the Kashmir region TO much more active hotspots in the western fringe of their country. India, as well as the rest of the planet, has no interest in seeing Pakistan completely collapse into chaos, and thus stood down in Kashmir in order for the Pakistanis to go get the anti-terrorist job done elsewhere in order to hold their country together. Which brings us to the next section of conflicts in the world, those brought about by failing/failed states, so let's stick with Pakistan for a page more . . .

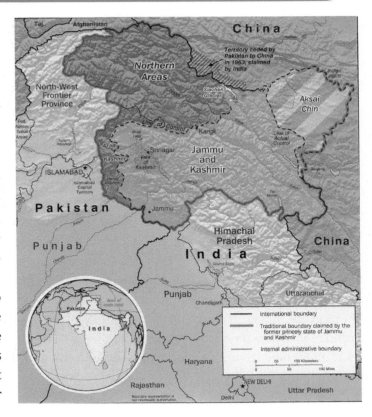

FLAILING, FAILING STATES

You think you are a state? Well, for this category of conflict, you receive a big fat F for your attempt to prove it. Or you are on the brink of getting that F. Or there are a lot of us that think you deserve an F, even if you officially do not have it on your statehood college transcripts yet. This section will deal with those supposedly sovereign spaces whose very existence is questionable, or who are falling down on the job, badly. Failing at the very function for which they were formed. . . .

So how do we determine this status? Well, it is a bit fuzzy. A "failed state" is one perceived by the rest of the world as having failed at some of the basic conditions and responsibilities of a sovereign government. There is no single general consensus on an exact definition what makes for a failed state, but it likely has one or more of the following failing conditions:

→ "the government" has lost control of all or parts of its territory

→ "the government" has lost its monopoly on the legitimate use of physical force within the state (meaning warlords, paramilitary groups, or terrorist use force or kill people in that state, and the government can't stop them)

→ "the government" is no longer taken seriously by all its citizens to make and enforce laws and run courts

→ "the government" has lost the ability to provide public services

→ "the government" has lost its ability to interact with other states as a full member of the international community

→ there is widespread corruption and criminality that "the government" cannot stop

→ there is a steep economic decline that "the government" cannot stop

If any one of these things is happening in a state, it still may be considered legitimate in the eyes of other sovereign entities, but once you start stacking up two, or three, of five of these factors, then the very existence of the state is dubious. And that is when all of the rest of us on the planet start debating if the place is officially "failed" and we all expect chaos, civil war, or international intervention to unfold at any second. Below is a short list of those places that are in this questionable conflict-riddled category in the coming years . . . which brings us back to Pakistan . . .

PAKISTANI PROBLEMS WITH TERRORISTS & TALIBAN

Wow, that's a mouthful! And it's a mouthful that Pakistan may not be able to swallow, and still survive. But hang on a second; maybe some of you are questioning the placement of Pakistan here in the failed state section. Can it really be all that bad? Yes my plaid friends, it really is that bad. In fact, it's worse. There is so much going wrong here that I am going to have to wildly over-summarize and over-generalize just to get to the main points. So here it goes. . . .

Pakistan has been on the brink of full-on catastrophe and chaos for some time now, and most look to the woes of the struggling democracy as reason for the calamity. However, the Plaid Avenger wants you to consider the indecisive nature of the the military and the intelligence communities within Pakistan as the primary problem behind the current calamity in Afghanistan, the Swat Valley, and the entire western side of Pakistan, which taken together may spell the possible future failure of the state itself.

Pakistani leadership is trying to pacify the wants and desires of 170 million citizens—that's the 6th biggest population in the world. It's also a devoutly Islamic society, including the whole spectrum of religious views from the mainstream to the seriously extreme. There are a slew of extremist factions and separatists groups pulling the country apart, especially all around the Afghan border (look up Waziristan, Balochistan, the Taliban—man, that sounds like a Dr. Seuss book).

Look at the map opposite, and understand this: the areas in blue are pretty much a no-man's land for government control. It consists of ethnic groups that have never considered themselves part of any country (like Balochistan), the Federally Administered Tribal Areas (FATA), and the North-West Frontier Province, which sounds like a section of Disney World to me. Taken collectively, these are all nice names for areas that have one thing in common: they are too wild for the government to really control, so the Pakistani government gave them some sort of autonomy in exchange for not causing chaos. Today's big problem? The deal is off. The Taliban and al-Qaeda have convinced the locals that they don't need any "deal" with the Pakistani government, since they exert true control on the ground. They have a point.

Some of these extremist types (like al-Qaeda and the Pakistani Taliban) are fighting for establishment of a theocratic Muslim state, while others are fighting for an independent state within Pakistan (i.e., Balochistan, Waziristan). Some are just fighting against the government because they don't like the fact that the Pakistani leadership is so cozy with the US when it comes to foreign policy. Remember, Pakistan receives plenty of foreign aid from the US, primarily to keep up the War on Terror. As you can imagine, the terrorists don't like that idea much, so total destabilization of Pakistan by any means is one of their ongoing goals. And they are winning.

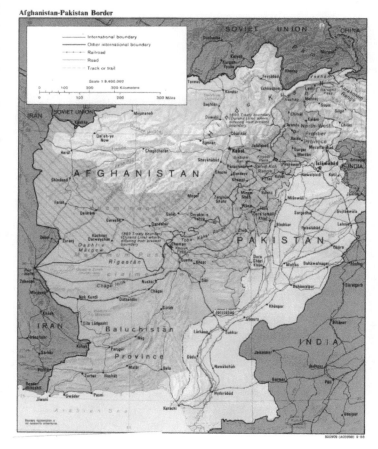

Afghanistan-Pakistan Border

Need proof? Not only did most folks believe that Osama bin Laden and his crew are somewhere in Pakistan (proven to be true!), and that most terrorist attacks worldwide involve Pakistani elements (UK subway bombing, Mumbai terror attacks, etc.), but now there are open attacks in Pakistan on US and NATO supply convoys heading to Afghanistan. A specific insult? In February 2009, the Pakistani government essentially surrendered the Swat Valley region of their country to the Taliban. That is inside their own country, man! 2010 Update: Pakistan's military finally got serious in late 2009 about holding their country together and organized a massive redeployment of troops to fight the Taliban and reclaim control over the western half of their country. They wrestled back control of the Swat Valley, and have made significant gains in the hottest spots of the FATA.

This place is in dire straits due to well-organized and passionate terrorist/separatist groups fighting with the home field advantage on extremely tough terrain. But let me finish with what I started with: the failure of Pakistan to hold itself together can be attributed to its own faults just as much as outside forces. Pakistani politics are hopelessly corrupt and gridlocked, with politicians bickering about judicial control and foreign policy while their state is imploding around them. The abysmal state of

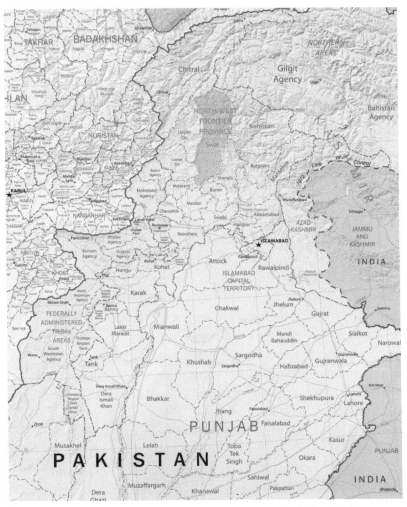

Swat Valley highlighted in pink. Notice proximity to capital of Islamabad.

their economy, widespread energy shortages, and collapsing infrastructure ain't helping much either.

What's worse, the Pakistani military has been fighting what can at best be described as a totally half-hearted campaign against these insurgents. Why would they do that? Because the military and intelligence communities are the ones who actively built and helped the Afghan Taliban come to power in the first place, so now they have found it difficult to go out and kill the guys that used to be their allies. On top of that, the whole Pakistani military structure has been so focused on an eventual war with India, that they can not seem to re-adjust their priorities to accept the fact that they are losing the whole country from the inside out! They have more of their military stationed on the border with India than actually out fighting the dudes who are ripping their country apart!

2011 Update: The US 'invasion' of Pakistan to kill Osama bin Laden certainly has made the world even more critical and mistrusting of the Pakistani military and/or ISI. That, in combination with the unpopular war on terrorism and the US drone strikes into Pakistani territory, has created extremely low public opinion of the US and lack of trust on both sides. But the US needs Pakistan to continue to help with this war and the Pakistanis need the billions in US aid, so an uneasy relationship must go forward. Pakistan's current huge crisis is the rise of the home-grown Pakistan Taliban (a different group than the Afghan Taliban), a group intent on the overthrow of Pakistan itself! And they are pulling out the stops in terms of terrorist attacks on civilian targets inside of Pakistan to push their agenda . . . scary stuff in an already challenged state!

US IN AFGHANISTAN

Well, let's just stay in the neighborhood, shall we? U.S. involvement in Afghanistan is still fresh in our memories—oh right, because they're still there . . . but not for much longer! The story, however, is a long one. In 1996, the Taliban, a fanatic Islamic extremist group, took control of Afghanistan, which had before then been a mishmash of warlords fighting for power. On October 7, 2001 NATO (led primarily by the US) began a military campaign against Afghanistan known as *Operation Enduring Freedom*. The invasion was in direct response to the 9/11 terrorist's attacks and under the belief that Afghanistan was harboring the same terrorists that planned the attacks. With the help of a rebel group called the Northern Alliance (which gains much of its funding through opium sales), America and its allies established a new, secular government in Afghanistan. This invasion kicked off what is now known as the Global War on Terror.

The main problem in Afghanistan has been the unsuccessful implementation of its established government. Afghanis are arguing that the ruling government is comprised of ethnic Tajiks, while the majority of Afghanistan is Pashtun. The Pashtuns are freaking out cause they think the Tajiks are too Westernized with their skintight jeans and pink Armani t-shirts, creating a secular government. The Pashtuns would prefer a Pashtun-dominated Islamic based government, which is strongly supported by its neighbor to the south, Pakistan. A reoccurring problem the U.S. is seeing is the failure of its exported form of democracy in the Middle East (*See: U.S. in Iraq*).

A recent surge in Taliban activity has many speculating that the U.S. and its NATO allies are failing in reconstructing Afghanistan. A clear example of that was on April 2008, when

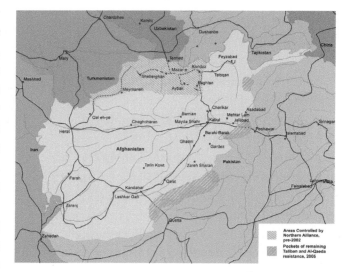

Afghan President Hamad Karzai survived a failed assassination attempt. Even though the attempt failed, it was a victory for the insurgents, as they were at least capable of getting close to Karzai. This puts immense pressure on Karzai's administration to successfully establish and proclaim an effective government.

A particular problem posing itself against the Afghan government is the sharp increase in production of opium, which can be processed into heroin. Can't be that big of a problem, right? Afghanistan is a war torn country; when are you ever going to have time to grow opium? Apparently there's something in the air, because Afghanistan produces 92% of the world's opium and 80% of the world's heroin. Let's get down and dirty with the problem itself. The Taliban and al-Qaeda have used the opium trade to re-equip, re-supply, and re-organize themselves, and have turned the tide of the battle back in their favor.

Yes, you are reading this correctly: the Taliban is currently gaining ground and beating the NATO coalition. The US has been outright begging other countries to contribute more to the cause, but unfortunately, more folks are pulling out than putting in: Australia is out, and Canada and UK have been downsizing troop numbers as well, preparing for an eventual total withdrawal. As just expressed in the last section on Pakistan, this whole area on both sides of the border is a no-man's land for control, from either government. NATO is losing just as bad on the Afghan side as the Pakistani military is losing on the Pak side. It's all one big interconnected mess that no one in the world wants to see disintegrate further, but no one in the world wants to send more troops or get involved with at all. Geez! Why does the US have to do every thing?

2014 UPDATE: This is it! This year the US has picked to pull out the rest of their fighting troops, leaving behind only a small force of security advisors' to continue to help train Afghan military and security forces. A hundred thousand troops are on their way out this year, leaving behind a shaky democracy (at best), a non-existent economy, and a fully charged

Taliban who is now a decade-old, battled-hardened group who for years has been playing the long game knowing that eventually the US has to leave. The Plaid Prediction for this place is not good: civil war is imminent, and/or direct Taliban takeover in short order is also quite possible.

We call Afghanistan a country because it occupies a space on the map of the world, but it isn't really one. For years, Afghanistan has been a battlefield for rival warlords and ethnic groups. Every time somebody has tried to set up a strong central government, such as the Soviet Union in the 1980s and the US/NATO in 2001, non-stop war has been the standard operating procedure for the locals. People in Afghanistan are used to war and are good at defending their homeland.

I almost hesitate to call this place a failed state, only because I'm not really sure it's ever been a state to begin with! And if you need any further convincing that Afghanistan will plummet back into chaos once the US leaves, one need look no further than Iraq, where this exact scenario is already playing out. . . .

US IN IRAQ

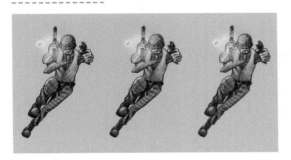

Ever since Saddam Hussein invaded Kuwait in 1990, he has been one of America's worst enemies. In the first Gulf War of 1990–91, aka Persian Gulf War, , American forces drove Hussein out of Kuwait and destroyed much of Iraq's army. When asked why he didn't pursue Hussein to Baghdad to finish the job and remove him from power, former President George H.W. Bush told a group of Gulf War veterans, "Whose life would be on my hands as the commander-in-chief because I, unilaterally, went beyond the international law, went beyond the stated mission, and said we're going to show our macho? We're going into Baghdad. We're going to be an occupying power—America in an Arab land—with no allies at our side. It would have been disastrous. We're American soldiers; we don't do business that way."

After George W. Bush was elected president in 2000, he decided that American soldiers **did** do business that way. As a part of the US led 'War on Terror,' the US started the 2003 Second Gulf War aka the Iraq War, when America unilaterally attacked Iraq, went into Baghdad, and occupied the country—America in an Arab land. But a funny thing happened on the way to Baghdad. Hussein's forces did not even show up. The war was a cakewalk; it was almost too easy. People wondered what happened to Hussein's famous Republican Guard.

Then the sinking truth set in. Hussein had intentionally told his army to stand down. They knew they never would win a traditional ground war with US forces, so they dispersed and began an insurgency, fighting a guerrilla war with bombings and sneak attacks. Hussein was eventually captured, but since America declared "Mission Accomplished" more than 4500 American soldiers died in Iraq, with new attacks happening daily. Insurgents have even kidnapped foreigners and beheaded them on television. So what to do in the face of a vaguely-established new democracy with a load of civil strife and tensions mounting? Why leave of course! By October 2011, then US President Barack Obama announced full troop withdrawal by year's end, and in December they declared victory, and war over. Ummm . . . yeah . . . about that. . . .

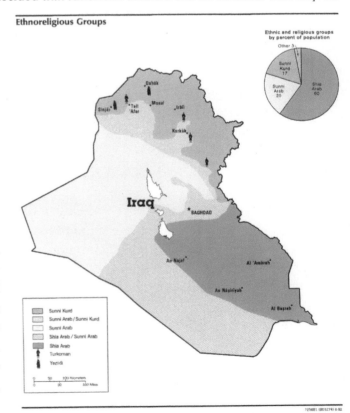

Ethnoreligious Groups

Ethnic and religious groups by percent of population

Other 3

Sunni Kurd 17

Sunni Arab 20

Shia Arab 60

Iraq

Dahūk
Sinjār Tall 'Afar Mosul Irbil
Karkūk
BAGHDAD
An Najaf Al 'Amārah
An Nāşiriyah
Al Başrah

Sunni Kurd
Sunni Arab / Sunni Kurd
Sunni Arab
Shia Arab / Sunni Arab
Shia Arab
Turkoman
Yezidi

0   50   100 Kilometers
0   50   100 Miles

Why would the mighty US have left? Public support in America for the war decreased because nobody wants to subject their children to guerilla warfare. Furthermore, one of America's justifications for invading Iraq was that Saddam had weapons of mass destruction, but when American forces found no WMDs in Iraq, many people felt misled and bamboozled. In addition, many Americans never supported the war in the first place, and an overwhelming majority of citizens in the countries of European allies totally were against the war from the start as well. It was wildly unpopular the world over. So getting out of there was the easiest thing ever for most politicians in Team West. Not so much for the locals. . . .

Because this place is a mess. America tried to establish a democracy in Iraq, but was doomed from the start because Iraq consists of three rival ethnic groups that do not get along: The Kurds in the north, the Sunnis in the middle, and the Shi'ites in the south. Good luck getting those guys to agree on a government. Good luck indeed. As foreign influence diminished, the likelihood of total civil war and implosion increased. These guys aren't out of the woods yet. By a long shot.

2014 INSANE BLOODY UPDATE: Unfortunately, I called this one exactly right years ago when I wrote those words you just read. As of June 30 2014: After organizing and fighting in Syria and conducting guerrilla warfare in Iraq, a group named ISIS (Islamic State of Iraq and Syria) started an open military campaign to take control of the entire state of Iraq. They took the second biggest Iraqi city of Mosul, are now consolidating territorial control over large swaths of Syria and Iraq, and are moving toward the capital of Bagdad itself. This fight could be over before this book even gets to print! This ISIS group is composed of radical Sunni Muslims Wahabbist and some al-Qaeda freaks, who are on a holy Jihad war to take over the entire Levant! This is full on civil war now! The Shia-led central Iraqi government has failed horrifically to date to repel this group, and now the US and others are debating about stepping back in to save the place from disaster. You heard it here first:

The black flag of the ISIS group . . . aka the Islamic State of Iraq! Yikes!

if the Kurds up in Kurdistan also declare independence (a strong possibility at this point, as they are successfully defending their territory from ISIS), this state we call Iraq is not just failed, it is finished! Too hot to handle or even speculate about right now, but no good can come of this!

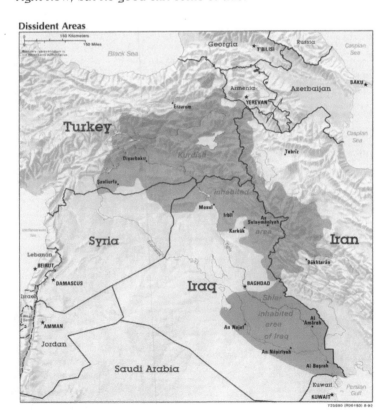

Dissident Areas

KURDISTAN WILL CUT OUT

Because of the disintegration of the Iraqi and Syrian states, it is highly likely that the Kurds will cut out of Iraq by declaring an independent state . . . which no one can do anything to prevent at this point in history. This will cause a potential showdown with Turkey. This one is definitely happening soon, and is not on anyone's radar. Iraq is actually already in a state of civil war, but the US and most of its press do not want to recognize that for political purposes. While the main source of friction and fighting is between Sunni and Shia groups, the actually full-on implosion of the Iraqi state will fully occur when the Kurds (who control the northern part of the state) will declare independence and a separate sovereign state of Kurdistan. While lighting the tinderbox for war in Iraq is in itself bad enough, what compounds this situation is that there are Kurdish people in other states like Iran, Syria, and—most importantly—Turkey, which may be incited to attempt to join this Kurdish state as well. Turkey has had a long repressive history with the Kurds, and Turkey actually invaded northern Iraq a few years back to root out the PKK, a Kurdish terrorist group responsible for attacks on Turkish soil. So if and when the Kurds declare a separate state in order to distance themselves from the Iraq debacle, it will immediately raise tensions with Turkey, and within Turkey.

SYRIA & LEBANON

Starting as a civil uprising back in March 2011, as a part of the broader 'Arab Spring' across the Middle East, this mess has continued to devolve, and devolve, and devolve into absolute bloody insurrection, dictatorship repression, civil war, and has now even spawned the newest terrorist organization on the planer (ISIS) which is carving out areas of the country to form a new extremist Islamic state. If this doesn't count as a failed state, I'm not sure what does . . . cuz there ain't no one in control of this chaos. As of this writing: 150,000 dead, millions displaced, insurgents of all stripes are growing in numbers and in bloodlust.

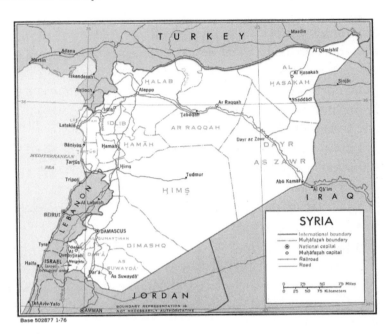

From the get-go, President Bashar al-Assad's government has attempted to smash all opposition with an iron fist, and is sporadically accused of using biological weapons and targeting civilian populations. The 'rebel' groups that have been fighting the government are a hodgepodge of pro-democracy reformers, moderate Sunni republicans, Hezbollah militia, Iranian-supported mercenaries, hardline Sunni Islamists, al-Qaeda inspired nut bags, and foreign fighters who are coming in just for the fun of it. Outside entities like the US and Europe have hesitated to get involved, because quite frankly they have no idea who to give money and guns to! Its such a confusing jumble of peeps fighting each other!

And it has now turned into a full-on proxy war, with a decidedly Sunni/Shia split between sides. Bashar al-Assad and his government are Alawite Muslims, which is a sect of Shia Islam . . . so the government is supported by Shia Iran and Shia Hezbollah (that political/terrorist group in Lebanon). Russia also supports Assad, because, well, Syria is their only ally left in the Middle East. The majority of Syrians are Sunni Muslims, so money and guns have been funneling in to them from the rich Gulf Arab states (including Saudi Arabia) whose leaders hate the Assad regime. The crazed al-Qaeda types are also Sunni, so their endgame is the end of Assad as well . . . and June 2014 update: the complete whack-ass crazy ISIS terrorist group that is sweeping across Syria and Iraq are 100% jihadist nuts who want to kill everyone who is not Sunni Muslim, so they are all about the destruction of Assad personally, but the destruction of the state of Syria as a whole. As Lebanon's fate is tied to Syria's fate, look for poor Lebanon to be completely enveloped into the chaos as well . . . and they have barely been a functioning state for decades already.

Future prediction for the coming year? Gonna. Get. Worse.

VENEZUELAN IMPLOSION

The great socialist experiment of our times, and in our hemisphere, is rapidly falling apart and it is more than likely that Venezuela will end up in a state of civil war, or at the very least massive civil unrest, by the end of this year. While the socialist firebrand and beloved leader Hugo Chavez was alive, he could effectively rally the majority and keep a lid on dissent in Venezuela, but he died in March 2013, and was succeeded by Nicolás Maduro . . . who honestly is just not up for the task of continuing the 'revolution,' nor of holding the country together. He ain't necessarily a bad dude, but certainly ain't no Chavez, and does not command the required presence to keep this state afloat. Because the pro-Chavez and anti-Chavez peoples of the state are starting to spiral out of control. . . .

Since the beginning of 2014, a series of protests, political demonstrations, and civil unrest have been occurring throughout Venezuela. The protests initially erupted largely as a result of Venezuela's high levels of violence, inflation, and chronic shortages of basic goods. There were two specific sparks: (1) actress and former Miss Venezuela Monica Spear and her husband were killed on 6 January 2014 during a roadside robbery, while their five-year-old daughter was in the car; and (2) In February an attempted rape of a young student on a university campus in San Cristobal led to protests from students over crime. The government's response? Arrest the protestors. Thus fueling the anti-government anger, thus fueling more protests, thus fueling more arrests and crackdowns . . . see how this cycle of violence works?

The protesters claim that the horrific state of the Venezuelan economy and crime rates are caused by the economic policies of Venezuela's government, including strict price controls, which allegedly have led to one of the highest inflation rates in the world. However, government supporters claim that government economic policies, especially that of under previous president Hugo Chávez (1999–2013), improved

the quality of life of Venezuelans, and blame external factors for ongoing problems. This is an ideological showdown for the very soul of this state, and I am afraid it may lead to a full-on civil insurrection/civil war/state collapse.

As a major oil producer and exporter, this has major economic repercussions for the globe, but on top of that, Venezuela has been the poster child/leader of the leftist shift all across Latin America for the last decade and a half. Its devolution changes the entire ideological game for South America, Cuba, and Latin America as a whole. So watch this one closely in the coming year, as it could explode fully at any second. . . .

DEMOCRATIC REPUBLIC OF CONGO (DRC)

HOT SPOT: EASTERN CONGO

The giant super-state of Africa has basically been in a state of self-destruction since its independence . . . and it wasn't doing very well even before that. The Democratic Republic of Congo mystifies even those of us that follow world events, due to the complexity and convoluted history of conflict. The site of "Africa's World War" in which 5–6 MILLION have died, and which most westerners have never even heard of, this state will definitely soon be crumbling apart under the sheer strain of its size, of its ethnic diversity, and its mismanagement . . . but mostly because of its vast mineral resources! Say what? Dig this:

The history of the Democratic Republic of the Congo has been a complicated one, full of instability, coups d'etat, violence, and name changes. When the Republic of the Congo gained independence from Belgium in 1960, a guy named Patrice

Lumumba was elected president, but was overthrown in a coup d'etat in 1965 by Mobutu Sese Seko. Sese Seko, a dictator, was supported by the United States and Belgium, and controlled every aspect of life in the country. He even changed the name of the country to Zaire in 1971. After the Cold War was over, the United States did not think it was necessary to support Mobutu anymore, so he was overthrown by Laurent Kabila in 1997, who renamed the country back to Democratic Republic of The Congo-Kinshasa. The name of the capital city is added to the name of the country to distinguish it from the Republic of the Congo-Brazzaville which is a different country that is next door to Democratic Republic of the Congo. Get it? Good. Someone explain this stuff back to me.

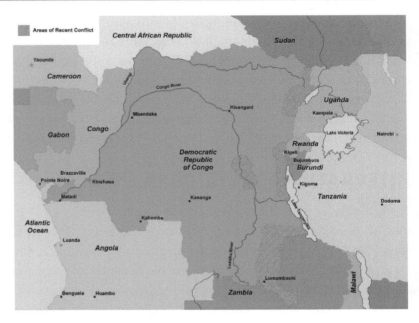

This new Kabila government immediately came under fire by rebel groups and has yet to establish any sort of stability. Kabila quickly lost control of parts of eastern Congo to Ugandan, Rwandan, and Burundian backed rebels while still receiving support from Angolan, Zimbabwean, and Namibian troops. The organization Human Rights Watch wrote that Congo is a human rights disaster area and "soldiers of the national army and combatants of armed groups continue to target civilians, killing, raping, and otherwise injuring them, carrying out arbitrary arrests and torture, and destroying or pillaging their property. Hundreds of thousands of persons have fled their homes, tens of thousands of them across international borders." In fact, the DRC conflict is responsible for the establishment of a new category of war crime: institutionalized rape. Roaming groups of soldiers raping women with total disregard for laws or morals has left an entire generation of women physically and psychologically wounded and another generation fatherless. In fact, the conflict in the Democratic Republic of the Congo has been called "Africa's World War" and 5–6 million people have died as a result of it and the subsequent chaos resulting from it, to this day. A peace treaty was signed in 2003, but the situation remains unstable and could devolve back into war at any time. In a few years, you could be watching "Hotel Congo."

Why the fighting? Quite simply: money. This area is huge, massively huge, and filled with gold, diamonds, uranium and practically every commodity the developed world wants. Also, there is no sense of nationhood and the leaders of the different factions are war criminals and thieves who are only looking to line their own pockets. The natural wealth of the Congo is taken out of the country, leaving a poor, uneducated population embroiled in turmoil. Things here are really bad and the near future holds few foreseeable good changes.

Some areas of the country are effectively controlled by the armies of other neighboring states, rebel, or terrorist groups from neighboring states, or local rebel groups, or local warlords, or . . . I think you are starting to get the point.

To sum up the DRC: huge country with a weak central government; a valuable and tempting resource base; a hideout for rebel groups of all sorts; a place where perhaps seven different states have forces in and out of the territory. What a mess, and it could implode soon, turning into an all-out territorial grab by all the involved parties: The DRC, Uganda, Rwanda, Burundi, Angola, etc. This DRC mess has the distinction of having the largest UN peace-keeping force stationed in it already. However, given the huge size of the country, the small size of the UN force, and the general warlord-driven politics of the region, don't think for one minute that the UN or anyone else is really in control of this part of the planet.

2012 Update: Just to give you a sense of the on-going nature of this Congo mess, the UN just released a report accusing Rwanda of fomenting internal rebellion in the DRC by training men to penetrate the Congolese military, and then mutiny against the DRC government itself. Talk about some internal shenanigans!

Of course Rwanda vehemently denies this, but the report comes on the heels of mutineer activity in eastern Congo, in what appears as an attempt to tease away those resource-rich provinces by undermining the central DRC government. Looks like the mad scramble for land and resources is still alive and well in the Great Lakes region, and could become even more violent soon. There are over a dozen 'rebel' groups operating in this part of the world, and 3 of them are large enough to be considered full-fledged paramilitary groups that can inflict serious damage on each other, and the local populations.

All of these groups have targeted civilians, government forces, and each other . . . just when I think this mess can't get an worse, it usually does. Watch for full-on warfare involving surrounding states in the next few years.

INTERNATIONAL INTRIGUES

Our final classification of clashes involve those disputes that involve major world powers or multiple sovereign states, or span whole regions . . . and also have the unique character of being waged over strange spaces: be they over oceans, over air spaces, or in the case of North Korea and Iran, over possible intentions. Many of these potential battles also are intriguing in that when/if they blow up, they have the potential for world war . . . as they will be bringing at least two of the world powers into the fray. Even the ones that don't have world war potential are fascinating in that they will be re-defining and shaping regional power structures, affecting dozen of states, and having repercussions well beyond the physical places where the fight may be taking place. These spats will shape world policy on the future of the oceans, world policy on terrorism, and will define the reach of particular world super-powers, and rising second tier powers like Iran. So follow these fights as they unfold in the coming 365. . . .

SENKAKU SHOWDOWN!

Tensions between the two Asian rivals have worsened in recent years over a group of uninhabited islands in the East China Sea called Senkaku Islands by the Japanese, Diaoyu Islands by the Chinese, and the Diaoyutai Islands by the Taiwanese. No matter what you call it, they are just a small group of rocks that are technically owned by Japan, that no one lives on, nor have any proven resource value. They are located roughly due east of Mainland China, northeast of Taiwan, west of Okinawa Island, and north of the southwestern end of the Ryukyu Islands.

The Chinese claim the discovery and control of the islands since the 14th century, but 'lost' control of them during their chaotic 19th and 20th centuries, when Japan was on the rise and China was sucking. Specifically, the Japanese took possession after their successful rout of China in the 1895 Sino-Japense War, in which Japan gained the Korean Peninsula, parts of Machuria, the island of Taiwan and the surrounding seas . . . in which these rocks in question are located.

So who cares about a handful of rocks that changed hands over a century ago? That would be the 2nd and 3rd largest economies on planet earth right now: China and Japan. What's at stake? Perhaps their is oil under them. Perhaps its about fishing rights. Perhaps its about China not wanting a foreign power located so closely to their coast. But is is DEFINITELY about national pride, and establishment of who is actually in charge of Pacific Rim in the coming centuries. And it is a big deal.

Why? Well, if you read this book, you now know that China and Japan pretty much still hate each other since the Japanese imperial phase that saw them take over, and decimate. large parts of China in the WW2 era. China is now back, and assertive, and wants to make sure that Japan gets back in line as a second their power, behind China's dominant power in this region. The Japanese for their part are also beating the national pride drum, and are tired of apologizing for the WW2 era, and don't want to lose face in front of their citizens. So in both sides of this equation are millions and millions of Chinese and Japanese citizens who want their governments to restore legitimacy to their proper place in the Pacific, and who get instantly infuriated at the opposing country any time boats bump into each other around these rocks! Insanity, isn't it?!? What's more, since the USA has military and strategic alliances with Japan, if a war were to break out over the rocks, America might get sucked into it . . . making it a world war status conflict!

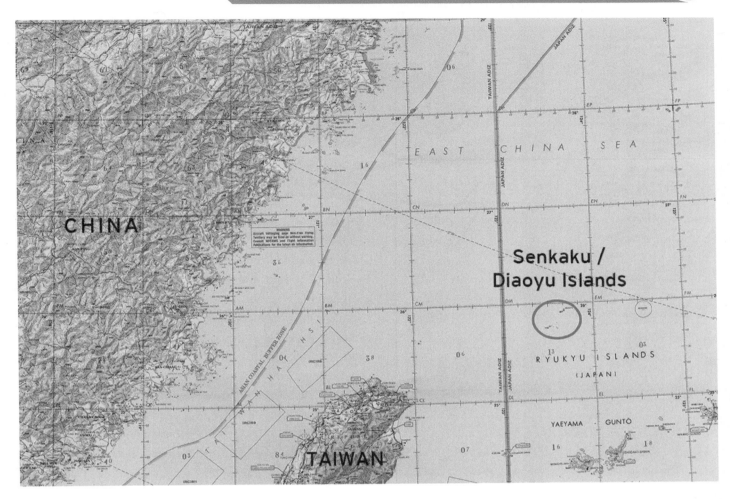

You will see public protests in both countries in the coming year about this issue, and likely see fishing and military boats from both countries floating dangerously close to each other in the waters around this area. For its part, China already unilaterally declared all the airspace above the islands and a wide swatch around it as officially Chinese airspace . . . essentially daring anyone to test their resolved on the issue. Which the USA promptly did within 12 hours when it flew several bomber straight across the Pacific, over the Senkaku, and then home again! It's a high-stakes chess game afoot my friends! Watch this one closely in the coming year! And that is not the only ocean patch that the Chinese are flexing their muscles over . . . there is also. . . .

SOUTH CHINA SEA & THE SPRATLY SPAT

Another great episode that explains the rising power of China in yet another maritime dispute, but this one has even more parties involved! Again, in it's now more comfortable role as regional hegemon, China is exerting itself and its claimed ownership/control of more maritime areas. This battle is for a way bigger area than the Senkaku Islands, and may contain vast oil and fishing reserves. This is the enormous South China Sea, which is entirely located completely south of China . . . thus the name! . . . but also thus the confusion of how China can claim the entire thing! But they do, and 6 other countries claim parts of it too, which makes it a hot spot to watch big time. Fishing vessels have already bumped into each other, China has already planted flags on dink-ass little rock outcrops all over the place, and Vietnamese ships have been fired out, China has told everyone, outright, that it claims it all and will not tolerate any discussion about the waters at all! WIth anyone! Ever! This affects not only the nations that may be going up directly against Chinese military, but also explains the rapid rise of

South China Sea Islands

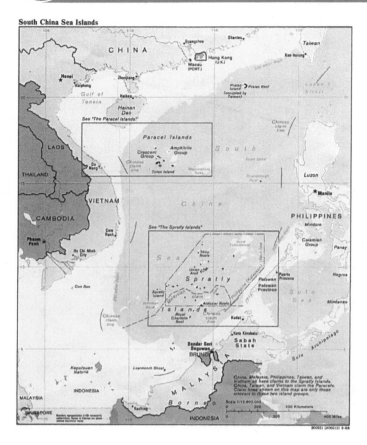

the Chinese Navy (including multiple aircraft carriers they are building), the US's shift to the Pacific (we are establishing bases in Australia, and most recently are re-establishing military ties with the Philippines), and may even result in a US military base in Vietnam! Crazy stuff! All in an effort to build US alliances in the region, while containing Chinese ambitions. It also affects economic activities to the entire Pacific realm, as these waters are tremendously important for their shipping lanes, and all of the states with partial claims on the South China Sea are members of the ASEAN economic alliance, which may become a spoiler to China. Mari-time, military, and economic mischief abounds!

Spanning from the Singapore and Malacca straits to the Strait of Taiwan, the South China Sea is one of the world's most hotly disputed bodies of water. It also has these huge group of tiny islands—named the Spratly Islands—which are closer to rock outcroppings than habitable land, splattered across the entire area . . . and a handful of countries claim parts of those rocks as well. China lays claim to nearly the entire sea, overlapping with the maritime claims of Taiwan, Vietnam, Malaysia, Brunei, and the Philippines. With sovereign territory, natural resources, and national pride at stake, this dispute threatens to destabilize the region and even draw the United States into a conflict.

Exercising sovereignty over the South China Sea would be a strategic boon for China given that more than half of the world's merchant tonnage, a third of crude oil trade, and half of liquefied natural gas trade travel through the contested waters. And, with its waxing political, economic, and military weight, China seems to be taking a harder line on the issue. Chinese officials continue to emphasize the protection of China's territorial integrity, which of course includes the South China Sea in their view. A Chinese defense white paper, released in April 2013, declares that China will "resolutely take all necessary measures to safeguard its national sovereignty and territorial integrity."

Thus, Beijing's public proclamations that the South China Sea as an outright core interest of theirs, and one they will defend by force if necessary. They WILL NOT back down from this fight . . . and I assure you that Vietnam won't either, as that scrappy little nation has already defeated a super power in the past, and is the state most likely to challenge China directly in this watery claim. Watch out! This one could blow at any time as well!

THE SAHEL REGION OF AFRICA

The Sahel: A transition zone in every sense of the word "transition," this belt of climatic cross-over is also a strip of savagery that spans the continent of Africa, and is the scene of scores of conflicts. The Sahel originally refers to a scattered scrub vegetative biome: a zone of transition between the Sahara desert to the north and the savanna grasslands to the south . . . but is also a cultural interaction space where nomadic/cattle country meets agricultural/urbanized life, and most importantly, where Islam meets Christianity!!! And frequent cultural fracases are the result! This unstable strip stretches 3,400 miles due west/east across parts of Senegal, Mauritania, Mali, Algeria, Niger, Chad, Sudan, South Sudan and Eritrea. Herders clash with farmers, Christians clash with Muslims, "Arab" clashes with "African," and nomadic clashes with urban. Political parties even within single sovereign states tend to polarize the peoples, and have torn states asunder. As a result, civil wars erupt, terrorist groups organize, and militias roam the region trafficking in drugs and arms, seizing hostages for ransom, and trading livestock.

Just to name a few of the most recent active actions across this lethal ribbon:

→ In Mali, the Tuareg Rebellion of 2012 separated the entire northern half of the country from the central government, and was to become a new separate state . . . for a whole 5 minutes before radical Islamists took over the territory for themselves. A military coup toppled the Mali government in March, while separatists and al Qaeda-linked fundamentalists took over the country's north. The end game resulted in the French intervening to help restore stability and take back control from the jihadists, and they still have an active force there still.

→ The Central African Republic (CAR) which by all measures is now described as a failed state in permanent crisis due to a civil war between Muslim minority Séléka rebel coalition and the François Bozizé government forces . . . the rebels captured the capital in March of 2013, Bozizé fled the country, and the UN and AU forces were called in to help restore order. As of this writing in 2014: 300,000 refugees, massive crimes against humanity have been perpetrated, full-on genocide is occurring, mostly by the Christian majority purging out Muslims.

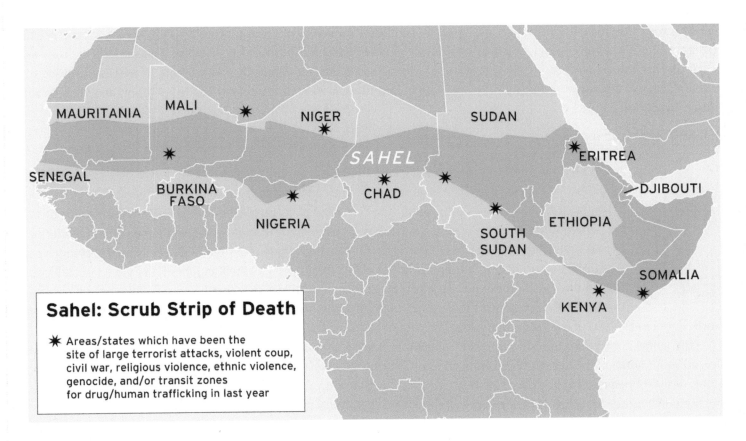

Sahel: Scrub Strip of Death

✴ Areas/states which have been the site of large terrorist attacks, violent coup, civil war, religious violence, ethnic violence, genocide, and/or transit zones for drug/human trafficking in last year

→ The most recent Chadian Civil War which began in December 2005 and ran to 2010. Since its independence from France in 1960, Chad has been overwhelmed by the civil war between the Arab-Muslims of the north and the Sub-Saharan-Christians of the south . . . which is the exact thing that happened in . . .

→ Sudan: After the First Sudanese Civil War (1955–72) and then the Second Sudanese Civil War (1983–2005) between the mostly Arab/Muslim north and the mostly Christian/African south, the country split in two . . . right along the Sahel zone lines! Of course, the 5 decades of war resulted in two million dead and four million displaced . . . which is the highest civilian death toll since World War II. And it ain't over yet, as sporadic fighting between Sudan and South Sudan is still a reality of the present and future. As is still is in. . . .

→ Darfur: this region of western Sudan made international headlines back in 2003 for acts of genocide perpetrated by government sponsored Arab/Muslim groups agains non-Arab populations. Yep. Same story, in the same state! And even though the world is no longer paying attention to it, the UN is still there, and atrocities still abound.

→ Go google "Boko Haram" to get the whole inside scoop on this despised and deadliest of new terror groups hoping to rip the northern part of Nigeria away from the south and make it into a separate Islamic state that follows strict Sharia law. Linked to Al-Qa'ida in the Islamic Maghreb (AQIM) and located in Nigeria, Niger and Cameroon, the group is known for attacking churches, schools, and police stations; kidnapping western tourists; and assassinating members of the Islamic establishment who have criticized the group. Violence linked to the Boko Harām insurgency has resulted in an estimated 10,000 deaths between 2002 and 2013. You probably heard about them in the news in 2014 for kidnapping and enslaving hundreds of school girls which (finally) gained international attention to the havoc they are reeking.

→ The Islamist group al-Shabaab sent gunmen to attack the upmarket Westgate shopping mall in Nairobi, Kenya in September 2013. The 48 hour siege resulted in 67 deaths, including four attackers, and 175 wounded. al-Shabaab claimed the attack was retribution for a Kenyan military's deployment in Somalia which attacked the terrorist group who had been operating across the Kenyan-Somali border for decades.

Starting to see the trend? This belt, which, again, is actually defined by physical geography, is the transition zone key to understanding the cultural geography, ethnic genocide, and terrorism across all of North Africa. This increase in civil war, in cultural friction, and increased terrorist activity have increased interest in outside intervention, as well as increased US military interest, and increased US military presence in new bases across the continent. The Sahel is hot, and likely to get hotter this decade. . . .

NORTH KOREA VS. ?????? THE US? SOUTH KOREA? JAPAN? EVERYBODY?

North Korea is America's worst enemy in the Far East. Ok, the world's worst enemy. They possess about 5,000 tons of biological and chemical weapons. They administer death camps where 200,000 political and religious criminals are worked to death. They are also run by an inexperienced and possibly insane dictator, Kim Jong-Un, who is not averse to funding his burgeoning **nuclear program** using money laundering, heroin smuggling, and counterfeit currency. Former US President George W. Bush has said of Kim Jong-Il (the former North Korean leader and father of Kim Jong-Un), "I loathe Kim Jong-Il. I've got a visceral reaction to this guy, because he is starving his people. And I have seen intelligence of these prison camps—they're huge—that he uses to break up families and to torture people. It appalls me." Bush even included North Korea in his "axis of evil" speech in 2002.

The whole conflict started in 1950, when communist North Korea invaded South Korea and America sent troops to defend South Korea in the Korean War. Ever since then, North and South Korea have endured an uneasy peace. The United States has stationed more than 30,000 American troops in South Korea, in case North Korea tries to invade again. However, the thing that scares American policymakers most about North Korea is the fact that they are building nuclear weapons.

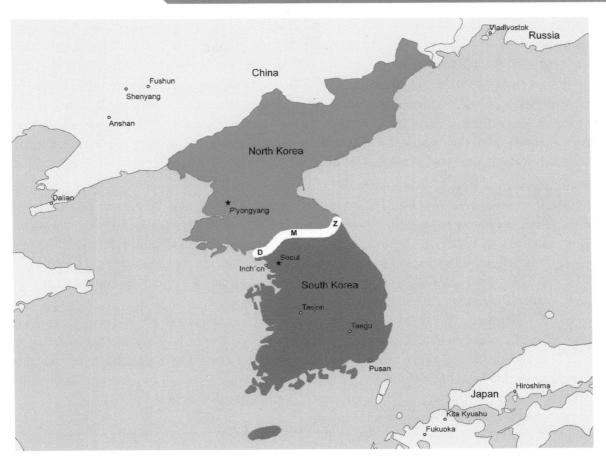

   To resolve the nuclear issue, there has been some diplomacy, but the Korea issue is complicated. At first, North Korea wanted the US to sign a peace treaty that would formally end the Korean War, which never happened, but now they just want security. Unfortunately, Kim Jong-Il and his advisors have decided that the best way to gain security is to threaten their neighbors with nuclear attacks and boast that they are not averse to preemptive war. The US wanted to resolve this whole issue in the form of six-party talks involving themselves, North Korea, South Korea, Japan, China and Russia. In November of 2005, the last round of these talks was held and an agreement was reached to provide North Korea with financial aid if they stop their nuclear program. But Kim won't finalize the deal unless the US drops its demands regarding his country's trade activities. The US responded to this by saying that the trade and nuclear issues are separate and they refuse to merge the two.

   North Korea is not much of a threat militarily and would probably get their butts kicked by South Korea in a war—or really, a war with anybody. Keep in mind: They would lose, but after possibly killing millions of Koreans on both sides of the border; the leadership does have death potential, but no hopes of actually 'winning' long term as their state is totally insane and totally defunct. In fact, much of their population is starving. But with atomic weapons, and the crazy rhetoric for which the Kims are famous, many people are scared of North Korea. Nukes, nuts, and nothing to lose: not a good combination.

   In 2010, North Korea did everything in their power to infuriate the US, South Korea, Japan, the IAEA, the UN, and even China! They tested multiple rockets which could carry a nuclear payload, as well as at least two underground atomic explosions which seem to indicate that they are back in full swing trying to produce nuclear weaponry. They pretty much threatened the entire planet with annihilation if anyone messes with them or their ships, which may be exporting illegal nuclear cargo. The erratic behavior of the regime has totally enraged China, pretty much their only ally on Earth. Harsh UN sanctions have been slapped on the Hermit Kingdom, and we are all now curiously watching to see what these nut-bags will do next.

2012 Update: With the totally inexperienced and clueless Kim Jong-Un taking charge of this psychotic state in 2012, it is really anybody's guess what direction they will go in next. A lot of us were hopeful that he would turn over a new leaf for the country and open up a bit to the outside world, but those dreams are fading fast. North Korea flipped off the entire world by launching a rocket into space (it failed), and by all accounts the regime is now moving fast to do another yet another nuclear underground explosion as a show of its savvy and strength. Meanwhile the people starve and the military continues its aggressive rhetoric. Unfortunately, there is no way this mess can end well. The only option that the Plaid Avenger can see that would end this nightmare without the use of nuclear weapons would be a ground invasion from China. They did it once during the Korean War . . . so let's rally President Xi Jinping for a repeat performance! Go get 'em guys!

SOMALIAN SKULLDUGGERY

Well, the end is near . . . for this chapter my friends, not for us! What better way to finish a section on failed states than to reference, at least briefly, the poster child of all poster children of failed states, and that is the stinking cesspool that is Somalia. Don't be offended, my Somali friends! I am not talking about you personally or your ethnic group generally; I love you all! But the political backwash of a country called Somalia just sucks like a shop-vac. On high speed.

However, I am not even going to go into the entire back-story of why this state disintegrated. A combination of colonial incompetence, ethnic infighting, civil strife, civil war, foreign meddling and general corruption have served to turn this state into a warlord-dominated and Islamic-fundamentalist breeding ground. Even when outsiders have made an attempt to make a positive impact, it has turned out horrifically bad. Go watch *Black Hawk Down* see the US's ill-fated go at it . . . and nobody has tried to help since.

No, I simply wanted to bring up Somalia for a completely different reason altogether . . . pirates, of course! Hoist the Jolly Roger! Swab the decks! Shiver me timbers! Okay, enough of that nonsense. The point is, you have heard a tremendous amount of news in the last few years, (and even a Tom Hanks movie) about the exponential increase in pirate attacks and hostage situations and skullduggery on the high seas, all happening off the Somali coast, perpetrated by Somali pirates. While it's quite fascinating and makes for flashy news stories, the bigger picture is being missed by most. What picture is that?

Just this: allowing states to fail ultimately affects the whole global community. Now, perhaps I am getting too preachy and not simply covering the facts, as all good textbooks do. Oh, I mean as all boring textbooks do. But the thing is, these are facts. All of the failed and/or failing states around the globe are now having global repercussions on all the rest of us, since we are now all connected in the globalized world. Did I use the word globe enough for you yet?

Just dig this: whenever we turn our backs on a part of the globe that we feel does not affect us in any

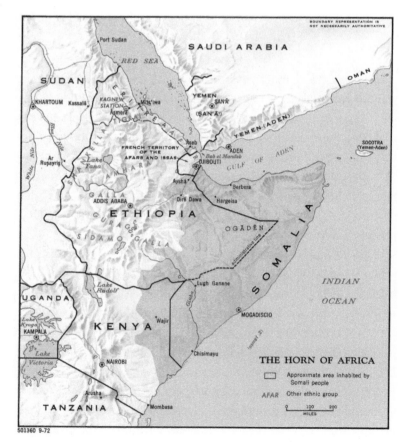

THE HORN OF AFRICA

☐ Approximate area inhabited by Somali people

AFAR Other ethnic group

0   100   200
MILES

501360 9-72

way, we get bit in the ass. Yeah, that's what happens when you turn your back. North Korea's nuclear issue, Afghanistan's/Pakistan's no-man's-land issue, Somalia's pirate issue, even Zimbabwe's AIDS/cholera issue are all bound by one common theme: all these problems were spawned in areas that the world gave up on, and now the world has to deal with the consequences. Because AIDS spreads. Because extremist militants attack. Because nuclear bombs can be launched. And yes, because pirates will climb out of the hole that the world left them in, and come and steal your booty.

The point of this rant at the end of our hot-spots chapter? Heck, I don't know. Maybe it's to keep you engaged in what's happening in the world so that the next hot-spot can be avoided. But for sure it is to alert you to a stone cold fact of life in our times: world ignorance may be bliss, but it won't stop world problems from reaching your shore. Even when it comes ashore dressed as a pirate. Speaking of the next imminent hot-spot, let's finish the chapter with a current conflagration that may end up re-defining our times and the world order if/when it goes down. Let's head back to Persia. . . .

IRAN VS. THE US, OR THE WORLD IF NECESSARY

Iran and the United States have been enemies since the 1979 Iranian Revolution, when Islamic extremists overthrew the US-backed Shah, and took control of the government. When the fanatics took over, one of the first things they did was raid the American embassy in Tehran and take 52 Americans hostage for 444 days. The images of American hostages on American TV made Americans furious at Iran. Iran has also funded terrorist groups worldwide, including Hezbollah, which was responsible for bombing a Marine barracks in Lebanon in 1983 and killing 241 American marines.

Iran has the dubious distinction as being listed by the State Department as the top state sponsor of terror worldwide, making it a natural target in the war on terror. To make things worse, Iranian officials are hostile to America's main ally in the Middle East, Israel. The current President of Iran, Mahmoud Ahmadinejad, has even said that Israel should be wiped off the map, and other clerics are famous for their anti-Semitic rhetoric about Israel. No wonder George W. Bush called Iran part of the "axis of evil" in 2002.

Despite years of talking shit, the US may finally have an excuse to attack Iran . . . and others have been getting on board for the action as well. In 2006, Iran admitted that it was enriching uranium to build nuclear power plants. Iranian officials claim that they are just doing this for nuclear power, but some American officials are concerned that this is just the first step in developing a nuclear bomb. Game on.

Obviously, American policy-makers are scared to death of Islamic extremists, known for supporting terror and threatening Israel, having nuclear weapons. There has been speculative debate for years now about a possible US invasion or a US air-strike to shut down this nuclear nonsense. Before that happens, the US has been pushing hard for crippling UN economic sanctions against the regime. Both avenues of punishment have been resisted big-time by China and Russia, both of which are doing big deals with Iran for energy importing and exporting, so this situation has some beyond-the-Middle-East, whole world repercussions as well.

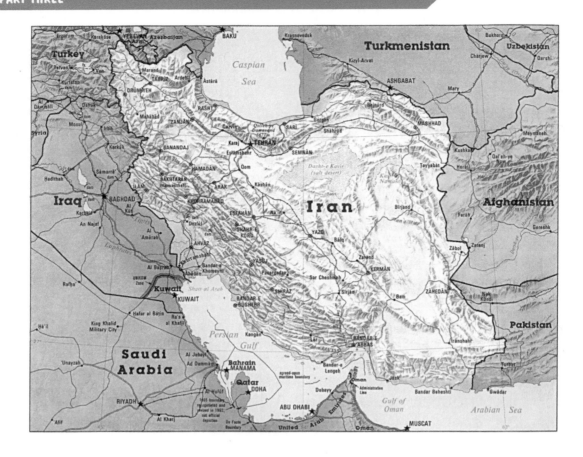

However, even China and Russia have been re-thinking their protection of Iran as Iran has become even more ballsy with their nuclear ambitions in the last two years: they failed to allow IAEA inspectors full access to crucial areas, were busted with a secret underground uranium processing base, and even made proud claims that they are enriching uranium fuel to levels higher than needed for nuclear energy, but not quite high enough to be used for bombs. They continue to intentionally stay in this murky middle ground, as if they actually are daring outsiders to attack.

Think I'm exaggerating? Both the Ayatollah and former Iranian President Ahmadenijad made fiery speeches about how it is Iran's "God-given right" to have a nuclear industry, and that nothing on the planet will stop them. (I forget what part of the Bible God specifically addresses fissile material in, but I'm sure it's in there.) No amount of wheeling and dealing seems to be able to push Iran off of this path, which is why they may be bombed off it instead.

Iran's air force and navy is shaky at best, and the US could take it in a couple hours with some surgical bombing. A ground war is another story. Unlike Arab Muslims, who usually consider themselves more Muslim than nationalistic, Iranian people are very nationalistic and would resist an attack violently. I cannot conceive of anybody in the US government being dumb enough to attempt a ground war in Iran, but we should never underestimate the capacity for ignorance, I suppose.

I believe the worst case scenario that we might see would involve surgical bombing of the nuclear production facilities, and the US probably won't even be the ones to do it! Israel has publicly stated many times now that a nuclear Iran is unacceptable and that they will not let it happen. The Iranian situation is now the #1 most worried about foreign policy issue to nearly every single Israelite.

It will probably be Israel that conducts a bombing campaign, unless some diplomatic breakthrough occurs. The US, the EU, most of Team West and, strangely enough, most Arab states will secretly support such a strike, while publicly distancing themselves from it. This will serve to infuriate the Iranians, and piss off the Russians and Chinese as well. The Iranian government has also said that it will step up terrorist attacks, especially in US occupied Iraq, if force is used against them.

Do you think that's all the Iranians can do if they get ticked off? Consider this: Iran could, in less than a day, drop sea mines into, or start picking off ships in, the critical Strait of Hormuz, where a huge daily dose of the world's oil supply passes through. In effect, they could halt the entire world economy by clogging up this small artery of oil flow, which would spike oil prices exponentially and instantaneously. This one could get ugly, quickly—and everyone is going to be involved one way or another.

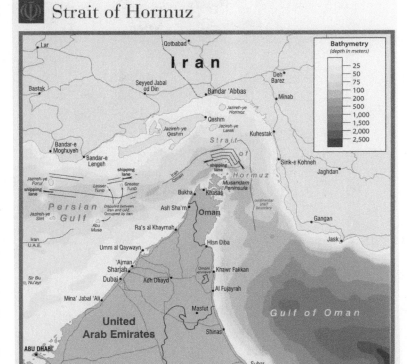

Strait of Hormuz

Strait of Hormuz: Potential Iranian Target to log-jam.

Eastern Europe in 1815 . . . where is Ukraine?

Team Play: Evolving Power Centers of the 21st Century

18

CHAPTER OUTLINE:

→ Let's Meet the Political Power Players

→ Military Machinations

→ Economic Power Players

→ Finally the Final Final Final Summary

MY friends, it's long past due to wrap up this blundering batch of buffoonery I call a book. What better way to round up a world regions textbook than to back up from the entire planet and assess exactly how these world regions are interacting with each other—either working with each other, against each other, or forming up teams to cancel each other out. It's such a delicious stew of international intrigue! Think of it as different clicks in a worldwide high school, except these states have nuclear weapons and control the planet. Other than that, it's the same old same old human nature. But I digress.

Let's once again look at the major players that are affecting the course of events on our planet strategically, politically and economically. Then we'll call it a day.

LET'S MEET THE POLITICAL POWER PLAYERS

No sense trying to pretend otherwise: the United States is still the undisputed heavyweight when it comes to a single entity with the absolute most political power. To put it simply, what the US wants, the US gets. An example: There was never any credible threat that any other country was going to attack the US because of its most recent invasion of Iraq. Most countries, institutions, and people on the planet may have been staunchly opposed to that US move, but no one was actually going to do anything about it. That is raw political power, and the US is the only state that has been able to wield that sword solo for a couple of decades.

But the world is changing fast, my friends. The US still has the power, but other states and groups are rising fast to match their diplomatic skills with Uncle Sam. With no reservations, I tell you that we are moving from a uni-polar world (only one power, namely the US) to a multi-polar world (with many major powers). This is the setting of transformation that we find ourselves in here at the beginning of the 21st century. Politically speaking, let's take a look at the major forces that shape decision-making across the globe, and who teams up with who in order to get their agendas pushed. I will identify the major players, their motivations, their agendas, who they like and who they hate, and see how that is translating into current events that are going down around the globe.

SAMMY SOLO

America has long been a team of one. Its physical isolation far from the rest of the world has forever insulated it from the global events in which it participates, or in which it declines to participate. Its entire history could be viewed as a wavering cycle of either isolation (in which it does not get involved in global events) or full-on engagement (in which it is full-on the predominate player.) Take a wild stab which side of the spectrum it has been on for the last couple of decades . . . although with the US finished in Iraq, pulling out of Afghanistan, and refusing to get involved in Syria, one could argue the the pendulum is swinging back to self-absorbed stance.

Sammy sez: step off!

But when Uncle Sammy decides to go for the global gusto, he has seemingly unlimited energy, money, military, and ambition to get the job done, regardless of world opinion. For the last hundred years, the US usually had a global coalition of participants that backed it up. Let's call them Team West for right now. In WWI, WWII, the Korean War, the Cold War, the first Iraq War and countless others, the US served as a team leader that rallied the world to victory, or at least something close to it. Point is, that mostly the US didn't have to stand alone, because the UK or the EU or NATO or the UN or Team West was there to back them up.

However, since the end of the Cold War—circa 1990—Sammy has increasingly worked on his own to achieve his agenda. Why? During most of the last fifty years, it was the US and the USSR in a bi-polar world of competition, giving the world at least two options to choose from. But now it has been a couple of decades that the US has stood completely alone at the top of the heap, alone at the top of this uni-polar world. In this time, the US quickly grew accustomed to the role of sole political power. Therefore, it has increasingly acted alone whenever it really wanted something that the rest of the players would not support.

I'm not just picking on the former Bush administration, either; the Clinton administration acted unilaterally on several different occasions as well, either to stop perceived terrorism in Africa or perceived genocide in the Balkans. The dynamic duo of Dick Cheney and Donald Rumsfeld just took the US even exponentially further down this unilateral path (willing to act alone), with the most extreme example being the aforementioned second Iraq War. Beyond even a willingness to act alone, the US for over a decade has actually preferred to act alone, so it does not have to waste time with negotiation and organization and cooperation that is required by international diplomacy.

The times are a-changing, though. Under the US President Barack Obama administration, the US is already rapidly leaning back to diplomacy and coalition-building as the most effective tool for international action. We can see this unfolding currently as the US is working exceptionally hard to rally world support on the growing threats of both the North Korean and the Iranian nuclear industry, and/or the situation in Syria, and it is having some success. The entire EU backs them on it, and even Russia and China are starting to come around to the US point of view after lots of wrangling and back-room debates. But that does not mean that any concrete multi-lateral actions have been the result . . . as of course it has not. The NATO-led Afghan mission has also been recast as needing more international support, as opposed to being primarily a US military force, but it is winding down in either case, so it won't matter for much longer how it is labeled. Because either way it will be labeled as chaos shortly after the US departure.

Uncle Sam had a solo run in the sun for some time, but it appears that in this newly forming multi-polar world, his future role may be as primary leader of Team West. Okay, who the heck are they?

TEAM WEST

Throughout this text, I have thrown around the concept of a core world culture that is primarily steeped in western traditions and values. It's finally time to identify just exactly who is on Team West. Obviously, the core of the team is Western Europe, the place where Western Civilization was born and evolved, thus the team name. *"Western civilization"* encompasses the big ideas on which our lives are based. I'm talking about the big things here: things like philosophies, legal systems, economic systems, medical practices, writing systems, and particularly religions like Christianity and all its sub-sects.

However, the true adhesive of this team is based on common ideologies. All the members of Team West also share particular values such as liberal democratic governance, emphasis on individual human liberties and rights, and the free market/capitalist system. These are more important to consider than the other stuff, since there are some non-European actors on this team; while they are not Christian or European or even white, they are staunch believers in the team's principle values. That's why they are on the team. Okay, so who?

Joining Western Europe, we have most of Eastern Europe, and let's just go ahead and refer to the team elements of both regions as the EU. The EU countries may have been the historic core of Team West, but the all-star quarterback of the squad is now the US. But there are even more players on the team! Canada and Australia always show up to the game, as does Japan. That's right, Japan is definitely part of Team West even though it is in the east, and Asian to boot! It shares all the West's values and has staunchly supported the US and the team ever since the end of WWII. One other surprise entry: Turkey. As a staunch secular democracy, capitalist country, and NATO member, Turkey also primarily sides up with Team West, even though it is Muslim. Primarily, but not always. We could even count Israel on Team West, although for various reasons, it does not usually openly participate in most of the Team's competitions.

How does it work in today's world? Well, when the Team comes up with a desired policy or goal—say, the War on Terror or even the War on Drugs, the US usually leads the charge, and will often be supported on the UN Permanent Security Council by having the UK and France back them up. The EU as a whole typically is on board as well. Then when it is time for the real action to go down, Canada, Australia and Turkey will jump in to join with the EU and US troops, and Japan will voice its support for the endeavor, and perhaps even supply money or materials to aid the effort, since Japan does not have offensive military capacity. Then the team hits the field and tries to score the goal. Got it? Good.

Team West is a potent force today, mostly because it consists of most of the planet's richest countries, some of the most powerful and technologically advanced militaries, as well as having three nuclear powers (US, UK, France . . . and covertly throw in Israel to make four). Can you feel the power yet? No? Consider this: all NATO countries are on Team West, which essentially means that NATO itself is a tool of Team West's arsenal. And NATO kicks butt.

An example as referenced above, the team's quarterback, Uncle Sam, is currently leading the charge against Iran to embargo the state and possibly have eventual military strikes against them. The EU is already on board, as is Canada, Japan, Australia, and even perhaps Israel, which has been conducting military exercises as part of the propaganda campaign designed to scare the Iranians. See how it all works, my friends?

But I must give you this 2014 update, my friends: Team West, and the US in particular, have really taken it on the chin since this whole global economic meltdown started. Most places on planet Earth now squarely put the blame on the capitalist excesses of Team West. This has started a whole re-think in terms of following the capitalist path of the West, with some folks suggesting that capitalism as a whole is on the way down. And some think the same about democracy too: authoritarianism is on the rise again, with active military coups taking power in Thailand, Egypt, and most of the Arab Spring countries have still not made it to the democracy finish line . . . and of course Russia is leaning back in that direction too, and China never left it. Many countries now see democracy itself as inherently unstable, unpredictable, and messy . . . which it is! So some states are shying away from it in these chaotic days. Couple the democratic gridlock and economic crap with what is now seen by many as over-extension of political and military power in the Middle East, Central Asia, and Eastern Europe, and you have the makings of a good old fashioned smear campaign against the West. I can't lie: they are taking a hit right now in the global scheme of things and will lose some power. But the Team is resilient, and rich, so they are still a global playa' for some time to come. Had enough of the West yet? Then let's get on to some other entities which are not part of this team.

THE BEAR

We already had a whole chapter on Russia, so I can keep it brief here. It was largely assumed that Russia would inevitably join Team West after the collapse of the Soviet Union in the early 1990s. It does share most of the common cultural characteristics of Western Civilization, but somehow it has always remained just on the fringe of that grouping, always slightly different, with a distinct "not European" component. After the demise of the USSR, Russia was a second rate power, with even less prestige, that was really at the mercy of international business and banking and stronger world powers like the US, the EU, and NATO. Russia pretty much just had to go with the flow, even when events were inherently against its own national interests, like the growth of NATO, for example. But all that has reversed in the last fourteen years under the long judo-chopping arm of Vladimir Putin. He, along with tons of revenue from the Russian oil industry, has brought the Bear politically back to a position of strength. He plays hardball in international affairs now, and has reinvigorated a Russian national pride that has long laid dormant. In short: the Bear is back. Grrrr! Russia's full-frontal invasion of Georgia in 2008 was essentially a calling card that they are back, ready to wrestle, and more than willing to engage militarily if pushed on issues of their national security.

2014 UPDATE: WOW, DID I CALL THAT ONE, OR WHAT? Not only was the Plaid Avenger dead on the mark predicting that Russia was no longer afraid to exert itself militarily, it of course actually went full-out blitz on Ukraine, grabbing back the Crimean Peninsula, knowing full well it would infuriate Team West and most of the world. And they did it anyway, and may do more, and don't appear to care about world opinion much at all anymore. Yep. That's power.

Now that the Bear is strong again, it has been re-thinking the whole joining Team West business, and it appears it is declining the invitation. Russia prefers to be an independent player in order to be a global power again, as well as to achieve its own unique foreign policy goals, and it is in a prime position to do this.

Since the Russians supply one-third of European energy demand, this gives them all sorts of economic and political leverage over their Eurasian neighbors. Get Russia upset at you, and you might not have heating oil next winter. But the Europeans aren't the only ones who need oil. Look eastward and you see Japan and China both vying for petroleum resources as well, so Russia is really sitting pretty right now. The Russians are forging economic and strategic political ties to their east as well, mostly with China. They are at a historical east/west pivot point and they are straddling the fence about which relationships to make or break. I think they're being extremely savvy and are going to avoid taking sides in anything. **2014 UPDATE:** May 2014, Russia and China ink a 30 year gas/oil deal worth billions. The die has been cast. Russia has turned East. They will not be joining Team West in our lifetime.

We can see this rift between the Bear and Team West in a variety of ways in today's world. The US and many Europeans express concern or outright criticize Russia for becoming too authoritarian under Putin, thus detracting from their democracy and human liberties, which is, of course, one of the foundations of Team West. The US and EU also accuse Russia of using oil as a political weapon to get its way; a not entirely false charge, but is that illegal?

In return, Russia usually counters most US movements at the UN Permanent Security Council, where it holds veto power. One need look no further than the current Iranian issue, where the Russians have been quick to dismiss Team West's fears of Iran's nuclear ambitions, has promised to veto any severe embargoes or military action against Iran, and is even helping the Iranians build their nuclear power industry. Russia has also propped up the regime in Syria, and protects them from any UN involvement via their veto power. Russia is increasingly working together with China economically, diplomatically, and militarily, which is perhaps unsettling to the other world teams.

The Bear is a pole of political power not to be taken lightly, but is there really ever a situation that a bear would be taken lightly? While I would never brawl with a bear, I would be even less likely to disagree with a dragon . . .

ENTER: THE DRAGON

Just as with the Bear, the Dragon has been resurrected as of late and is on the fast track to becoming a major world political power. As you undoubtedly deduced from the chapter on China, the state had a rough go of things for most of the 20th century, from internal destruction to civil wars to colonialization to wartime atrocities to counter-productive communist policies. Since the capitalist reforms of the 1970s, China has really turned things around and become an economic powerhouse of the globe. Remember, it jumped two whole slots a few yearsa ago to become the 2nd largest economy on the planet, and will likely take the top slot from Uncle Sam in the coming decades. Or maybe even this decade.

With that economic clout has come global political power. The Dragon has been very focused on rebuilding itself for the last few decades and therefore was not that interested in exerting itself in the global political structure, except of course when it was in its own interest to do so; think of any issue dealing with the status of Taiwan. China also is the staunchest supporter of sovereignty in the world, so its attitude towards global issues and conflicts outside of its own territory has typically been "live and let live," although "live and let die" may be a more appropriate descriptor. China does not like to get involved in the affairs of other states, no matter how horrific the situation, and in return wants no one poking their noses into Chinese business. You dig?

Now this would all be just an asinine academic excursion but for one thing: China is one of the veto-wielding members of the UN Permanent Security Council. When international action is called for by Team West to punish Iraq or Syria or North Korea, the Chinese position is usually to play the spoiler and veto any actions by the UN, or at least dumb-down the UN action to the point that it is ineffective or useless. Need an example?

Take Sudan, Zimbabwe, and North Korea. Most of the world does not like, does not support, and does not want to trade with these countries because they have fairly brutal records of persecuting their own people. The morally and ethically right thing to do is to refuse to support that state in any capacity. But China says, "We don't care. We will trade with you. Can we buy oil from you, Sudan? Thanks, buddy!" Perhaps their open policy is not such a great thing. This causes friction between China and other rich countries around the world, especially "Team West" countries. Developed countries say, "You can't do that," and then China says, "Hell yeah we can. We are a sovereign state." The issue is a source of friction for a lot of the Team West countries. Both China and Russia are very good at this game. Speaking of which . . .

I just can't help myself but to over-exaggerate a subtle theme that has been forming globally for some time now, and has burst into full flower just this year: China and Russia are totally in love, dudes! These two titans have pretty much agreed on every single global item of consequence in the last year, and have met more times than I can even count anymore. From heads of state meetings on trade and security, to a joint space program, to meetings of the SCO and the BRIC; China and Russia have pretty much thrown down the gauntlet and created a new power block for the planet.

What do I mean by this? Well, in addition to all the joint stuff they are doing, both President Putin and former Chinese Premier Wen Jaibao have openly body-slammed Team West for causing the current financial meltdown. They both are also pushing for way more security and cooperation via the SCO structure . . .not through the UN or any other Western device. These guys made a joint declaration after their SCO meeting in June 2009 that they favoring scrapping the US dollar as the world standard. Dang! This is total cage-match smack-down. But back to China now . . .

With the number two economy on the planet and increasing international pressure to pony up and accept some global responsibilities, China's star is on the rise politically too. China's steadfast respect for the power of sovereignty still trumps all, and thus, the Dragon will be very hesitant to throw its weight around in the international arena. China will hesitate from any unilateral action on its own part, and also mostly stalls multilateral actions even when sponsored by the UN. While the Dragon does (in theory) support the US-led War on Terror, it will continue to stymie any major efforts by Team West to transform other countries, either by invasion or persuasion. With its UN veto-wielding power, it more often than not will be siding with Russia to counter major moves by Team West, while strengthening its hand at home and in its immediate neighborhood. That is done with an eye towards countering another global power in the Dragon's 'hood: India.

TEAM HINDU

I would be remiss to not include the other significant growing power center on our planet, and that would be our Indian friends. Like China, Team Hindu is a burgeoning economy here in the 21st century and that economic power will increasingly be translated into political power. That's right, India will soon be another major axis of planetary political power. Especially if their new main man Modi, the Hindu head cheese, has success transforming the economy into a capitalist powerhouse! But I don't want to exaggerate the circumstances! India is on the rise for sure, but currently is nowhere near the economic, political, or military power of a Russia or China—yet. Stress on the "yet."

India is soon to be a major power player because it has a whole lot of unique attributes which will soon force it to be a player. Let's quickly list them: India will soon be the most populous state on the planet; China is now, but not for too much longer. India has a vibrant and growing economy, and it is simultaneously diversifying nicely. It is a nuclear power, and, yes, it has nuclear weapons. India has a signi-ficant military, certainly the strongest in the region. As part of that, India has an aircraft carrier. That means it has a means of projecting power abroad. The only other countries in the neighborhood that fit that description are China and Russia, who are both power players already.

However, unlike their Chinese and Russian counterparts, Team Hindu actually shares a lot of attributes, ideals, and goals with Team West. India is a strong, established democracy—the biggest in the world, actually—and a supporter of free market principles and global trade. India, like Team West, has human rights and individual liberties as a foundation stone of its society. India is an avid supporter of the War on Terror and, as such, does quite a bit at home to thwart extremism of all sorts. They even speak English, for Pete's sake . . . how much more Western could they get?

Well, I'm pushing it a bit far now—they are not Western, and do maintain their own distinct culture, as well as their own distinct take on foreign policy and global affairs. But look for them to become a stronger and stronger voice in the near future. Former US President Bush met with the former Indian Prime Minister Manmohan Singh multiple times during his presidency in an effort to warm relations between the two titans. First time that has ever happened between these two titan countries. The US was actually working very hard to get a deal signed with India in order to help them out with their nuclear power industry, and that deal finally went through in 2009. It's a really big deal, too! The US and India are now in a friendship that is just now starting to blossom, all because of nukes!

Why do I keep bringing up the nuclear issue? The US and Team West want to legitimize the Indian nuclear position—India never signed the nuclear non-proliferation treaty—so as to further isolate and de-legitimize any future Iranian one. See how this works?

Team Hindu may still be a minor world player right now, but that status won't last for too much longer. They are one of the five potential additions to the UN Permanent Security Council, and given their western tendencies, I would look for their membership to be supported by Team West. I would also speculate that Team Hindu will likely be increasingly siding with Team West on global issues of the future, albeit with a more reserved tendency to use force. Gandhi taught them way too well.

POLITICAL TEAM SUMMARY

Uncle Sam has been the sole political superpower on the world stage, but that brief era is coming to a close. The rise of China, India, and the rebirth of a strong Russia preclude the US from acting unilaterally in the future. The likely US response is to rebuild its base of allies and become a stronger leader of Team West. And Team West will be strong.

China, and increasingly Russia, has a tendency towards authoritarianism and an emphasis on the trump card of sovereignty; therefore, they do not completely share in the value systems of Team West. As such, these two power players are a likely alliance that will balance/counteract the Team West influence on global affairs. Team Hindu does actually have much more in common with Team West, and will likely be an ally of the West in future global issues.

This section has merely been a summary of the **MAJOR** power players on the planet. Of course, there are other states and entities that will affect global events and influence the direction of the planet in a myriad of ways. Brazil, South Africa, Islam, and ASEAN all spring to mind. We simply focused on the most powerful, most organized, and most influential teams that are or will be the major shapers of global political policy for the near future. These are the teams that will decide how to conduct the War on Terror and the global War on Drugs. These are the teams that will decide all major UN actions, including the possible expansion of the UN Security Council. These are the teams that will shape all major policies on global warming, fighting global crime, and conducting global war.

They are the majors of political power. But what about other types of power?

MILITARY MACHINATIONS

You might think that political power immediately equals military power, and you might be right. Might does sometimes make right. There are many individual powerful states on the planet right now. However, the alliances of military power that have occurred between states has been accelerating greatly in the last decade, and these military power centers have themselves become an important component of understanding how the planet works. Let's briefly look over the major military entities on the planet and how they interact with one another.

THE BIG BOYS WITH THE BIG TOYS

The United States is the undisputed heavyweight of military power, technology, and spending. Look back at the table in chapter 1 to check out military expenditures for 2011 and you can quickly calculate that the US alone accounts for half of the total world spending on all things military. Boom! That's some power! However, China actually has the largest standing army, and is spending fast to catch up with the US. But even these two they are not alone.

Let's just call a spade a spade here. The most powerful state military entities in terms of global reach are the US, the UK, France, Russia, China, and India. Why those? Those are all the states with mostly modern militaries, the money to make them stronger and keep them up-to-date, the ability to project that power outside of their own countries, and they all have nuclear weaponry to boot. Places like Israel and Pakistan may have many of those qualities, but not all. Certainly Israel, Pakistan, Turkey and Brazil are powerful regional entities, but I would not necessarily call them powerful global ones. That make sense?

All the power players, except India, are also on the UN Permanent Security Council, which means they all have a big voice in how that organization is run. For this section, just know this: these are all countries that have, or could, act unilaterally from a military standpoint. But that's not really the way things go down anymore.

ARMED ALLIANCES

Mostly, use of military force in global conflicts occurs as a coordinated effort between states as part of a bigger institution or entity. Most of these groups have already been discussed throughout this text, but let's hit them up one last time and make some predictions on not only future use of military power across the globe, but also how these entities will be working with, or against, each other in shaping global events.

THE UN

Most member states contribute troops, supplies, or money to the United Nations for its active military missions. Since any active mission must clear the UN Security Council, the countries described above get to decide where those missions will be allowed to happen. Have you figured out where virtually all UN missions have occurred? If you said "Africa," then give yourself a gold star! Why there? Because it's the only place that all the major powers can agree on with regard to having a UN presence. China would never allow a UN military mission to Burma (their ally), Russia would never allow a UN mission to Kazakhstan (their ally), and the US would never allow a UN mission to Canada—okay, maybe that one would fly, but you get the picture.

UN Secretary General Ban Ki-Moon.

The only places that UN troops end up in are typically poor, developing countries in which no major power has a vested interest, and therefore no one on the Security Council will veto. The Democratic Republic of the Congo, Liberia, Haiti, Lebanon, Côte d'Ivoire, Chad, Ethiopia, East Timor and Kosovo are all excellent examples of current UN deployments that lack vested interest of major power players. However, Kosovo should throw up a red flag to you, as I have explained earlier in this text how it has become a hot-button issue between world powers.

Dig this: that mission was agreed upon by Russia back in the late 1990s when it was in a significantly weakened state. Russia would never let something like that go down now! Same thing goes in reverse though: the UN mission to poor, undeveloped Sudan has been held up by China for years, because it formed a vested interest in the place when Chinese companies started developing Sudan's oil fields. Only increased international pressure has forced the Chinese to allow a mission to happen, but in a much more limited capacity; that is, a smaller force with more constraints on what it can do.

That brings us to the last point about military maneuvers of the UN: they are largely weak and ineffectual. The big powers can never all agree on any single issue, and therefore any mission which gets passed through the UN is typically watered down to the point of nothingness, if there is any action taken at all. Remember the Rwandan genocide? What a freakin' debacle! The major powers couldn't even form a plan of action on that huge mess, even though no one had a vested interest in the place. The entanglements of getting military missions passed through the UN system is the major reason that the US typically bypasses the UN altogether and goes straight for NATO.

NATO

NATO is still the most powerful and effective military arrangement on the planet right now, and the preferred choice of the US, since the US is the primary player and most influential member in the club. In other words, the US has a much easier time getting actions sanctioned by NATO than by the UN. You know all about Article 5 and how the group works, so let's cut to the chase.

NATO is essentially the military arm of Team West—yes, the whole Team, not just the US. Think about it. Makes sense, doesn't it? All members of NATO are part of Team West, and even Japan and Australia (non-NATO members) often help out NATO missions with supply and support services. This is what makes Team West so potent: a combination of common ideologies with an organized and powerful military structure. They can get the job done, and more often than not, can agree about what the job is.

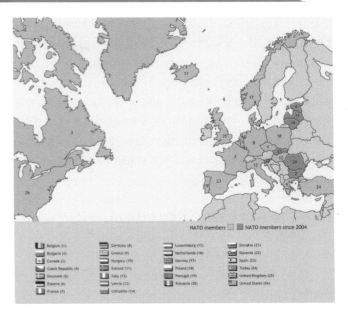

This is significant in today's world because NATO has done a major mission shift to now incorporate the War on Terror. As you know, after 9/11, the US invoked NATO Article 5 in order to get the club to help out with the Afghanistan mission, which they did. The current war in that state is a NATO war, not a US one. That sets the stage for many more NATO engagements in this global war on terrorism in the future. Granted, NATO will not dive head first into every US anti-terror effort; it is important to note that NATO did not support the current US war in Iraq. However, they may eventually come to the US's assistance in this situation as well. NATO has become stronger over the years and has increased its mission to a global reach, which is fitting since they are the most effective global military in the world. 2011 Update: NATO acts as the arm of Team West once again by organizing the charge against Muammar Qaddafi's Libya! Stay tuned for the climactic conclusion of that confusion! 2012 Update: Mission accomplished: Muammar now worm food.

What's not to love about NATO? If you don't necessarily agree with all the foundations of Team West, then you might be a bit leery of NATO's power. Russia is the obvious antagonist of NATO and continues to be irate about the US/NATO missile defense shield in Eastern Europe, which was officially launched in May 2012. In the past, Vladimir Putin openly threatened to target missiles at the Ukraine if they attempted to join NATO, and also to more generally target the nuclear arsenal at Europe as a whole if NATO proceeded with its missile-defense systems being built in Eastern Europe. Then there was that small matter of Russia invading Georgia before it could become a NATO member. And now Russia has absorbed the Crimean Peninsula from Ukraine, and has destabilized the country enough to warrant further invasion at Vladimir Putin's whim. In other words, this resurgent Russia's physical power plays back into Eastern Europe have proven to the world that Russia fully intends to re-exert its influence in, if not outright control of, its 'hood.

Why am I talking so much about Russia in this NATO section? Good question. Just this: just a year ago I would have told you that NATO is done expanding, and was being slowly de-funded by its broke European members, and that the group was receding in power and influence. But the Ukrainian unraveling and Russia's chest-thumping and expansionary tactics have once again breathed life into NATO, and is a born-again entity as well. NATO's sole reason for being created in the first place—to protect against Russian aggression—has been re-established and has reinvigorated the group. Even the European members are now anxious to use NATO to project a powerful front by increasing funding, increasing military exercises, and possibly even increasing NATO troop presence in the most threatened Eastern European member states, like Estonia, Latvia, Lithuania, and Poland.

Yep. Cold War is back on. But this time it is not an ideological battle between the US and the USSR, it is actually a more simplistic battle for territorial influence, and a drive by Russia to re-establish its street credibility, while also protecting/promoting its self-interests in the vicinity. Which is what big powers do. Long story short: Russia and NATO are once again full antagonists, and both will be spending more money on military maneuvers . . . and there is no hope whatsoever that Russia will be joining Team West, nor working with NATO anymore . . . because it appears that Russia has other things in mind. It is not looking to ever join NATO, because perhaps it has an alternative. . . .

SCO: SHANGHAI COOPERATION ORGANIZATION

Of course! Russia + China + all of Central Asia. That's the SCO: the anti-NATO! Okay, that is a bit too strong of a descriptor, but the Plaid Avenger wants you to be smart about the world, not politically correct. The heck with that! Let's get to the goods.

SHANGHAI COOPERATION ORGANISATION

■ Member States ■ Observer States ■ Dialogue Partners

Though the declaration on the establishment of the Shanghai Cooperation Organization contained a statement that it "is not an alliance directed against other states and regions and it adheres to the principle of openness," most observers believe that one of the original purposes of the SCO was to serve as a counterbalance to the US and NATO, and, in particular, to avoid conflicts that would allow the US to intervene in areas near both Russia and China. How are they pulling that off?

SCO states are strengthening their defense pacts with each other to ensure that no interference or invasion will be allowed from outsiders. In essence, they have formed an Asian version of NATO. The movement in this direction is why most US military bases have now been kicked out of SCO states, and also why the US request to become an observer state of the SCO has been declined every single year. It's also why the SCO has started doing the joint military exercises.

However, it is a very young organization, so I wouldn't call them a military powerhouse on par with NATO just yet. But the alliance does consist of two nuclear powers, Russia does have the second largest nuclear arsenal on the planet, and China does have the largest standing army. While I do not predict this entity doing anything aggressive anytime soon, I also do not think it's a stretch of the imagination that the US and/or NATO would be extremely hesitant to ever attack any of the SCO members, even the dinky ones, for fear that the defense alliance might strike back. I mean, that is the whole point of a defense alliance after all.

Let's pretend the US/NATO strikes a terrorist cell in Uzbekistan, which then forces the SCO to counter-attack. Sounds like a WWIII scenario to me! Don't get too worked up yet, though; these guys are still organizationally pretty weak, and their mettle is untested. Would Russia and China really counter a US attack? Only time will tell. Actually, let's hope we never have to find out.

The more important issues have to do with those observer states of the SCO, particularly Iran. Iran would just love to be a full-fledged SCO member so that it could have an insurance policy against western aggression. Iran thinks that the SCO membership will somehow shield it from any UN or NATO or US attack. But it's not a full member yet, and Russia and China are not likely to grant it that status, because no one involved wants to get into a global power battle over the antics of Persian Presidents or agitated Ayatollahs. Again, let's hope that is the case.

The SCO is a fascinating study on the future of Eurasia, and the world. Be sure to keep up with them as the entity continues to evolve. We already have two major power players in the club, as well as the entire region of Central Asia. India is also an observer member: a position which sets it up as a future power player on the globe. Iran really wants to be part of the club, which effectively moves SCO power and influence into the Middle East as well. These guys are just all over the place!

OUT OF AMMO

Summary: the UN is weak, NATO is strong and has a new-found energy since the Ukrainian debacle, and the SCO is evolving. All three will be the primary military alliances which shape future conflicts around the globe, both the small ones and the big ones. The UN is likely to expand the Permanent Security Council, which will make them even weaker from a military standpoint; more voices at the table means less things that can be agreed upon by all. NATO is nearing the end of its physical expansion period, having pulled in most of Eastern Europe, and stands more powerful than ever given its new objectives of countering Russia once more. The SCO is young and ineffectual so far, but that means it has the most growing to do, and possibly the greatest future potential of them all.

Of course, each individual state has its own military might, and the power players will continue to dominate the game in the global competitions. Smaller regional powers like Israel, Brazil, Pakistan, Saudi Arabia, Turkey, and Iran will continue to build up their military might and will affect regional politics in their prospective neighborhoods, but won't have global reach. I suppose I should also throw in smaller and smaller military entities like rebel and terrorist groups, which can shape regional or even global policy in general by their actions. I know they are the ones that seem to be the big deal every single day when you read news headlines, but the reality is that they are, militarily, just a pin-prick on world affairs. Many pin-pricks can add up though.

I did want to give a final shout-out to my AU military brothers as an up-and-coming force for regional power as well. The African Union as a whole may be largely ineffective in terms of economics or politics, but the AU military is starting to gain some steam. It is increasingly being used for peace-keeping efforts all around Africa—it is a critical part of the UN mission in Sudan—and is getting a lot of support from the big power players to keep on getting bigger and better. Indeed, perhaps it will be a strong and professional pan-African army that will help the region as a whole put itself back together, fight corruption, and build a stability that will allow for economic investment and growth. Wouldn't that be grand? Speaking of economics . . .

ECONOMIC POWER PLAYERS

Another facet of global competition is the quest for more dollars—more dollars for your country, more dollars for your country's businesses, more dollars for your country's citizens. Mo' money, mo' money, mo' money! Do we have Team play at the global scale for economics? Sure we do!

To be sure, economies are made up of businesses, which are made up of people, and all people have different personal incentives and goals and ambitions and opportunities to make money. I'm not going to get down to the abstract, philosophical reasons for why people do what they do, or even why countries do what they do. When the rubber hits the road, it's really every man for himself, or in this case, every state for itself, since states are first and foremost beholden to their citizens to make sure that their economy is as strong and rich as possible. Some states are obviously way better at this than

others, sometimes at the expense of other states. But that is the way the proverbial cookie crumbles.

However, even though every state looks out after itself first, some states have figured out that they get richer when working with other states. The economic centers of the planet do shift according to who is doing the best, producing the best, or growing the fastest. We can look at what entities and Teams are major forces for economic activity in the world, how they compete with each other, and what the future might bring in terms of cooperation or collision. Sounds like fun, huh? Then let's expound about some economies!

BIG BOYS WITH BIG TOYS: PART DEUX

Why not stick with the pattern? We can identify the individual states with the biggest economies, for starters. The US is by far and away the biggest economy in the world, larger than the next three economies combined. You know the next three: China, then Japan, then Germany. The rest of the top ten is rounded out with mostly rich European countries, but watch out! India and Brazil have jumped into top ten status too, and a simmering South Korea is making its way up the ranks, as is Indonesia, Mexico, and Turkey. Why should we care about these economies with big numbers?

Because money is power! Rich societies are typically stable and happy societies—typically. Who doesn't want to be in a rich society? Levels of richness effects all sorts of things like political power, military might, standards of living, but also global patterns like commodity chains, trade routes, distribution of economic activities, and even immigration/emigration patterns. I know, big shock: people migrate from poor states to rich ones. Go figure.

Where these super-economies are located also has a lot to do with how economic alliances go down, how competitive these states are with each other, and how powerful they can become. Don't believe me? Then let's revisit the EU.

EUROPEAN UNION

The EU has a dozen of the top world economies within its ranks, but individually, none of them can compete effectively with the US economically. But hold the phone! Take all of the EU states together as a single entity, and you are suddenly looking at the richest organization on the planet, even richer than the US in terms of GDP and GDP per capita. The EU as a singular entity has increased trade among its member states, increased trade abroad, and become much more effective at attracting international investment. That's what everybody wants to do! Which is why these supranational organizations based on trade have become so popular lately—and rich.

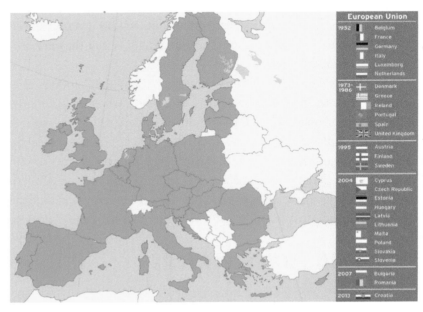

Before we get to that, a few more points on the EU. Like NATO, the EU seems to be at the end of its expansion streak. It has pretty much absorbed all of Eastern Europe, and the only likely future members are the states that were formerly Yugoslavia (no real economic gain for the EU here), the Ukraine (big potential gain here) and finally Turkey (big gain with some big risks here). The former Yugoslavian states are economically weak and have little to offer even when strong. The Ukraine is an agricultural and manufacturing powerhouse with a huge land area and population that would radically affect the EU. Turkey is almost exactly like Ukraine for all the reasons just listed, but promises to be divisive within Europe as a whole because of its Islamic culture. Look back to chapters on Western Europe and Turkey for more details.

The EU is at a breaking point where lots of people think they can simply not get any bigger and maintain any sort of effective control and economic cohesion, but they do have a lot to work with already.

2012 Update: Wow. Could that previous sentence be any more prophetic? The recession in the EU, followed by the complete economic debacle in Greece, now appear to be shaking the very foundations of the EU and especially the smaller sub-grouping of the Eurozone countries. With Greece likely to be kicked out of the Eurozone, and a handful of other EU states on the brink of bankruptcy, it seems all too apparent right this second that maybe the group has grown too big too fast and the vast economic differences between some of its member states are causing a huge rift in the fabric of the block itself. Stay tuned for new EU assessments as they unfold!

Creating this union has been the only thing not just keeping Europe afloat, but making it a global powerhouse to compete with the US and a rising China. The EU will continue to bust ass to fine tune its EU program—because it knows it has to in order to continue to compete with the economic prowess of Asia as a whole. Asia is on fire with economic growth, from Korea to Indonesia to India. The center of the economic game is shifting. For hundreds of years, the Europeans were the center of the global economy, eventually forfeiting their top slot to the powerhouse Americans who have held onto it for about a century. But nothing ever stays the same forever, and the 21st century promises to be the Asian century in terms of becoming the economic center of the universe.

This transition is already well underway, and the Europeans do not want to get left totally behind as the Asians start kicking economic ass and taking names. In the 1990s, when the USSR folded up shop and the EU expansion began in earnest, there was even talk of eventually incorporating Russia into the EU. What a continental powerhouse that would have been! Russia is such an important part of the EU economy that it is already invited to major summits and consultations on EU policy, even though it is not a member. **2014 UPDATE:** Yep, Russia is now out of any possible partnership with the EU, due to the Ukrainian situation which has pitted the old Cold War teams against each other once more, even economically. But the Europeans know they need Russian energy, so there is no way anyone is actually fully severing ties or anything drastic like that . . . it is more of a cooling of future investments, deterioration of trust between the parties, and a lack of any future potential for further integration. So, Russia in the EU? Right this second it doesn't appear that it will ever be, by choice, because the Bear is straddling the fence, already looking to the other side of the continent for alternative companionship, which brings us . . .

BACK TO THE SCO . . . AND A NEW UNION OF NOTE

What? Back to the SCO! Remember them? I've already told you that this group may eventually be an anti-NATO, but they also started the organization to be a regional trade block. With the Eurasian titans of Russia and China as the bookends, this is increasingly looking like the team to beat.

As you already know, the SCO is building infrastructure like roads and pipelines across the member states, mostly to facilitate natural resource extraction. Remember I told you about the possibility of them becoming a natural gas cartel akin to OPEC? That may very well be the building block that turns into increased trade among the nations, but more importantly will facilitate trade to other parts of the world. Central Asia has done well to throw in its chips with the SCO. Otherwise, they would be marooned in the middle of the continent fending for themselves. That trade block stuff really works for some states! Instead of being on their own, those states are now a central hub of export supplying petroleum products to the eastern and western sides of the Eurasian continent. Damn! Silk Road back in action!

More important than salvaging Central Asia is the blossoming relationship between Russia and China. You know from the political power players section above that both Team Bear and Team Dragon typically find themselves as the opposing end of a lot of Team West initiatives, so they have a common alliance of sorts already. Both states are avid supporters of sovereignty, and don't want anyone messing in their backyards. The evolution of a friendship between the two, and then the creation of the SCO, is a predictable development. Team West and the EU were hopeful that Russia was going to join their squads eventually, but it doesn't look like that is going to happen, perhaps ever.

Russia is strong again, but it's no powerhouse economy on its own. Hanging out with China certainly is beneficial to Russia, especially since the Chinese demand for natural resources and energy supplies is insatiable. The Russians find themselves in the enviable position of supplying energy to the powerhouse EU and also the powerhouses China and Japan. Their economy is thriving, and it probably will be getting a lot better, because as that world economic focus shifts to Asia, Russia stands to continue to benefit even more. The Bear will be increasingly dining on Moo Goo Gai Pan and sake shots as it rolls over to economically embrace the Dragon.

But don't let me suggest that Putin's Russia is just going to rely on selling more stuff to China to secure their economic future Oh no, my friends! He has much grander designs on the Asian 'hood than just that! As mentioned in several places throughout this titillating textbook, Putin's grand ambitions of restoring the old soviet order have now become reality! The Eurasian Union has been officially born!

While originally an idea posed by Kazak President Nursultan Nazarbayev, the Eurasian Union, is without a doubt the wet dream of Russian President Vladimir Putin, who now openly seeks to reassert a resurgent Russia's influence and outright power into areas it lost during the collapse of the Soviet Union in 1991. It is being proposed as a economic/political trade block alternative to the EU . . . and one that has Russia at the center. Signed into existence on 29 May 2014, between Russia, Belarus and Kazakhstan; Armenia has requested to join, Azerbaijan is debating it; other states are being pressured hard to join, or possibly face the wrath of Russia.

Think that is an exaggeration? Well, the whole unraveling of Ukraine hinged upon them getting ready to initiate talks to join the EU, which were then intentionally tanked by pro-Russian President Viktor Yanukovych, who of course wanted Ukraine to join this Russian-led Eurasian Union group. Of course this incited riots that led to his ejection . . . and then Russia's power play to take over parts of the country. Could this exact scenario happen in other states that Russia wants to pull back into its orbit? Maybe. Maybe not. But Russia has certainly put everyone on alert that they are not afraid to use force to get their way, and that has everyone in the vicinity thinking twice before siding up with Team West on any new initiative . . . even economic ones!

Putin's grand gamble here is that his union (and yes, it is referred to as his personal grand strategy at this point) will become a successful alternative to the European Union, and that states of Central Asia, Eastern Europe, and even the Caucuses Region will want to sign up for it. I suppose it is being marketed as a great way to boost economic growth, without having to bother with all that pesky democracy and human rights stuff. And it is definitely being pushed as an alternative to joining Team West. And it definitely is Russia trying to regain "influence" in areas it previously held as its own. It is not a stretch of the imagination to see that Putin is trying to reassemble—at a minimum—the old Commonwealth of Independent States, which was an economic working group of previously SSR's of the old USSR after it collapsed.

So even economically, Russia is pulling its old team back together to counter the west, and pushing its new team to integrate even further to counter the west. See what I mean now? For many countries in Eastern Europe and Eurasia, they will be hard pressed to jump into the Eurasian Union, or the SCO . . . both of which are alternative power structures to the Team West tool-kit. Will it work? I am still reserved about the chances of a true Russian revival, but no one is disputing that the Dragon will continue to breath fire for centuries to come . . .

THE DRAGON AS ASIAN LEADER

I've been stressing that Asia as a whole is on fire with economic activity, and with no doubts you know that China is the main engine for all this explosive exponential economic energy. Team Dragon has been cranking out double digits in terms of increases in GDP growth year after year for over a decade. That is crazy talk

growth! Their manufacturing, industrial and service sectors have been blowing up as they produce everything from toothpaste to Nike shoes to rocket parts for NASA spacecraft. They have become the workshop of the world.

Why am I bothering to repeat this? Because with no reservations, I tell you that China will become the world's largest economy in the next fifty years. It will displace the US from the top slot, although you have learned enough from this book already to know that this shift does not equate to China being truly richer or better off than the US. They will have the biggest economy, and more importantly, will continue to be a focus for international investment, as well as increasingly a center for evolution of new technologies.

Here is the real deal about being a world economic power player and the focus of world economic energies: more things happen in your neighborhood. Huh? You know what I'm sayin'! With this increase in economic energy comes an increase in interest, an increase in investment, an increase in education, and an increase in technologies. The best and smartest people are attracted to where the flame is burning the brightest. The newest innovations in all sorts of technologies almost invariably occur in hot, happening societies where the action is, and that center of action is increasingly Asia!

Team Dragon is being looked upon as not only a very desirable trading partner among other Asian states, but as a model of success to be duplicated. Most Southeast Asian economies are now tied into China, as are the Koreas, and even the Central Asian countries via the SCO as described above. China is truly regaining its historical legacy as the center of the Asian universe.

Tigers back with the Dragon!

Asian countries are not alone. Japan (okay, they are Asian too, but very western!) is turning more to trade, invest, and compete with China, as is Russia, Australia, and the US. Japan and Australia in particular have publicly pronounced that they know they must work together more economically with China if they are to continue to thrive; both countries' foreign policy initiatives have shifted immensely to suck up to China more, even though both are still part of Team West. Hey man, money talks!

I call it Team Dragon because I want you to know that it really has become more of a team affair. Yes, China is the center of the action—but remember, there are other economic powerhouses in this 'hood. Taiwan, South Korea, Singapore, and Hong Kong were once titled the "Asian Tigers" because they underwent decades of phenomenal industrialization and economic growth. Those tigers have now come home to roost under mother Dragon. This place is a veritable smorgasbord of animated animals, and they are getting rich! Throw Japan into the lot, and you have a whole cluster of top world economies here—with one big difference:

Most top twenty economies are still in Europe, and Europe is about played out. The rest of the top twenty world economies are Asian, and they are just getting warmed up! Look for Asian states to totally dominate the world economic measures of wealth in the coming decades. For sure, China will soon be number two, Japan will be number three, and South Korea will be number ten. And they are all next door neighbors! Indonesia is a rising regional titan, India is about to be reincarnated as a vibrant economy, and Vietnam is a dark horse with a big stride ahead of it as well. Are you starting to get a fever for the flava' of East Asia?

To sum up: there will be major growth, competition, and shifts in economic focus and economic policy around the world as Asia rises. Don't get all sad on me—your region is still good! The US and Europe will still be rich, don't worry about that! There are just a few other entities I want mention before escaping this escapade.

SOME OTHER ECONOMIC ENTITIES OF NOTE

I've mentioned ASEAN back in the chapter on Southeast Asia; it is a supranationalist organization with more potential for economic growth than I can imagine right now. Smack dab in the middle of the powerhouse China and Japan to the north and growing power India to its west, all of which are associate members, the ASEAN core might well become the epicenter of an Asian EU. Wow. That would be insane. It's still some ways off, but the point is that this group is in a position of it being almost impossible for them to NOT grow economically right now. Southeast Asia is starting to absorb a lot of manufacturing jobs not just from abroad, but also directly from China. It's good to be close to the top dogs. Speaking of Asian top dogs . . .

As with the rise in political clout, Team Hindu stands to become an economic powerhouse in its own right, although it is going to take a bit longer than their Asian brothers. It's that mondo big population that makes things problematic for them to advance as fast! But advance they will nonetheless, but for different reasons than others. Namely, India is fast becoming a high-end technological center for growth in Asia and the world.

China and other Asian nations have focused first on becoming agri-cultural and manufacturing giants, then incorporating more and more service sector and quaternary activities as they get richer. India is on rather the reverse course: it focuses heavily on increasing its service sector, and especially on new technology stuff. Computer programming, information technology, and engineering applications are fields that India is specializing in immediately. While international business may be attracted to China or Indonesia for their low wages, international business is attracted to India for their lower wages AND highly educated and skilled service sectors. Why pay a European programmer when you can pay an Indian programmer half as much? Look for Team Hindu's star to rise fast in our technologically-focused world.

Finally, I will reference the Latin Lefties. Team Lefty if you like. You can love him, you can hate him, or you can depose him in an illegal coup, but Hugo Chavez represents this Latin America shift to the left that has taken on a bit of a revolutionary flavor.

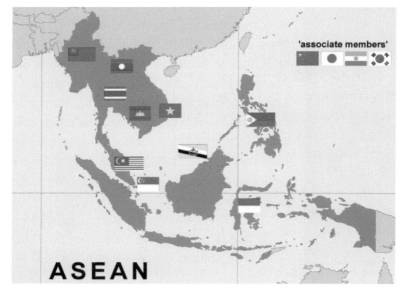

'associate members'

ASEAN

While the leftward swing is a political thing, it also heavily incorporates an economic component, which is why I include it here in this section. I won't go so far as to suggest that Team Latin Left is an economic power of global significance, but it is one of global interest, since it represents a significant shift in thinking and strategic economic alliances.

What am I talking about? Team Lefties have a common denominator: a desire for greater social services and wealth redistribution via the state structure. They aren't communist, but some are pretty far to the left side of the Socialist spectrum. The so-called Bolivarian Revolution has resulted in a couple of things I want you to note: a movement away from outside free trade unions, namely the FTAA, which is currently dead in the water. That is related to the second big thing: a Team Lefty shift away from the US as its sole economic influence, and shifting toward China, but also the Middle East and Sub-Saharan Africa.

Interesting turn of events. Need proof? In May 2008, China joined the board of the Inter-American Development Bank, which is a major source of funds for Latin American countries

Just look at all that leaning!

to finance projects for the social and economic development for the region. Bam. First Asian member of that group. Uncle Sammy ain't happy about losing influence in his backyard, and Team Dragon doesn't mind spreading its love around.

FINALLY THE FINAL FINAL FINAL SUMMARY

We have identified the major forces that shape the future of our planet by battling it out every day politically, militarily, and economically, or sometimes cooperating with each other. Either on their own or within international organizations, the most powerful sovereign states plan, buy, sell, trade, agree, disagree, fight, make peace, plot, unify, embargo, or smack-down each other in various ways to achieve their objectives. It's not a pretty process, but it's the only one we have until smarter folks like you make it better.

Does it appear that I have forgotten large parts of the globe? Oh no, my friends, I have not forgotten anyone! This chapter was about major movers and shakers, and that list is not all inclusive. Sub-Saharan Africa is at the whim of the international system until they straighten themselves out internally . . . and watch out, because that is actually happening now! Sub-Saharan Africa is actually heating up economically, and has had consistent growth for the last decade! Go for it guys! The Middle East as a region is doomed for eternal conflict and a variety of debilitating internal problems until they get their business straight, and it looks like they got a ways to go before that sun rises. Most of the sub-regions of Latin America are doing pretty good with consistent economic growth and political stability, but do not impact world events in a major way. Eastern Europe has effectively become part of Western Europe. Regions like Turkey, Japan and Australia play more of a role within the ranks of Team West than they do on their own.

But history is not done, my friends! Unpredictable events can shake things up quickly in today's world and change the course of states, or regions, or the whole planet. Even with my super-human powers of insight and heightened fashion sense, I could not have predicted the attacks of 9/11, the massive Japanese quake and nuclear disaster of 2011, the "Arab Spring" still unfolding, or the horrible outcome of the final Indiana Jones movie release. Events like these have huge repercussions for the regions in which they take place, but are also felt throughout the world in the form of global policies and actions that are implemented due to the event itself.

That's why it is important to keep up with your planet. You have taken a very large step in the right direction by reading this book. Congratulations on that feat of endurance and patience. But the game is far from over. Stay involved and educated about is going on in the world, so that when the time comes, you can make the biggest impact for positive change, no matter what it is you decide to do in life.

Lots of people ask me how I keep up with the events and actions of the planet. Well, I do a lot of on-the-spot super hero action within these regions on a daily basis, but when I'm at home in the underground lair, here are a few of the Web sites I keep up with, so that I can keep up with the world:

→ http://news.bbc.co.uk/ for daily lightweight international news

→ http://www.iht.com/ for more in-depth daily international news

→ http://www.worldpress.org/ for deep in-depth weekly assessments of major news events. On this site, you will also find links to every single digital newspaper on the planet from every single country that has them.

You can tune in to CNN, MSNBC, and Fox News Channel anytime to check out the daily arcane domestic political party finger pointing nonsense, and up to the minute status on whatever celebrity drama is making headlines that day. Mainstream American news sources are not really worth much else.

Always read at least two different news sources for every story you are learning about; make sure one news source is from inside the country where the news happens, and one is outside the country where the news happens. It's the only way to get a balanced view of what's actually happening on the planet.

For those of you who want to do more, here are some ideas and things you can do.

Ever heard of micro-finance? It is an awesome and easy way to help the poor for those of you who may not have time to volunteer. Sometimes charities and organizations get so big and wrapped up in politics that the money never actually gets to those who really need it. What microfinance does, is supplies small loans (sometimes it's even as little an amount as $25) to individuals (usually women, 'cause let's face it guys, when it comes right down to it, women do tend to be the most responsible sex) to help them start small businesses. For example, a source close to the Plaid Avenger was working with a group of Haitian women in the Dominican Republic who were trying to support themselves by making and selling candles. These women were making candles one at a time and it was taking them forever. But, if they had someone who could loan them the money to buy a candle mold, they could make a lot more candles in a lot less time and increase the revenue as a result. Get it? With microloans, a small amount of money can go a long way in helping change and improve lives but with dignity—because it's not a handout, the recipient has to pay the loan back. What my friend in the DR did, was have them pay the loan back to a community account, so that money could stay within the community and be there for the next person needing a loan—recycling money for good—I love it!

If you are interested in micro-lending, check out this site—you can even donate under Team Plaid!

→ http://www.kiva.org/team/team_plaid

Not into micro-lending and want to choose a cause instead? Here's another cool website that allows you to give to a grassroots project of your choice.

→ http://www.globalgiving.org/dy/fundraiser/helpaddfund/gg.html?projid=4185

To be unaware and ill-informed about the world around you is to be a passive player in the game of life. Keep your head in the game. Stay informed. Read. Pay attention. Write a letter. Protest. Help someone in need. Start a revolution. Or just do anything to make the world a better place. Peace.

And party on!

. . . to be globally engaged and informed!!!

Image Credits for Selected Chapters

CHAPTER 1

Page 21: © 2007 JupiterImages Corporation; **Page 22:** map courtesy of Katie Pritchard, flags from *The World Factbook*; **Page 23: top:** map courtesy of Katie Pritchard; **bottom:** map courtesy of Katie Pritchard; **Page 24:** map courtesy of Katie Pritchard; **Page 25:** map courtesy of Katie Pritchard; **Page 26:** maps courtesy of Katie Pritchard; **Page 27:** OECD logo used by permission; **bottom:** © 2007 JupiterImages Corporation; **Page 28: top:** © NATO; **bottom:** © Royalty-free/CORBIS; **Page 29:** NATO logo used by permission of NATO; **bottom:** © NATO; **Page 30:** map courtesy of Katie Pritchard; **Page 21:** courtesy of Katie Pritchard; **Page 32** *The World Factbook*; **Page 33: top:** © AP Photo/Hussein Malla; **middle:** Organization of American States; **Page 34: top left & right:** Copyright © African Union, 2003. All rights reserved. Used by permission; **middle right:** U.S. Air Force photo by Tech. Sgt. Jeremy T. Lock; **flags at bottom:** *The World Factbook*; **Page 35:** flags: *The World Factbook*; **bottom:** © European Communities, 2009; **Page 36: top:** © European Communities, 2009; **bottom:** Roberto Stuckert Filho/PR/ABr; **Page 37:** WTO logo used with permission; **Page 40:** map courtesy of Katie Pritchard; **Page 41:** Government of Argentina

CHAPTER 3

Page 82: photo ca. 1867, gift of Oliver Wendell Homes, from Library of Congress; **Page 83:** maps by CIA, courtesy of University of Texas Libraries; **Page 84:** all maps: from CIA, courtesy of University of Texas Libraries; **Page 85: top:** map courtesy of Katie Pritchard; **bottom:** Library of Congress; **Page 86: top:** map source: *The World Factbook*; **bottom:** photo from the Frank & Frances Carpenter Collection, Library of Congress; **Page 87: top:** © 2008 JupiterImages Corporation; **bottom:** © Jose Miguel Hernandez Leon, 2012. Used under license of Shutterstock, Inc.; **Page 88: top:** Library of Congress; **bottom:** scanned from Helmolt, J.F. ed. *History of the World* (New York, Dodd, Mead & Co., 1902); **Page 89: top:** courtesy of Brazilian Embassy; **bottom left:** Library of Congress; **left center:** Fabio Rodrigues Pozzebom/ABr; **center:** courtesy of Secretaria de Imprensa, Brasil; **right center:** photo by Warren K. Leffler, from Library of Congress; **right:** Government of Argentina; **Page 90: top:** photo ca. 1900, from Library of Congress; **bottom:** from *Judge, February 15, 1896*, Library of Congress; **Page 91: top:** photo ca. 1900, from Library of Congress; **bottom:** photo by Elias Goldensky (1933), from Library of Congress; **Page 92:** photo by Warren K. Leffler, from Library of Congress; **Page 93:** © Shutterstock, Inc.; **Page 95:** map courtesy of Katie Pritchard; **Page 96:** Government of Argentina; **Page 98:** maps courtesy of Katie Pritchard; **Page 99:** all photos courtesy of Secretaria de Imprensa, Brasil, except the far right photo, which is courtesy of Wilson Dias/ABr; **Page 100: left:** painting by Ricardo Acevedo Bernal; **left center:** The White House Historical Association (White House Collection); **right center:** National Archives & Records Administration; **right:** Library of Congress

CHAPTER 5

Page 151: top: © 2009 JupiterImages Corporation; **bottom:** © 2007 JupiterImages Corporation; **Page 152: top:** © Shutterstock, Inc.; **bottom:** map courtesy of Katie Pritchard; **Page 153: top:** CIA map, courtesy of University of Texas Libraries; maps from *The World Factbook*; **Page 154:** maps from *The World Factbook*; **right:** by Elihu Vedder, 1896, Library of Congress; **bottom:** stereo by Underwood & Underwood, 1907, Library of Congress; **Page 155:** maps from *The World Factbook*; **bottom:** map from *The Cambridge Modern History Atlas*, 1912, courtesy of University of Texas Libraries; **right:** © 2007 JupiterImages Corporation; **Page 156:** from George Grantham Bain Collection, Library of Congress; **Page 157: top:** from George Grantham Bain Collection, Library of Congress; **bottom:** map source: *The World Factbook*; **Page 158:** from George Grantham Bain Collection, Library of Congress; **Page 159: top:** © JupiterImages Corporation; **bottom:** Dept. of Defense photo by R.D. Ward; **Page 160:** Dept. of Defense photo by Tech Sgt. Jim Varhegyi; **Page 161:** map courtesy of Katie Pritchard; **Page 162: top:** Government of Argentina; **bottom:** CIA map, courtesy of University of Texas Libraries; **Page 163:** Dept. of Defense photo by PH2 (AC) Mark Kettenhofen, U.S. Navy; **Page 164: top:** © NATO; **bottom:** © NATO; **Page 165: top:** Dept. of Defense photo by SSgt. Jeremy T. Lock, U.S. Air Force; **bottom:** © 2007 JupiterImages Corporation; **Page 166:** © NATO; **Page 167: left:** from George Grantham Bain Collection, Library of Congress; **center:** © European Communities,

2004; **right:** © European Communities, 2009; **Page 168: top:** Dept. of Defense photo by TSGT Jim Varhegyi; **bottom:** map scanned from *Atlas of the Middle East* (CIA, 1993), courtesy of University of Texas Libraries; **Page 169:** photo from the George Grantham Bain Collection, Library of Congress; **Page 170:** maps from *The World Factbook*; **Page 171:** maps from *The World Factbook*; **bottom:** photo from Frank & Frances Carpenter Collection, Library of Congress; **Page 172:** map courtesy of Katie Pritchard; **Page 173:** White House photo; **Page 174: top:** photo from George Grantham Bain Collection, Library of Congress; **bottom:** map from *Historical Atlas* (1911) by William Shepherd, courtesy of University of Texas Libraries; **Page 175:** CIA map, courtesy of University of Texas Libraries; **Page 176: left:** © Bettmann/Corbis; **right:** © Bettmann/Corbis; **Page 177: top:** CIA map adapted by Katie Pritchard, courtesy of University of Texas Libraries; **bottom:** © Daniella Zalcman; **Page 178: top:** ca. 1917, from Library of Congress; **bottom:** Dept. of Defense photo; **Page 180:** CIA map, courtesy of University of Texas Libraries; **bottom:** Dept. of Defense photo; **Page 182: top left:** National Archives & Records Administration; **top left center:** Dept. of Defense photo; **top right center:** © Bettmann/Corbis; **top right:** Dept of Defense photo; **middle left:** Saudi Information Office; **middle left center:** © European Communities, 2010; **middle right:** © European Communities, 2004; **bottom left:** Dept. of Defense photo by R.D. Ward; **bottom right:** © European Communities, 2009; **Page 183: left:** © Mohsen Shandiz/Corbis; **center:** photo from Official Russian Presidential Press and Information Office; **right:** © DENIS BALIBOUSE/Reuters/Corbis

CHAPTER 7

Page 219: top: © 2008 JupiterImages Corporation; **bottom:** map source: *The World Factbook*; **Page 220:** map source: *The World Factbook*; **Page 221: top:** maps: *The World Factbook*; **bottom:** map source: *The World Factbook*; **Page 223: middle:** photo © 1941 by J. Russell & Sons, from Library of Congress; map source: *The World Factbook*; **Page 224: right:** map source: *The World Factbook*; **left:** map source: *The World Factbook*; **Page 225: bottom:** source: *Nuclear Weapons and NATO: Analytical Survey of Literature* by U.S. Dept. of the Army, courtesy of University of Texas Libraries Perry-Castaneda Map Collection; **Page 227:** maps source: *The World Factbook*; **Page 227:** map courtesy of Katie Pritchard; **Page 228: top:** courtesy of Katie Pritchard; **bottom:** map by U.S. Central Intelligence Agency, courtesy of University of Texas Libraries, Perry-Castaneda Map Collection; **Page 230: left:** Dept. of Defense photo by Tech. Sgt. Cedric H. Rudisill, US. Air Force; **left center:** © NATO; **right center:** © NATO; **right:** photo from Official Russian Presidential Press and Information Office; **Page 231 top:** © NATO; **bottom:** © Shutterstock, Inc.; **Page 232 top:** © 2008 JupiterImages Corporation; **bottom:** © 2008 JupiterImages Corporation; **Page 233:** © 2008 JupiterImages Corporation; **Page 234: top:** from *Former Yugoslavia: A Map Folio* (1992) by CIA, courtesy of University of Texas Libraries; **bottom: left:** White House Photo Office Collection (1971), from Library of Congress; **bottom right:** © Lucas Jackson/Reuters/Corbis; **Page 236:** CIA map, courtesy of University of Texas Libraries; Page 237: map by Katie Pritchard

CHAPTER 9

Page 293 top: map courtesy of Katie Pritchard and NOAA; **bottom:** Dept. of Defense photo by Mass Communication Specialist Seaman Bryan Reckard, U.S. Navy; **Page 294: top:** Dept. of Defense photo by Tech. Sgt. Rob Marshall; **middle:** *Blue Marble: Next Generation* image produced by Reto Stockli, NASA Earth Observatory (NASA Goddard Space Flight Center); **Page 295: top:** Tsuta-ya Kichizo (1858), from Library of Congress; **middle:** U.S. Navy photo by Photographer's Mate 2nd Class Nathanael T. Miller; **bottom:** Library of Congress; **Page 296: top:** Library of Congress; **bottom:** Dept. of Defense photo by PHI (AW) M. Clayton Farrington, U.S. Navy; **Page 297:** Heinoya (1870), from Library of Congress; **Page 298: top:** Dept. of Defense photo; **bottom:** art by Rivinger (1860), from Library of Congress; **Page 299:** map source: *The World Factbook*; **Page 300:** photo by U.S. War Department Signal Cops, from Library of Congress; **Page 301:** Office for Emergency Management, Office of War Information; **Page 302: top:** image created by James Montgomery Flagg for Office of War Information Domestic Operations Branch, from National Archives & Records Administration; **bottom:** from the George Frantham Bain Collection, Library of Congress; **Page 303: top left:** photo by Office of War Information Overseas Operations Branch (August 9, 1945), from National Archives & Records Administration; **top right:** Dept. of Defense photo, Dept. of the Navy, U.S. Marine Corps; **bottom right:** photo by U.S. Army, from Library of Congress; **Page 304: left:** photo from NASA; **left center:** photo ca. 1918, from Library of Congress; **right center:** © European Communities, 2006; **right:** Dept. of Defense photo by D. Myles Cullen; **Page 305:** all photos © 2007 JupiterImages Corporation; **Page 306: top left:** © 2008 JupiterImages Corporation; **middle right:** Dept of Defense photo; **bottom left:** Dept. of Defense photo; **Page 307:** Dept. of Defense photo; **Page 308: left:** Dept. of Defense photo by Helen C. Stikkel; **right:** Dept. of Defense photo by D. Myles Cullen; **Page 309:** both are © Shutterstock, Inc.; **Page 310: top:** © Shutterstock, Inc.; **bottom:** Library of Congress; **Page 311: left:** Harris & Ewing Collection, Library of Congress; **left center:** Library of Congress; **right:** Dept. of Defense photo; **Page 312: left:** © European Communities, 2004; **right:** Dept. of Defense photo by D. Myles Cullen

CHAPTER 10

Page 246: top: © 2007 JupiterImages Corporation; **bottom:** adapted from *Blue Marble: Next Generation* image produced by Reto Stockli, NASA Earth Observatory (NASA Goddard Space Flight Center); **Page 247:** created by Strobridge Lithography Co. (ca. 1896), from Library of Congress; **Page 248: top:** © Shutterstock, Inc.; **center left:** scanned from Helmolt, J.F. ed. *History of the World* (New York, Dodd, Mead & Co., 1902); **center right:** © Shutterstock, Inc.; **bottom:** map source: *The World Factbook*; **Page 249: top:** created by W. Holland (1803), from Library of Congress; **center top:** engraving by A. Muller (1879), Library of Congress; **center bottom:** scanned from Helmolt, J.F. ed. *History of the World* (New York, Dodd, Mead & Co., 1902); **bottom:** map source: *The World Factbook*; **Page 250:** from the George Grantham Bain Collection, Library of Congress; **Page 251:** lithography by M.A. Striel'tsova (ca. 1918), from Library of Congress; **Page 252:**

bottom left: *New York Times*, 1919, Library of Congress; **left center:** *Tsar Nicholas II* (1915) by Boris Kustodiyev; **right center:** Library of Congress; **right:** photo ca. 1909; **Page 253: left:** Library of Congress; **right:** photo from *Liberty's Victorious Conflict: A Photographic History of the World War* by The Magazine Circulation Co., Chicago, 1918; **Page 254:** map source: *The World Factbook;* **bottom:** from New York World-Telegram & the Sun Newspaper Photograph Collection, Library of Congress; **Page 255:** U.S. Signal Corps photo, from Library of Congress; **Page 256:** photo by U.S. Office of War Information Overseas Picture Division, Library of Congress; **Page 257:** map source: *The World Factbook;* **Page 258:** photo by U.S. Office of War Information Overseas Picture Division, Library of Congress; **Page 259 top:** © Shutterstock, Inc.; **bottom:** © 2007 JupiterImages Corporation; **Page 260:** official White House portrait; **Page 261: top left:** Library of Congress; **top center:** U.S. Signal Corps photo, from Library of Congress; **top right:** photo from Franklin D. Roosevelt Library, Library of Congress; **bottom left:** White House Photo Collection, Library of Congress; **bottom left center:** © Bettmann/Corbis; **bottom right center:** © Bettmann/Corbis; **bottom right:** White House Photo Collection, Library of Congress; **Page 263:** White House Photo Collection, Library of Congress; **Page 264:** photo by Earle D. Akin Co, 1909, from Library of Congress; **Page 267:** map courtesy of Katie Pritchard; **center:** photo from Official Russian Press and Information Office; **bottom:** © NATO; **Page 268 top:** © NATO; **middle:** photo from Official Russian Press and Information Office; **bottom:** photo from Official Russian Press and Information Office; **Page 269:** map from U.S. Dept. of Energy; **Page 270:** © NATO; **Page 271:** Produced by the Office of The Geographer and Global Issues, Bureau of Intelligence and Research, US Dept. of State, courtesy of University of Texas Libraries; **bottom:** © Shutterstock, Inc. **Page 272 top:** © Shutterstock, Inc.; **bottom:** cover of *The Great Train Robbery* by Scott Marble, 1896, from Library of Congress; **Page 273: left:** George Grantham Bain Collection, Library of Congress; **right center:** U.S. Signal Corps photo, from Library of Congress; **right:** © NATO; **Page 274: left:** © European Communities, 2009; **left center:** © European Communities, 2009; **right center:** © NATO; **right:** © NATO

CHAPTER 13

Page 366: left: *Blue Marble: Next Generation* image produced by Reto Stockli, NASA Earth Observatory (NASA Goddard Space Flight Center); **top right:** NASA/Goddard Space Flight Center Scientific Visualization Studio; **middle:** © Regien Paassen, 2012. Used under license of Shutterstock, Inc.; **Page 367:** © Pichugin Dmitry, 2012. Used under license of Shutterstock, Inc.; **Page 358:** all images: © 2007 JupiterImages Corporation; **Page 369:** © 2007 JupiterImages Corporation; **Page 370: top:** ca. 1915, Library of Congress; **bottom:** © Shutterstock, Inc.; **Page 371: middle:** © Debra James, 2012. Used under license of Shutterstock, Inc.; **bottom:** © Shutterstock, Inc.; **Page 372 top:** © 2007 JupiterImages Corporation; **bottom:** © Gavran333, 2012. Used under license of Shutterstock, Inc.; **Page 373:** © 2007 JupiterImages Corporation; **Page 374:** stereograph (1919), from Library of Congress; **Page 375:** lithography by Cincinnati Lithography Co., from Library of Congress; **Page 376:** all images © Shutterstock, Inc.; **Page 377: left:** Dept. of Defense photo by Robert D. Ward; **center:** Dept. of

Defense photo by Cherie Cullen; **right:** Dept. of Defense photo by Cpl. Mark Doran, Australian Defense Force

CHAPTER 16

Page 433: CIA map (2014), courtesy of University of Texas Libraries; **Page 434: left:** *The Historical Atlas* by William R. Shepherd, 1923, courtesy of University of Texas Libraries Perry-Castaneda Map Collection; **right:** *The Public Schools Historical Atlas* edited by C. Colbeck, published by Longmans, Green, and Co. 1905, courtesy of University of Texas Libraries Perry-Castaneda Map Collection; **Page 435:** CIA, courtesy of University of Texas Libraries Perry-Castaneda Map Collection; **Page 436: top:** CIA, courtesy of University of Texas Libraries Perry-Castaneda Map Collection; **bottom:** *The Ottoman Empire, 1801–1913* by William Miller. Published by Cambridge University Press, 1913, courtesy of University of Texas Libraries Perry-Castaneda Map Collection; **Page 437:** CIA map, (2014), courtesy of University of Texas Libraries; **Page 438:** map courtesy of Katie Pritchard; **Page 439:** maps courtesy of Katie Pritchard; **Page 441:** © Shutterstock, Inc.; **Page 442:** map courtesy of Katie Pritchard; **Page 443: top left:** Dept. of Defense photo by Tech. Sgt. Cedric H. Rudisill, US. Air Force; **top left center:** © NATO; **top right center:** © NATO; **top right:** photo from Official Russian Presidential Press and Information Office; **bottom:** map courtesy of Katie Pritchard; **Page 444:** © European Community; **Page 445:** both images © Shutterstock, Inc.; **Page 446:** both images © Shutterstock, Inc.; **Page 447:** © Shutterstock, Inc.; **Page 448:** © Shutterstock, Inc.; **Page 450:** CIA map, (2014), courtesy of University of Texas Libraries; **Page 451:** map courtesy of Katie Pritchard; **Page 452:** both images © Shutterstock, Inc.; **Page 453:** © Shutterstock, Inc.

CHAPTER 17

Page 456: © 2007 JupiterImages Corporation; **Page 457:** map courtesy of Katie Pritchard; **Page 458:** map source: *Former Yugoslavia: A Map Folio*, 1992 (CIA), courtesy of University of Texas Libraries; **Page 459:** Produced by the Office of The Geographer and Global Issues, Bureau of Intelligence and Research, US Dept. of State, courtesy of University of Texas Libraries; **Page 461:** map courtesy of Katie Pritchard; **Page 462:** CIA map (1988), courtesy of University of Texas Libraries; **Page 463:** CIA map, courtesy of University of Texas Libraries; **Page 464:** from *Global Trends 2015: A Dialogue About the Future With Nongovernment Experts* [page 73], National Intelligence Council, 2000, courtesy of University of Texas Libraries; **Page 465:** CIA map (2003), courtesy of University of Texas Libraries; **Page 466:** CIA map (1988), courtesy of University of Texas Libraries; **Page 467:** CIA map (2008), courtesy of University of Texas Libraries; **Page 468:** map courtesy of Katie Pritchard; **Page 469:** CIA Map (1992), courtesy of University of Texas Libraries; **Page 470: top:** © Daniel Leal-Olivas/Corbis; **bottom:** CIA map (1992), courtesy of University of Texas Libraries; **Page 471:** CIA map (1976), courtesy of University of Texas Libraries; **Page 472:** CIA map (2007), courtesy of University of Texas Libraries; **Page 473:** map courtesy of Katie Pritchard; **Page 475:** CIA map, courtesy of University of Texas Libraries; **Page 476:** both CIA maps, courtesy of University of Texas Libraries; **Page 477:** map courtesy of Katie Pritchard; **Page 479:** map courtesy of Katie Pritchard; **Pages 480 and 481:** both CIA maps, courtesy of University of Texas Libraries; **Page 482:** CIA map, courtesy of University of Texas Libraries; **Page 483:** CIA map, courtesy of University of Texas Libraries; **Page 484:** © Shutterstock, Inc.

CHAPTER 18

Page 486: lithograph by Illinois Co., Chicago (1917), from Library of Congress; **Page 487:** ca. 1917, from Library of Congress; **Page 488:** © 2008 JupiterImages Corporation; **Page 489:** © 2008 JupiterImages Corporation; **Page 490:** © 2008 JupiterImages Corporation; **Page 491: top:** Dept. of Defense photo; **bottom:** © 2008 JupiterImages Corporation; **Page 492: top:** © NATO; bottom: NATO logo used by permission of NATO; **Page 493:** map courtesy of Katie Pritchard; **Page 494:** courtesy of Katie Pritchard; **Page 495: bottom:** © 2007 JupiterImages Corporation; **Page 496:** map courtesy of Katie Pritchard; **Page 498:** both maps courtesy of Katie Pritchard; **Page 499: top:** © 2008 JupiterImages Corporation; **bottom:** © 2007 JupiterImages Corporation; **Page 500:** map courtesy of Katie Pritchard; **Page 501:** map courtesy of Katie Pritchard; **Page 503:** © Shutterstock, Inc.

Glossary

adiabatic lapse rate—The loss of energy in the atmosphere as a function of elevation.

AIDS—Acquired immune deficiency syndrome. A deadly disease due to viral infection.

attenuated—An extended or "long" bordered state. An example is Chile.

atmospheric flows—Directed wind or ocean currents.

autocracy—A government where an individual has total power.

Barbadian Slave Code—The legal basis in English law, which ensured perpetual (permanent) slavery until abolished in 1833 throughout the British Empire in large measure due to the work of Christian activists.

bellicosity—The act of being war-like or threateningly hostile.

capital—Any city exercising political authority and hegemony or control over the surrounding areas within a state.

capitalism—An economic system whereby the means of production or ownership of wealth is controlled by private individuals.

caste system—A system of hereditary social status in India and associated with Hinduism.

centrifugal—Cultural forces that act to pull people apart.

centripetal—Cultural forces acting to pull people together.

counter movement—A movement of people usually in a direction opposite of an earlier movement due to related circumstances.

cultural diffusion—The spreading of a way of life generally considered to be in any direction from a cultural hearth.

cultural hearth—The origin of a particular aspect of culture such as a cultural complex (graduation ceremony), ritual (marriage) or tradition (saying prayers at the dinner table).

culture—How humans live, work, worship and act. The transmission of culture from one generation to another is education. The development of culture, its diffusion and the display of rituals and complexes are often survival mechanisms developed to meet the challenges of the physical environment.

curse of South America—Term used to describe the difficulties of establishing networks due to the challenges of physical geography and landforms.

deciduous—Refers to forests where trees generally lose their leaves in fall.

democracy—A system of government where sovereignty rests ultimately with the people.

developing economy—The creation of wealth generally by harvesting resources such as timber, coal or farming.

diaspora—The historical term used to describe the dispersal of the Jewish people from their homeland by the Babylonians.

drawn borders—Artificially constructed lines drawn on maps typically reflective of the western imperial powers after the age of exploration and reflects their attempts to organize and administratively control territory.

dredge—A verb that describes an attempt to channel or deepen a waterway to aid navigation and transport.

epidemic—A disease spreading or diffusing within a region.

escarpments—High-walled canyons often the result of tectonic activity.

failed state—A former nation-state unable to control a portion of its own territory.

feudalism—A political system whereby loyalty is rewarded by the dispensing of land for a fee or an oath of loyalty.

forward capital—A capital city located in an area to encourage movement of people into the area for the purpose of solidifying control.

gateway city—Cities located near rivers allowing access to the oceans. Particularly important during the age of sail.

GNP—Gross national product. The total value of goods and services within a state.

holocaust—A policy used to exterminate a particular people group.

hydroelectric power—The use of flowing water to create electric power.

hydrography—The study of surface waters on the surface of the Earth.

industry—The production of goods or services within an area.

insurgency—A rebellious attempt to gain power over a formally recognized authority.

kudzu—A naturalized species of East Asian legume that is a vine.

leeward—The side of a topographic feature such as a hill or mountain opposite of the direction of prevailing winds or currents where air tends to draw moisture from or dry as a result of a rain shadow effect.

leaching—The phenomenon attributed to heavy rains drawing soil components deeper into the Earth. Particularly important in tropical wet climates where heavy rainfall and slash and burn agricultural techniques can create conditions catastrophic for plant growth.

nationalization—The act of causing ownership to be public property.

natural borders—Areas of the Earth's surface that tend to isolate or act as barriers to the exchange of culture and movement. Usually associated with the border of a state.

networks—A group of interconnecting transportation or communications centers, for example, a radio antenna and a radio receiver in your car. That exchange of information would be an example in communications. Parking your car and getting on a train would be a transportation network.

orographic effect—The tendency for energy to be released at higher elevations where the result is often precipitation such as snow or rain on the side of a mountain facing the oncoming front or current.

pandemic—A disease that spreads from one region to another.

periphery—The parts of the world with less developed economies and less technology than the more industrial or core regions of the world.

permafrost—Soils frozen for indeterminate amounts of time in the arctic and sub-arctic climates (see introduction).

plaza—A public square prominently placed in an urban center.

population pyramid—A schematic diagram used to demonstrate the age of a population by using the total numbers of population on an x-axis while demonstrating the numbers of male and females which are placed in an ascending pattern on the y-axis usually in five year increments.

prairie—Generally shorter grasslands denoting increasing aridity or dryness. Typical of a semi-arid climate such as the southern plains of North America where few if any trees exist.

region—An area of similarity from the standpoint of physical geography or in terms of culture.

shantytown—A generally impoverished area denoted by temporary or expedient structures outside of the cities in most peripheral nations.

slavocracy—A view held in the early to mid 1800's that held to the belief that the slave states of the south would continue to grow in influence into the Caribbean and Middle America with resulting power to slave owners.

smelting—An industrial process whereby metal is derived from ore.

specie—Coinage used as a medium of exchange or money.

tectonic activity—The energy resulting in changes to the Earth's surface as a result of tectonic or plate dynamics.

windward—The side of a hill or mountain facing the oncoming prevailing wind or current and generally moister as a result of adiabatic lapse rates resulting in precipitation.

Political Map of the World, August 2013

AUSTRALIA   Independent state
Bermuda   Dependency or area of special sovereignty
Sicily / AZORES   Island / island group
★   Capital

Scale 1:35,000,000
Robinson Projection
standard parallels 38°N and 38°S

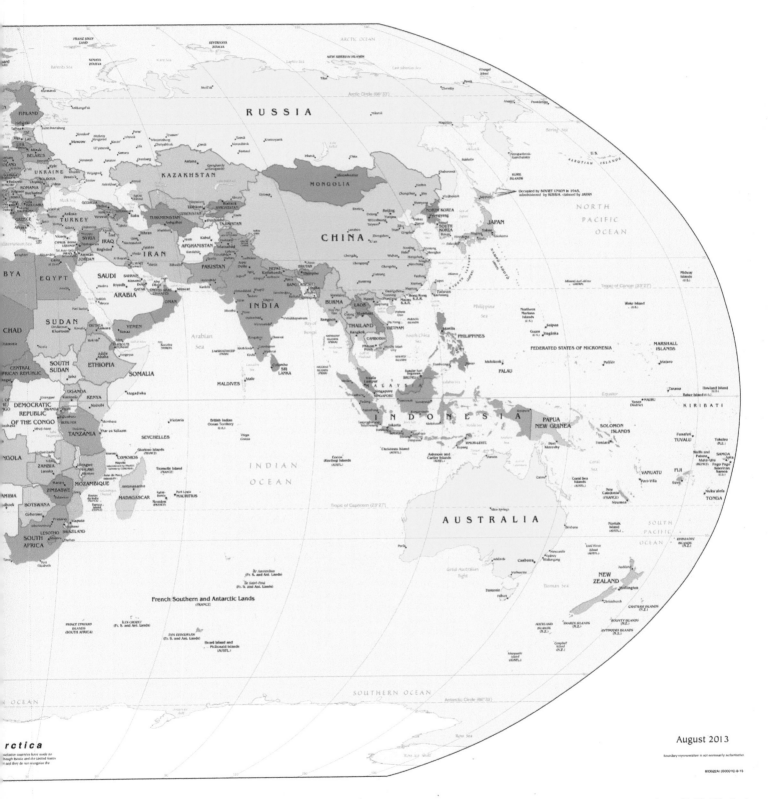

Map Source: *The World Factbook*

August 2013